HORMONAL CARCINOGENESIS IV

Hormonal Carcinogenesis IV

Proceedings of the
Fourth International Symposium

HORMONAL CARCINOGENESIS IV

edited by

JONATHAN J. LI
Professor of Pharmacology, Toxicology and Therapeutics
Director, Hormonal Carcinogenesis Laboratory
The University of Kansas Medical Center
Kansas City, KS, USA

SARA A. LI
Associate Research Professor of Pharmacology, Toxicology and Therapeutics
Hormonal Carcinogenesis Laboratory
The University of Kansas Medical Center
Kansas City, KS, USA

ANTONIO LLOMBART-BOSCH
Director, Departamento de Patologia
Facultad de Medicina y Odontologia
Universidad de Valencia
Valencia, Spain

Springer

Jonathan J. Li
Division of Etiology and Prevention
of Hormonal Cancers
The University of Kansas Medical Center
3901 Rainbow Boulevard, 2010 Breidenthal
Kansas City, KS 66160-7417

Sara A. Li
Division of Etiology and Prevention
of Hormonal Cancers
The University of Kansas Medical Center
3901 Rainbow Boulevard, 2010 Breidenthal
Kansas City, KS 66160-7417

Antonio Llombart-Bosch
Departmento de Patologia
Facultad de Medicina y Odontologia
Camino de Vera s/n
Valencia 46071 Spain

Library of Congress Cataloging-in-Publication Data

Hormonal Carcinogenesis IV / edited by Jonathan J. Li, Sara A. Li, Antonio Llombart-Bosch.
p.; cm.
Includes bibliographical references and index.
ISBN 0-387-23783-6 (alk. paper)
1. Hormones--Carcinogenicity—Congresses. 2. Hormones-Carcinogenicity-Congresses.
3. Generative organs-Cancer-Congresses. I. Title: Hormonal Carcinogenesis 4. II. Title: Hormonal Carcinogenesis four. III. Li, Jonathan J. IV. Li, Sara Antonia. V. Llombart Bosch, Antonio. VI. International Symposium on Hormonal Carcinogenesis (4th : 2003: Valencia, Spain)
[DNLM: 1. Neoplasms, Hormone-Dependent—Congresses. 2. Breast Neoplasms—etiology—Congresses. 3. Gonadal Steroid Hormones—adverse effects—Congresses. 4. Neoplastic Processes– Congresses. 5. Prostatic Neoplasms—etiology—Congresses. 6. Receptors, Cell Surface—physiology– Congresses.
QZ 200 H81131 2005]
RC268.7.S7H67 2005
616.99'4071—dc22

Printed in the United States of America.

9 8 7 6 5 4 3 2 1 SPIN 11014911

springeronline.com

In Memoriam to

Charles Bretton Huggins, M.D.

Ben May Laboratories
for Cancer Research
University of Chicago
Nobel Laureate, 1960

This volume is dedicated to the following individuals who unstintingly lent their resources and/or facilities to these Symposia since its inception in 1991, thus contributing importantly to our collective understanding of hormonal carcinogenesis and hormonal cancer research.

Janet Daling, Ph.D.

Jan-Åke Gustafsson, M.D., Ph.D.

Manuel Llombart-Bosch, M.D.

George Lucier, Ph.D.

James Pickar, M.D.

Acknowledgements

Patrons

Fundación Instituto Valenciano Oncología (IVO)
Office of Science and Culture of the Valencia Community
GlaxoSmithKline
National Institutes of Health
National Cancer Institute
National Institute of Child Health and Human Development
National Institute of Diabetes and Digestive and Kidney Diseases
National Institute of Environmental Health Sciences
Office of Research on Women's Health
Wyeth-Ayerst

Sponsors

DAKO
Zenyaku Kogyo

Contributors

GTx
Hormone Pellet Press

Executive and Scientific Advisory Boards

L to R: F. Waldman‡, S. Chen‡, A. Llombart-Bosch*, G. Mueller‡, D. Tindall‡, J.J. Li*, S.A. Li*, J. Daling‡, H. Rochefort‡, S. Nandi‡, A. Weisz‡, V. Beral‡, S. Liao‡, J. Pickar‡

Absent SAB members: S. Grisolia, E.V. Jensen, and S. Mohla

* Executive Board
‡ Scientific Advisory Board

Participants of the Fourth International Symposium on Hormonal Carcinogenesis

Preface

It has been over a decade since the First International Symposium on Hormonal Carcinogenesis convened in 1991. Since then, the field has rapidly expanded with considerable progress in both breast and prostate cancers; while ovarian and endometrial cancer have been hampered, in part, due to the absence of suitable hormone-mediated animal models. While knock-out, transgenic, and cell-culture systems have been extremely useful in identifying specific gene/protein alterations and the ensuing pathways affected, the precise molecular mechanisms whereby sex hormones elicit their oncogenic effects still remain elusive. Moreover, despite the considerable progress made in breast cancer research, the exact role of progestins in the presence or absence of estrogen in breast growth, differentiation, and malignant transformation is lacking. Elucidating the incipient molecular alterations in early/pre-invasive lesions elicited by these hormones is a growing important focus of this field. The main purpose of these Symposia has been to address vital questions that impact our understanding of the causation, dependency, progression, resistance, and prevention of hormonally-associated cancers.

We are indebted to the Scientific Advisory Board members who worked with us reviewing and offering suggestions to finalize the scientific program. We offer special thanks for the guidance and support of *Dr. Gerald Mueller*. His wisdom played an indispensable role in maintaining the excellence of these Symposia. We also acknowledge the numerous external reviewers that worked diligently to revise and improve the quality of the manuscripts. We are very grateful to *Ms. Tandria Price*. Her enthusiasm for the project, her effective and diplomatic interactions with contributors and administrators, and her superb organizational skills were evident during the Symposium and in the preparation of this volume. We are deeply grateful to the Fundación Instituto Valenciano Oncología (IVO), Universidad de Valencia, and the Office of Science and Culture of the Valencia Community, who hosted this Symposium and provided funds and gracious staff that worked efficiently to make this a most memorable Symposium. We are indebted to *Ms. Paula Callaghan*, our Springer-Verlag editor, for her support and highest publication standards. We appreciate the financial support of the NIH institutes and companies, listed separately, which have been indispensable to the success of this Symposium.

Kansas City, Kansas, USA — *Jonathan J. Li, Ph.D.*
Sara Antonia Li, Ph.D.

Valencia, SPAIN — *Antonio Llombart-Bosch, M.D, Ph.D.*

Contents

Dedication ... v

Acknowledgments ... vi

Preface ... ix

Participants ... xix

Symposium Address

Hormones, Centrosomes, and Genetic Instability in Mammary Carcinogenesis
William R. Brinkley, David L. Stenoien, and Thea Goepfert ... 1

STATE OF THE ART LECTURES ... 17

1 **Chromosomal Instability: A New Paradigm for Estrogen-induced Oncogenesis**
Jonathan J. Li, Jeffrey Salisbury, and Sara Antonia Li ... 19

2 **The Molecular Pathogenesis of Human Prostate Cancer**
William G. Nelson, Angelo M. DeMarzo, Theodore L. DeWeese, and William B. Isaacs ... 34

PART 1. BREAST CANCER I: PROGESTERONE ACTION 45

3 **Pathologic, Clinical, and Epidemiologic Characteristics of Invasive Lobular Breast Carcinoma and a Review of Studies Evaluating its Association with Hormone Replacement Therapy**
Christopher I. Li 47

4 **Progestin-Regulated Genes and Breast Cancer Risk: Good or Bad?**
Henri Rochefort, Majida Esslimani-Sahla and Danny Chalbos.. 65

5 **Role of Progesterone Receptors in Mammary Development and Carcinogenesis**
Gopalan Shyamala.. 77

PART 2. BREAST CANCER II: MECHANISMS OF GENOMIC INSTABILITY 85

6 **C-Myc Deregulation Promotes a Complex Network of Genomic Instability**
Sabine Mai, Amanda Guffei, Thierry Fest, and Frederic Mushinski 87

7 **Deregulation of Cyclin E and Genomic Instability**
Charles H. Spruck, Adrian P. L. Smith, Susanna Ekholm Reed, Olle Sangfelt, Jaimie Keck, Heimo Strohmaier, Juan Mendez, Martin Widschwendter, Bruce Stillman, Anders Zetterberg and Steven I. Reed 98

8 **Centrosome Amplification and the Origin of Chromosomal Instability in Breast Cancer**
Jeffrey L. Salisbury 106

PART 3. BREAST & PROSTATE: EARLY *IN-SITU* LESIONS 119

9 **Premalignant Breast Disease: Anatomic Lesions and Hormonal Associations**
David Page 121

10 **Aromatase Overexpression: Effect of Tissue Estrogen on Phenotypic and Biochemical Changes in Aromatase Transgenic Mice**
Rajeshwar Rao Tekmal, Nameer Kirma, Usha Mandava, and Roopa Luthra 130

11 **Prostate Epithelial Carcinogenesis: Putative and Controversial Precursor Lesions**
Cristina Magi-Galluzzi and Angelo M. De Marzo 141

PART 4. PROMOTION/PREVENTION OF HORMONE CANCERS 151

12 **Estrogen Can Prevent Breast Cancer by Mimicking the Protective Effect of Pregnancy**
Satyabrata Nandi, Raphael C. Guzman, Gudmundur Thordarson, and Lakshmanawamy Rajkumar 153

13 **Aromatase Inhibition and Breast Cancer**
William R. Miller 166

PART 5. PROSTATE I: ANDROGEN/ESTROGEN ACTION 181

14 **Androgens and Prostate Cancer Etiology: Sorting Through the Evidence**
Ronald K. Ross, Leigh Pearce, Juergen Reichardt, and Gerhard Coetzee 183

15 **Id-1 Protein as a New Marker for Prostate Cancer**
Y.C. Wong, M.T. Ling, X.H. Wang, X.S. Ouyang, A.L.M. Cheung, and Franky Chan 197

PART 6. PROSTATE II: ANDROGEN RECEPTOR: DEPENDENCE/RESISTANCE 209

16 **Four Stages of Prostate Cancer: Suppression and Eradication by Androgen and Green Tea Epigallocatechin Gallate**
Shutsung Liao, John M. Kokontis, Chih-pin Chuu, Stephen Hsu, Junichi Fukuchi, Mai Dang, and Richard A. Hiipakka 211

17 **Androgen Receptor and Interleukin-6 Signaling in Prostate Cancer Progression**
Zoran Culig, Hannes Steiner, Sonia Godoy-Tundidor, Barbara Comuzzi, Georg Bartsch, and Alfred Hobisch 221

18 **Role of the Androgen Receptor and P13K/Akt in the Survival of Androgen-Refractory Prostate Cancer Cells**
Haojie Huang and Donald J. Tindall 233

PART 7. ENDOMETRIUM/OVARIAN/COLON 245

19 **The BDII Inbred Rat: A Model for Genetic Analysis of Endometrial Carcinoma**
Karin Klinga-Levan and Göran Levan 247

20 **Potential Role of Gonadotropin-Releasing Hormone and Estrogen in Ovarian Cancer**
Peter C.K. Leung, Kyung-Chul Choi, and Nelly Auersperg....... 258

COMMUNICATIONS

Session I. Epidemiology/Human Derived Studies 271

Epidemiology:

Dose of Progestogen in Postmenopausal Combined Hormone Therapy and Risk of Endometrial Cancer
Susan D. Reed, Lynda F. Voigt, Shirley A. Beresford, Deirdre A. Hill, Jennifer A. Doherty, and Noel S. Weiss 273

Mammographic Densities and Urinary Hormones in Healthy Women with Different Ethnic Backgrounds
Gertraud Maskarinec, Andrew E. Williams, Sabina Rinaldi, and Rudolph Kaaks 277

Polymorphism of the Estrogen Receptor Beta Gene in Breast Cancer
Cristina Caraion, Nadine Vincent, Claude Lambert, Constantin Caraion, Guorong Li, Pierre Seffert, Jean-Marc Dumollard, and Christian Genin . 287

Characterization of Genetic Polymorphism of Glycine N-Methyltransferase Gene in Hepatocellular Carcinoma
Yu-Chuen Huang, Yi-Ping Shih, Chun-Chih Li, Li-Ying Liao, and Yi-Ming A. Chen 293

Human-Derived Studies:

BREAST CANCER

Role of Soy Phytoestrogens Genistein and Daidzein in Focal Adhesion Assembly and Focal Adhesion Kinase (FAK) Activity in Breast Cancer Cells
Nicolas G. Azios and Suranganie F. Dharmawardhane 300

Identification of Genes Regulated by the Antiprogestin, Onapristone, in Breast Cancer Cells Using Microarray Analysis
Simon J. Crook, J. David Brook, Anette Sommer, Joern Kraetzschmar, John F.R. Robertson 308

ENDOMETRIAL/CERVICAL CANCER

Expression of Estrogen α and β Variants, Androgen and Progesterone Receptors in Human Normal and Neoplastic Endometrium
Maciej Skrzypczak, Sylwia Zagulska-Szymczak, Rosette Lidereau, Ivan Bieche, Sebastian Lewandowski, Katarzyna Radwanska, Cezary Szczylik, Michael Vidaud, Jerzy A. Jakowicki, and Leszek Kaczmarek 314

PROSTATE CANCER

Toremifene (ACAPODENE™) Reduces High Grade Prostatic Intraepithelial Neoplasis in a Phase IIa Clinical Trial
Mitchell S. Steiner 321

Stereological Study of Mean Nuclear Volume Weighted by Volume in Normal Prostate, Prostatic Intraepithelial Neoplasia, and Adenocarcinoma
Fernando Teba, Rocio Martín, Vicente Gómez, and Luis Santamaría 329

Constitutively Active Androgen Receptor Variant Detected in a Human Prostate Cancer
Jocelyn Céraline, Marion D. Cruchant, Eva Erdmann, Philippe Erbs, Jean-Emmanuel Kurtz, Brigitte Duclos, Didier Jacqmin, Dominique Chopin, Patrick Dufour, and Jean-Pierre Bergerat 336

Cancer Prevention by Green Tea via EGCG-Mediated Inhibition of Fatty Acid Synthase
Koen Brusselmans, Ellen De Schrijver, Walter Heyns, Guido Verhoeven, and Johannes V. Swinnen 343

Silencing of the Fatty Acid Synthase Gene by RNA Interference Inhibits Growth and Induces Apoptosis of LNCaP Prostate Cancer Cells
Ellen De Schrijver, Koen Brusselmans, Walter Heyns, Guido Verhoeven, and Johannes V. Swinnen 350

Androgens Stimulate the SREBP Pathway in Prostate Cancer Cells by Inducing a Shift in the SCAP-Retention Protein(s) Balance
Hannelore Heemers, Walter Heyns, Guido Verhoeven, and Johannes V. Swinnen 357

Session II. Hormone Metabolism & Cell/Molecular Biology 365

Hormone Metabolism

Metabolism of 17β-Estradiol in ACI Rat Liver and Mammary Gland After Chronic Estradiol Treatment
Sonia Mesia-Vela, Rosa I Sanchez, Kenneth R. Reuhl, Allan H. Conney, and Frederick C. Kauffman 367

Modulation of Transforming and Clastogenic Activities of Catechol Estrogens by a Catechol-*O*-Methyltransferase Inhibitor in Syrian Hamster Embryo Fibroblasts
Takeki Tsutsui, Takeo W. Tsutsui, Yukiko Tamura, and J. Carl Barrett . 375

Cell/Molecular Biology:

BREAST CANCER

Bi-directional Regulation of Human Progesterone Receptors and the Mitogen Activated Protein Kinase Pathway in Breast Cancer Cell Models
Emily Faivre, Ming Qiu, and Carol A. Lange 381

Abnormal Properties of Mutants in the Hinge Region of Er∀: Implications in Breast Cancer
Carlos Martínez-Campa, Pedro Zuazua, Juana María García-Pedrero, Pedro Casado, Pedro Sánchez Lazo, and Sofía Ramos 391

The Effects of Eicosapentaenoic Acid Upon Proliferation and 17β-Dehydrogenase Activity in MCF-7 Breast Carcinoma Cells
Alison S. Whitehouse, Eftychia Oikonomou, and Eric F Adams 398

Effect of Dietary Genistein on Estradiol-Induced Mammary Carcinogenesis in the ACI Rat
Valerie K. Turan, Kenneth R Reuhl, and Paul E. Thomas 405

Cytogenetic Analysis of Genomic Destabilization in Solely Estrogen-Induced Female ACI Rat Mammary Neoplasms
Dan Papa, Jonathan J. Li, and Sara Antonia Li 412

Microarray Analysis of Estrogen-Induced Protection Against Breast Cancer
Lakshmanaswamy Rajkumar, Demi-Nhung Dang, Mark D. Hartnett, David L. Hirschberg, Kenneth C. Loh, Raphael C. Guzman, Gudmundur Thordarson, and Satyabrata Nandi 419

Pregnancy Levels of Estrogen Prevents Mammary Cancer
Raphael C. Guzman, Lakshmanaswamy Rajkumar, Gudmundur Thordarson, and Satyabrata Nandi 426

Hormonal Dependence of Mammary Premalignant Progression
Daniel Medina, Frances S. Kittrell, Anne Shepard, Jamel Hill, and Powel Brown 431

The Possible Role of IGF-I and Androgens in the Development of Canine Inflammatory Mammary Carcinoma
Juan Carlos Illera, Gema Silván, M. Dolores Pérez-Alenza, Ana R. Sánchez-Archidona, Ana Nieto, and Laura Peña 436

ENDOMETRIAL CANCER

Hormonal Activation of the Gab-1 Docking Protein in Uterine Cells
Diane M. Klotz, Jon A. Proctor, David K. Walmer, R. Gregg Richards, and Richard P. DiAugustine 443

Endometrial Adenocarcinoma in Syrian Hamsters Treated with Diethylstilbestrol, Tamoxifen and N-Ethyl-Nitrosourea
Jaime Ferrer, Faustino Pérez-Mínguez, Antonio Leal, Amando Peydró, and Antonio Llombart-Bosch 450

OVARY/PITUITARY/KIDNEY/LIVER CANCERS

Hormonal Regulation of ZEB-1 and Implication for Progression of Human Reproductive Cancers
Bynthia M. Anose, Michael P. Linnes, and Michel M. Sanders........ 455

Evaluation of Messenger RNA of Pituitary Tumour-Transforming Gene-1 (PTTG1) as a Molecular Marker for Micrometastasis
Manuel Valladares Ayerbes, Lourdes Calvo, Guillermo Alonso, Pilar Iglesias, Maria J Lorenzo, Inmaculada Brandón, Mar Haz, Marga Reboredo, Silvia Antolín, and Luis Antón Aparicio........ 462

Presence of CCK-B Receptor mRNA in Human Functionless Pituitary Tumours
Eftychia Oikonomou, Alison Whitehouse, Rosalind Mitchell, and Eric F. Adams 468

Estrogen-induced Mutations and its Role in the Development of Tumorigenesis
Kamleshwar P. Singh, Jose Antonio López-Guerrero, Antonio Llombart-Bosch, and Deodutta Roy 475

Englitazone Delays Fetal Growth in Late Gestation in the Rat
Julio Sevillano, Inmaculada C. López-Pérez, Emilio Herrera, M. Pilar Ramos, and Carlos Bocos 480

PROSTATE CANCER

The Tumour Suppressor Gene PTEN Plays Role in Cell Cycle Regulation and Apoptosis in Prostate Cancer Cell Lines
Alice Hlobilková, Michaela Šváchová, Jana Knillová, Eva Pimrová, Petra Řiháková, Per Guldberg, and Zdeněk Kolář 487

The Coactivators CBP and p300 in Androgen Independent Prostate Cancer
Jose D. Debes, Zoran Culig, and Donald J. Tindall 494

Expression Study of Estrogen Receptor-related Receptors and Steroid Hormone Receptors in Human Prostatic Cells
C.P. Cheung, Lung-Wai Chan, Ki Lui, Uwe Borgmeyer, Shiuan Chen, and Franky L. Chan 501

Immunohistochemical and In Situ Detection of Sex Hormone-Binding Globulin (SHBG) Expression in Breast and Prostate Cancer: Implications for Hormone Regulation
Scott M. Kahn, Daniel J. Hryb, Atif M. Nakhla, Saeed M. Khan, Nicholas A. Romas, and William Rosner 508

Activation of Androgen Receptor in Prostate Cancer: Role of Protein Kinase A and Extracellular Signal-regulated Kinases
Yehia Daaka, Elizabeth Kasbohm, and Charles Yowell 515

Cadmium and Zinc Chloride-induced Preneoplastic Changes in the Rat Ventral Prostate: An Immunohistochemical and Molecular Study
Riánsares Arriazu, José M. Pozuelo, Rosario Rodríguez, Nuno Henriques-Gil, Teresa Perucho, Rocío Martín, and Luis Santamaría 522

Index

Participants

C. MARCELO ALDAZ The University of Texas, M.D. Anderson Cancer Center, Smithville, TX, USA

LORENZO ALONSO Hospital Universitario de Málaga, Madrid, SPAIN

LEONCIO ARRIBAS ALPUENTE Hospital Universitario de Málaga, Madrid, SPAIN

BYNTHIA MARIA ANOSE Department of Biochemistry, Molecular Biology, and Biophysics, University of Minnesota, Minneapolis, MN, USA

MANUEL VALLADARES AYERBES Medical Oncology, Pathology and Research Units, Medicine Department, Juan Canalejo University Hospital, La Coruña, SPAIN

NICOLAS G. AZIOS Section of Molecular Cell and Developmental Biology and the Institute for Cellular and Molecular Biology, The University of Texas at Austin, Austin, TX, USA

SANDRA BALLESTER Departmento de Patología, Universidad de Valencia, Facultad de Medicina y Odontología, Valencia, SPAIN

ELISA BANDERA The Cancer Institute of New Jersey, New Brunswick, NJ, USA

FREDERICK BELAND National Center for Toxicological Research, Jefferson, AR, USA

VALERIE BERAL Cancer Research UK Epidemiology Unit, Oxford, UNITED KINGDOM

ISABELLE M. BERQUIN Wake Forest University, Winston-Salem, NC, USA

WILLIAM R. BRINKLEY Department of Cellular and Molecular Biology, Baylor College of Medicine, Houston, TX, USA

KOEN BRUSSELMANS Laboratory for Experimental Medicine and Endocrinology, Department of Developmental Biology, Gasthuisberg, Catholic University of Leuven, Leuven, BELGIUM

CRISTINA CARAION Clinical Immunology Laboratory, University Hospital of Saint Etienne, FRANCE

CARMEN CARDA Departamento de Patología, Universidad de Valencia, Facultad de Medicina y Odontología, Valencia, SPAIN

JOCELYN CÉRALINE Laboratoire de Cancérologie Expérimentale et de Radiobiologie, IRCAD, Strasbourg, FRANCE

FRANKY L CHAN Department of Anatomy, The Chinese University of Hong Kong, Shatin, N.T. Hong Kong, CHINA

SHIUAN CHEN Beckman Research Institute of the City of Hope, Duarte, CA, USA

YI-MING A. CHEN Institute of Public Health, School of Medicine, National Yang-Ming University, Taipei, TAIWAN

YONG Q. CHEN Comprehensive Cancer Center, Wake Forest University School of Medicine, Winston-Salem, NC, USA

SIMON J. CROOK Institute of Genetics, University of Nottingham, Queen's Medical Centre, Nottingham, UNITED KINGDOM

ZORAN CULIG Department of Urology, University of Innsbruck, Innsbruck, AUSTRIA

YEHIA DAAKA Department of Surgery and Pharmacology and Cancer Biology, Duke University Medical Center, Durham, NC, USA

JANET R. DALING Fred Hutchinson Cancer Research Center, Seattle, WA, USA

ELLEN DE SCHRIJVER Laboratory for Experimental Medicine and Endocrinology, Department of Developmental Biology, Gasthuisberg, Catholic University of Leuven, Leuven, BELGIUM

JOSE D. DEBES Department of Urology, Mayo Clinic and Foundation, Rochester, MN, USA

RICHARD P. DIAUGUSTINE Hormones and Cancer Group, Laboratory of Molecular Carcinogenesis, National Institute of Environmental Health Sciences, National Institutes of Health, Research Triangle Park, NC, USA

SURANGANIE F. DHARMAWARDHANE Section of Molecular Cell and Developmental Biology and the Institute for Cellular and Molecular Biology, The University of Texas at Austin, Austin, TX, USA

MANUEL ENCISCO Departamento de Patología, Universidad de Valencia, Facultad de Medicina y Odontología, Valencia, SPAIN

JAIME FERRER Departamento de Patología, Universidad de Valencia, Facultad de Medicina y Odontología, Valencia, SPAIN

JAMES FIORICA Gyn Oncology Program, University of South Florida, H. Lee Moffitt Cancer Center, Tampa, FL, USA

MATIAS GARCIA-FANTINI Monforte de Lemos, Lugo, SPAIN

CECILIA GAROFALO Kimmel Cancer Center, Thomas Jefferson University, Philadelphia, PA, USA

SONIA GODOY-TUNDIDOR Department of Urology, University of Innsbruck, Innsbruck, AUSTRIA

JAN-ÅKE GUSTAFSSON Karolinska Institute, Department of Medical Nutrition, Huddinge, SWEDEN

RAPHAEL C. GUZMAN Cancer Research Laboratory, University of California, Berkeley, CA, USA

HANNELORE HEEMERS Laboratory for Experimental Medicine and Endocrinology, Faculty of Medicine, Onderwijs en Navorsing (9), Gasthuisberg, K.U. Leuven, Leuven, BELGIUM

ALICE HLOBILKOVÁ Laboratory of Molecular Pathology and Institute of Pathology, Faculty of Medicine, Palacký University, Hněvotínská 3, Olomouc, CZECH REPUBLIC

HAOJIE HUANG Departments of Urology, and Biochemistry and Molecular Biology, Mayo Clinic, Rochester, MN, USA

JUAN CARLOS ILLERA Department of Animal Physiology, Veterinary School, Complutense University, Madrid, SPAIN

SCOTT M. KAHN Departments of Medicine and Urology, St. Luke's/Roosevelt Hospital Center, and College of Physicians and Surgeons, Columbia University, New York, NY, USA

KARIN KLINGA-LEVAN Department of Natural Sciences, Skövde University, Skövde, SWEDEN

RAJKUMAR LAKSHMANAWAMY Cancer Research Laboratory, University of California, Berkeley, CA, USA

CORAL A. LAMARTINIERE Department of Pharmacology/Toxicology, University of Alabama at Birmingham, Birmingham, AL, USA

CAROL A. LANGE Department of Medicine, Division of Hematology, Oncology, and Transplant, University of Minnesota Cancer Center, Minneapolis MN, USA

LOIC LE MARCHAND Cancer Research Center of Hawaii, University of Hawaii, Honolulu, HI

ANTONIO LEAL Departamento de Patología, Universidad de Valencia, Facultad de Medicina y Odontología, Valencia, SPAIN

PETER C. K. LEUNG Department of Obstetrics and Gynaecology, BC Children's and Women's Hospital, University of British Columbia, Vancouver, British Columbia, CANADA

GÖRAN LEVAN CMB-Genetics, Göteborg University, Göteborg, SWEDEN

KRISTINA LEVAN Su/Sahlgrenska, Gotenborg, SWEDEN

CHRISTOPHER I. LI Fred Hutchinson Cancer Research Center, Seattle, WA, USA

JONATHAN J. LI Department of Pharmacology, Toxicology and Therapeutics, The University of Kansas Medical Center, Kansas City, KS, USA

SARA ANTONIA LI Department of Pharmacology, Toxicology and Therapeutics, The University of Kansas Medical Center, Kansas City, KS, USA

SHUTSUNG LIAO The Ben May Institute for Cancer Research and Department of Biochemistry and Molecular Biology, University of Chicago, Chicago, IL, USA

MICHAEL PAUL LINNES Department of Biochemistry, Molecular Biology, and Biophysics, University of Minnesota, Minneapolis, MN, USA

ANTONIO LLOMBART-BOSCH Departamento de Patología, Universidad de Valencia, Facultad de Medicina y Odontología, Valencia, SPAIN

ANTONIO LLOMBART-CUSSAC Department of Medical Oncology, Fundación Instituto Valenciano de Oncología, Valencia, SPAIN

BEATRIZ LLOMBART-CUSSAC Departamento de Patología, Universidad de Valencia, Facultad de Medicina y Odontologia, Valencia, SPAIN

JOSE ANTONIO LOPEZ-GUERRERO Departamento de Patología, Universidad de Valencia, Facultad de Medicina y Odontología, Valencia, SPAIN

CRISTINA MAGI-GALLUZZI Department of Pathology, the Cleveland Clinic Foundation, Cleveland, OH, USA

SABINE MAI Manitoba Institute of Cell Biology, CancerCare Manitoba, The University of Manitoba, Winnipeg, MB, CANADA

M. MATILDE MARQUES Centro de Química Estructural, Instituto Superior Técnico, Lisboa, PORTUGAL

CARLOS MARTÍNEZ-CAMPA Departamento de Bioquímica y Biología Molecular. Instituto Universitario de Oncología Principado de Asturias. Universidad de Oviedo, Oviedo, SPAIN

GERTRAUD MASKARINEC Cancer Research Center of Hawaii, Honolulu, HI, USA

JUAN MENDEZ Cold Spring Harbor Laboratory, Cold Spring Harbor, NY, USA

DANIEL MEDINA Department of Molecular and Cellular Biology, Baylor College of Medicine, Houston, TX, USA

SONIA MESIA-VELA Laboratory for Cellular and Biochemical Toxicology, Department of Chemical Biology, Ernest Mario School of Pharmacy, Rutgers, The State University of New Jersey, Piscataway, NJ, USA

WILLIAM R. MILLER Edinburgh Breast Unit Research Group, University of Edinburgh, Western General Hospital, Edinburgh, SCOTLAND

GERALD MUELLER McArdle Lab, University of Wisconsin, Madison, WI, USA

SATYABRATA NANDI Cancer Research Laboratory, University of California, Berkeley, CA, USA

WILLIAM G. NELSON The Sidney Kimmel Comprehensive Cancer Center and Brady Urological Institute, The Johns Hopkins University School of Medicine, Baltimore, MD, USA

EFTYCHIA OIKONOMOU Pharmaceutical Sciences Research Institute, Aston University, Birmingham, UNITED KINGDOM

CARLOS ORTIZ DE SOLÓZANO Escuela Politécnica Superior, Universidad Autónoma de Madrid, Madrid, SPAIN

DAVID L. PAGE Vanderbilt University Medical Center, Nashville, TN, USA

DAN PAPA Department of Pharmacology, Toxicology and Therapeutics, The University of Kansas Medical Center, Kansas City, KS, USA

FAUSTINO PEREZ-MINGUEZ Departamento de Patología, Universidad de Valencia, Facultad de Medicina y Odontología, Valencia, SPAIN

JAMES H. PICKAR Clinical Research and Development, Wyeth-Ayerst, Collegeville, PA, USA

KARL C. PODRATZ Mayo Clinic, Rochester, MN, USA

ANDRES POVEDA Departamento de Patología, Universidad de Valencia, Facultad de Medicina y Odontología, Valencia, SPAIN

NAFIS RAHMAN Gynecology & Obstetrics Section, Department of Physiology, University of Turku, Turku, FINLAND

DAVID RAMOS Departamento de Patología, Universidad de Valencia, Facultad de Medicina y Odontología, Valencia, SPAIN

SOFÍA RAMOS Departamento de Bioquímica y Biología Molecular. Instituto Universitario de Oncología Principado de Asturias, Universidad de Oviedo, Oviedo, SPAIN

STEVEN I. REED The Scripps Research Institute, La Jolla, CA, USA

SUSAN D. REED Department of Obstetrics & Gynecology and Epidemiology, University of Washington, Division of Public Health Sciences, Fred Hutchinson Cancer Research Institute, Seattle, WA, USA

SUSANNA EKHOLM REED The Scripps Research Institute, La Jolla, CA USA; The Karolinska Institute, Stockholm, SWEDEN

HENRI ROCHEFORT Inserm U.540 and Faculty of Medicine, Montpellier, FRANCE

RONALD K. ROSS, Flora L. Thornton Chairman of Preventive Medicine, USC/Norris Comprehensive Cancer Center, Los Angeles, CA, USA

AMPARO RUIZ Departamento de Patología, Universidad de Valencia, Facultad de Medicina y Odontología, Valencia, SPAIN

JEFFREY L. SALISBURY Tumor Biology Program, Mayo Clinic, Rochester, MN, USA

LUIS SANTAMARÍA Department of Morphology. School of Medicine, UAM, Madrid, SPAIN

JULIO SEVILLANO Facultad de Ciencias Experimentales y de la Salud, Universidad San Pablo-CEU, Boadilla del Monte, Madrid, SPAIN

GOPALAN SHYAMALA Department of Cell and Molecular Biology, University of California, Lawrence Berkeley Lab, Berkeley, CA, USA

MACIEJ SKRZYPCZAK Second Department of Gynaecology, University School of Medicine, Lublin, POLAND

EDUARDO SOLSONA-NARBONA Department of Medical Oncology, Fundación Instituto Valenciano de Oncología, Valencia, SPAIN

MEIR J. STAMPFER HSPH Epidemiology, Harvard Medical School, Boston, MA, USA

MITCHELL S. STEINER Department of Urology, University of Tennessee, Memphis, TN, USA

EVA SURMACZ Kimmel Cancer Center, Thomas Jefferson University, Philadelphia, PA, USA

OSSAMA TAWFIK Department of Surgical Pathology, The University of Kansas Medical Center, Kansas City, KS, USA

RAJESHWAR RAO TEKMAL Division of Reproductive Research, Department of Obstetrics and Gynecology, The University of Texas Health Science Center, San Antonio, TX, USA

PAUL E. THOMAS Joint Graduate Program in Toxicology; Laboratory for Cancer Research, Ernest Mario School of Pharmacy, Rutgers, the State University of New Jersey, Piscataway, NJ, USA

DONALD J. TINDALL Departments of Urology, and Biochemistry and Molecular Biology, Mayo Clinic, Rochester, MN, USA

PAOLO TONIOLO Ob/Gyn and Environmental Medicine, Population Sciences, NYU Cancer Institute, New York University School of Medicine, New York, NY, USA

TAKEKI TSUTSUI Department of Pharmacology, The Nippon Dental University, School of Dentistry at Tokyo, Tokyo, JAPAN

VALERIE K. TURAN Joint Graduate Program in Toxicology, Ernest Mario School of Pharmacy, Rutgers, the State University of New Jersey, Piscataway, NJ, USA

FRED M. WALDMAN University of California San Francisco, Cancer Center, San Francisco, CA, USA

ALESSANDRO WEISZ Dipart. di Patolgia Generale, Seconda Universita di Napoli, Napoli, ITALY

S. JOHN WEROHA Department of Pharmacology, Toxicology and Therapeutics, The University of Kansas Medical Center, Kansas City, KS, USA

ALISON S. WHITEHOUSE Pharmaceutical Sciences Research Institute, Aston University, Birmingham, UNITED KINGDOM

Y.C. WONG Cancer Biology Laboratory, Department of Anatomy; Central Laboratory of Institute of Molecular Technology for Drug Discovery and Synthesis, The University of Hong Kong, Hong Kong, CHINA

GABRIEL ZANGO Endocrine Laboratory, Dept. of Clinical Biochemistry, Faculty of Health Sciences, Ben Gurion University, Beer-Sheva, ISRAEL

ZHU ZHE PL Group, Institute of Molecular and Cell Biology, Singapore, SINGAPORE

Symposium Address

Hormones, Centrosomes, and Genomic Instability in Mammary Carcinogenesis

William R. Brinkley, David L. Stenoien, and Thea Goepfert

Introduction

The centrosome, a cytoplasmic organelle acquired at fertilization via paternal inheritance, plays a vital role in cell division and in the maintenance of cell polarity throughout development. The importance of the centrosome in achieving proper orientation and segregation of duplicated chromosomes and assuring stability of the eukaryotic genome is well established. The centrosome can be recognized in the cytoplasm of most eukaryotic cells as a discrete, microscopic domain that functions as the cell's principal microtubule organizing center (MTOC). Like the genome, it undergoes duplication during late S-phase, producing a pair of centrosomes that function to organize the bipolar spindle in mitosis.

The centrosome duplication cycle, like the cell cycle, is regulated by a vital system of checkpoint-signaling proteins, about which little is currently known. Errors in centrosome duplication and/or distribution can result in aberrant daughter cells that either lack centrosomes (acentric cells) or receive a single centrosome, producing mono-polar spindles, or they can receive more than two centrosomes, resulting in the assembly of multipolar spindles during mitosis. These aberrations can lead to catastrophic errors in chromosome distribution resulting either in cell death or transformation to become tumor cells. Using antibodies specific for centrosome associated proteins, we and others have noted that many tumors, both *in-vitro* and *in-vivo*, display cells with a variety of centrosome aberrations (1-7). The most commonly reported defect causes cells to develop supernumerary centrosomes (greater than the expected 1-2 centrosomes/cell), a process identified as "centrosome amplification." Retention and maintenance of centrosome duplicity is critical in the cell cycle to assure bipolar spindles and the maintenance of a diploid genome. Centrosome amplification could therefore account, in part, for the genomic instability and aneuploidy commonly found in cancer cells. Although bipolar spindles can form in the absence of centrosomes in some germ cells of eukaryotes this organelle plays a prominent role in spindle assembly and function in somatic cells (reviewed in 5). Thus, it becomes essential to identify and characterize the molecular components involved in maintaining the correct centrosome number in diploid cells and to understand the cause of aberrations.

Although little is known about the role of hormones in the biology of centrosomes, they serve an important role in normal cell growth and developments, and are implicated in tumorigenesis. In a later section of this report, we review recent studies in our laboratory of centrosomes and their involvement in hormone mediated aneuploidy and cancer in an experimental rat mammary model for carcinogenesis.

The Resurrection

The implication of centrosome anomalies in cancer were first identified at the dawn of the 20th Century. In a 1914 treatise, *Zur Frage der Enstehung Maligner Tumoren*, Theodore Boveri (8) proposed that sea urchin zygotes containing more than two centrosomes, segregated their chromosomes abnormally due to the presence of extra spindle poles induced by polyspermy (Figure 1). He noted that although multipolarity was generally lethal, occasional segregants survived to produce embryos with tumor-like outgrowths. As one of the early experts on the role of centrosomes in cell division, Boveri was the first to propose a direct link between oncogenesis and the presence of multipolar spindles, aneuploidy and loss of tissue architecture in these embryos. Although he never studied cancer cells *per se*, his astute observations of this simple invertebrate system allowed him to derive many important postulates that still apply to cancer, including the concept of oncogenes, cell cycle checkpoints, tumor-suppressor genes, genetic instability, the clonal origin of tumors, chromosome specific "weakness" (telomeres), loss of cell adhesion, and genetic mosaicism. Moreover, he achieved this monumental task with little more than a primitive light microscope, a keen sense of observation and a truly remarkable intuition.

Boveri's novel hypothesis that centrosome anomalies could be responsible for aneuplody and the ontogeny of cancer, created a brief but spirited debate at the time, but was never widely accepted by the cancer establishment and it was essentially ignored until recently. The development and rise of *Drosophila* genetics by the Morgan school of genetics (9) and the discovery by Muller, *et al.* (10) that x-rays were potent mutagens and carcinogens, led to the widely accepted view that cancer is caused by a somatic gene mutation. The subsequent discovery of cellular oncogenes and tumor suppressor genes in the latter half of the 20th Century added considerable reinforcement to the somatic mutation idea (11-12). Despite the complex genetic basis of cancer, mutations in somatic genes remains the accepted hypothesis for oncogenesis by most cancer researchers today (for alternative view, see 13). However, recent reports of centrosome anomalies associated with many tumor cells (3-6), along with the well-established role of this organelle in mitotic spindle assembly and chromosome segregation (6, 7, 14), has revived Boveri's 96 year-old hypothesis that aberrant centrosomes and aneuploidy (genomic instability) are incipient events in oncogenesis.

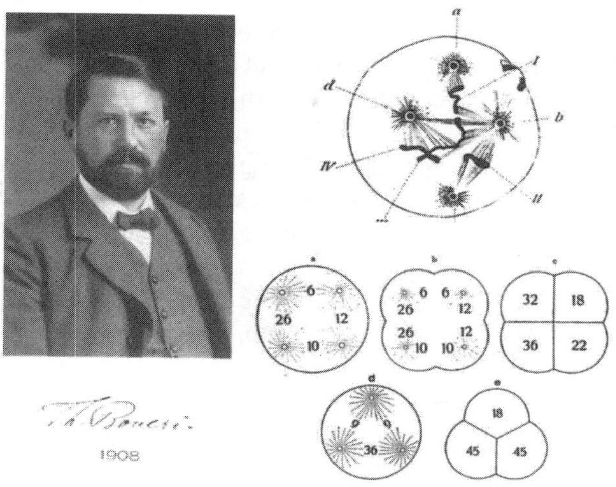

Figure 1. Theodore Boveri, whose portrait is shown here in 1908, proposed that aneuploidy and alterations in cellular polarity, distinguishing characteristics of most cancer cells, resulted from defects in the function of the centrosome. His early drawings, shown here, depict possible arrangements of chromosome and patterns of segregation in dividing cells with three (tripolar) or four (tetrapolar) spindle poles resulting from double fertilization of sea urchin eggs. (Reproduced with permission from University of California Press Berkeley and Los Angeles, CA. Cambridge University Press London, England. Copyright 1967, by Regents of The University of California).

The role of centrosomes in cancer resurfaced again in 1996 when Fukasawa, *et al.* (1) reported centrosome amplification in mouse embryonic fibroblast null for the tumor suppressor p53. This group used anti-γ tubulin antibodies to detect and count centrosomes by immunofluorescence and reported that cells with the p53 null phenotype displayed more than the normal complement of centrosomes, whereas wild type and heterozygous cells displayed normal numbers of centrosomes. Three additional manuscripts published in 1998 from the laboratories of Roop and Brinkley (2), Salisbury (3), Pihan, Doxy, *et al.* (4) reported centrosome anomalies, especially centrosome amplification, in tumor cells *in vivo.* These reports were followed by additional findings that centrosomal abnormalities are common to many human cancers (reviewed in 6) and have sparked a lively resurrection of Boveri's original hypothesis (14).

The Enlightenment: The Aurora Family of Serine/Theronine Kinases

The remarkable re-discovery of centrosome amplification in many common cancers led to a renewed interest in aneuploidy in neoplasia and initiated a search for a cellular and/or biochemical mechanism responsible for this phenomenon. It was immediately obvious; however that progress would be hampered by a dearth of knowledge about the molecular basis of centrosome maturation and replication in eukaryotic cells (15, 16). Theoretically, centrosome amplification (more than the normal 1-2 centrosomes/cell) can occur by one of several pathways: (a) through a defect in a checkpoint pathway that controls centrosome duplication in late S-phase resulting in the over duplication , (b) via failure to partition duplicated centrosomes into daughter cells due to arrested or aberrant cytoplasmic cleavage of cytokinesis (6, 17), or (c) due, possibly, to fragmentation of pericentriolar material into small dispersed bodies that retain their capacity serve as MTOCs.

New light was cast onto the mechanism of centrosome amplification with the discovery of mitotic serine/threonine kinases representing the Aurora kinase family that included the prototypic yeast *Ipl1* and the *Drosophila Aurora* kinases (reviewed in 18, 19). Two groups from the laboratories of Sen and Brinkley at Houston (20) and the Bischoff group at Los Angeles (21) reported elevated expression of Aurora A (AurA) in many human cancers. Moreover, antibodies made against AurA were found to localize to the centrosomes of both interphase and mitotic cells (Figure 2).

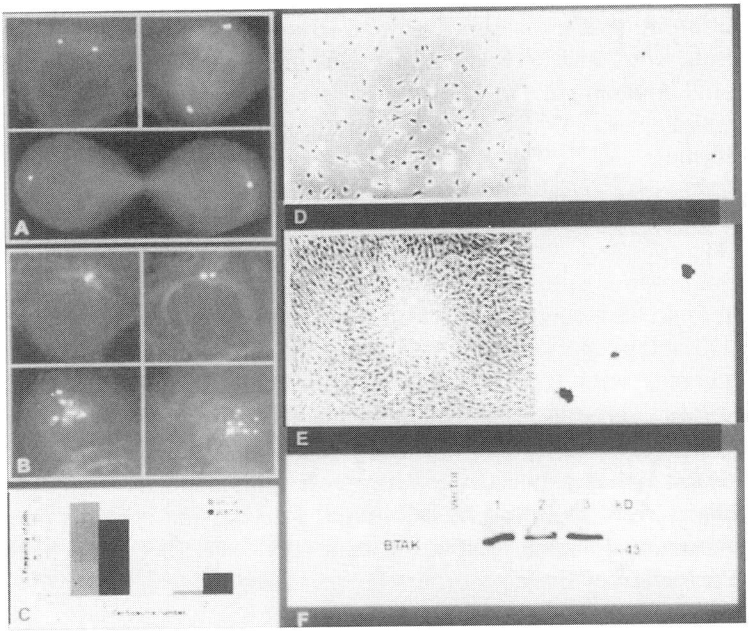

Figure 2. Localization and expression of Aurora kinase in mammalian cells. **(A)** Antibodies against AurA are localized in the centrosomes of HeLa cells and when the gene, STK15 is overexpressed by transfection, multiple centrosomes appear. Two centrosomes appear in cells transfected with the vector, while 20% of the STK15 transfected cells displayed > 3 centrosomes/cell. **(B)** Centrosomes counts in 200 vector and STK15-transfected cells are shown in **(C)** Figures 1D-E show growth of cells in agar of stable transfected cells with vector **(D)** and NIH 3T3 cell transfected with STK 15 and grown in 0.5% bovine calf serum. Micrographs were taken at a total magnification of X100. F, Western blot analysis of STK 15-transfected 3T3 clones showing expression of STK15. [From Figure 5 in Zhou, *et al.*, 1998 (20)].

Although mRNA levels and AurA expression remain low during G_1 and S phase of the cell cycle, message levels peak during G2/M phase and fall as cells exit mitosis (18, 21, 22). Even though AurA kinase was found to be expressed in normal cells, especially those progressing through the cell cycle, experimental overexpression of AurA kinase was shown to cause transformation of human and rodent cells *in vitro* and *in vivo* (20, 21). Thus, this kinase appears to be oncogenic and centrosomes appear to be its primary target organelle. Much less is known, however, about the mutational status of the STK15 allele and its natural substrates.

The Aurora kinase family is highly conserved and has been isolated from a variety of species including *Saccharomyces cerevisiae*, *Drosophila melanogaster*, *C. elegans*, *Xenopus laevis*, *Mus musculus*, *Rattus Norvegicus*, and *Homo sapiens* (reviewed in 18). Three members of the family are found in human cells including AurA, B, and C. For comparison, the structure and functional domains of these three kinases are shown in Figure 3. Recent studies indicated that an orthologue of human AurA is also overexpressed in rat mammary tissues in virgin females exposed to a carcinogen, NMU, and the expression levels, along with centrosome amplification, appear to be early markers for tumorigenisis in this animal model (19).

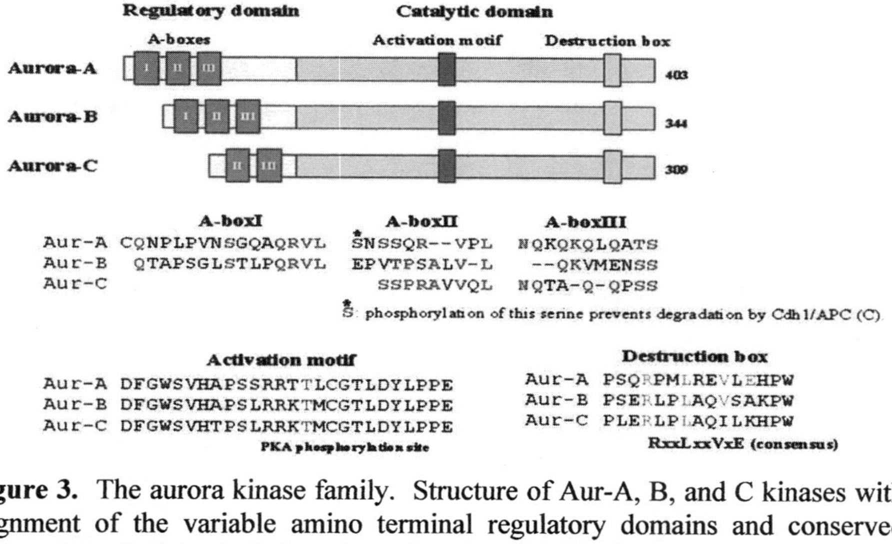

Figure 3. The aurora kinase family. Structure of Aur-A, B, and C kinases with alignment of the variable amino terminal regulatory domains and conserved carboxyl terminal catalytic domains. Three putative Aur boxes (A-box I, II, and III) with sequence motifs are shown. Sequence motifs are also for the activation motif and destruction box. [From Katayama, *et al.*, 2003 (18)].

Although the search continues for natural substrates and the nature of binding sites within cells, we have investigated AurA binding properties at centrosomes and spindles of human cells and determined it to be highly dynamic in comparison to other centrosome-associated proteins (23). This was accomplished by expressing a GFP-AurA in living cells to mark the centrosomes and by analyzing the cells in various stages of division by fluorescence microscopy. GFP-AurA expression was induced by doxycycline treatment in stably transfected HeLa Tet-on cells and detectable levels appeared as early as 2-4 hours following induction (Figure 4). As shown in Figures 5A-O, it was possible to detect the kinase at centrosomes throughout the cell cycle and to observe a rapid migration from spindle poles to the mitotic spindle at prometaphase (Figure 5C, D). The dynamics of AurA kinase were apparent at two levels: first within the duplicated centrosomes themselves, where dynamic oscillations and the unique tumbling motions of the entire organelle were detected (Figure 6), and secondly, at the molecular level as detected by fluorescence recovery after photobleach (FRAP). The latter procedure demonstrated a rapid exchange into an out of the centrosome and spindle ($t_{1/2}$~3sec), indicating a dynamic state of equilibrium between bound kinase and unbound component in the microenvironment (Figure 7). In contrast, the tubulins (α and γ) were found to be relatively stable and immobile (23). These novel observations provide considerable evidence for a kinetic vs structural role for AurA kinase in regulating spindle activity.

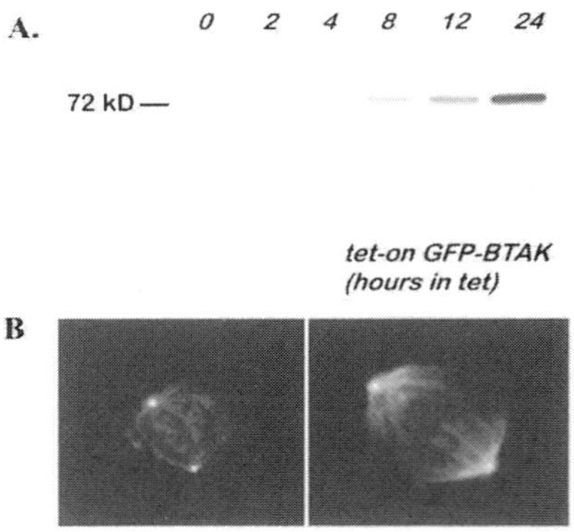

Figure 4. Regulated expression of GFP-AurA. **(A)** GFP-AurA subcloned into the pTRE vector (Clontech). A stable cell line was generated in HeLa Tet-on cells. GRP-AurA expression was induced by doxycycline addition and detectable levels of AurA were evident at 2-4 h. Following induction, GFP-AurA became localized to centrosome during interphase and prometaphase **(B)** before moving into the spindle **(C)** during prometaphase. [From Stenoien, *et al.,* 2003 (23)].

Specific domains of AurA are needed to target the kinase to the centrosome and other sites in the mitotic spindle as indicated by mutational analysis (23). As shown in Figure 8, AurA targeting in human cells requires a minimum of amino acids 1-193 for targeting, but amino acids 1-310 assure a more pronounced association. Deletion of the amino terminal 129 amino acids results in a protein that shows a more stabilized centrosomal localization as indicated by reduced recovery after photobleaching (FRAP). These experiments are still ongoing, but it is clear from our initial FRAP analysis that AurA kinase is highly dynamic with respect to its association with the centrosome, indicating a tendency to reside at the functional sites for very brief periods, in contrast to other proteins such as tubulins, NuMA, and other spindle proteins. The ability to analyze fusion protein bioluminescence and protein dynamics in live cells, during real time, has provided remarkable new insight into the micro-environment, both in the nucleus and in the cytoplasm (23, 24).

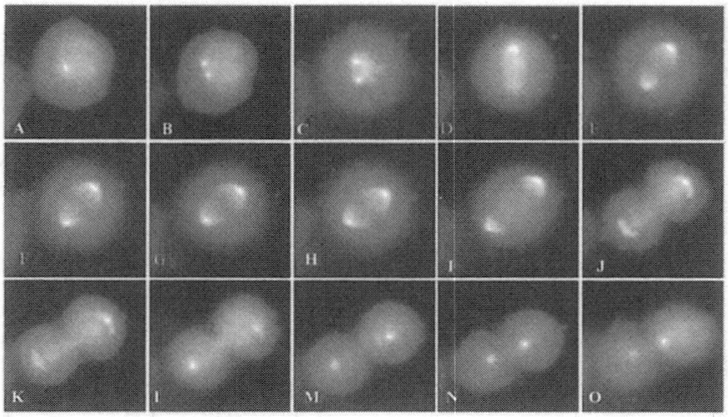

Figure 5. GFP-AurA localization during the cell cycle. Note the abrupt progression of GFP-AurA from the centrosomes to the spindle in frames C and D. Images were recorded live over a 2 h time-frame [From Stenoien, *et al.*, 2003 (23)].

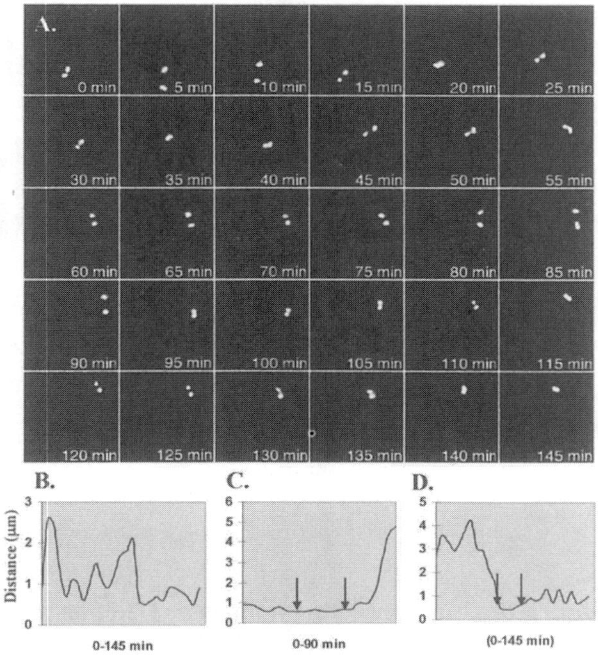

Figure 6. Analysis of GFP-AurA-marked centrosomes during the G_2 phase of a HeLa cell cycle. Note that the two centrosomes undergo oscillatory movements and appear to rotate with respect to the plane parallel to the monolayer axis (compare frames at 45 min with frames at 55 and 65 min). The distance between centrosomes changes over time as shown by the graph (B). Depolymerization of microtubules by nocodazole (C) or depletion of ATP levels (D) markedly reduces the oscillatory motion [From Stenoien, *et al.*, 2003, (23)].

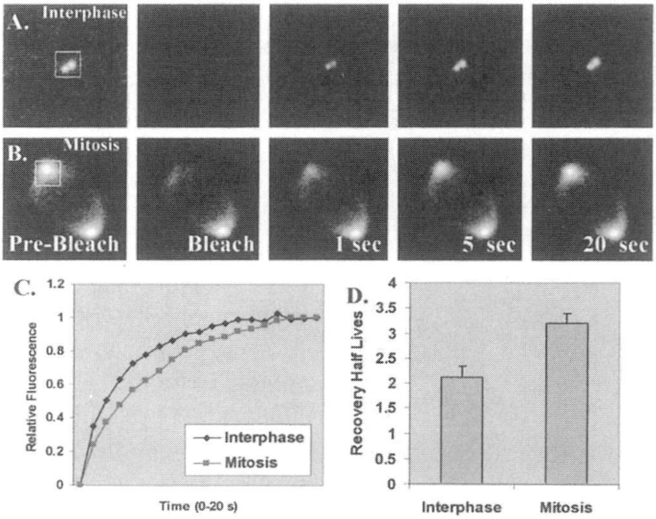

Figure 7. GF-AurA is a dynamic component of the centrosome and mitotic spindle. FRAP analysis was performed on HeLa cells stably transfected with GFP-AurA. Following a short photobleach in the region denoted by the box, GFP-AurA recovers very rapidly in both the centrosome (**A**) and spindle (**B**). The recovery curves from 10 cells are shown in (**C**). (**D**) The calculated $t_{1/2s}$ of GFP-AurA in the interphase centrosome and mitotic spindle are 2.1 ± 0.2 sec, respectively [From Stenoien, *et al.*, 2003, (23)].

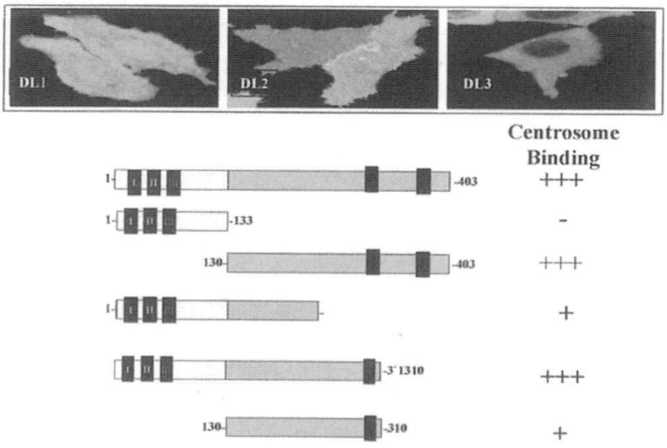

Figure 8. AurA targeting domains. Deletion analysis of centrosome targeting comparing the full-length AurA to deleted domains. AurA deletions were generated as GFP fusions to analyze the effects on AurA domains on centrosomeal targeting. DL1 (1-133) did not target to the centrosome whereas DL2 and DL3 showed moderate to strong association (inset photos). [From Stenoien, *et al.*, 2003 (23)].

Centrosome Amplification and AurA Expression in Rat Mammary Carcinogenesis: Hormone Mediation

Although AurA kinase expression and centrosome amplification have been investigated in cells *in vitro* and has been reported in a number of human tumors in vivo, few studies have utilized animal models where experimental manipulation is possible. For this reason, we adapted a novel rat mammary model, developed in the Medina laboratory (25), to explore the effects of carcinogens and hormones on centrosome amplification and AurA expression *in vivo*.

The rat mammary model used in our studies was initially established to investigate the paradigm of the hormone-induced refractory state imposed by the exposure of virgin female rats to elevated levels of estrogen and progesterone (E + P) at a critical period in development (26). It is our working hypothesis that in the mammary gland of the immediate post-pubescent female, hormones induce a molecular switch in a population of stem cells that renders the descendent mammary cells refractory to carcinogens such as methylnitrosourea (MNU) and others (25). A schema illustrating this hypothesis is diagrammed in Figure 9. Cells progressing through pathway 1 are susceptible to the carcinogen and become neoplastic, but mammary cells from females exposed to elevated E + P, via pregnancy or experimental manipulation (pathway 2), at the appropriate stage of development, become resistant to MNU exposure. Using this model, we analyzed centrosome amplification and AurA expression in cells progressing through each of the pathways. The details of these experiments and the results have been published elsewhere (27), but they are reiterated here to establish that centrosome amplification and overexpression of AurA are early events in carcinogenesis and sensitive to hormone action. Thus, E + P treatment under these experimental conditions, suppresses carcinogenesis via pathways that are, as yet, poorly understood.

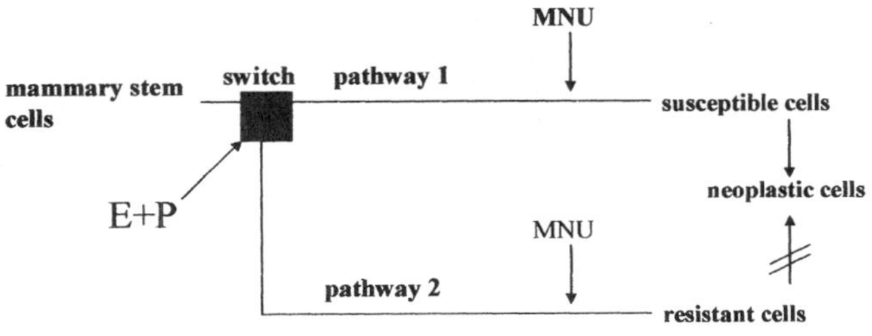

Figure 9. The rat mammary model. Scheme illustrates the proposed E + P switch (black box) showing pathway 1 of MNU-susceptible rats and pathway 2 animals that are resistant.

Centrosome counts were obtained from both epithelial and stromal cells exposed to MNU according to schemes 1 and 2 shown in Figures 10 A-D. Mammary epithelia and associated stromal cells provide an excellent built-in control for centrosome counts. Thus, proliferating cell populations (mammary epithelium) generally display cells with 1-2 centrosomes/cell, where as the non-dividing stromal cells in G_0, contain mostly single centrosomes (Figure 10B). The same populations of mammary cells from animals exposed to MNU at 97-104 generally display elevated numbers of epithelial cells with 3 or more centrosomes/cell (Figure 10D). Centrosome amplification is considered significant when 10% or more of the cells counted contained more than 2 centrosomes per cell. As shown in Figure 10D, the incidence of centrosome amplification in cells from MNU-treated rats varied from a low of 10% to a high of 46% (300 cells were scored in each mammary gland). The nonproliferating populations of stromal cells from both normal glands as well as tumors displayed mostly single centrosomes, as anticipated.

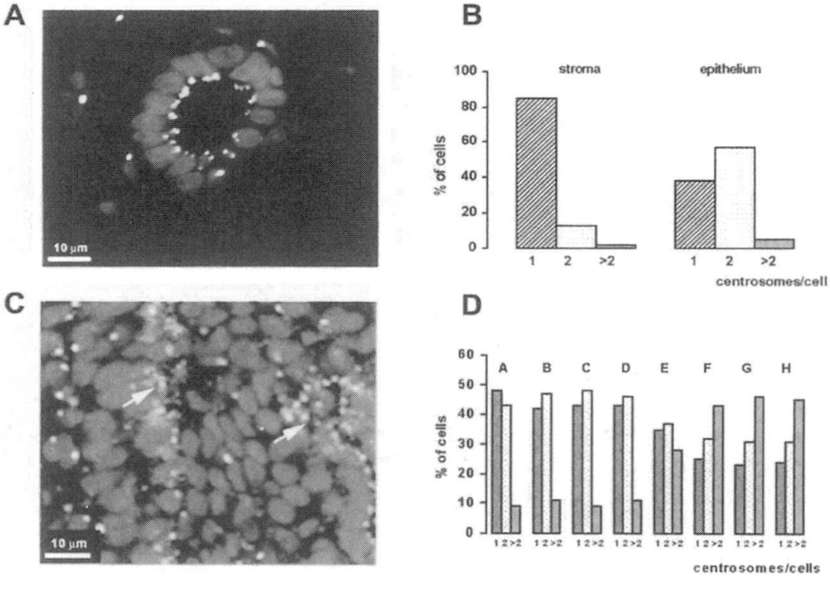

Figure 10. **(A)** Confocal microscope optical sections of mammary epithelium showing centrosome at the basal region below nuclei of normal mammary gland. The centrosome plot in **(B)** shows centrosome distribution in both stroma and normal epithelium from age matched virgin rats (AMV). Confocal image for mammary tumor **(C)** and centrosome plot **(D)** show centrosome plots of eight different mammary tumors (A-H). Details on the centrosome quantitation age given in Goepfert, *et al.*, 2002 (27).

In order to determine when the first indications of centrosome amplification occurred after exposure to MNU, we examined mammary glands at various intervals after exposure. We found mostly normal centrosome profiles, but also detected small foci of cells where centrosome amplification was occurring as early as 40 days post treatment. Histological examination of these samples revealed no indication of pre-malignant or malignant pathology. Mammary tumors arise within 15 weeks after exposure to MNU, and AurA mRNA levels and centrosome amplification correlate with tumorigenesis (Figure 11).

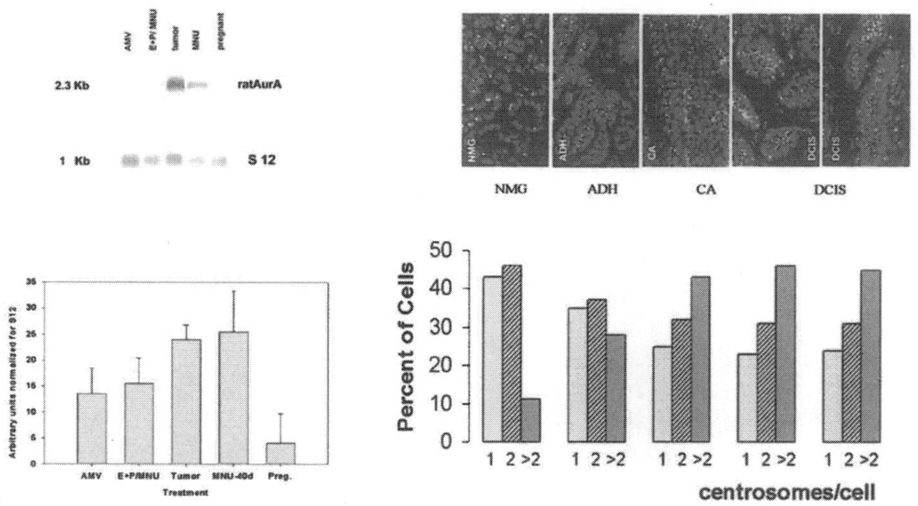

Figure 11. AurA mRNA expression and centrosome amplification correlated with tumorigenesis in rat mammary gland. Northern blot containing PolyA$^+$ RNA (5 ug) from rat mammary gland tissue of different regimens. Tissue of AMVs of animals subjected to hormonal treatment prior to MNU (E + P/MNU), of animals subjected to MNU treatment. The confocal images show sections of immunostained tissues from the normal mammary gland (NMG) and from ADH, CA, and DCIS. The corresponding centrosome plots are shown below each image. [From Goepfert, *et al.,* 2002 (27)].

To correlate centrosome amplification with overexpression of AurA, we analyzed the same rat mammary tissue samples using our human probe. However, we soon found it necessary to clone a rat orthologue to human AurA/STK15. The rat AurA (rAurA) displayed 85% identity to human kinase and 96% to the mouse orthologue (27). Northern analysis was applied to examine AurA gene expression following various treatments. As shown in Figure 11, AurA mRNA expression levels were significantly elevated [compared to control aged-matched virgin (AMV glands)] in all rat mammary tumors examined, as well as in mammary gland examined as early as 40 days after treatment. Tissues from pregnant rats, and from rats pretreated with E + P, prior to exposure, displayed control levels of AurA expression and mammary epithelial cells displayed normal centrosomes.

Rat AurA gene expression appeared to be specifically influenced by hormone treatment as indicated by mRNA levels throughout pregnancy. Early in pregnancy (6 days), message levels were elevated (compared to tissues from the same time points from E+P treatment), but later (18 days) and during lactation (10 days) message levels decreased significantly to levels comparable to AMVs (Figure 12). Such fluctuation in the expression of this kinase gene is likely due to a natural elevation in mitotic index known to occur in proliferating mammary epithelia at early pregnancy. Proliferation levels remained high initially, but decreased at the end of pregnancy and the onset of lactation. Such mitotic activity clearly influences the normal ebb and flow of AurA expression in populations of tissue cells during developmental stages of growth and differentiation and must be taken into account when analyzing mRNA expression in normal or neoplastic cells *in vivo*.

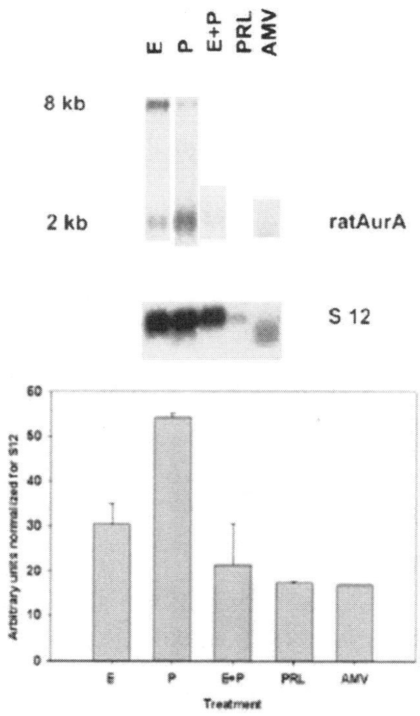

Figure 12. Northern blot containing poly A^+ RNA (5 µg) from mammary gland tissue of animals subjected to E; 20 µg), P; 20 mg, and prolactin (PRL;10 µg/g body weight) stimulation for 6 days. RNA samples were probed with radioactive labeled *ratAurA* cDNA as indicated in the figure. The blot was exposed to a Phosphorlmager, and the results were quantitated using the software Image-Quant (Molecular Dynamics). S12 ribosomal protein was used as the internal control. The data for rat AurA were plotted in arbitrary relative volume units/ug RNA normalized for S12 ribosomal protein expression. The columns show the range of variance of two treatments (n = 2); bars, SD. [From Goepfert, *et al.,* 2002 (27)].

Future Directions

Much more can be learned about the Aurora kinase family of kinases and their involvement in cell transformation when investigators more clearly identify and define molecular substrates for these catalytic enzymes, especially AurA. Until then, the role of centrosomes in cell transformation and tumorigenesis remains a "chicken vs egg" paradox. *Which comes first, centrosome amplification or some prior condition that leads to genomic instability, with subsequent, downstream dis-regulation of centrosomes?* Our results with the rat mammary model strongly suggest that both overexpression of AurA and centrosome amplification represent early, perhaps incipient events in cell transformation. Since centrosomes are responsible for achieving spindle bipolarity, a condition absolutely required for proper segregation of chromosomes, disregulation of centrosome number may be the "Rosetta stone" for genomic instability. Thus, an aberration in the process of centrosome duplication, as illustrated in Figure 13, may switch cells to an errant pathway causing aneuploidy, and initiating a selection process that results in clonal outgrowth of tumor cells. Alternatively, a similar fate could arise from a defect in cytokinesis causing misdivision of the cytoplasm and nucleus, resulting in supernumerary centrosomes and tetraploidization, as reported by Meraldi, *et al.* (17). The later pathway (Figure 13) argues that AurA overexpression does not necessarily disregulate by over-duplicating centrosomes, but produces cells with an extra burden of centrosomes through misdivision that will initially double the centrosome number. Further variations in centrosome numbers could arise in subsequent divisions causing multipolar spindles to form leading to aneuploidy and apoptosis, in most cells, or perhaps a favorable partitioning of the genome in a rare segrant, favoring survival, clonal expansion from daughter cells and tumor production.

If the initial lesion that gives rise to genomic instability involves targets within the fundamental machinery of cell division, such as those that maintain correct centrosome number and distribution, the surviving cancer cells must ultimately compensate for this harmful defect. Otherwise mitotic chaos would persist. Mature cancer cells divide as virtual "mitotic machines," however, ultimately overgrowing and out-competing normal diploid cells for space and nutrients, leading to the death of the host. Thus, it seems highly likely that a natural selection process exists to set the mitotic process back on track, rescuing tumor cells from an initial chaotic state of division, to one where mitotic efficiency reigns supreme (7).

Cancer, like all living forms, is thus, a product of Darwinian evolution that capitalizes on the time-honored process of natural selection that leads to diversity in all life forms. The successful development of future anti-cancer drugs and treatment regimens not only requires additional knowledge of the molecular basis of cell division, but also must take into account the basic concepts of evolution and natural selection within tissue cells. Accordingly, a cancer cell is a "survivor" of the harsh natural ecosystem of the host, and, in some cases, the pharmaceutical armamentarium of modern medicine, --and becomes a species all its own (13).

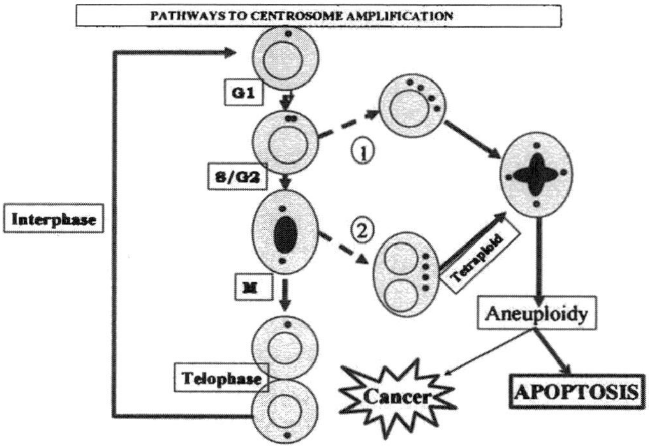

Figure 13. Centrosome pathways. Centrosomes are duplicated once, and only once during a normal cell cycle, producing a pair of centrosomes that separate and migrate to opposite poles of the mitotic spindle during mitosis (**M**). The replicated chromosomes are aligned on the metaphase plate of the bipolar spindle and partitioned equally to each daughter cell producing diploid progeny. At telophase, each daughter receives a single centrosome that becomes the major MTOC at G_1-phase. Centrosome amplification occurs when errors occur, either in the process of duplication (**1**), or partitioning of centrosomes (**2**) such that the cells receive extra copies of centrosomes. In **pathway 1,** centrosomes undergo multiple rounds of replication during the cell cycle, possibly due to failure of a centrosome-associated checkpoint, producing supernumerary centrosomes and a multipolar spindle at the subsequent M. In **pathway 2,** centrosomes accumulate in a cell that has undergone endoreduplication and failure to complete cytokinesis. Either pathway produces dis-regulation of chromosome segregation, resulting in aneuploidy followed by cell death (apoptosis). According to this scheme, a rare daughter cell receives a favorable complement of genes, growth factors, etc., (the mutator phenotype) producing progeny that proliferate and survive to produce a cancer.

Acknowledgement

The authors are grateful to Janet Hom for administrative support, to Kevin Brinkley for editorial assistance and to Rebecca Moore and Joshua Newton for technical assistance. This research was supported in part by grants from NIH CA 64255 and CA 41434. *We dedicate this work to the memory of Professor T.C. Hsu.*

References

1. Fukasawa K, Choi T, Kuriyama R, et al (1996) Abnormal centrosome amplification in the absence of p53. Science 271:1744-1747.
2. Wang X-J, Greenhalgh DA, Jiang A, et al (1998) Analysis of centrosome abnormalities and angiogenesis in epidermal-targeted p53^{172H} mutant and p53-knockout mice after chemical carcinogenesis: Evidence for a gain of function. Molec Carcinogenesis 23:185-192.
3. Ingle WL, Lutz WH, Ingle JN, et al (1998) Centrosome hypertrophy in human breast tumors implications for genomic stability and cell polarity. Proc Natl Acad Sci USA 95:2950-2955.
4. Pihan GA, Purohit A, Wallace J, et al (1998) Centrosome defects and genetic instability in malignant tumors. Cancer Res 58:3974-3985.
5. Sen S, Zhou H, Zhang R-D, et al (2003) Amplification/overexpression of a mitotic kinase gene is associated with aneuploidy and aggressive behavior of human bladder cancer. J Natl Cancer Inst 94:1320-1329.
6. Nigg EA (2002) Centrosome aberrations: cause or consequence of cancer progression? Nature Rev Cancer 2:815-825.
7. Brinkley BR (2001) Managing the centrosome numbers game: from chaos to stability in cancer cell division. Trend Cell Biol 11:18-21.
8. Boveri T (1914) The Origin of Malignant Tumours. Baltimore:Williams and Wikins.
9. Morgan TH, Bridges CB, Sturtevant AH (1925) The genetics of Drosophila. Bibliogr Genet 2:3-262.
10. Muller HJ (1927) Artificial transmutation of the gene. Science 66:84-87.
11. Bishop JM (1981) Enemies within: genesis of retrovirus oncogenes. Cell 23:5-6.
12. Bishop JM (1995) Cancer: the rise of the genetic paradigm. Genes Development 9:1300-1315.
13. Duesberg P, Resnick D (2000) Aneuplody, the somatic mutations that makes cancer a species of its own. Cell Motility Cytoskeleton 47:81-107.
14. Brinkley BR, Goepfert TM (1998) Supermunerary centrosomes and cancer; Boveri's hypothesis resurrected. Cell Motility Cytoskeleton 41:281-288.
15. Palazzo RE (2003) Centrosome and spindle pole body Dynamics; a review of the EMBO/EMBL Conference on centrosomes and spindle pole bodies, Heidelberg, Sept. 13-17. 2002. Cell Motility Cytoskeleton 54:148-154.
16. Sluder G, Hinchcliffe EH (2001) The coordination of centrosome reproduction with nuclear events during the cell cycle. Curr Top Dev Biol 49:267-289.
17. Meraldi P, Honda R, Nigg EA (2002) Aorora-A overexpression reveals tetraploidization as a major route to centrosome amplification in p53 -/- cells. EMBO J 21:483-492.

18. Katayama H, Brinkley WR, Sen S (2003) The aurora kinases: role in cell transformation and tumorigenesis. Cancer Metastasis Rev 22:451-464.
19. Goepfert TM, Brinkley BR (2000) The centrosome-associated aurora/IP-like kinase family. Curr Top Dev Biol 49:331-342.
20. Zhou H, Kuang J, Zhong L, et al (1998) Tumour amplified kinase STK15/BTAK induces centrosome amplification, aneuploidy and transformation. Nature Genet 20:189-193.
21. Bichoff RJ, Anderson L, Zhu Y, et al (1998) A homologue of Drosophila aurora kinase is oncogenic and amplified in human colorectal cancers. EMBO J 17:3052-3065.
22. Farruggio DC, Townsley FM, Ruderman JV (1999) Cdc20 associates with the kinase aurora2/AIK. Proc Natl Acad Sci USA 96:7306-7311.
23. Stenoien DL, Sen S, Mancini MA et al (2003) Dynamic association of a tumor amplified kinase, aurora A with the centrosome and mitotic spindle. Cell Motility Cytoskeleton 55:134-146.
24. Stenoien DL, Patel K, Mancini MG, et al (2001) FRAP reveals that mobility of oestrogen receptor-alpha is ligand-and protesosome dependent. Nature Cell Biol 3:15-23.
25. Silvaraman L, Stephens LC, Markaverich BM, et al (1998) Hormone – induced refractoriness to mammary carcinogenesis in wistar-furth rats. Carcinogenesis 19:1573-1581.
26. Russo IH, Russo J (2000) Hormonal approach to breast cancer prevention. J Cell Biochem Suppl. 34:1-6.
27. Goepfert TM, Adigun YE, Zhong L, et al (2002) Centrosome amplification and overexpression of Aurora A are early events in rat mammary carcinogenesis. Cancer Res 62:4115-4122.

STATE OF THE ART LECTURES

1

Chromosomal Instability: A New Paradigm for Estrogen-induced Oncogenesis

Jonathan J. Li, Jeffrey Salisbury, and Sara Antonia Li

Introduction

Human sporadic breast cancer (BC) comprises >90% of all BC cases whereas familial BC is less than 10% (1). Despite its likely multifactorial origin, there is now pervasive evidence from epidemiological and animal studies, developed over the past several decades, that the causation of human sporadic BC primarily involves female sex hormones, particularly estrogens (Es) (1-8). This view is consistent with long standing epidemiological data relating extended exposure to Es and elevated BC risk, such as early first menarche, late age at menopause, nulliparity, late age at full-term pregnancy, and absence of lactation (5, 6). These BC risk factors are all related to pre-menopausal women. Moreover, all of the well-established BC risk factors are associated with elevated circulating E levels. Even lesser risk factors such as obesity and alcohol ingestion are known to significantly increase serum E concentrations in women (9, 10). These earlier studies are buttressed by results of the recent Breast Cancer Prevention Trial in which tamoxifen (TAM) treatment markedly reduced (44-55%) BC risk in women considered at increased risk for the disease (11). Since it is evident that Es are crucial to our understanding of sporadic BC etiology, it is surprising that so little is known about the involvement of Es in oncogenic processes in target tissues such as the breast, other than its ability to elicit cell proliferation. Our laboratory has utilized two seemingly disparate animal models in which malignant tumors are induced solely by Es in the absence of any other exogenous intervening carcinogenic or promotional agent (12-15). Based on our recent findings, it has been concluded that the early detection of chromosomal instability (CIN) and aneuploidy, in early lesions and primary neoplasms in both systems, is a hallmark characteristic of E-induced oncogenesis.

General Considerations

Animal models are most useful when they closely resemble the biologic and molecular aspects of the human disease they intend to study. Although the two experimental models to be discussed have many fundamental differences, they have in common the ability of certain cells within each of these tissues (*i.e.,* kidney, breast) to undergo neoplastic transformation in response to E treatment alone. Moreover, the neoplasms induced by Es in the castrated hamster kidney and in the intact female ACI rat mammary gland (MG) are both prevented by concomitant anti-estrogen treatment (Table 1) (15, 16). These data strongly indicate that estrogen receptor (ERα) is central in mediating the oncogenic response of Es. Although the minimum oncogenic dose of E to elicit tumors in the kidney is not known, typically, serum 17β-estradiol (E_2) levels of 1.6-2.4 ng/ml are used to induce a 100% tumor incidence. In contrast, only 60-130 pg/ml serum E_2 levels are sufficient to result in a 100% mammary tumor (MT) incidence in female ACI rats (15). The lower serum E_2 concentrations are well within the physiological range of serum E_2 levels reported in cycling female rats (17-88 pg/ml) (17). On the other hand, serum E_2 levels in normal cycling women have shown to be in the range of 40-360 pg/ml (18). Importantly, in normal breast and primary BC tissues, the E_2 concentrations were found to be extremely low (Table 1) (19-22) and approximate that seen in the hamster kidney during chronic E treatment.

Table 1. Serum and Tissue E_2 Concentrations During E-induced Oncogenesis.

Tissue site	Serum (pg/ml)	Tissue (pg/mg protein)
Hamster EUTK	1670 - 2480	4.6
ACI Rat MT	60 - 130	---
Human Breast	40 - 350	5.7
Human BC		8.9

Ectopic Uterine Tumors in the Kidney (EUTK)

While the tumors in the hamster kidney have been the foremost animal model studied in E-induced oncogenesis, there are a number of unusual aspects to this system which may be now understood. Unlike the breast, prevention of these tumors can be effectively blocked by the concomitant treatment with either progesterone or androgen (16). This finding provides a clue as to the cell of origin of these tumors arising in the kidney. Since the hamster reproductive and urinary systems arise from a common germinal ridge of multi-potential stem cells, we have postulated that some of these germinal/stem cells, normally destined to the uterus, migrate and establish themselves in the cortico-medullary area of the kidney, the earliest region formed in the developing kidney (23). These ectopically located "uterine" germ cells remain dormant unless exposed to a sustained level of E. This contention is supported by the following findings: 1. Resemblance of early renal interstitial lesions to blastema, with positive staining of mesenchymal markers (*e.g.* vimentin, desmin) but only a trace of cytokeratin (24, 25). The cytokeratin expression markedly rises during tumor progression. 2. Only a subset of these renal interstitial stem cells in the cortico-medullary region express progesterone receptor (PR) after only two to three weeks of E treatment (Figure 1) (23). 3. After one mo, this subset of interstitial stem cells co-express ERα (Figure 2) (23). 4. The lack of ERβ expression in these E-sensitive renal interstitial stem cells, in early tumorous lesions, and primary tumors is consistent with the established selective expression and proliferative role of ERα in uterine tissue (26). 5. The ability of P to completely block these now designated ectopic uterine stem cell tumors in the kidney (EUTK) is consistent with the opposing action of P in the uterus. 6. The essentially identical isoform profiles of ERα and PR in primary EUTK to those of corresponding receptor profiles in the hamster uterus (23). 7. The specific high proliferative activity in this same subset of interstitial stem cells in response to E and the overexpression (OE) of early E response genes (12) further supports our contention. An attractive aspect of this experimental model is that EUTK occur in the absence of any sequential intervening morphological stages, but develop as a continuous progression of a subset of E-sensitive interstitial stem cells leading to tumor formation.

Table 2. Prevention of E-induced Oncogenesis of Hamster EULTK and ACI Rat MT by Tamoxifen.

Species	Treatment	No. Animals	% Animals w/Tumors	Tumor Multiplicity/Animal
Hamster[1]	Untreated	8	0	0
	TAM	8	0	0
	E_2	10	100	16.7 ± 1.1
	E_2 + TAM	8	0	0
ACI Rat[2]	Untreated	10	0	0
	TAM	8	0	0
	E_2	12	100	15.6 ± 1.6
	E_2 + TAM	8	0	0

[1] Castrated hamsters were treated with a 20-mg pellet of E (DES or E_2) alone or in combination with a 20-mg pellet of TAM, every 3.0 mo for 8 mo.

[2] Intact female ACI rats were treated with a single 20-mg pellet containing 3 mg of E_2 + 17 mg of cholesterol alone or in combination with a single 20-mg pellet of TAM for a 6.0-mo period.

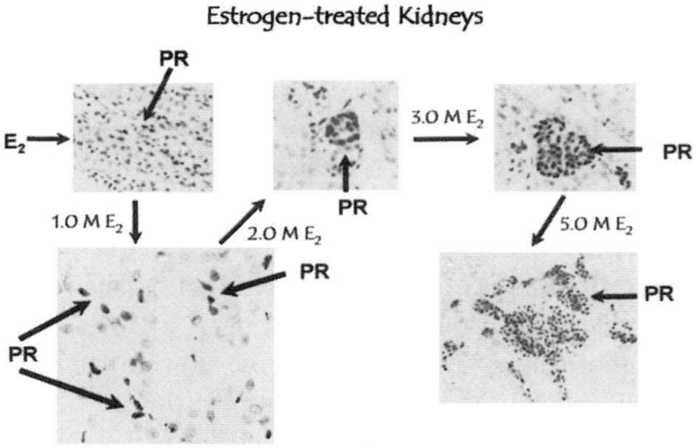

Figure 1. Immunohistochemical detection of E-induced PR expression in castrated male hamster after 1.0, 2.0, 3.0, and 5.0 months of E treatment. Note the PR^+ interstitial cells (arrows) at the cortical-medullary junction in the animals treated for 1.0 mo, PR^+ cells in a nascent tumor foci after 2.0 mo, in an early renal tumor foci after 3.0 mo, and in a moderate size EUTK foci, after 5.0 mo (200x).

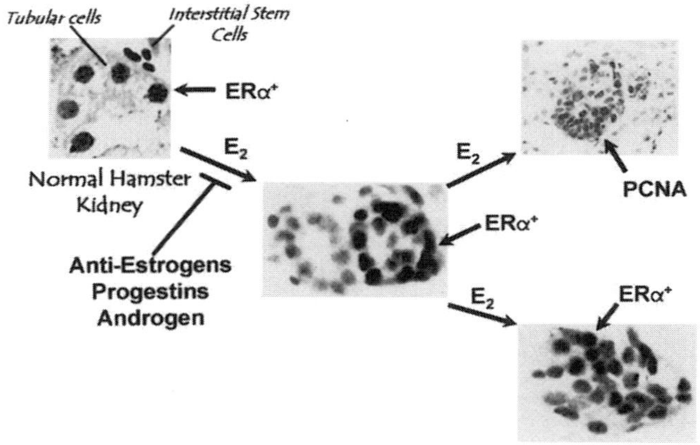

Figure 2. ERα expression during E-induced oncogenesis. Normal hamster kidney cortical tubule cells (arrows) exhibited $ER\alpha^+$ $ER\alpha^+$ expression in normal proximal tubular cells are downregulated upon E-treatment. Thereafter, individual and small groups of $ER\alpha^+$ interstitial cells are seen after 3.0-mo, and moderate size renal tumor foci after 4.0-mo E-treatment (250x). In addition, a small tumor foci is shown stained for PCNA.

Mammary Tumors in Female ACI Rats

Numerous rat strains exhibit variable susceptibility to E-induced MTs including ACI, Noble, and Long Evans (60-100% tumor incidence) and to a lesser extent,

August, Wistar-Wag, and SD (36-42%) (14, 15, 27-32). Recently, we have shown that >40% MTs may be induced in female SD rats after only 5.0-mo of estrone treatment (El-Bayoumy, K and Li, JJ, unpublished data). An attractive feature of the E-induced female ACI rat model to study human sporadic BC causation is the ability of this strain to elicit 100% multiple MTs employing E_2 serum levels of only 60-180 pg/ml (Figure 4) (15). Moreover, we have established that <60 pg/ml of E_2 was also effective in producing essentially a 100% MT incidence (Li, SA, unpublished data). An E_2 dose/response relationship between MT multiplicity and incidence was evident. Below 35 pg/ml of E_2, no MTs were detected within a 6.0-mo treatment period. Additionally, above an E_2 serum level of 200 pg/ml, there was a reduction in MT multiplicity and incidence (Figure.3). These results indicate that within a relatively narrow range of serum E_2 concentrations (all at low E_2 pg/ml). MT multiplicity and incidence may be enhanced. Below and above this serum E_2 range, MTs are either absent or inhibited, respectively. These data are consistent with epidemiological findings that only modest elevations in serum E_2 levels may increase human BC risk and high levels of E_2, such as found during pregnancy, may reduce BC risk (33, 34).

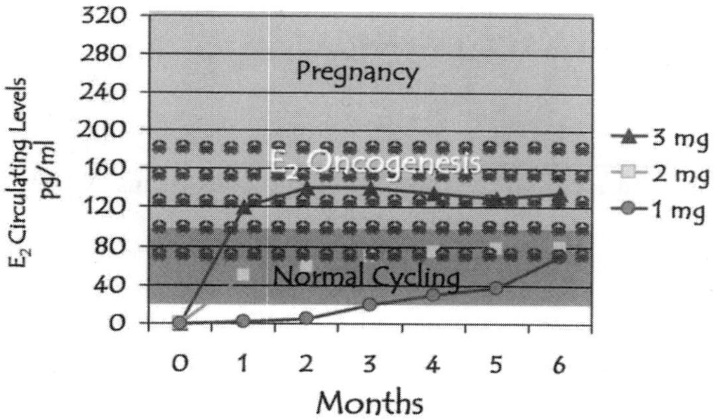

Figure 3. Serum levels determined at monthly intervals in female ACI rats treated with either 1, 2, and 3 mg E_2. The data represent the mean of four individual samples measured in duplicate. 100% MT incidence was elicited at a E_2 serum concentration range between 60-180 pg/ml.

Synthetic chemical carcinogens (dimethylbenz(a)anthracene, DMBA and nitrosomethylurea, NMU) induced MT models have dominated experimental BC research for nearly a half century. Recent studies in Long Island, NY, however, provide compelling evidence that environmental carcinogens, including polycyclic aromatic hydrocarbons, do not have an appreciable role in BC etiology (35, 36). In a blinded study, employing nuclear image cytometry (NIC), we have shown that two synthetic chemical carcinogens (*i.e.*, DMBA and NMU) and one environmental carcinogen, 6-nitrochrysene (6-NC) yielded primarily diploid (> 85%) MTs in female rats (Table 3) (14). In contrast, both DCIS and primary MTs induced in either female ACI or Noble rats by E treatment alone resulted in MTs which were highly aneuploid (> 84%). These results provide very strong evidence that the induction of experimental breast tumors by potent mammary chemical carcinogens occurs by distinctly different molecular mechanisms utilizing different pathways compared to those breast tumors induced solely by Es, and it is the latter mammary neoplasms which more closely resemble those seen in human DCIS and invasive ductal BCs (37, 38).

Table 3. NIC Assessment of Aneuploid Frequency in MTs Induced by Hormones or Chemical Carcinogens in a Variety of Rat Strains.

		% Aneuploid Cells[1]	
	Strain	**CIS**	**MGT**
HORMONES			
17β-E_2	ACI	84.5 (6)	90.9 (9)
17β-E_2 + TP	Noble	85.2 (5)	89.3 (5)
CHEMICAL CARCINOGENS			
DMBA	BuF/N		14.6 (5)
NMU	BuF/N		12.8 (5)
6-NC	SD		10.5 (5)

[1] Number in parentheses indicates number of individual scans (x 100 cells).

Steroid Hormone Receptors (ERα, PR)

Generally, similar hormone receptor responses to sustained E exposure were obtained in the kidneys of male hamsters and mammary glands of female ACI rats (15, 23). In kidneys of untreated control hamsters, a predominate 50-kDa ERα variant and a slightly lower expression of a 64-kDa ERα isoform was found (Figure 4). In contrast, untreated normal mammary glands contained only a single major 56-kDa ERα variant (Figure 4). Following chronic treatment with E_2, the 64- kDa ERα variant level was mainly elevated in hamster kidneys. With further E_2 treatment, a 58-kDa isoform appeared after 4.0- to 5.0-mo and was present in all primary EUTK examined whereas the 64-kDa ERα isoform was evidently loss (Figure 4). In addition to the presence of both 50- and 58-kDa ERα variants, both the hamster uterus and EUTK samples invariably expressed the 66-kDa form, presumably the full length ERα. Based on the immunohistochemical analyses in renal tissue sections, changes in these ERα isoforms following E exposure were evidently confined to the E sensitive renal interstitial cells which subsequently undergo neoplastic development and multiply. Similarly, after low dose E_2

treatment, a number of ERα forms were detected in female ACI rat mammary glands. In addition to the full length 66-kDa ERα form, the major ERα variants were 56- and 47-kDa (Figure 4). Moreover, two lesser ERα variants, a 55- and 72-kDa, were also seen as well as a new 54-kDa ERα form. The dominant form of ERα in MTs, however, was this 54-kDa ERα. This latter finding may be important in mediating E-dependent MT growth advantage. E elevated PR both in treated male hamster kidneys and in female ACI rat mammary glands and in their respective primary tumors (Figure 4). In primary EUTK, PR-B, -A, and–C expressions were markedly increased whereas in primary ACI rat MTs, only PR-A and –C exhibited strong expression following sustained E exposure.

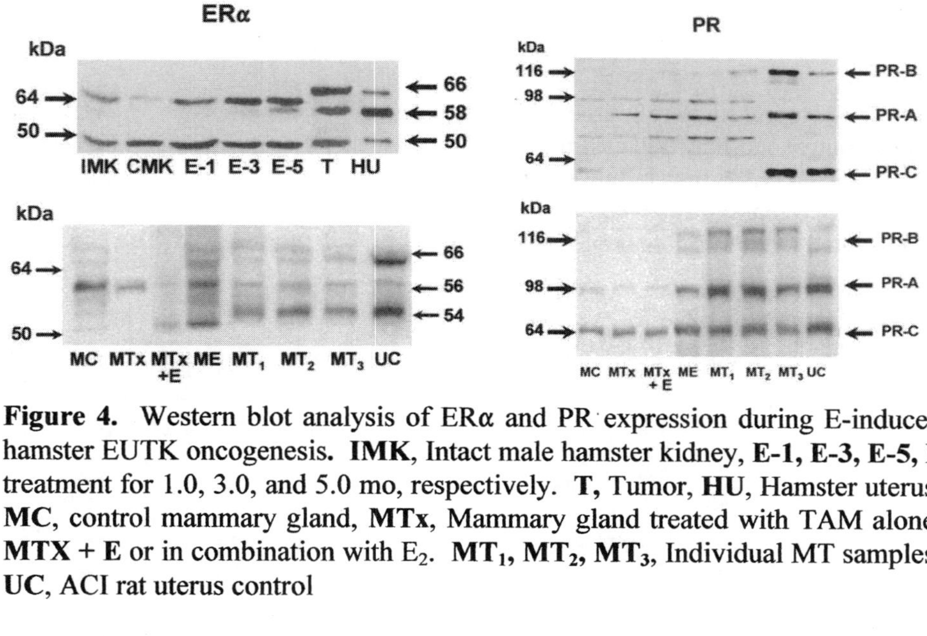

Figure 4. Western blot analysis of ERα and PR expression during E-induced hamster EUTK oncogenesis. **IMK**, Intact male hamster kidney, **E-1, E-3, E-5,** E treatment for 1.0, 3.0, and 5.0 mo, respectively. **T,** Tumor, **HU**, Hamster uterus. **MC**, control mammary gland, **MTx**, Mammary gland treated with TAM alone, **MTX + E** or in combination with E_2. **MT_1, MT_2, MT_3**, Individual MT samples. **UC**, ACI rat uterus control

c-*Myc* Gene Overexpression and Amplification

In addition to a similarity in steroid receptor responses, characteristic of E-dependent cells, both developing E-induced hamster EUTK and ACI MTs (*i.e.*, DCISs) overexpress c-*myc* and MYC protein (12, 14, 39). This upstream c-*myc*/Myc response to E treatment is a characteristic response of this hormone on its target tissues. Recently, it has been shown that OE of c-*myc* elicits CIN (40, 41). However, it is likely, that downstream cell cycle genes, mediated by c-*myc* OE, are more intimately involved in eliciting the loss of mitotic stability resulting in CIN and subsequent aneuploidy. Southern blot analyses were performed on primary hamster EUTK and ACI rat MTs taken from individual E_2-treated animals (Figure 5) (14, 39). In EUTK, 67% (8/12) exhibited c-*myc* amplification (range 2.4-3.6).

A 66% (6/9) amplification of c-*myc* (range 3.4-6.9) was also found in ACI rat MTs. In untreated control hamster kidneys and ACI rat mammary glands, the mean densitometric level of c-*myc* had a range between 0.7-1.5. The amplification of c-*myc* in E_2-induced EUTK and ACI rat MTs is due, in part, to a consistent gain in the number of copies of chromosome 6 (hamster) and 7 (rat), where the c-*myc* gene resides (Figure 6).

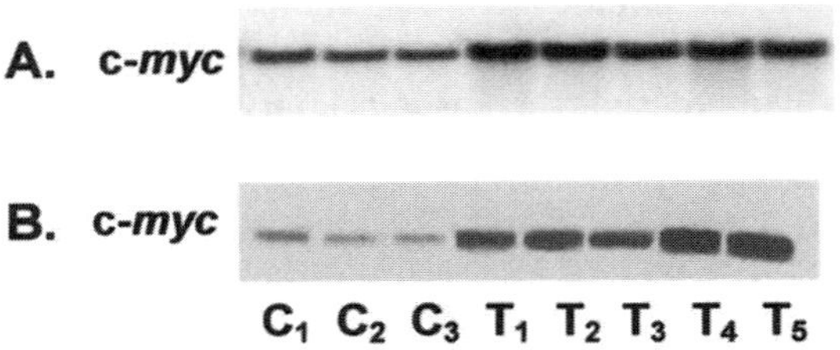

Figure 5. Representative southern blot analyses of c-*myc* expression. **A.** Hamster. Three control age-matched castrated male kidneys (C_1-C_3) and five EUTKs (T_1-T_5) isolated from individual animals after 7 mo E-treatment. B. ACI rat. Three control age-matched mammary glands from intact female ACI rats (C_1-C_3) and 5 MT (T_1-T_5) isolated from individual animals after 6.0 mo E_2 treatment.

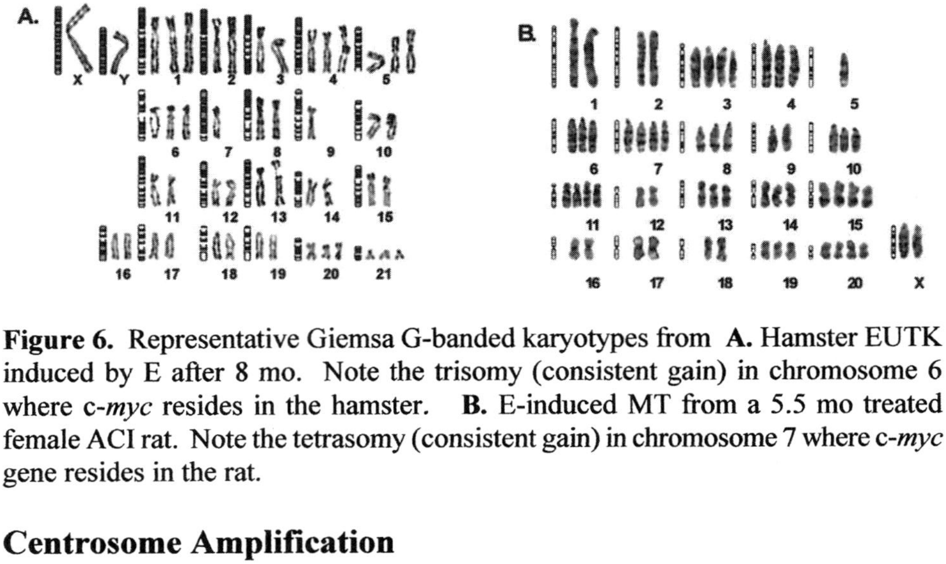

Figure 6. Representative Giemsa G-banded karyotypes from **A.** Hamster EUTK induced by E after 8 mo. Note the trisomy (consistent gain) in chromosome 6 where c-*myc* resides in the hamster. **B.** E-induced MT from a 5.5 mo treated female ACI rat. Note the tetrasomy (consistent gain) in chromosome 7 where c-*myc* gene resides in the rat.

Centrosome Amplification

The centrosome is the major microtubule-organizing center. Centrosomes are essential to the control of spindle bipolarity, spindle positioning, and cytokinesis. Abnormalities in centrosome number, common in human DCISs and primary invasive ductal BCs (42-44), can interfere with bipolar spindle formation and chromosome segregation, thus leading to CIN and aneuploidy. Normal hamster renal tissues exhibit regular comet-like staining for pericentrin (red) which occurs apically and low levels of centrin staining (Figure 7). EUTK sections show large aggregates of pericentrin and elevated levels of centrin staining (green), typical of centrosome amplification (Figure 7). Centrosome amplification was also commonly detected in primary ACI rat MTs (Lingle, W., Salisbury, J, Li, JJ, and Li SA, unpublished data). It is concluded from these findings that centrosome defects may be a characteristic feature of solely estrogen-induced oncogenic processes and a primary causative event leading to E_2-induced tumor formation.

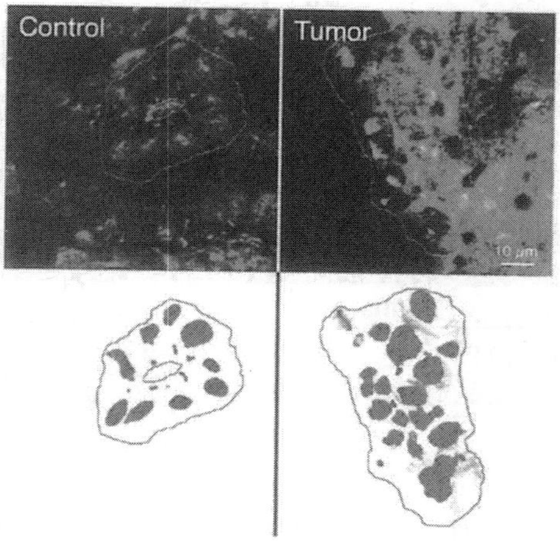

Figure 7. Normal control and EUTK specimens stained with a cocktail of antibodies against pericentrin (rabbit Ig) and centrin (mouse monoclonal ascites) followed by the appropriate secondary antibodies (Alexa 568 - red for pericentrin and Alexa 488 – green for centrin) and the DNA intercalating dye DAPI. Sections were observed using a 63X 1.2 NA objective and recorded digitally on a Zeiss 510 confocal microscope. The images represent maximum projections of five 0.2 µm optical sections. The schematic representations (lower figures) illustrate nuclear (blue), pericentrin (red), and centrin (green) signals. Bar = 10 µm.

Chromosomal Instability and Aneuploidy

In addition to the nonrandom occurrence of trisomies and tetrasomies in chromosome 6 and 7 in E_2-induced hamster EUTK and ACI rat MTs, respectively, where the c-*myc* gene resides (Figure 6), other chromosomes were also consistently gained or lost (14, 45). Nonrandom gains in chromosomes 3, 6, 20, and 21 were detected in primary hamster EUTK employing a stringent criterion. Consistent chromosome losses, however, were not found. In E_2-induced ACI rat MTs, chromosomes 11, 13, 19, and 20 exhibited high frequencies of trisomies and to a lesser extent tetrasomies and a monosomy in chromosome 12 using the same criterion. In these E_2-induced neoplasms, consistent or nonrandom numerical chromosome gains or losses were considered to have a frequency of occurrence $\geq 30\%$. Recurrent numerical chromosome alterations (>20 to <29.9%) were also detected in both E_2-induced primary neoplasms, as well as random chromosome changes were seen. A major challenge is to discern of the mechanisms whereby consistent, recurrent, and random numerical chromosomal alterations are generated together.

Summary and Conclusions

The cellular and molecular resemblance of E_2-induced primary hamster EUTK and ACI rat MTs to human DCIS and invasive ductal BC (IDC) is striking (Table 4).

Table 4. Similarities in Salient Molecular Parameters Between Primary Human Ductal Breast Carcinomas and E-induced Tumors.

	Hamster		**Human**		**ACI Rat**	
	EarlyFoci (%)	**EUTK (%)**	**DCIS (%)**	**IDC (%)**	**DCIS (%)**	**IDC (%)**
ER^+	97	100	60-84	73	100	100
$ER^+ + PR^+$	100	100	66	71	100	100
Aneuploid	94	100	55-78	85-92	84	91
MYC Protein OE	100	100	100	71-100	--	100
c-*myc* Gene Ampl	--	67	--	12-86	--	66
Cyclin D1 Protein OE	--	100	72	43	--	100
Cyclin D1 mRNA OE	--	--	76-87	83	--	100
Centrosome Ampl	--	100	100	80	--	100
AurA Protein OE	--	100	--	94	--	100

In particular, the presence of ERα and PR and the histopathology of the ACI rat MT are analogous to human ductal BC. Other similarities to human ductal BC and these solely E-induced neoplasms are: MYC protein OE, c-*myc* amplification, and cyclin D1 OE (46-49). Most remarkable, however, is the coincident high frequency of centrosome amplification and aneuploidy in these solely E-induced rodent neoplasms, and human DCIS and primary invasive ductal BCs (Table 4) (42-45). The OE of Aurora A kinase [Aur A, breast tumor amplified kinase (BTAK)] was also detected in both primary hamster EUTK and ACI rat MTs (Table 4). While no direct association between Aur A OE, centrosome amplification, and CIN have been shown, they always occur together (50). Aur A OE has been implicated in eliciting these latter events (51-53).

Based on our current knowledge, the scheme in Figure 7 is proposed for E-induced oncogenesis in the two mammalian models studied. E interacts with its receptor (ERα) in susceptible cells to elicit c-*myc*/MYC OE. The OE of MYC turns on specific down stream cell cycle-related genes (*i.e.*, cyclins D1, D2, and E). It is also possible that E-ERα may directly activate the cyclin D gene family. The persistent OE of these cell cycle related proteins lead to elevated levels of centrosomal proteins (*i.e.*, pericentrin, centrin, γ tubulin), and centrosome-associated kinases (Aur A) located in the lattice network/spindle regions of centrosomes. Thus, disturbances in centrosome regulation result in defects in duplication/separation, resulting in centrosome amplification, considered a mutational event. Importantly, our findings for the first time directly link E-action to the deregulation of the cell cycle, Aur A OE, CIN, and aneuploidy as causative events leading to solely E-induced tumor development in two *in-vivo* systems. It is noted that these aforementioned similarities in molecular characteristics between E_2-induced hamster EUTK and ACI rat MTs and their respective early lesions occur despite the absence of P in the E-induced hamster EUTK model, and the presence of this hormone in the E-induced ACI MT system.

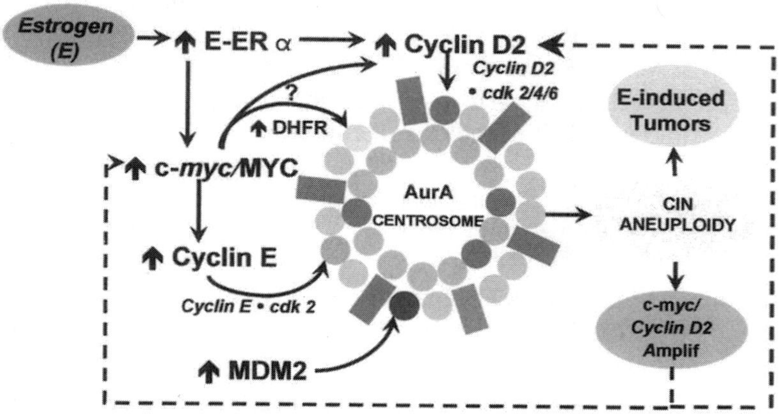

Figure 8. Scheme proposed for the development of EUTK in the hamster kidney and ACI MT-induced solely by chronic E exposure. E binds to its overexpressed ERα as a result of E treatment. The E-ERα complex transactivate c-*myc* leading to its OE and subsequent amplification. Also, E-ERα interactions lead to a rise in cyclin E, cyclin D2, and MDM2. The cyclin•cdk complexes and other regulatory proteins (MDM2, DHFR) bind to specific centrosome proteins located in the lattice network of centrosomes. The concerted interactions of these complexes cause chromosomal instability and eventual malignant tumor formation.

Acknowledgements

This investigation was supported by NIH Grants CA58030 and 33519. We wish to thank Dan Papa, Ph.D., and S. John Weroha for their able technical assistance in these studies.

References

1. King M-C, Rowell S, Love SM (1993) Inherited breast and ovarian cancer. What are the risks? What are the choices? J Am Med Assoc 269:1975-1980.
2. Clemons M, Gross P (2001) Estrogen and the risk of breast cancer. New Engl J Med 344:276-285.
3. Toniolo PG (1997) Endogenous estrogens and breast cancer risk. The case for prospective cohort studies. Environ Health Perspect 105:587-601.
4. Felgelson HS, Henderson BE (1996) Estrogens and breast cancer. Carcinogenesis 17:2279-2284.
5. Adami HO, Persson I, Ekbom A, et al (1995) The aetiology and pathogenesis of human breast cancer. Mutation Res. 333:29-35.
6. Henderson BE, Ross R, Bernstein L (1988) Estrogens as a cause of human cancer: the Richard and Linda Rosenthal Foundation Award Lecture. Cancer Res 48:246-253.
7. Rothschild TC, Boylan ES, Calhoon RE, et al (1987) Transplacental effects of diethylstilbestrol on mammary development and tumorigenesis in female ACI rats. Cancer Res 47:4508-4516.
8. Hammouche N, Samperez S, Riviere M-R, et al (1982) Estrogen and progesterone receptors in mammary tumors induced in rats by simultaneous administration of 17β-estradiol and progesterone. J Steroid Biochem 17:415-419.
9. Wu AH, Pike MC, Stram DO (1999) Meta-analysis: Dietary fat intake, serum estrogen levels and the risk of breast cancer. J Natl Cancer Inst 91:529-534.
10. Sarkola T, Makisal OH, Fakunaga T, et al (1999) Acute effect of alcohol on estradiol, estrone, progesterone, prolactin, cortisol, and luteinizing hormone in premenopausal women. Alcoholism Clin Exp Res 23:976-982.
11. Fisher B, Costantino JP, Wickerham DL, et al (1998) Tamoxifen for prevention of breast cancer: Report of the National Surgical Adjuvant Breast and Bowel Project P-1 Study. J Natl Cancer Inst 90:1371-1388.
12. Hou X, Li JJ, Chen WB, et al (1996) Estrogen-induced protoncogene and suppressor gene expression in the hamster kidney: Significance for estrogen carcinogenesis. Cancer Res 56:2616-2620.
13. Li JJ Papa D, Li SA (2003) Ectopic uterine interstitial cell tumors in the hamster kidney. Minerva Endocrinologica 28:321-328.
14. Li JJ, Papa D, Davis MF, et al (2002) Ploidy differences between hormone- and chemical carcinogen-induced mammary neoplasms: Relation to human breast cancer. Molec Carcinogen 33:56-65.
15. Li SA, Weroha SJ, Tawfik O, et al (2002) Prevention of solely estrogen-induced mammary tumors by tamoxifen: Evidence for estrogen receptor mediation. J Endocrinol 175:297-305.

16. Li JJ, Cuthbertson TL, Li SA (1980) Inhibition of estrogen carcinogenesis in the Syrian golden hamster kidney by antiestrogens. J Natl Cancer Inst 64:795-800.
17. Butcher RL, Collins WE, Fugo NN (1974) Plasma concentration of LH, FSH, progesterone and estradiol-17β throughout the 4-day estrous cycle of the rat. Endocrinology 94:1704-1708.
18. Schiff I, Walsh B (1995) Menopause. In: Becker KL (editor) Principles and Practice of Endocrinology and Metabolism, 2nd ed. New York:J.B. Lippincott, 916.
19. Dorgan JF, Longcope C, Stephenson Jr HE, et al (1997) Serum sex hormone levels are related to breast cancer risk in postmenopausal women. Environ Health Perspect 105:583-585.
20. Kuller LH, Couley JA, Lucas L, et al (1997) Sex steroid hormones, bone mineral density, and risk of breast cancer. Environ Health Perspect 105:593-598.
21. Thijssen JHH, van Landeghem AAJ, Poortman J (1986) Uptake and concentration of steroid hormones in mammary tissues. Ann NY Acad Sci 484:106-116.
22. Bonney RC, Reed MJ, Davidson K, et al (1983) The relationship between 17β-hydroxysteroid dehydrogenase activity and oestrogen concentrations in human breast tumours and in normal breast tissue. Clin Endocrinol Oxf 19:727-739.
23. Li JJ, Weroha SJ, Davis MF, et al (2001) Estrogen and progesterone receptors in renomedullary interstitial cells during Syrian hamster estrogen-induced tumorigenesis: Evidence for receptor-mediated oncogenesis. Endocrinology 142:4006-4014.
24. Gonzalez A, Oberley TD, Li JJ (1989) Morphological and immunohistochemical studies of the estrogen-induced Syrian hamster renal tumor: Probable cell of origin. Cancer Res 49:1020-1028.
25. Oberley D, Gonzalez A, Lauchner LJ, et al (1991) Characterization of early lesions in estrogen-induced renal tumors in the Syrian hamster. Cancer Res 51:1922-1929.
26. Gustafsson JA (2000) Novel aspects of estrogen action. J Soc Gynecol Invest 7:58-59.
27. Harvell DME, Strecher TE, Tochacek M, et al (2000) Rat strain-specifications of 17β-estradiol in the mammary gland: Correlation between estrogen-induced lobulo alveolar hyperplasia and susceptibility to estrogen-induced mammary cancers. Proc Natl Acad Sci USA 97:2779-2784.
28. Blankenstein MA, Broerse JJ, van Zwieten MJ, et al (1984) Prolactin concentration in plasma and susceptibility to mammary tumors in female rat from different strains treated chronically with estradiol-17β. Breast Cancer Res Treat. 4:137-141.

29. Blankenstein MA, Broerse JJ, de Vries JB, et al (1977) The effect of subcutaneous administration of oestrogen on plasma oestrogen levels and tumor incidence in female rats. Eur J Cancer 13:1437-1443.
30. Cutts JH, Noble RL (1964) Estrone-induced mammary tumors in the rat. I. Induction and behavior of tumors. Cancer Res 24:1116-1130.
31. Dunning WF, Curtis MR, Segaloff A (1952) Strain differences in response to estrone and the induction of mammary gland, adrenal and bladder cancer in rats. Cancer Res 13:147-152.
32. Nelson WO (1944) The induction of mammary carcinoma in the rat. Yale J Biol Med 17:217-228.
33. Sivaraman L, Medina D (2002) Hormone-induced protection against breast cancer. J Mamm Gland Biol Neopl 7:77-92.
34. Guzman RC, Yang J, Rajkamar L, et al (1999) Hormonal prevention of breast cancer: Mimicking the protective effect of pregnancy. Proc Natl Acad Sci USA 96:2520-2525.
35. Gammon MD, Neugut AI, Santella RM, et al (2002) The Long Island Breast Cancer Study Project: description of a multi-institutional collaboration to identify environmental risk factors for breast cancer. 74:235-254.
36. Gammon MD, Wolff MS, Neugut AI, et al (2002) Environmental toxins and breast cancer on Long Island . II. Organochlorine compound levels in blood. Cancer Epidemiol Biomarkers Prevent 11:686-697.
37. Killeen JL, Namiki H (1991) DNA analysis of ductal carcinoma in situ of the breast. A comparison with histologic features. Cancer 68:2602-2607.
38. Arnerlov C, Emdin SO, Cajander S, et al (2001) Intratumoral variations in DNA ploidy and s-phase fraction in human breast cancer. Anal Cell Path 23:21-28.
39. Li JJ, Hou X, Banerjee SK, et al (1999) Overexpression and amplification of c-myc in the Syrian hamster kidney during estrogen carcinogenesis: A probable critical role in neoplastic transformation. Cancer Res 59:2340-2346.
40. Felscher DW, Bishop JM (1999) Transient excess of Myc activity can elicit genomic instability and tumorigenesis. Proc Natl Acad Sci USA 96:3940-3944.
41. Mai S, Hanley-Hyde J, Rainey GJ, et al (1999) Chromosomal and extrachromosomal instability of the cyclin D2 gene is induced by Myc overexpression. Neoplasia 1:241-252.
42. Lingle WL, Barrett SL, Negron VC, et al (2002) Centrosome amplification drives chromosomal instability in breast tumor development. Proc Natl Adad Sci USA 99:1978-1983
43. Lingle WL, Salisbury JL (1999) Altered centrosome structure is associated with abnormal mitoses in human breast cancer. Am J Pathol 155:1941-1951.

44. Lingle WL, Lutz WH, Ingle JN, et al (1998) Centrosome hypertrophy in human breast tumors: Implications for genomic stability and cell polarity. Proc Natl Acad Sci USA 95:2950-2955.
45. Papa D, Li SA, Li JJ (2003) Comparative genomic hybridization of solely estrogen-induced extopic uterine-like stem cell neoplasms in the hamster kidney: Nonrandom chromosomal alterations. Mol Carcinog 38:97-105.
46. Spandidos DA, Pintzas A, Kakkanas M, et al (1987) Elevated expression of the myc gene in human benign and malignant breast lesions compared to normal tissue. Anticancer Res 7:1299-1304.
47. Rummukainen JK, Salminen T, Lundin J, et al (2001) Amplification of c-myc oncogene by chromogenic and fluorescence in situ hybridization in archival breast cancer tissue array samples. Laboratory Invest 11:1545-1551.
48. Umekita Y, Yoshida H (2000) Cyclin D1 expression in ductal carcinoma in situ, atypical ductal hyperplasia and usual ductal hyperplasia: an immunohistochemical study. Pathol Int 50:527-530.
49. Weinstat-Saslow, Merino MJ, Manrow RE, et al (1995) Overexpression of cyclin D mRNA distinguishes invasive and in situ breast carcinomas from non-malignant lesions. Nature Med 12:1257-1260.
50. Dutertre S, Decamps S, Prigent C (2002) On the role of aurora-A in centrosome function. Oncogene 21:6175-6183.
51. Katayama H, Brinkley WR, Sen S (2003) The Aurora kinase: Role in cell transformation and tumorigenesis. Cancer Metast Rev 22:451-464.
52. Tanaka T, Kimura M, Matsunago K, et al (1999) Centrosomal kinase AIKI is overexpressed in invasive ductal carcinoma of the breast. Cancer Res 59:2041-2044.
53. Zhou H, Kuang J, Zhong L, et al (1998) Tumour amplified kinase STK15/BTAK induces centrosome amplification, aneuploidy, and transformation. Nature Genet 20:189-193.

2

The Molecular Pathogenesis of Human Prostate Cancer

William G. Nelson, Angelo M. DeMarzo, Theodore L. DeWeese, and William B. Isaacs

Introduction

Prostate cancer (PCA) has become the most commonly diagnosed cancer among men in the USA, with an estimated 189,000 cases diagnosed in 2002 (1). Encouragingly, over the past several years, increased use of serum prostate-specific antigen (PSA) screening has increased the fraction of men diagnosed with PCA confined to the prostate gland, leading to more effective use of surgery and radiation therapy for treatment, and to a decline in PCA mortality (2, 3). Despite these improvements, some 30,200 men will likely died of progressive metastatic cancer in 2002 (1). Furthermore, even though men with early PCA can be cured using surgery or radiation therapy, the side effects of treatment frequently include erectile dysfunction, urinary incontinence, or rectal irritation (4-6). New insights into the etiology of PCA are needed so that new strategies for its prevention can be developed.

Recent studies of the earliest molecular steps in the development of human PCA have generated new evidence supporting causative roles for prostate inflammation and diet in prostatic carcinogenesis. These new findings have provided new clues as to how PCAs likely arise, and new insights into how the disease might be prevented. A new lesion, termed proliferative inflammatory atrophy (PIA), in which prostate epithelial cells undergo regenerative proliferation in response to inflammatory damage, appears to be a precursor to prostatic intraepithelial neoplasia (PIN) and to PCA (7). PIA lesion cells exhibit many signs of stress, including the induction of carcinogen-detoxification enzymes such as glutathione S-transferases GSTA1 and GSTP1 (7). Somatic inactivation of *GSTP1*, encoding the human π-class GST, renders prostate epithelial cells vulnerable to suffer genome damage mediated by reactive chemical species generated by inflammatory cells, or ingested as part of the diet (8). By leading to more somatic genome alterations, loss of *GSTP1* function leads to PIN or PCA.

Thus, *GSTP1* likely acts as a "caretaker" gene in the prostate (9). When induced, as in PIA lesion cells, it affords protection against cell and genome damage; when its function is lost, genomic instability, driven by genome damage, ensues. New PCA prevention strategies can target this vulnerability to genome damaging, perhaps by attenuating prostatic inflammation, buttressing carcinogen defenses, or by both approaches.

PCA Epidemiology: Prostate Inflammation and Diet

PCA is a disease of Western lifestyle. PCA incidence and mortality are known to vary widely between different geographic regions, with high rates in the USA and Western Europe, and low rates in Asia (10). Asian immigrants to North America adopt higher PCA risks, especially after more than 25 years exposure to a Western lifestyle, and Asian men born in North America have high PCA risks (11-13).

Chronic (or recurrent) prostatitis is one etiological factor that is increasingly suspected to lead to PCA (14). Prostate inflammation is ubiquitously present in prostates removed by radical prostatectomy for PCA in the USA, however, the prevalence and age distribution of asymptomatic prostatitis, in the USA, or elsewhere, is not known. About 9% of men between 40 and 79 years of age report suffering of symptomatic prostatitis, with half of them having repeated episodes (15-17). Although many of these inflammations may be triggered by infections, the infectious cause is most often not identified. Since prostatitis is so common in the USA, often asymptomatic, and of uncertain etiology, causative associations between prostatitis and PCA have been difficult to assess in epidemiology studies. Despite these limitations, an increased PCA risk has been associated with sexually transmitted infections, independent of the specific pathogen, hinting that the inflammatory response to infection, rather than the infectious agent itself, may lead to PCA (18, 19). In addition, host responses to prostate infections may underlie some familial PCA clusters. Genetic studies of familial PCA have identified two candidate PCA susceptibility genes, 2'-5'-oligoadenylate-dependent ribonuclease L (*RNASEL*), and macrophage scavenger receptor 1 (*MSR1*). These genes are thought to encode proteins with critical functions in host responses to a wide variety of infectious agents (20-23). Finally, an inflammatory lesion in the human prostate, PIA, may be a precursor to PIN and to PCA (7).

A key feature of Western lifestyle that may promote PCA development is the diet. Several epidemiology studies have implicated various dietary components, such as animal fat and charred meat, rich in the Western diet, as high PCA risk factors; while vitamins, fruits, and vegetables, poor in the Western diet, as dietary factors that decrease PCA risk (24-31). However, whether the high PCA risk diet represents an error of *commission* (*i.e.,* over-consumption of animal fats and charred meats), *omission* (*i.e.*, under-consumption of fruits and vegetables), or *both* has not been proven. Nevertheless, there are carcinogens present in the Western diet. Male rats fed the heterocyclic aromatic amine 2-amino-1-methyl-6-phenylimidazo[4,5-β] pyridine (PhIP), an intermediate metabolite formed while preparing "charred" or "well-done" meat, develop DNA mutations in prostate cells that result in PCAs (32, 33).

Somatic Genome Alterations in PCA Cells

Autopsy studies suggest that PCA begins in USA men at quite a young age. Small cancer foci have been reported in about 29% of men aged 30-40 years (34). Not surprisingly, by the time the disease is diagnosed, age 60-70 years, the DNA of the neoplastic cells has accumulated a large number of somatic genome alterations, including gene mutations, amplifications, and deletions, chromosomal rearrangements, and changes in DNA methylation (Figure 1). Not only a large number of somatic genome alterations arise in each individual PCA patient, but among different PCA patients, and between different PCA lesions from a single patient, indicating a striking heterogeneity of genome changes. This suggests that the alterations may arise as a consequence of prolonged exposure to genome damaging stresses, such as those associated with inflammation or ingestion of heterocyclic amine carcinogens, given rise to poor maintenance of genome integrity, by either one of these processes or both.

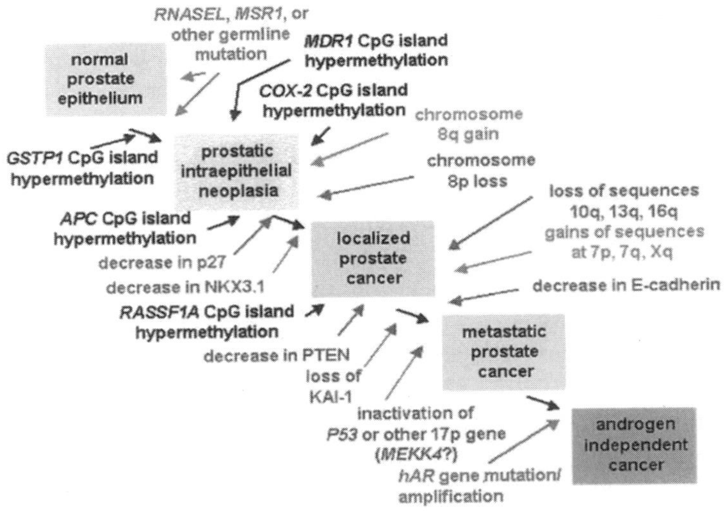

Figure 1. The molecular pathogenesis of human PCA: *GSTP1* CpG island hypermethylation initiates prostatic carcinogenesis.

Table 1. *GSTP1* CpG Island Hypermethylation in Prostate Cancer.

Study	Detection Technique	Results
Lee, *et al.* (36)	SB[1]	**Tissue:** 100% PCA[5], 0% BPH[6]
Lee, *et al.* (49)	RE-PCR[2]	**Tissue:** 91% PCA[5]
Brooks, *et al.* (38)	RE-PCR[2]	**Tissue:** 70% PIN[7]
Santourlidis, *et al.* (50)	RE-PCR[2]	**Tissue:** 75% PCA[5], 0% TCC[10]
Millar, *et al.* (51)	BGS[3]	**Tissue:** 83% PCA[5]
Suh, *et al.* (52)	RE-PCR[2]	**Ejaculate:** 44% PCA[5]
Goessl, *et al.* (53)	MS-PCR[4]	**Tissue:** 94% PCA[5], 0% BPH[6] **Plasma:** 71% PCA[5] **Ejaculate:** 50% PCA[5]
Goessl, *et al.* (54)	MS-PCR[4]	**Urine:** 73% PCA[5], 29% PIN[7], 2% BPH[6]
Cairns, *et al.* (55)	MS-PCR[4]	**Tissue:** 79% PCA[5] **Urine:** 27% PCA[5]
Lin, *et al.* (37)	SB[1], RE-PCR[2],BGS[3]	**Tissue:** 100% PCA[5]
Goessl, *et al.* (56)	MS-PCR[4]	**Tissue:** 90% PCA[5], 0% BPH[6] **Plasma:** 72% PCA[5] **Urine:** 76% PCA[5] **Ejaculate:** 50% PCA[5]
Jeronimo, *et al.* (57)	MS-PCR[4]	**Tissue:** 91% PCA[5], 54% PIN[7], 29% BPH[6]
Chu, *et al.* (58)	MS-PCR[4]	**Tissue:** 100% PCA[5], 7% BPH[6]
Jeronimo, *et al.* (59)	MS-PCR[4]	**Tissue:** 91% PCA[5], 29% BPH[6] **Urine/Plasma:** 54% PCA, 3% BPH[6]
Goessl, *et al.* (60)	MS-PCR[4]	**Prostate Biopsy Washings:** 100% PCA[5], 67% PIN[7], 0% BPH[6]
Harden, *et al.* (61)	MS-PCR[4]	**Tissue:** 73% PCA[5], 0% BPH[6]
Gonzalgo, *et al.* (62)	MS-PCR[4]	**Post-Biopsy Urine:** 58% PCA[5], 33% non-PCA, 67% atypia/PIN
Nakayama, *et al.* (63)	MS-PCR[4]	**Tissue (LCM[8]):** 91% PCA[5], 69% PIN[7], 6% PIA[9], 0% normal
Gonzalgo, *et al.* (64)	MS-PCR[4]	**Prostate Secretions:** 76% PCA[5]

[1] SB, southern blot analysis [2]RE-PCR, methylation-sensitive restriction enzyme-PCR
[3] BGS, bisulfite genomic sequencing [4] MS-PCR, methylation-specific-PCR
[5] PCA, prostate cancer [6] BPH, benign prostate hypertrophy
[7] PIN, prostate intraepithelial neoplasia [8] LCM, laser capture microdissection
[9] PIA, proliferative inflammatory atrophy [10] Transitional cell carcinoma

GSTP1, encoding the π-class GST, likely acts as a "caretaker" gene in the prostate, preventing the acquisition of somatic genome changes in response to exposure to genome damaging agents. In mice, targeted disruption of *GSTP* genes leads to an increased susceptibility to skin carcinogenesis after exposure 7,12 dimethylbenz[a]anthracene (DMBA) (35). In the human prostate, GSTP1 is consistently expressed in normal basal epithelial cells, but not in normal columnar epithelial cells (36, 37). This enzyme expression is highly induced in cells comprising PIA lesions (7). In contrast, cells in PIN lesions or in PCA rarely ever express GSTP1 (36, 38). In fact, loss of GSTP1 expression often marks the transition between PIA, PIN, or PCA (7). The lack of GSTP1 expression can be attributed to somatic *GSTP1* gene "CpG island" DNA hypermethylation, resulting in silencing of *GSTP1* transcription via recruitment of $^{5\text{-m}}$C-binding domain proteins (37, 39). *GSTP1* "CpG island" DNA hypermethylation is now recognized to be the most common somatic genome change in PCAs (Table 1).

Loss of *GSTP1* "caretaker" function early during prostatic carcinogenesis may link PCA epidemiology with the molecular pathology of PCA. Lack of *GSTP1* activity has been shown to render PCA cells vulnerable to genome damage mediated by the heterocyclic amine carcinogen PhIP (8) and by oxidants (41). In response to oxidative genome damaging stresses, the absence of *GSTP1* leads not only to an increase in oxidized DNA bases in genomic DNA, but also to improved cell survival (41). This form of "tolerance" to oxidative genome damaging stresses is reminiscent of the "oxidation tolerance" exhibited by cells with DNA mismatch repair enzyme deficiency, which are prone to suffer increased oxidized DNA base damage, increased mutations, and decreased cell death following exposure to oxidative stresses (40). In cells with both *GSTP1* and DNA mismatch repair enzyme deficiency, "oxidation tolerance" may lead to an increased vulnerability to mutation and neoplastic transformation upon exposure to oxidative genome damaging stresses, including those inflicted by chronic (or recurrent) inflammation (41).

Proliferative Inflammatory Atrophy (PIA)

GSTP1 is typically expressed at high levels in PIA lesions (7). These prostate lesions are composed of proliferating, atrophic-appearing, prostate epithelial cells that are often juxtaposed to activated inflammatory cells (7). The epithelial cells present are distinct in several ways from the atrophic epithelial cells seen after androgen deprivation or anti-androgen treatment. First, PIA lesions are focal, not diffuse. PIA cells are often proliferating quite actively, rather than lying quiescently. Moreover, the PIA cells show many molecular signs of stress, including induction of GSTP1, GSTA1, and COX-2 expression (7, 42, 43). PIA lesions are readily evident in radical prostatectomy specimens containing PCA, often present directly adjacent to PIN lesions or PCA (44). Inflammatory lesions in other organ sites, including the liver, stomach, and colon, are known cancer precursors. In the prostate, rare PIA lesions exhibit somatic genome alterations, such as *GSTP1* "CpG island" hypermethylation and other changes, reminiscent of PIN lesions and PCAs (63). The mechanism by which PIA cells acquire such changes, particularly *de-novo* hypermethylation of *GSTP1* "CpG islands" DNA sequences, has not been elucidated.

New Opportunities for PCA Prevention

The convergence of PCA epidemiology, indicating a possible role for prostate inflammation, and a significant role for the diet, in PCA development, with molecular pathology, revealing that neoplastic prostate cells may have acquired an increased vulnerability to carcinogen damage, provides an opportunity for the discovery and development of rational new approaches to PCA prevention. Possible strategies include reduced exposure to genome damaging oxidants and other carcinogens, and intake of antioxidant micronutrients, including vitamin E, selenium, and carotenoids such as lycopene, which may be able to intercept reactive oxygen species before they inflict genome damage in the prostate. Administration of anti-inflammatory agents, when distributed into prostate tissues, may reduce oxidant production by prostate inflammatory cells. Consumption of cruciferous vegetables, containing the isothiocyanate compound, sulforaphane, an inducer of GSTs and other carcinogen-detoxification enzymes, may raise carcinogen defenses in the prostate and in other tissues to compensate for acquired defects in *GSTP1* "caretaker" gene function (45-47).

Thus, human PCA itself, featuring ongoing threats to genome integrity associated with prostate inflammation and with high-risk dietary practices, may be the most rational "disease" that needs to be targeted for prevention. To discover and develop new agents to treat prostatic carcinogenesis, new clinical trial strategies featuring new "disease" biomarkers will likely be required. For the near future, the most promising PCA prevention strategies under consideration may be the use of anti-oxidant micronutrients (the SELECT trial) and anti-inflammatory agents (48).

References

1. Jemal A, Thomas A, Murray T, et al (2002) Cancer statistics. CA Cancer J Clin 52:23-47.
2. Hankey BF, Feuer EJ, Clegg LX, et al (1999) Cancer surveillance series: interpreting trends in PCA-- Part I: Evidence of the effects of screening in recent PCA incidence, mortality, and survival rates. J Natl Cancer Inst 91:1017-1024.
3. Bartsch G, Horninger W, Klocker H, et al (2001) PCA mortality after introduction of prostate-specific antigen mass screening in the Federal State of Tyrol, Austria. Urology 58:417-424.

4. Walsh PC, Marschke P, Ricker D, et al (2000) Patient-reported urinary continence and sexual function after anatomic radical prostatectomy. Urology 55:58-61.
5. Walsh PC, Lepor H, Eggleston JC (1983) Radical prostatectomy with preservation of sexual function: anatomical and pathological considerations. Prostate 4:473-485.
6. Horwitz EM, Hanlon AL, Hanks GE (1998) Update on the treatment of PCA with external beam irradiation. Prostate 37:195-206.
7. De Marzo AM, Marchi VL, Epstein JI, et al (1999) Proliferative inflammatory atrophy of the prostate: implications for prostatic carcinogenesis. Am J Pathol 155:1985-1992.
8. Nelson CP, Kidd LC, Sauvageot J, et al (2001) Protection against 2-hydroxyamino-1-methyl-6-phenylimidazo[4,5- b]pyridine cytotoxicity and DNA adduct formation in human prostate by glutathione S-transferase P1. Cancer Res 61:103-9.
9. Kinzler KW, Vogelstein B (1997) Cancer-susceptibility genes. Gatekeepers and caretakers [news; comment] [see comments]. Nature 386:761, 763.
10. Hsing AW, Tsao L, Devesa SS (2000) International trends and patterns of PCA incidence and mortality. Int J Cancer 85:60-67.
11. Haenszel W, Kurihara M (1968) Studies of Japanese migrants. I. Mortality from cancer and other diseases among Japanese in the United States. J Natl Cancer Inst 40:43-68.
12. Shimizu H, Ross RK, Bernstein L, et al (1991) Cancers of the prostate and breast among Japanese and white immigrants in Los Angeles County. Br J Cancer 63:963-966.
13. Whittemore AS, Kolonel LN, Wu AH, et al (1995) PCA in relation to diet, physical activity, and body size in blacks, whites, and Asians in the United States and Canada. J Natl Cancer Inst 87:652-661.
14. Gardner WA, Bennett BD (1992) The prostate overview: recent insights and speculations. In: Weinstein RS, Garnder WA (eds) Pathology and Pathobiology of the Urinary Bladder and Prostate. Baltimore:Williams and Wilkens, 129-148.
15. Roberts RO, Lieber MM, Rhodes T, et al (1998) Prevalence of a physician-assigned diagnosis of prostatitis: the Olmsted County study of urinary symptoms and health status among men. Urology 51:578-584.
16. Giovannucci E (2001) Medical history and etiology of PCA. Epidemiol Rev 23:159-162.
17. Hoekx L, Jeuris W, Van Marck E, et al (1998) Elevated serum prostate specific antigen (PSA) related to asymptomatic prostatic inflammation. Acta Urol Belg 66:1-2.
18. Hayes RB, Pottern LM, Strickler H, et al (2000) Sexual behaviour, STDs and risks for PCA. Br J Cancer 82:718-725.

19. Dennis LK, Dawson DV (2002) Meta-analysis of measures of sexual activity and PCA. Epidemiology 13:72-79.
20. Zhou A, Paranjape J, Brown TL, et al (1997) Interferon action and apoptosis are defective in mice devoid of 2',5'-oligoadenylate-dependent RNase L. Embo J 16:6355-6363.
21. Carpten J, Nupponen N, Isaacs S, et al (2002) Germline mutations in the ribonuclease L gene in families showing linkage with HPC1. Nat Genet 30:181-184.
22. Xu J, Zheng SL, Mychaleckyj J, et al (2002) Germline mutations of the macrophage scavenger receptor 1 gene are associated with PCA risk in Caucasian and African American men. Nature Genetics, 32:321-325.
23. Platt N, Gordon S (2001) Is the class A macrophage scavenger receptor (SR-A) multifunctional? - The mouse's tale. J Clin Invest 108:649-654.
24. Gann PH, Hennekens CH, Sacks FM, et al (1994) Prospective study of plasma fatty acids and risk of PCA. J Natl Cancer Inst 86:281-6.
25. Gann PH, Ma J, Giovannucci E, et al (1999) Lower PCA risk in men with elevated plasma lycopene levels: results of a prospective analysis. Cancer Res 59:1225-1230.
26. Giovannucci E, Rimm EB, Colditz GA, et al (1993) A prospective study of dietary fat and risk of PCA [see comments]. J Natl Cancer Inst 85:1571-1579.
27. Le Marchand L, Kolonel LN, Wilkens LR, et al (1994) Animal fat consumption and PCA: a prospective study in Hawaii. Epidemiology 5:276-282.
28. Clark LC, Combs GF, Jr., Turnbull BW, et al (1996) Effects of selenium supplementation for cancer prevention in patients with carcinoma of the skin. A randomized controlled trial. Nutritional Prevention of Cancer Study Group [see comments] [published erratum appears in JAMA 1997 May 21;277(19):1520]. JAMA 276:1957-1963.
29. Clark LC, Dalkin B, Krongrad A, et al (1998) Decreased incidence of PCA with selenium supplementation: results of a double-blind cancer prevention trial. Br J Urol 81:730-734.
30. Cohen JH, Kristal AR, Stanford JL (2000) Fruit and vegetable intakes and PCA risk. J Natl Cancer Inst 92:61-68.
31. Heinonen OP, Albanes D, Virtamo J, et al (1998) PCA and supplementation with alpha-tocopherol and beta- carotene: incidence and mortality in a controlled trial. J Natl Cancer Inst 90:440-446.
32. Stuart GR, Holcroft J, de Boer JG, et al (2000) Prostate mutations in rats induced by the suspected human carcinogen 2- amino-1-methyl-6-phenylimidazo[4,5-b]pyridine. Cancer Res 60:266-268.

33. Shirai T, Sano M, Tamano S, et al (1997) The prostate: a target for carcinogenicity of 2-amino-1-methyl-6-phenylimidazo[4,5-b]pyridine (PhIP) derived from cooked foods. Cancer Res 57:195-198.
34. Sakr WA, Grignon DJ, Crissman JD, et al (1994) High grade prostatic intraepithelial neoplasia (HGPIN) and prostatic adenocarcinoma between the ages of 20-69: an autopsy study of 249 cases. In Vivo 8:439-443.
35. Henderson CJ, Smith AG, Ure J, et al (1998) Increased skin tumorigenesis in mice lacking pi class glutathione S-transferases. Proc Natl Acad Sci USA 95:5275-5280.
36. Lee WH, Morton RA, Epstein JI, et al (1994) Cytidine methylation of regulatory sequences near the pi-class glutathione S-transferase gene accompanies human prostatic carcinogenesis. Proc Natl Acad Sci USA 91:11733-11737.
37. Lin X, Tascilar M, Lee WH, et al (2001) GSTP1 CpG island hypermethylation is responsible for the absence of GSTP1 expression in human PCA cells. Am J Pathol 159:1815-1826.
38. Brooks JD, Weinstein M, Lin X, et al (1998) CG island methylation changes near the GSTP1 gene in prostatic intraepithelial neoplasia. Cancer Epidemiol Biomarkers Prev 7:531-536.
39. Lin X, Nelson WG (2003) Methyl-CpG-binding domain protein-2 mediates transcriptional repression associated with hypermethylated GSTP1 CpG islands in MCF-7 breast cancer cells. Cancer Res 63:498-504.
40. DeWeese TL, Shipman JM, Larrier NA, et al (1998) Mouse embryonic stem cells carrying one or two defective Msh2 alleles respond abnormally to oxidative stress inflicted by low-level radiation. Proc Natl Acad Sci USA 95:11915-11920.
41. DeWeese TL, Nelson WG (2003) Inadequate "caretaker" gene function and human cancer development. Methods Mol Biol 222:249-268.
42. Parsons JK, Nelson CP, Gage WR, et al (2001) GSTA1 expression in normal, preneoplastic, and neoplastic human prostate tissue. Prostate 49:30-37.
43. Zha S, Gage WR, Sauvageot J, et al (2001) Cyclooxygenase-2 is up-regulated in proliferative inflammatory atrophy of the prostate, but not in prostate carcinoma. Cancer Res 61:8617-8623.
44. Putzi MJ, De Marzo AM (2000) Morphologic transitions between proliferative inflammatory atrophy and high-grade prostatic intraepithelial neoplasia. Urology 56:828-832.
45. Zhang Y, Talalay P, Cho CG, et al (1992) A major inducer of anticarcinogenic protective enzymes from broccoli: isolation and elucidation of structure. Proc Natl Acad Sci USA 89:2399-2403.

46. Zhang Y, Kensler TW, Cho CG, et al (1994) Anticarcinogenic activities of sulforaphane and structurally related synthetic norbornyl isothiocyanates. Proc Natl Acad Sci USA 91:3147-3150.
47. Fahey JW, Zhang Y, Talalay P (1997) Broccoli sprouts: an exceptionally rich source of inducers of enzymes that protect against chemical carcinogens. Proc Natl Acad Sci USA 94:10367-10372.
48. Klein EA, Thompson IM, Lippman SM, et al (2001) SELECT: the next PCA prevention trial. Selenium and vitamin E cancer prevention trial. J Urol 166:1311-1315.
49. Lee WH, Isaacs WB, Bova GS, et al (1997) CG island methylation changes near the GSTP1 gene in prostatic carcinoma cells detected using the polymerase chain reaction: a new prostate cancer biomarker. Cancer Epidemiol Biomarkers Prev 6:443-450.
50. Santourlidis S, Florl A, Ackermann R, et al (1999) High frequency of alterations in DNA methylation in adenocarcinoma of the prostate. Prostate 39:166-174.
51. Millar DS, Ow KK, Paul CL, et al (1999) Detailed methylation analysis of the glutathione S-transferase π (*GSTP1*) gene in prostate cancer. Oncogene 18:1313-1324.
52. Suh CI, Shanafelt T, May DJ, et al (2000) Comparison of telomerase activity and *GSTP1* promoter methylation in ejaculate as potential screening tests for prostate cancer. Mol Cell Probes 14:211-217.
53. Goessl C, Krause H, Müller M, et al (2000) Fluorescent methylation-specific polymerase chain reaction for DNA-based detection of prostate cancer in bodily fluids. Cancer Res 60:5941-5945.
54. Goessl C, Muller M, Heicappell R, et al (2001) DNA-based detection of prostate cancer in urine after prostatic massage. Urology 58:335-338.
55. Cairns P, Esteller M, Herman JG, et al (2001) Molecular detection of prostate cancer in urine by GSTP1 hypermethylation. Clin Cancer Res 7:2727-2730.
56. Goessl C, Muller M, Heicappell R, et al (2001) DNA-based detection of prostate cancer in blood, urine, and ejaculates. Ann NY Acad Sci 945:51-58.
57. Jeronimo C, Usadel H, Henrique R, et al (2001) Quantitation of GSTP1 methylation in non-neoplastic prostatic tissue and organ-confined prostate adenocarcinoma. J Natl Cancer Inst 93:1747-1752.
58. Chu D-C, Chuang C-K, Fu J-B, et al (2002) The use of real-time quantitative polymerase chain reaction to detect hypermethylation of the cpg islands in the promoter region flanking the gstp1 gene to diagnose prostate carcinoma. J Urol 167:1854-1858.

59. Jeronimo C, Usadel H, Henrique R, et al (2002) Quantitative GSTP1 hypermethylation in bodily fluids of patients with prostate cancer. Urology 60:1131-1135.
60. Goessl C, Müller M, Heicappell R, et al (2002) Methylation-specific PCR for detection of neoplastic DNA in biopsy washings. J Pathol 196:331-334.
61. Harden SV, Guo Z, Epstein, JI, et al (2003) Quantitative gstp1 methylation clearly distinguishes benign prostatic tissue and limited prostate adenocarcinoma. J Urol 16:1138-1142.
62. Gonzalgo ML, Pavlovich CP, Lee SM, Nelson WG (2003) Prostate cancer detection by GSTP1 methylation analysis of postbiopsy urine specimens. Clin Cancer Res 9:2673-2677.
63. Nakayama M, Bennett CJ, Hicks JL, et al (2003) Hypermethylation of the human glutathione S-transferase-pi gene (GSTP1) CpG island is present in a subset of proliferative inflammatory atrophy lesions but not in normal or hyperplastic epithelium of the prostate: a detailed study using laser-capture microdissection. Am J Pathol 163:923-933.
64. Gonzalgo ML, Nakayama M, Lee SM, et al (2004) Detection of GSTP1 methylation in prostatic secretions using combinatorial MSP analysis. Urology 63:414-418.

PART 1. BREAST CANCER I: PROGESTERONE ACTION

3

Pathologic, Clinical, and Epidemiologic Characteristics of Invasive Lobular Breast Carcinoma and a Review of Studies Evaluating its Association with Hormone Replacement Therapy

Christopher I. Li

Introduction

Sex hormones have been implicated in the etiology of breast cancer (BC) since the eighteenth century, when Ramazzini, an Italian physician, observed that nuns were more likely to develop BC than other women. Since that time, observational studies have documented that exposures with a hormonal component, including reproductive factors, use of exogenous hormones, and anthropometric characteristics, are among the most important and consistent BC risk factors. Thus, a further understanding of the hormonal basis of BC has been a topic of considerable interest to basic scientists, epidemiologists, and clinicians.

As our knowledge of BC has grown, so has our appreciation for its complexity. It is clear that BC is a heterogeneous disease at the clinical, pathologic, and molecular levels. For example, evaluation of the role that estrogen (E) exposure plays in BC concurred with the identification of the estrogen receptor (ERα), and now tumors are routinely classified as either ERα^+ or ERα^- since ERα status influences treatment recommendations. Epidemiologic studies also suggest that tumors that are hormone receptor positive are different from those that are hormone receptor negative, as they have different risk factors and seemingly different etiologies (1).

In addition, there are different histologic types of BC. The two most common are invasive ductal carcinoma (IDC) and invasive lobular carcinoma (ILC), which account for approximately 80% and 5-15% of all BCs, respectively (2, 3). In this chapter a summary of the pathologic, clinical, and epidemiologic characteristics of ILC, in contrast to IDC; recent work investigating associations between hormone replacement therapy (HRT) use and risks of ILC compared to IDC; and the accumulated evidence supporting the role of progesterone (P) in the etiology of BC in general, and ILC in particular are presented.

Characteristics of ILC Compared to IDC

Pathologic, Immunohistochemical and Genetic Characteristics. Classic ILC was first identified and distinguished from other types of BC by the pathologists Foote and Stewart in 1941. Based on their histopathologic descriptions, classic ILC is characterized by slender linear strands of cancer cells in single file, which are present diffusely throughout the tissue (4). Lobular tumor cells are also typically small and uniform in size, and often have a signet-ring shape. Alternatively, IDC has a more heterogeneous microscopic pathology, and unlike ILCs, IDCs typically grow as solid masses of cancer cells.

One of the molecular characteristics of ILC that may account for its linear growth pattern is that nearly all ILCs do not express e-cadherin, which is a cell-cell adhesion molecule. A study comparing e-cadherin expression in 42 ILCs and 100 IDCs found that 41/42 ILCs had complete loss of expression, while all 100 IDCs had moderate to strong e-cadherin expression (5).

Other immunohistochemical differences between ILC and IDC have also been observed. ILCs are more likely to be both $ER\alpha^{+}$ and PR^{+} compared to IDC (6), and to differ with respect to the expression of a variety of other molecular tumor markers including cyclin D1 (7), p53 (8), and vascular endothelial growth factor (VEGF) (9, 10). Studies have also evaluated differences in genome-wide alterations between ILC and IDC using comparative genomic hybridization (CGH) (11, 12). A study of 19 ILCs and 29 IDCs noted numerous differences between these two types of BC in their DNA copy numbers of various chromosomes (12). For example, 90% of ILCs had a loss of chromosome 16q compared to only 37% of IDCs. Since the gene encoding the transcription of e-cadherin is located on 16q, loss of 16q by ILCs may, to some extent, explain the differences in e-cadherin expression observed in ILCs compared to IDCs. Many of these same differences were observed in another CGH comparison of ILC and IDC (11).

Thus, beyond differences in their pathologic features, ILC and IDC appear to have different molecular and genetic profiles, suggesting that the etiologies of these two carcinomas are distinct.

Clinical Characteristics. Clinical differences between ILC and IDC are somewhat reflective of their underlying pathologic differences. Since ILC tends to present as diffuse linear slender strands of cells, rather than as the more discrete solid masses of cancer cells that characterize IDC, ILC is more difficult to palpate on clinical breast examination and more likely to be missed by mammography compared to IDC (13-15). One study reported that 19% of mammograms from patients ultimately diagnosed with ILC were initially reported as negative, 46% of which showed no evidence of ILC when re-examined after the diagnosis of ILC was made (16). These characteristics may partly explain why ILCs tend to be larger in size at diagnosis than IDC (3). Despite being more difficult to detect and being larger in size at diagnosis, ILC does have a better prognosis than IDC (17, 18).

The proportions of women with these cancers do differ by race/ethnicity. In the USA, compared to non-Hispanic whites, African Americans are 40%, Native Americans and Asian Americans/Pacific Islanders are 50%, and Hispanic whites are 30% less likely to be diagnosed with ILC (19). These differences may be explained by differences in the proportions of women exposed to factors related to ILC risk, such as HRT (as described in detail below) by race/ethnicity. For example, in the USA, African American women are less likely to use HRT than white women (20).

Time Trends in ILC and IDC Incidence Rates

From the 1970's to the mid-1980's, about 80% of all invasive breast carcinomas diagnosed in the USA had a ductal histology, and 5-10% had a lobular histology (2,3). However, ILC incidence rates have been increasing steadily among US women (21, 22). The most recent of these studies documented a 65% increase in ILC incidence rates from 1987-1999, while incidence rates of IDC have remained essentially constant, increasing only 3% over this same time period (Figure 1) (22).

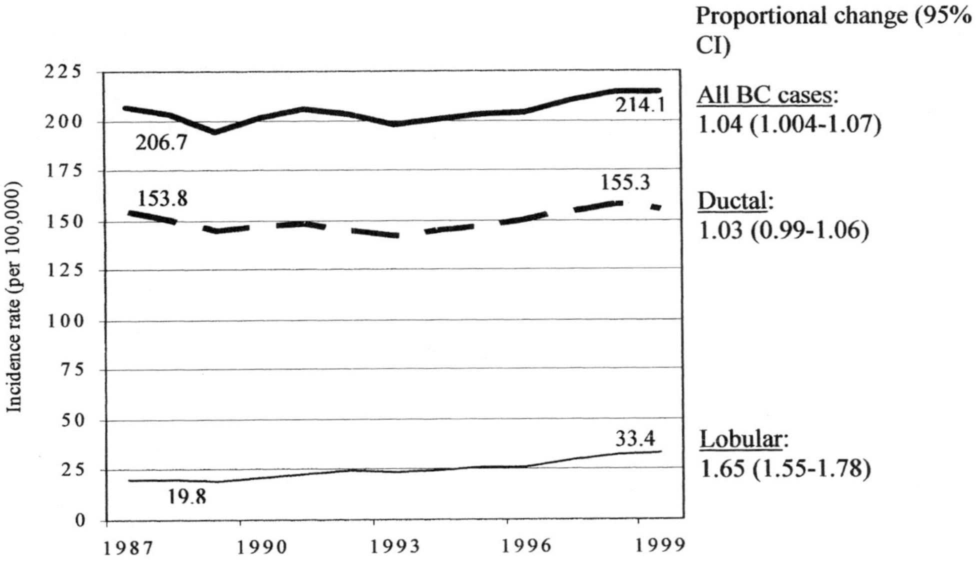

Figure 1. Incidence rates of all BC cases, IDC, and ILC in the USA, 1987-1999.

The proportion of all BC cases with a lobular histology increased from 9.6% in 1987 to 15.4% in 1999. ILC rates have also been increasing more rapidly than those for IDC in Geneva, Switzerland (23). In Geneva, ILC rates increased 14.4% per year from 1976-1999 while rates of IDC increased only 1.2% per year.

The Relationship Between HRT and Risk of Lobular Carcinoma

As a result of the observations described above, attention has been paid to potential risk factors that may be more strongly related to ILC risk than to IDC risk. In particular, there is a growing interest in the relationship between combined E and P HRT (CHRT) and ILC risk. Two main observations have driven this research. First, CHRT use in the USA increased over the same time period that ILC rates increased and IDC rates remained constant. From 1982-1992 the number E and P prescriptions increased 2.3-fold and 4.9-fold, respectively (24). Among controls from a recent USA multi-center case-control study of postmenopausal women spanning 1994-1998, 45% were current HRT users (25). Second, CHRT use has been shown to be associated with an elevated BC risk in numerous studies, and more recently in the Women's Health Initiative (WHI), a randomized controlled trial. The pooled analysis conducted by the Collaborative Group on Hormonal Factors in Breast Cancer of 51 observational studies found that current use of CHRT or P alone for 5 years or longer increased BC risk 53% (26). Consistent with this result, the WHI reported that CHRT is associated with a 26% increase in BC risk after 5.2 years of follow-up (27). Since ILCs are also more likely than IDCs to be hormone receptor positive, there is reason to suspect that ILCs may also be more responsive to hormonal exposures, like CHRT, than are IDCs.

Six case-control studies have now evaluated the relationship between CHRT use and risk of ILC compared to IDC (Table 1) (25, 28-32). Each of these studies were conducted using different USA populations. The number of lobular cases included in these studies ranged from 58 to 408. Three of the studies included a broad range of postmenopausal women (29-31), two focused on younger postmenopausal (<65 years of age) (25, 28), and the other on older postmenopausal women (65-79 years of age) (32). Each study evaluated ever and/or current use of ERT and CHRT separately. Three of these studies also evaluated different patterns of CHRT use (25, 29, 32). Two defined continuous CHRT (CCHRT) use as the use of the P component for 25+ days/month along with a daily dose of E, and sequential CHRT (SCHRT) use as the use of the P component for <25 days/month along with a daily dose of E (25, 32). The third study defined CCHRT use as the use of the same number of E and P pills in each prescription, and SCHRT use as that prescribed with different number of E and P pills in each prescription (29).

Ever Use of ERT and CHRT. Five of the six studies evaluated ever use of ERT and CHRT and risk of ILC and IDC (Table 2) (25, 28, 30-32). In general these studies were quite consistent. Ever use of ERT was associated with 1.1- to 1.6-fold increases in risk of ILC (all of which were within the limits of chance) and 0.5- to 1.2-fold changes in risk of IDC in these studies. The two studies that evaluated duration of use were also consistent with each other, indicating that duration of ERT use was not associated with either ILC or IDC, including use for as long as 25 or more years according to one of these studies (25, 32).

Table 1. Studies of Use of ERT and CHRT and Risk of BC by Histologic Type.

Authors	Setting	Study Design	Age, Years	HRT Exposures Reported
Li, *et al.* (28) (2000)	Seattle (King County)	Population-based case-control study 58 lobular cases 370 ductal cases 492 controls	50-64	Ever and current use of ERT and CHRT
Chen, *et al.* (29) (2002)	Group Health Cooperative of Puget Sound	Nested case-control study 91 lobular cases 4 non-lobular cases 92 controls	50-74	Current use and durations of recent use of ERT and CHRT (including SCHRT and CCHRT)
Newcomb, *et al.* (30) (2002)	Wisconsin, Massachusetts, and New Hampshire	Population-based case-control study 408 lobular cases 105 ductal cases 571 controls	50-79	Ever use of ERT and CHRT
Daling, *et al.* (25) (2002)	Atlanta, Detroit, Los Angeles, Philadelphia, and Seattle	Population-based case-control study 263 lobular cases 1,386 ductal cases 1,953 controls	35-64	Ever use, current use, and durations of ever and current use of ERT and CHRT (including SCHRT and CCHRT)
Newcomer, *et al.* (31) (2003)	Wisconsin, Massachusetts, New Hampshire, and Maine	Population-based case-control study 219 lobular cases 2,172 ductal cases 3,179 controls	<75	Ever use and recent use of ERT and CHRT
Li, *et al.* (32) (2003)	Seattle-Puget Sound Area (King, Pierce, and Snohomish counties)	Population-based case-control study 196 lobular cases 656 ductal cases 1,007 controls	65-79	Ever use, current use, and durations of ever and current use of ERT and CHRT (including SCHRT and CCHRT)

Table 2. Ever Use of ERT and CHRT and Risk of BC by Histologic Type.

Authors	HRT Regimen	Proportion of Controls who were Ever Users	Histology	Type of Use/ Duration of Use	Odds Ratio (95% CI)
Li, *et al.* (28) (2000)	ERT	35%	Lobular	Ever use	1.2 (0.4-3.1)
			Ductal	Ever use	0.5 (0.4-0.8)
	CHRT	18%	Lobular	Ever use	2.0 (0.8-4.8)
			Ductal	Ever use	0.7 (0.5-1.1)
Newcomb, *et al.* (30) (2002)	ERT	Not reported	Lobular	Ever use	1.2 (0.9-1.6)
			Ductal	Ever use	1.2 (1.1-1.4)
	CHRT	Not reported	Lobular	Ever use	2.0 (1.3-3.2)
			Ductal	Ever use	1.4 (1.1-1.8)
Daling, *et al.* (25) (2002)	ERT	38%	Lobular	Ever use	1.1 (0.7-1.8)
				6 mo-5 y	1.0 (0.6-1.7)
				5+ y	1.3 (0.8-2.2)
			Ductal	Ever use	0.8 (0.6-1.0)
				6 mo-5 y	0.8 (0.6-1.1)
				5+ y	0.7 (0.6-1.0)
Daling, *et al.* (25) (2002)	CHRT	27%	Lobular	Ever use	1.8 (1.2-2.6)
				6 mo-5 y	1.6 (1.0-2.4)
				5+ y	2.0 (1.3-3.2)
			Ductal	Ever use	1.0 (0.8-1.3)
				6 mo-5 y	1.0 (0.8-1.3)
				5+ y	1.0 (0.8-1.3)
Newcomer, *et al.* (31) (2003)	ERT	19%	Lobular	Ever use	1.6 (1.0-2.4)
			Ductal	Ever use	1.1 (0.9-1.3)
	CHRT	4%	Lobular	Ever use	3.0 (1.6-5.7)
			Ductal	Ever use	1.1 (0.8-1.5)

Table 2. Continued.

Authors	HRT Regimen	Proportion of Controls who were Ever users	Histology	Type of Use/ Duration of Use	Odds Ratio (95% CI)
Li, *et al.* (32) (2003)	ERT	39%	Lobular	Ever use	1.3 (0.8-2.0)
				6 mo-5 y	1.1 (0.6-2.0)
				5-15 y	1.5 (0.8-2.9)
				15-25 y	1.3 (0.6-2.6)
				25+ y	1.3 (0.7-2.4)
			Ductal	Ever use	1.0 (0.8-1.4)
				6 mo-5 y	0.9 (0.6-1.3)
				5-15 y	1.2 (0.8-1.8)
				15-25 y	1.4 (0.9-2.2)
				25+ y	0.9 (0.6-1.3)
	CHRT	16%	Lobular	Ever use	2.7 (1.7-4.3)
				6 mo-5 y	1.8 (0.9-3.6)
				5-15 y	3.7 (2.0-6.6)
				15+ y	2.6 (1.3-5.3)
			Ductal	Ever use	1.5 (1.1-2.0)
				6 mo-5 y	1.3 (0.9-2.1)
				5-15 y	1.5 (1.0-2.3)
				15+ y	1.6 (1.0-2.6)

Ever use of CHRT was associated with 1.8- to 3.0-fold increased risks of ILC, but with 0.7- to 1.5-fold changes in risk of IDC. Further, CHRT appears to increase BC risk only when used for 5 years or longer (26). In the two studies evaluating the duration of CHRT, 5-year or longer users had higher risks of ILC than did users for less than 5 years in both studies (25, 32).

Together, these five studies indicate that ever use of CHRT is associated with an elevation in ILC risk, and possibly, higher risk of IDC when used longer, but ever use of ERT was not associated with either ILC or IDC.

Current/Recent Use of ERT and CHRT. Five of the six studies also evaluated associations between current/recent use of HRT and risk of ILC and IDC (Table 3) (25, 28, 29, 31, 32). In three studies, current ERT use was not associated with ILC risk, with odds ratios (ORs) ranging from 0.9 to 1.3; while two found that current/recent use of ERT was associated with 1.8- to 2.0-fold increases in ILC risk (29, 31).

Table 3. Current Use of ERT and CHRT and Risk of BC by Histologic Type.

Authors	HRT Regimen	% of Controls who were Current Users	Histology	Type of Use/ Duration of Use	Odds Ratio (95% CI)
Li, *et al.* (28) (2000)	ERT	24%	Lobular	Current use	0.9 (0.3-3.0)
			Ductal	Current use	0.7 (0.4-1.1)
	CHRT	14%	Lobular	Current use	2.1 (0.8-5.8)
			Ductal	Current use	0.7 (0.4-1.1)
Chen, *et al.* (29) (2002)	ERT	16%	Lobular	Current use	2.0 (1.0-3.8)
			Nonlobular	Current use	1.1 (0.8-1.5)
	CHRT	11%	Lobular	Current use	3.9 (2.1-7.4)
			Nonlobular	Current use	1.3 (0.9-1.8)
Daling, *et al.* (25) (2002)	ERT	27%	Lobular	Current use	1.2 (0.7-2.1)
			Ductal	Current use	0.8 (0.6-1.0)
	CHRT	18%	Lobular	Current use	2.2 (1.4-3.3)
			Ductal	Current use	1.2 (0.9-1.5)
Newcomer, *et al.* (31) (2003)	ERT	7%	Lobular	Recent use	1.8 (1.0-3.4)
			Ductal	Recent use	0.9 (0.7-1.2)
	CHRT	2%	Lobular	Recent use	3.6 (1.8-7.6)
			Ductal	Recent use	0.9 (0.6-1.3)
Li, *et al.* (32) (2003)	ERT	27%	Lobular	Current use	1.3 (0.8-2.2)
				6 mo-5 y	1.1 (0.4-2.9)
				5-15 y	1.8 (0.8-3.9)
				15+ y	1.4 (0.8-2.6)
			Ductal	Current use	1.0 (0.7-1.3)
				6 mo-5 y	0.5 (0.3-1.0)
				5-15 y	1.3 (0.8-2.1)
				15+ y	1.1 (0.8-1.7)
Li, *et al.* (32) (2003)	CHRT	11%	Lobular	Current use	3.1 (1.9-5.2)
				6 mo-5 y	1.3 (0.5-3.6)
				5-15 y	4.6 (2.5-8.5)
				15+ y	3.0 (1.4-6.3)
			Ductal	Current use	1.7 (1.2-2.4)
				6 mo-5 y	1.3 (0.8-2.3)
				5-15 y	1.7 (1.1-2.7)
				15+ y	2.0 (1.2-3.4)

However, the latter two studies included the smallest proportions of control women who were current users of ERT of the five studies (7% and 16% compared to 24-27% in the other three studies), and thus had limited statistical power. None of the studies found an association between ERT use and risk of ductal, or in one study, "non-lobular" carcinoma (29).

With respect to CHRT, these five studies reported similar findings: current/recent CHRT use was associated with 2.1- to 3.9-fold increases in ILC risk. Four of the five studies also found that current/recent use of CHRT was not associated with IDC risk (ORs ranging from 0.7 to 1.3) (25, 28, 29, 31). The one study that found an association between current CHRT use and IDC risk (OR = 1.7; 95% CI: 1.2-2.4) was the one that focused on older women with longer durations of CHRT use (32). In that study, CHRT use for five years or longer was associated with elevations in risk of both ILC and IDC, but use for less than five years was not (32). As a group, these studies suggest that current/recent CHRT use is associated with an elevated ILC risk, and that perhaps current/recent users of CHRT for more than five years may also have an elevated IDC risk.

Regimens of CHRT Use: Sequential vs. Continuous Use of Progestin. Three of the studies mentioned also evaluated associations between different regimens of CHRT use and BC risk by histologic type (25, 29, 32). Such evaluations are important because while the WHI study provided strong evidence that CCHRT is associated with an increased BC risk, limited data on the association between SCHRT and BC risk are available. The results of two of the studies summarized in Table 4 suggest that both SCHRT and CCHRT use are associated with an increased ILC risk, and with an increased risk of IDC/non-lobular carcinomas (29, 32). The other study, which was restricted to women younger than 65 years of age, observed that only CCHRT was associated with an increased ILC risk (25). Therefore, based on the limited evidence from these studies, SCHRT use appears to be associated with the same risk for ILC and IDC as is CCHRT use; however, further studies evaluating these associations are needed.

Summary. In the aggregate, the results of the studies described here suggest that neither ever use or current use of ERT is associated with changes in BC risk overall, IDC or ILC. This is inconsistent with the results of the meta-analysis by the Collaborative Group on Hormonal Factors in Breast Cancer, which found that current use of ERT for ≥ 5 years was associated with a 1.34-fold increased risk of BC (26). This meta-analysis was limited because data on the type of HRT used were only available for 39% of the eligible women, and the analysis was not restricted to women who were exclusive ERT users. As a result, some of the associations observed may have been due to combined effects of ERT and CHRT use, since this study and others have found that CHRT use is a stronger BC risk factor than is ERT use (33, 34).

Table 4. Current Use of Sequential and Continuous CHRT and Risk of BC by Histologic Type.

Authors	CHRT Regimen	% of Controls who were Current Users	Histology	Type of Use/ Duration of Use	Odds Ratio (95% CI)
Chen, *et al.* (29) (2002)	SCHRT	16%	Lobular	Recent use	
				<12 mo	1.2 (0.4-3.6)
				13-35 mo	2.0 (0.8-5.4)
				36+ mo	2.6 (1.1-6.1)
			Nonlobular	Recent use	
				<12 mo	1.4 (0.9-2.3)
				13-35 mo	0.9 (0.5-1.6)
				36+ mo	1.5 (0.9-2.4)
	CCHRT	5%	Lobular	Recent use	
				<11 mo	1.2 (0.3-6.0)
				11+ mo	6.1 (2.1-17.3)
			Nonlobular	Recent use	
				<11 mo	1.0 (0.5-2.0)
				11+ mo	1.2 (0.6-2.4)
Daling, *et al.* (25) (2002)	SCHRT	7%	Lobular	Current use	1.4 (0.8-2.5)
			Ductal	Current use	0.9 (0.7-1.3)
	CCHRT	10%	Lobular	Current use	2.4 (1.5-3.8)
			Ductal	Current use	1.3 (1.0-1.7)
Li, *et al.* (32) (2003)	SCHRT	2%	Lobular	Current use	2.6 (1.0-7.1)
			Ductal	Current use	1.8 (1.0-3.6)
	CCHRT	9%	Lobular	Current use	3.1 (1.8-5.3)
			Ductal	Current use	1.6 (1.1-2.3)

In a prior meta-analysis, use of ERT for ≥ 10 years was associated with a 15%-49% increased BC risk (35). It is noteworthy that data from the Nurse's Health Study from 1976 to 1986, a time period prior to the widespread use of CHRT, are consistent with more recent studies, including those described above, indicating lack of association between ERT use and BC risk (36). Specifically, this report found that use of ERT for ≥ 15 years was not associated with an increased

BC risk. This dataset was not included in the meta-analysis by Steinberg, *et al.* (35). They included data from the Nurse's Health Study that were published later (37), and included a larger proportion of ERT users who later also received CHRT. While recent data suggest that ERT use is not strongly associated with BC risk, completion of the ERT arm of the WHI randomized trial in the next few years will provide additional important information regarding this association. However, given the fact that the CHRT arm was terminated early, one could infer that any BC risk associated with ERT in the WHI study is likely to be smaller in magnitude than the BC risk found in the CHRT arm.

As previously indicated, there is convincing evidence from observational studies and the WHI randomized trial, that CHRT increase BC risk. Evidence is mounting regarding the adverse impact on BC risk by adding progestin to HRT. This adverse impact appears to manifest within several years of CHRT use, and is similar in magnitude irrespective of the type of CHRT (continuous or sequential use). Given the types of CHRT used most commonly by women in the latter part of the 20th century, an increased incidence of BC must be tallied as a consequence.

Also, the six studies described above consistently indicate that CHRT use is more strongly associated with an increased risk for ILC than it is for IDC (25,28-32). Thus, CHRT use may partially account for the increasing ILC rates observed in the USA and Europe (21-23). The biologic mechanisms underlying this association remain poorly understood, but one potential explanation may rest on the differences in ERα/PR expression between ILC and IDC, as 91% of ILCs are $ER\alpha^+$ compared to 79% of IDCs, and 75% of ILCs are PR^+ compared to 67% of IDCs (38).

P likely plays a key role in this association. We have observed that ERT use was not associated with BC risk of any particular ERα/PR tumor profile, but CHRT use was associated with an elevation in risk of $ER\alpha^+/PR^+$ tumors. The magnitude of the risk increased with increased duration of CHRT use, but not for $ER\alpha^+/PR^-$ or $ER\alpha^-/PR^-$ tumors (32). These data suggest that CHRT may promote BC through the stimulation of both receptors, and not through ERα alone, since CHRT use was associated with an elevated risk for $ER\alpha^+/PR^+$ tumors, but not for those $ER\alpha^+/PR^-$. While these findings require confirmation, they suggest that the P component of CHRT may be particularly important with respect to altering BC risk.

The Role of Progesterone in the Etiology of Breast Cancer

While E's role in the etiology of BC has been extensively studied, and E has been shown to play multiple roles in BC pathways, less is known about the relationship between P and BC etiology. Initially, P was thought to protect the breast from cancer based on its anti-proliferative effects on the endometrium. However, several recent studies suggest that P is involved in multiple pathways associated with BC. In 1994, Shi, *et al.* (39) summarized evidence documenting the role of P in the initiation and promotion of BC, including how P: 1. Stimulates growth in the normal human breast (in contrast to its effect on the endometrium). 2. Has mitogenic effects on BC cell lines. 3. Promotes mammary tumor growth in rodents. 4. Induces mitogenic growth factors and their receptors in hormone receptor positive BC cells. 5. Regulates receptors for adhesion molecules involved in metastases. 6. Tumorigenesis is inhibited by anti-progestins. Details regarding this evidence and more recent work documenting links between P and breast/mammary gland carcinogenesis are discussed below.

P's central role in the proliferation and development of the normal breast are well documented (40). In 1977, two studies evaluating cyclic variations in DNA synthesis in breast epithelium during the menstrual cycle were published (41, 42). These studies established that DNA synthesis is low during the follicular phase (E dominant) and high during the luteal phase (P dominant). Subsequent studies identified that this fluctuation in DNA synthesis specifically takes place in breast lobular epithelium, both during natural menstrual cycles (43) and in cycles regulated by oral contraceptive use (44). These studies also demonstrated that after lobular proliferation tapers off at the end of the luteal phase, a period of apoptotic activity occurs. It was hypothesized that because the increase in DNA synthesis occurs during the luteal phase, P, rather than E (long believed to be the primary promoter of breast proliferation), was primarily responsible for stimulating lobular epithelial cell proliferation in the breast. Experimental data ultimately substantiated this hypothesis (45), and now there is general agreement that DNA synthesis increases in the luteal phase of the natural menstrual cycle due to the peaking of P levels.

Developmentally, P is involved in the formation of lobular-alveolar structures that occurs during pregnancy. A study comparing normal mice to mice that lack the PR found that while the PR-null mice developed normal ductal structures, they did not form any lobular-alveolar structures when exposed to E and P (46). P's proliferative effects appear to be mediated through its regulation of genes associated with cell cycle progression, growth factors, and growth factor receptors. Treatment of cells arrested in G1 with progestin stimulates them to progress through the cell cycle and induces cyclins (including cyclin D1), cyclin-dependent kinases, and c-*myc* and c-*fos* (protooncogenes involved in cell proliferation) (47, 48). Further, the increases in cyclin D1 and c-*myc* can be blocked by administering the anti-progestin RU 38486, indicating the key role of P's interactions with the PR in stimulating cell cycle progression. P induces growth factors that stimulate proliferation, such as epidermal growth factor (EGF) (49-51) and tumor growth factor (TGF)-α, and suppresses growth factors that inhibit proliferation, such as TGF-β (52). Thus, there are multiple lines of evidence suggesting that P has proliferative effects on the normal breast. However, the mechanisms underlying these effects remain to be elucidated, and additional research in this area is required.

Though P's involvement in the proliferation and development of the normal human breast is well established, its effects on BC cells remain controversial because depending on the model, cell type, and duration of treatment, P can act as an inducer or repressor of BC cell proliferation. A detailed discussion of this literature is beyond the scope of this review, but it has been hypothesized by Lange, *et al.* that these apparent contradictory effects can be explained if P acts as a priming factor that mediates cross-talk between proliferative and anti-proliferative signals (53). Thus, the timing, dose, and route of P administration can alter the balance between these signals in different directions. However, these mechanisms and our perception of how P is related to human BC is not well-understood.

Our understanding of P's mechanism of action was further complicated by the identification of two distinct isoforms of the PR-A and PR-B (54). Both are transcribed from a single gene, but differ in that PR-A lacks the 164 N-terminal structure present in PR-B. Although both PRs have the same affinity for P, these two isoforms have different functions with respect to breast development and gene regulation. Studies using a transgenic mouse model carrying additional PR-A develop mammary glands with excessive lateral ductal branching, but lack lobular-alveolar development (55). Alternatively, mice with additional PR-B exhibit an excessive amount of lobular-alveolar growth, but exhibit an arrested development of branching ductal structures (56). With respect to transcriptional activity, PR-A does not activate transcription but is a strong repressor of PR-B, while PR-B is a transcriptional activator (57). Since PR-B is critical for lobular development and stimulates the transcription of genes implicated in BC, one potential mechanism by which the use of exogenous P could promote lobular, but not ductal, carcinoma is by its PR-B mediated effects. However, further research is required to test this hypothesis since no studies have been reported comparing the expression of PR's isoforms in IDC vs ILC. A study evaluating only IDC tumors reported that 66% expressed PR-A and 55% PR-B (58).

Conclusions

Incidence rates of ILC have been increasing steadily in the USA and Geneva, and likely in other parts of the world, while IDC rates have remained essentially constant. Based on its underlying pathology, ILC presents unique clinical challenges since it is more difficult to detect by both clinical breast exam and mammography. It is associated with better survival rates, and since the majority of ILCs are $ER\alpha^+$, most are amenable to treatment with anti-hormonal agents such as tamoxifen. However, until recently, the epidemiology of ILC has been poorly defined. Epidemiologic investigations aimed at understanding the increases in ILC incidence rates have consistently identified that CHRT use is more strongly associated with ILC risk rather than it is with IDC risk. CHRT may partly account for the increases in ILC rates observed based on these findings and the parallel rise in CHRT use and ILC incidence rates. CHRT also appears to act as a cancer promoter given that current use, but not former use, is positively associated with

ILC risk. P and PRs appear to play key roles in this process since ERT is not associated with risk of either ER^+ or PR^+ tumors, while CHRT is associated with an increased risk of $ER\alpha^+/PR^+$ tumors but not $ER\alpha^+/PR^-$ tumors. P is also known to stimulate the proliferation of lobules, but not ducts, particular through PR-B. Moreover, P's proliferative effects primarily target lobular rather than ductal tissue in the breast. Thus, one would expect that P would be more likely to promote neoplastic growth in lobular, but not in ductal, tissue, particularly through interactions with PR-B. However, the specific mechanisms by which P is involved in oncogenic pathways leading to the development of ILC remain poorly understood. The fact that progestins are commonly used by women world-wide, particularly as forms of contraception or HRT, further stress the importance of advancing our understanding of the roles that P plays in the etiology of BC.

The studies described here indicate that beyond their histologic differences, ILC and IDC appear to have different etiologies. Thus, it is important to take into account the heterogeneity of diseases such as breast cancer when assessing their risk factors. The failure to do so may result in important associations being missed, as illustrated by the results of studies showing a differential effect of CHRT use on risk of ILC and IDC. A greater understanding of risk factors for different histologic types of BC may aid in the development of improved management and treatment approaches for women with BC.

References

1. Potter JD, Cerhan JR, Sellers TA, et al (1995) Progesterone and estrogen receptors and mammary neoplasia in the Iowa Women's Health Study: how many types of breast cancer are there? Cancer Epidemiol Biomarker Prev 4:319-326.
2. Schnitt S, Connolly J, Recht A, et al (1989) Influence of infiltrating lobular histology on local tumor control in breast cancer patients treated with conservative surgery and radiotherapy. Cancer 64:448-454.
3. Sastre-Garau X, Jouve M, Asselain B, et al (1996) Infiltrating lobular carcinoma of the breast. Clinicopathologic analysis of 975 cases with reference to data on conservative therapy and metastatic patterns. Cancer 77:113-120.
4. Davis RP, Nora PF, Kooy RG, et al (1979) Experience with lobular carcinoma of the breast: Emphasis on recent aspects of management. Arch Surg 114:485.
5. Acs G, Lawton TJ, Rebbeck TR, et al (2001) Differential expression of e-cadherin in lobular and ductal neoplasms of the breast and its biologic and diagnostic implications. Am J Clin Pathol 115:85-98.

6. Stierer M, Rosen H, Weber R, et al (1993) Immunohistochemical and biochemical measurement of estrogen and progesterone receptors in primary breast cancer: Correlation of histopathology and prognostic factors. Ann Surg 218:13-21.
7. Oyama T, Kashiwabara K, Yoshimoto K, et al (1998) Frequent overexpression of the cyclin D1 oncogene in invasive lobular carcinoma of the breast. Cancer Res 58:2876-2880.
8. Domagala W, Markiewski M, Kubiak R, et al (1993) Immunohistochemical profile of invasive lobular carcinoma of the breast: predominantly vimentin and p53 protein negative, cathepsin D and oestrogen receptor positive. Virchow Archiv 423:497-502.
9. Coradini D, Pellizzaro C, Veneroni S, et al (2002) Infiltrating ductal and lobular breast carcinomas are characterized by different interrelationships among markers related to angiogenesis and hormone dependence. Brit J Cancer 87:1105-1111.
10. Lee AHS, Dublin EA, Bobrow LG, et al (1998) Invasive lobular and invasive ductal carcinoma of the breast show distinct patterns of vascular endothelial growth factor expression and angiogenesis. J Pathol 185:394-401.
11. Nishizaki T, Chew K, Chu L, et al (1997) Genetic alterations in lobular breast cancer by comparative genomic hybridization. Int J Cancer 74:513-517.
12. Gunther K, Merkelbach-Bruse S, Amo-Takyi BK, et al (2001) Differences in genetic alterations between primary lobular and ductal breast cancers detected by comparative genomic hybridization. J Pathol 193:40-47.
13. Dixon J, Anderson T, Page D, et al (1982) Infiltrating lobular carcinoma of the breast. Histopathology 6:149-161.
14. Yeatman TJ, Cantor AB, Smith TJ, et al (1995) Tumor biology of infiltrating lobular carcinoma. Implications for management. Ann Surg 222:549-559.
15. Silverstein MJ, Lewinsky BS, Waisman JR, et al (1994) Infiltrating lobular carcinoma: is it different from infiltrating duct carcinoma? Cancer 74:1673-1677.
16. Krecke K, Gisvold J (1993) Invasive lobular carcinoma of the breast: Mamographic findings and extent of disease at diagnosis in 184 patients. Am J Roentgenol 161:957-960.
17. Du Toit RS, Locker AP, Ellis IO, et al (1991) An evaluation of differences in prognosis, recurrence patterns and receptor status between invasive lobular and other invasive carcinomas of the breast. Eur J Surg Oncology 17:251-257.
18. Li CI, Moe RE, Daling JR (2003) Risk of mortality by histologic type of breast cancer among women aged 50 to 79 years. Arch Int Med 163:2149-2153.

19. Li CI, Malone KE, Daling JR (2002) Differences in breast cancer hormone receptor status and histology by race and ethnicity among women 50 years of age and older. Cancer Epidemiol Biomarker Prev 11:601-607.
20. Friedman-Koss D, Crespo CJ, Bellantoni MF, et al (2002) The relationship of race/ethnicity and social class to hormone replacement therapy: results from the Third National Health and Nutrition Examination Survey 1988-1994. Menopause 9:264-272.
21. Li CI, Anderson BO, Porter P, et al (2000) Changing incidence rate of invasive lobular breast carcinoma among older women. Cancer 88:2561-2569.
22. Li CI, Anderson BO, Daling JR, et al (2003) Trends in incidence rates of invasive lobular and ductal breast carcinoma. JAMA 289:1421-1424.
23. Verkooijen HM, Fioretta G, Vlastos G, et al (2003) Important increase of invasive lobular breast cancer incidence in Geneva, Switzerland. Int J Cancer 104:778-781.
24. Wysowski DK, Golden L, Burke L (1995) Use of menopausal estrogens and medroxyprogesterone in the United States, 1982-1992. Obstet Gynecol 85:6-10.
25. Daling JR, Malone KE, Doody DR, et al (2002) Relation of regimens of combined hormone replacement therapy to lobular, ductal, and other histologic types of breast carcinoma. Cancer 95:2455-2464.
26. Collaborative Group on Hormonal Factors in Breast Cancer (1997) Breast cancer and hormone replacement therapy: collaborative reanalysis of data from 51 epidemiological studies of 52,705 women with breast cancer and 108,411 women without breast cancer. Lancet 350:1047-1059.
27. Writing Group for the Women's Health Initiative Investigators (2002) Risks and benefits of estrogen plus progestin in healthy postmenopausal women: principal results from the Women's Health Initiative randomized controlled trial. JAMA 288:321-333.
28. Li CI, Weiss NS, Stanford JL, et al (2000) Hormone replacement therapy in relation to risk of lobular and ductal breast carcinoma in middle-aged women. Cancer 88:2570-2577.
29. Chen CL, Weiss NS, Newcomb P, et al (2002) Hormone replacement therapy in relation to breast cancer. JAMA 287:734-741.
30. Newcomb PA, Titus-Ernstoff L, Egan KM, et al (2002) Postmenopausal estrogen and progestin use in relation to breast cancer risk. Cancer Epidemiol Biomarker Prev 11:593-600.
31. Newcomer LM, Newcomb PA, Potter JD, et al (2003) Postmenopausal hormone therapy and risk of breast cancer by histologic type (United States). Cancer Causes Control 14:225-233.

32. Li CI, Malone KE, Porter PL, et al (2003) Relationship between long-term use of hormone replacement therapy and risk of breast cancer among women 65-79 years of age. JAMA 289:3254-3263.
33. Colditz GA, Hankinson SE, Hunter DJ, et al (1995) The use of estrogens and progestins and the risk of breast cancer in postmenopausal women. New Engl J Med 332:1589-1593.
34. Ross RK, Paganini-Hill A, Wan PC, et al (2000) Effect of hormone replacement therapy on breast cancer risk: estrogen versus estrogen plus progestin. J Natl Cancer Inst 92:328-332.
35. Steinberg KK, Smith SJ, Thacker SB, et al (1994) Breast cancer risk and duration of estrogen use: the role of study design in meta-analysis. Epidemiology 5:415-421.
36. Colditz GA, Stampfer MJ, Willett WC, et al (1990) Prospective study of estrogen replacement therapy and risk of breast cancer in postmenopausal women. JAMA 264:2648-2653.
37. Colditz GA, Stampfer MJ, Willett WC, et al (1992) Type of postmenopausal hormone use and risk of breast cancer: 12-year follow-up from the Nurses' Health Study. Cancer Causes Control 3:433-439.
38. Surveillance, Epidemiology, and End Results (SEER) Program (www.seer.cancer.gov) SEER*Stat Database: Incidence - SEER 9 Regs, Nov 2002 Sub (1973-2000), National Cancer Institute, DCCPS, Surveillance Research Program, Cancer Statistics Branch, released April 2003, based on the November 2002 submission.
39. Shi YE, Liu YE, Lippman ME, et al (1994) Progestins and antiprogestins in mammary tumor growth and metastasis. Human Reproduction 9:162-173.
40. Graham JD, Clarke CL (1997) Physiological action of progesterone in target tissue. Endocrine Rev 18:502-519.
41. Meyer JS (1977) Cell proliferation in normal human breast ducts, fibroadenomas, and other ductal hyperplasias measured by nuclear labeling with tritiated thymidine. Hum Pathol 8:67-81.
42. Masters JRW, Drife JO, Scarisbrick JJ (1977) Cyclic variation of DNA synthesis in human breast epithelium. J Natl Cancer Inst 58:1263-1265.
43. Ferguson DJP, Anderson TJ (1981) Morphological evaluation of cell turnover in relation to the menstrual cycle in the "resting" human breast. Br J Cancer 44:177-181.
44. Anderson TJ, Ferguson DJP, Raab GM (1982) Cell turnover in the "resting" human breast: influence of parity, contraceptive pill, age, and laterality. Br J Cancer 46:376-382.
45. Haslam SZ (1988) Progesterone effects on deoxyribonucleic acid synthesis in normal mouse mammary glands. Endocrinology 122:464-470.

46. Lydon JP, DeMayo FJ, Funk CR, et al (1995) Mice lacking progesterone receptor exhibit pleiotropic reproductive abnormalities. Genes Dev 9:2266-2278.
47. Musgrove EA, Lee CS, Sutherland RL (1991) Progestins both stimulate and inhibit breast cancer cell cycle progression while increasing expression of transforming growth factor α, epidermal growth factor receptor, c-fos, and c-myc genes. Mol Cell Biol 11:5032-5043.
48. Musgrove WA, Hamilton JA, Lee CS, et al (1991) Growth factor, steroid, and steroid antagonist regulation of cyclin gene expression associated with changes in T-47D human breast cancer cell cycle progression. Mol Cell Biol 13:3577-3587.
49. Murphy LC, Murphy LJ, Dubik D, et al (1988) Epidermal growth factor gene expression in human breast cancer cells: regulation of expression by progestins. Cancer Res 46:4555-4560.
50. Murphy LJ, Sutherland RL, Stead B, et al (1986) Progestin regulation of epidermal growth factor receptor in human mammary carcinoma cells. Cancer Res 46:728-734.
51. Murphy LC, Murphy LK, Shiu RPC (1988) Progestin regulation of EGF-receptor mRNA accumulation in T-47D human breast cancer cells. Biochem Biophys Res Commun 150:192-196.
52. Murphy LC, Dotzlaw H (1989) Regulation of transforming growth factor α and transforming growth factor β messenger ribonucleic acid abundance in T-47D human breast cancer cells. Mol Endocrinol 3:611-617.
53. Lange C, Richer JK, Horwitz KB (1999) Hypothesis: progesterone primes breast cancer cells for cross-talk with proliferative or antiproliferative signals. Molec Endocrinol 13:829-836.
54. Kastner P, Krust A, Turcotte B, et al (1990) Two distinct estrogen-regulated promoters generate transcripts encoding the two functionally different human progesterone receptor forms A and B. EMBO J 9:1603-1614.
55. Shyamala G, Yang X, Silberstein G, et al (1998) Transgenic mice carrying an imbalance in the native ratio of A to B forms of progesterone receptor exhibit developmental abnormalities in mammary gland. Proc Nat Acad Sci USA 95:696-701.
56. Shyamala G, Yang X, Cardiff RD, et al (2000) Impact of progesterone receptor on cell-fate decisions during mammary gland development. Proc Nat Acad Sci USA 97:3044-3049.
57. Richer JK, Jacobsen BM, Manning NG, et al (2002) Differential gene regulation by the two progesterone receptor isoforms in human breast cancer cells. J Biol Chem 277:5209-5218.
58. Ariga N, Suzuki T, Moriya T, et al (2001) Progesterone receptor A and B isoforms in the human breast and its disorders. Jpn J Cancer Res 92:302-308.

4

Progestin-Regulated Genes and Breast Cancer Risk: Good or Bad?

Henri Rochefort, Majida Esslimani-Sahla, and Dany Chalbos

Introduction

The assumed role of progesterone and progestins on mammary glands and human breast carcinogenesis has varied between laboratories and countries and is still controversial. Progestins inhibit estrogen (E) action in cell lines and provide protection against the mitogenic effect of Es in endometrium, which is why they are required to overcome the mitogenic activity of E in hormone replacement therapy (HRT) of non-hysterectomised postmenopausal women.

The effect of progestins on mammary carcinogenesis is much less clear since molecular and cellular studies diverge according to laboratories (1-2), and it was even proposed that progestins protect against breast cancer (BC) (3). The results of a large randomised NIH clinical trial in USA (4) involving post-menopausal women treated with medroxyprogesterone acetate (MPA) and the E conjugate, Premarin, vs placebo indicate that this association increases the risk of BC by 30%, as previously shown in several observational studies also attained with MPA, which is rarely used in France and other parts of Europe (5-7). Moreover, results obtained in the 1980s on estrogen receptor $(ER)^-$ and progesterone receptor $(PR)^+$ BC cell lines in our laboratory (8), and studies on pre-malignant human mammary glands (9), could be interpreted as due to beneficial effects of progestins. Since in France, contrary to USA, it has long been believed that progestins are protective or at least neutral in breast, the HRT currently practiced in this country has differed from that in the USA. In France, the percutaneous route of E administration is more frequently prescribed, along with various progestins other than MPA. Moreover, the proportion of French women treated for menopause with progesterone or a progestin associated with Es, instead of E alone, is higher (ratio $\approx$ 4.7 in the E3N cohort) than in USA (ratio $\approx$ 1.5 in WHI trial) where the higher frequency of hysterectomies allows women to take only E.

Controversies and Discrepancies Concerning Progestin Action on Mammary Glands

There is a relatively high consensus among laboratories on the mitogenic effect of Es on the mammary gland, and coherence between *in-vitro* studies on cell lines, and *in-vivo* epidemiological data, even though the detailed mechanism is not yet fully understood. By contrast, the effects of progestins after menopause, *i.e.,* whether good or bad, have been extensively debated according to the laboratory and country involved. This is mostly due to the fact that different progestins are being used and that most controlled clinical studies have been conducted using MPA, which is rarely prescribed in Europe. In Europe, and particularly in France, at least 15 different preparations are used, varying in their structure (natural progesterone, synthetic pregnanes, 19 nor-pregnane, and 19 nor-testosterone derivatives), and in their mode of administration (continuous or discontinuous, *per os* or transparenteral) (10). Consequently, the receptors interacting with progestins are not only the two PR isoforms A and B, but, in some cases, also the androgen receptor, ER, and the glucocorticoid receptor (Figure 1). The 19 nor-testosterone derivatives have been shown to display more estrogenic and androgenic activities than pregnane and nor-pregnane derivatives.

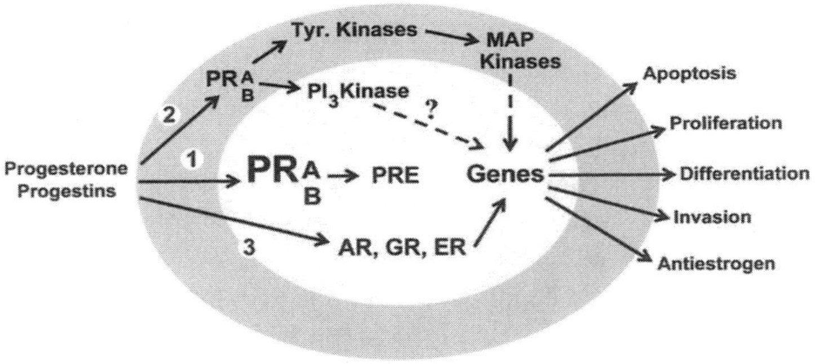

Figure 1. Schematic representation of progestin action in a BC cell. The relative importance of the pathways and end-points may vary according to the nature of the progestin used. **Pathway 1**- Classical legitimate pathway *via* nuclear PR-A/-B and transcriptional regulation. **Pathway 2**- Plasma membrane pathways activating the tyrosine, MAP or PI-3 kinases. **Pathway 3**- Illegitimate pathway *via* other receptors vary according to the affinity and transconformational activity of the different progestins (nor-testosterone, pregnane and nor-pregnane derivatives).

The other difficulty is that most reported studies were performed in culture on BC cell lines, with only a few controversial studies conducted with non-tumoral mammary cells. The results obtained on rodent models clearly indicate that progesterone stimulates the proliferation of lobulo-alveolar glands both in immature

(11) and ovariectomized mice (12). Moreover, PR-B gene knock-out, but not that of PR-A inhibits mammary gland development in mice (13, 14), and PR-B over expression facilitates endometrial carcinogenesis (15). However, the case may be different for aged human mammary glands exposed to hormone administered after menopause which may already be transformed.

A review of experimental data indicates that there are almost as many arguments favouring an adverse effect of progestins on mammary cells than those supporting a beneficial effect. By contrast, the published clinical trials point mostly adverse effects with MPA, which may display some androgenic and glucocorticoid activity. In fact, the WHI randomised clinical trial of menopaused patients undergoing HRT indicated an increased incidence of BC in women treated with E + MPA compared to placebo (4). The trial involving hysterectomized women treated by E alone is still in progress confirming additional MPA effects in mammary carcinogenesis.

Use of Human Breast Cancer Cell Lines to Detect Specific Receptor-Induced Proteins

Before the development of DNA micro-array and SAGE approaches, the use of [^{35}S]-methionine labelling of proteins synthesized in cell culture, before and after hormone stimulation, allowed us detection of several steroid hormone specific induced proteins, as defined by their molecular weight in SDS-PAGE, which could then be identified and studied in tumour samples to specify their significance in human carcinogenesis (16). Regarding progestins, using dose/response curves and anti-hormones, we could not discriminate between MPA and R5020 (promegestone) with respect to the induction of specific PR responses, such as the secreted 48 kD protein (16-17) and the cellular 250 kD protein, which were both inhibited by the anti-progestin RU486, but not by the anti-androgen flutamide (18 and D.Chalbos et unpublished).

The two progestins were also similarly active, but at higher concentrations, inducing a specific response to an androgen receptor such as that of a 43-kD androgen induced secreted protein (16), which has been identified as the Znα2 glycoprotein (19). The weaker activity of progesterone on the cell lines was probably due to its high metabolism of inactive products, while R5020 has been shown to be more stable. Both MPA and R5020 are pregnane derivatives, and, supposedly, both are less androgenic than 19 nor-testosterone progestins.

The Antiestrogenic Activity of Progestins: Good or Bad Significance?

It is clear that progesterone is a natural anti-E, thus partially explaining its protective effect in endometrium. We showed this anti-estrogenic activity on BC *in vitro*, since R5020, a pure progestin, inhibited the E-induced growth of MCF7 cells (8). However, proposals concerning mechanism of this anti-estrogenic activity have varied with time: Induction of 17β-OH-steroid dehydrogenase (9), increase of inactive E-sulfates (20), down regulation of the ER, and inhibition of transcription of ERE-controlled genes by squelching of limiting receptor co-factors (21). The decrease in the ERα level in mammary glands, after progestin treatment of pre-menopausal women with a benign breast disease was demonstrated in a population of 67 women by two fine needle biopsies performed before and after a 20-30 days treatment with lynestrenol, *i.e.*, a nor-testosterone progestin routinely used in France to treat benign breast disease in the 1980s. In the 20 women receiving only placebo, the ERα concentration measured by immunohistochemistry was not altered, while in the 47 women receiving lynestrenol, the ERα level was markedly reduced by this treatment in all patients (22). This finding was consistent with an anti-estrogenic action of progestin operating also *in- vivo*. Whether this effect will be good or bad in the long term remains controversial, since $ER\alpha^-$ BC cells (both *in-situ* and invasive) are more aggressive than $ER\alpha^+$ BC cells (23). Moreover, while the ERα complex stimulates cell growth, it inhibits invasion of ovarian and breast cancer cells (24, 25). This is consistent with a possible better prognosis of BC in postmenopausal women taking HRT.

In other tissues, such as bone or endothelial vascular cells, the anti-E activity of systemic progestins is certainly not beneficial by inhibiting the favourable effects of Es. This might explain the increased incidence of stroke and coronary disease observed in the WHI trial when MPA was associated with E.

The Significance of Fatty Acid Synthase: From a Marker of Differentiation to an Enzyme Involved in Carcinogenesis

Progestin regulation of some genes (Figure 1) such as the decrease in bcl-2 facilitating apoptosis (26) or the increase in alkaline phosphatase a marker of mammary gland differentiation (1, 2) suggest a favourable effect of. progestins. Others have suggested a higher aggressiveness such as the induction of the EGF receptor and VEGF (27).

We have extensively studied an abundant 250-kDa protein specifically induced by progestin at the gene level, in BC cells, *via* the PR. This protein was identified by several converging approaches (18, 28, 29) as fatty acid synthase (FAS, EC 2.3.1.85), the cytosolic enzyme which synthesizes C16 fatty acids from acetyl CoA and malonyl CoA (30).

The Mechanism of FAS Regulation by Progestin

The mechanism of FAS regulation by progestins was found to be direct, since it was not inhibited by cycloheximide, and rapid, with an increase of transcription within 20 min. It is also post-transcriptional since progestins increased FAS mRNA stability (29, 31). FAS was not induced by Es or glucocorticoids and different types of progestins used in HRT, including the "pure" progestin R5020 and MPA, were similarly active in the induction of FAS (18 and unpublished experiments). Progesterone however, was only active at higher concentrations.

In BC, FAS is specifically regulated *via* the PR since its induction by R5020 was inhibited by RU 486 but not by flutamide. Androgens such as dihydrotestosterone (DHT) were only active at μμM concentrations. Therefore, in $ER\alpha^+$ and PR^+ BC cell lines, progesterone was the major hormone to regulate FAS expression. Recently, other laboratories showed that mitogens such as the growth factors EGF, FGF-1, and HER-2/neu also induce FAS by stimulating PI-3 kinase and MAP kinase pathways leading to induction of the sterol regulatory element binding protein (SREBP1) transcription factor . This pathway has been documented in two $ER\alpha^-$ mammary cell lines: MCF10a transfected with ras (32), and H16 transfected with the HER-2/neu oncogene (33). In MCF7 and T47D BC cell lines, the classical pathway following activation of nuclear PRs by progestins was found to be the predominant or even exclusive regulation. Therefore *in-vivo,* in ERα and PR-positive BCs, FAS is most likely induced by progestins acting *via* their receptors. However, in ER^- and PR^- BCs, oncogenes coded proteins and growth factors *via* the PI-3, and MAP kinase pathways may be involved. Metabolic regulation by a lack of nutrients in cells is also possible.

Interestingly, in prostate cancer, androgens rather than progesterone induce FAS *via* the androgen receptor (34). Whether this effect is direct and transcriptional or mediated by the PI-3 kinase pathway is debated (35).

The Clinical Significance of FAS

The clinical significance of FAS has evolved with time. In our laboratory, pilot translational studies in patients, indicated that it was also associated with progesterone regulation since its mRNA level, as measured by *in-situ* hybridization (36), and the protein level, as estimated by immunohistochemistry (IHC) on frozen sections (37), were increased at the luteal phase mostly in lobulo-alveolar glands. In women treated for benign breast disease with synthetic progestins, FAS expression was increased in mammary lobules and ducts at a much higher level than that reached at the luteal phase with the endogenous progesterone (37).

Moreover, FAS staining in BC was generally greater in pre-menopausal patients than after menopause. These results indicated that progestins also induced FAS expression *in vivo*, in both malignant and non malignant mammary gland cells, and suggested that FAS is a marker of responsiveness to progestins. Based on the dominant paradigm concerning progestins in BC in France, we initially proposed that FAS is a differentiation marker: First, in normal mammary glands, FAS is known to be mostly active during lactation to produce milk lipids (30). Secondly, we observed high accumulation of lipid droplets in R5020-treated T47D cells with an increased cellular level of triglycerides and phospholipids (38). In parallel, R5020 decreased the secretion of lipids and glycoproteins (17). It was therefore tempting to propose in this context that FAS is a marker of mammary gland differentiation.

We did not consider that FAS was as potentially important as cathepsin D, a marker of aggressiveness in BC, for monitoring BC patients. However, interestingly, FAS content, as measured by IHC, was found to be correlated with the proliferation index (histone H4) and histological criteria of hyperplasia mostly in lobules (37), suggesting an opposite significance. This is consistent with the physiology of progesterone, which mostly stimulates the growth of lobular alveolar glands, and with clinical studies showing an increased proportion of lobular carcinomas among BCs developed under HRT (39, 40).

Ten years after our studies on this progesterone-induced enzyme, FAS was revealed to be a marker associated with high BC risk, and a potential target for cancer therapy. Different independent laboratories using various approaches came to this same conclusion. First, Pasternak, *et al.* at the John Hopkins Institute developed antibodies to haptoglobin and found that the corresponding antigen (OA-519) was of poor prognostic value in BC (41,42). The antibodies reacted with contaminating FAS, which was actually the antigen responsible for the increased staining detected in aggressive BCs. Later, Kuhadja proposed FAS as a new therapeutic target in cancer. FAS inhibitors, such as cerulenine and C75 (43, 44), have been proven efficient in blocking breast and ovarian cancer cell growth, at least *in vitro* and in nude mice. FAS mRNA and protein were shown by other laboratories to be increased in several solid tumour cells, including prostate cancer, compared to normal cells. The group of Dana Farber in Boston, using the SAGE approach, showed that among 50,000 detected transcripts, FAS was one of the most over-expressed genes in BC as compared to normal mammary cells (45). The cDNA micro-array approach also indicated high FAS over expression in prostate cancer compared to normal prostate (46).

What is the Significance of FAS Over-expression in Cancer Cells?

Is FAS over-expression a marker only associated with aggressive solid tumours or is it actively engaged in mammary carcinogenesis? FAS activity and endogenous fatty acid synthesis are required for normal embryonic development since FAS gene knock-out is lethal in mice (43). In normal adult cells, except in lactating mammary glands (30), FAS activity is low since fatty acids are mostly provided by food and imported from local vascularization (47).

The use of our specific FAS antibodies on formalin-fixed paraffin embedded sections of mammary tumours have shown that FAS is specifically overexpressed in BC cells and not in adjacent normal ducts lobules or stromal cells (M.Esslimani Sahla, *et a,l* in preparation).

We hypothesize that in solid tumours the bio-availability of fatty acids from the circulation is weak because their vascularization is generally poor, as also indicated by the lower extra-cellular pH. BC cells overexpressing FAS may have a growth advantage compared to normal peripheral cells in these solid tumours. Cancer cells require nutrients to grow, and not only mitogens. In addition to proteins, they also need fatty acids as precursors for synthesis of membranes, lipid mediators and lipid anchors and they may use FAS for endogenous synthesis of these fatty acids. FAS endogenous activity seems necessary for development of the embryo (48) and for growth of cancer cells, as supported by the specific effects of cerulenine or C75, which block proliferation of breast and ovarian cancer cells and induce apoptosis, while normal cells are insensitive (43, 44, 49, 50). This inhibitory effect is suppressed by adding an excess of palmitic acid, the major product of FAS activity. This clearly indicates that cancer cells require FAS activity for their sustained growth and suggests that FAS could be a potential target in the treatment of hormone-dependant and independent BCs.

Concluding Remarks

On the basis of recent epidemiological and clinical data on the increased incidence of BC in women treated with E + MPA, compared to placebo or E alone, we reviewed data on progestin action in human mammary cells from our laboratory and others. We propose that the increased risk of BC by progestin is due to the induction of master genes and that FAS is one of these genes. .

This review illustrates that the same experimental results can be interpreted differently according to the dominant hypothesis, which has varied with time in different countries. In France, in the 1980s and 90s, it was quite logical to concentrate our research on E action since Es were mitogenic *in vitro*, while progestins were not mitogens in our conditions. The reason why in post-menopausal women, the addition of progestins to E in HRT slightly increases the risk of human BC, while, *in vitro,* progestins are rather inhibitory is not known. FAS is induced by progestins and overexpressed in BC cells both *in vitro* and *in vivo*. While FAS activity appears to be involved in facilitating tumor growth *in vivo*, it is unknown why it does not appear to be active *in vitro*. This may be due to the different BC cell culture environments as monolayers on plastic or packed *in vivo* in a 3D solid tumor. The lack of other partners required for progestin mitogenicity is also possible.

The emergence of "evidence-based medicine", such as the large WHI randomized trial is crucial both for guiding us in basic research directions to understand mechanisms, and for reorienting the medical practice in this country. It is possible, on the basis of various hypotheses, that other progestins and routes of administration rather than oral would be better for women health. However, this will have to be proven through controlled and large scale epidemiological studies. The adverse role of progestins and FAS suggest new potential therapies. In ER^+

and PR^+ BC, resistant to anti-Es, anti-progestins - if non-toxic - could be useful, as we initially proposed for RU486 based on our *in-vitro* studies (51) and a pilot clinical trial (52).

Moreover, the most aggressive ER^- and PR negative BCs could be targeted by inhibiting the activity of enzymes, which are largely and specifically over expressed in these cancer cells. There is a striking parallel between estrogen and progesterone action in BC: the two ovarian hormones induce an enormous quantity of two enzymes, *i.e.,* cathepsin D and FAS, both of them facilitate the sustained growth of cancer cells *in vivo* by different synergic mechanisms. These two enzymes are also overexpressed in more aggressive ER^- BC via growth factors and oncogene encoded proteins. They both constitute new putative targets for cancer therapy, as shown by Kushada, *et al* for FAS and by our laboratory for cathepsin D (53). These two steroid and growth factor induced enzymes illustrate the synergy between E and progestins to facilitate the growth of hormone dependent solid tumors and the potential of enzyme inhibitors in cancer therapy.

Foot note added in proof: Since this review was completed (June 2003) results of a cohort of a million of British post menopausal women taking hormone replacement therapy with different preparations [Beral V and the MWS group (2003) Breast cancer and hormone-replacement therapy in the Million Women Study, Lancet 362: 419-427] confirmed the adverse effect of progestins on breast, including the progestins currently used in Europe

Acknowledgments

We thank Ms. Fabienne David for excellent secretarial assistance and Jean Yves Cance for the skilled preparation of the figure.

References

1. Graham JD, Clarke CL (1997) Physiological action of progesterone in target tissues. Endocr Rev 18:502-519.
2. King RJB (1993) Estrogen and progestin effects in human breast carcinogenesis. Breast Cancer Res Treat 27:3-15.
3. Plu-Bureau G, Lê MG, Sitruk-Ware R, et al (1994) Progestogen use and decreased risk of breast cancer in a cohort study of premenopausal women with benign breast disease. Br J Cancer 70:270-277.
4. Writing Group for the Women's Health Initiative Investigators (2002) Risks and benefits of estrogen plus progestin in healthy postmenopausal women. JAMA 288:321-333.
5. Magnusson C, Baron JA, Correja N, et al (1999) Breast cancer risk following long-term oestrogen- and oestrogen-progestin-replacement therapy. Int J Cancer 81:339-344.

6. Ross RK, Paganini-Hill A, Wan PC, Pike MC (2000) Effect of hormone replacement therapy on breast cancer risk: estrogen versus estrogen plus progestin. J Natl Cancer Inst 92:328-32.
7. Schairer C, Lubin J, Troisi R, et al (2000) Menopausal estrogen and estrogen-progestin replacement therapy and breast cancer risk. JAMA 283:485-491.
8. Vignon F, Bardon S, Chalbos D, et al (1983) Antiestrogenic effect of R5020, a synthetic progestin in human breast cancer cells in culture. J Clin Endocrinol Metab 56:1124-1130.
9. Gompel A, Malet C, Sprotzer P, et al (1986) Progestin effect on cell proliferation and 17 beta-hydroxysteroid dehydrogenase activity in normal human breast cells in culture. J Clin Endocrinol Metab 63:1174-1180.
10. Sitruk-Ware R (2000) Progestins in hormonal replacement therapy and prevention of endometrial disease in clinical practice. In: Sitruk-Ware R, Mishell Jr DR (eds) Progestins and Antiprogestins. New York:Marcel Dekker, Inc., 269-278
11. Haslam SZ (1988) Progesterone effects on deoxyribonucleic acid synthesis in normal mouse mammary glands. Endocrinology 122:464-470.
12. Raafat AM, Li S, Bennett JM, et al (2001) Estrogen and estrogen plus progestin act directly on the mammary gland to increase proliferation in a postmenopausal mouse model. Cell Physiol 187:81-89.
13. Lydon JP, Ge G, Kittrell FS, et al (1999) Murine mammary gland carcinogenesis is critically dependent on progesterone receptor function. Cancer Res 59:4276-84.
14. Conneely OM, Mulac-Jercevic B, Lydon JP, De Mayo F (2002) Distinct and overlapping reproductive activities of the PR-A and PR-B isoforms of the progesterone receptor. Abstract S28-4, Int Congress Hormones and Cancer, Fukuoka, Oct 2002.
15. De Vivo I, Huggins GS, Hankinson SE, et al (1999) A functional polymorphism in the promoter of the progesterone receptor gene associated with endometrial cancer risk. Proc. Nat. Acad. Sci. USA 99:12263-12268.
16. Rochefort H, Chalbos D (1984) Progestin specific markers in human cell lines: biological and pharmacological applications. Mol Cell Endocrinol 36:3-10.
17. Chalbos D, Rochefort H (1984) Dual effects of the progestin R5020 on proteins released by the T47D human breast cancer cells. J Biol Chem 259:1231-1238
18. Chalbos D, Westley B, May F, et al (1986) Cloning of cDNA sequences of a progestin-regulated mRNA from MCF7 human breast cancer cells. Nucleic Acids Res 14:965-982.

19. Chalbos D, Haagensen D, Parish T, et al (1987) Identification and androgen regulation of two proteins released by the T47D human breast cancer cells. Cancer Res 47:2782-2792.
20. Pasqualini JR, Chetrite G, Blacker C, et al (1996) Concentrations of estrone, estradiol, and estrone sulfate and evaluation of sulfatase and aromatase activities in pre- and postmenopausal breast cancer patients. J Clin Endocrinol Metab 81:1460-1464.
21. Chalbos D, Galtier F (1994) Differential effect of forms A and B of human progesterone receptor on estradiol-dependent transcription. J Biol Chem 269:23007-23012.
22. Maudelonde T, Lavaud P, Salazar G, et al (1991) Progestin treatment depresses estrogen receptor but not cathepsin D levels in needle aspirates of benign breast disease. Breast Cancer Res Treat 19:95-102.
23. McGuire WL, Clark GM (1992) Prognostic factors and treatment decisions in axillary-node-negative breast cancer. N Engl J Med 326:1756-1761.
24. Hayashido Y, Lucas A, Rougeot C, et al (1998) Estradiol and Fibulin-1 inhibit motility of human ovarian and breast cancer cells induced by fibronectin. Int J Cancer 75:654-658.
25. Rochefort H, Platet N, Hayashido Y, et al (1998) Estrogen receptor mediated inhibition of cancer cell invasion and motility: an overview. J Steroid Biochem Molec Biol 65:163-168
26. Gompel A, Somai S, Chaouat M, et al (2000) Hormonal regulation of apoptosis in breast cells and tissues. Steroids 65:593-598.
27. Hyder SM, Chiappetta C, Stancel GM (2001) Pharmacological and endogenous progestins induce vascular endothelial growth factor expression in human breast cancer cells. Int J Cancer 92:469-473.
28. Chalbos D and Rochefort H (1984) A 250-kilodalton cellular protein is induced by progestins in two human breast cancer cell lines MCF7 and T47D. Biochem Biophys Res Commun 121:421-427.
29. Chalbos D, Chambon M, Ailhaud G, et al (1987) Fatty acid synthetase and its mRNA are induced by progestins in breast cancer cells. J Biol Chem 262:9923-26.
30. Wakil S (1989) Fatty acid synthase, a proficient multifunctional enzyme. Biochemistry 28:4523-4530.
31. Chalbos D, Galtier F, Emiliani S, et al (1991) The antiprogestin RU486 stabilizes the progestin-induced FAS mRNA but does not stimulate its transcription. J Biol Chem 266:8220-8224.
32. Yang YA, Morin PJ, Han WF, et al (2002) Regulation of fatty acid synthase expression in breasr cancer by sterol regulatory element binding protein-1c. Exp Cell Res 282:132-137.
33. Kumar-Sinha C, Ignatoski KW, Lippman ME, et al (2003) transcriptome analysis of HER2 reveals a molecular connection to fatty acid synthesis. Cancer Res 63:132-139.

34. Swinnen JV, Esquenet M, Goossens K, et al (1997) Androgens stimulate fatty acid synthase in the human prostate cancer cell line LNCaP. Cancer Res 57:1086-1090.
35. Van de Sande T, De Schrijver E, Heyns W, et al (2002) Role of the phosphatidylinositol 3'-kinase/PTEN/Akt kinase pathway in the overexpression of fatty acid synthase in LNCaP prostate cancer cells. Cancer Res 62:642-646.
36. Chalbos D, Escot C, Joyeux C, et al (1990) Expression of the progestin-induce fatty acid synthetase in benign mastopathies and breast cancer as measured by RNA *in situ* hybridisation. J.Natl Cancer Inst 82:602-606.
37. Joyeux C, Chalbos D, Rochefort H (1990) Effects of progestins and menstrual cycle on fatty acid synthetase and progesterone receptor in human mammary glands. J Clin Endocrinol Metab 70:1438-1444.
38. Chambon M, Rochefort H, Vial HJ, et al (1989) Progestins and androgens stimulate lipid accumulation in T47D breast cancer cells *via* their own receptors. J Steroid Biochem 33: 915-922.
39. Daling JR, Malone KE, Doody DR, et al (2002) Relation of regimens of combined hormone replacement therapy to lobular, ductal, and other histologic types of breast carcinoma. Cancer 95:2455-2464.
40. Li CI, Stanford JL, Daling JR (2000) Hormone replacement therapy in relation to risk of lobular and ductal breast carcinoma in middle-ages women. Cancer 88:2570-2577.
41. Kuhajda FP, Piantadosi S, Pasternack GR (1989) Haptoglobin-related protein(Hpr) epitopes in breast cancer as a predictor of recurrence of the disease. N Engl J Med 321:636-641.
42. Alo PL, Visca P, Marci A, et al (1996) Expression of fatty acid syntase (FAS) as a predictor of recurrence in stage I breast carcinoma patients. Cancer 77:474-82.
43. Kuhajda FP, Jenner K, Wood FD, et al (1994) Fatty acid synthase: a potential selective target for antineoplastic therapy. Proc Natl Acad Sci USA 91:6379-6383.
44. Kuhadja FP, Pizer ES, Li JN, et al (2000) Synthesis and antitumor activity of an inhibitor of fatty acid synthase. Proc Natl Acad Sci USA 97: 3450-3454.
45. Porter DA, Krop IE, Nasser S, et al (2001) A SAGE (Serial Analysis of Gene Expression) view of breast tumor progression. Cancer Res 61:5697-5702.
46. Welsh JB, Sapinoso LM, Su AI, et al (2001) Analysis of gene expression identifies candidate markers and pharmacological targets in prostate cancer. Cancer Res 61:5974-5978.
47. Weiss L, Hoffmann GE, Schreiber R, et al (1986) Fatty-acid biosynthesis in man, a pathway of minor importance. Biol Chem Hoppe-Seyler 367:905-912.

48. Chirala SS, Chang H, Matzuk M, et al (2003) Fatty acid synthesis is essential in embryonic development: fatty acid synthase null mutants and most of the heterozygotes die in utero. Proc Natl Acad Sci USA 100:6358-6363.
49. Pizer ES, Wood FD, Heine HS, et al (1996) Inhibition of fatty acid synthesis delays disease progression in a xenograft model of ovarian cancer. Cancer Res 56:1189-1193.
50. Pizer ES, Jackisch C, Wood FD, et al (1996) Inhibition of fatty acid synthesis induces programmed cell death in human breast cancer cells. Cancer Res 56:2745-2747.
51. Bardon S, Vignon F, Chalbos D, et al (1985) RU486, a progestin and glucocorticoid antagonist, inhibits the growth of breast cancer cells via the progesterone receptor. J Clin Endocrinol Metab 60:692-697.
52. Romieu G, Maudelonde T, Ulmann A, et al (1987) The antiprogestin RU486 in advanced breast cancer: Preliminary clinical trial. Bulletin du Cancer 74:455-461.
53. Rochefort H, Glondu M, Esslimani Sahla M, et al (2003) How to target estrogen receptor negative breast cancer? Endocrine -Related Canc 10:1-6.

5

Role of Progesterone Receptors in Mammary Development and Carcinogenesis

Gopalan Shyamala

Introduction

It is well established that signaling through estrogen and progesterone through their cognate receptors, estrogen receptor α (ERα) and progesterone receptor (PR) are essential for both the development of mammary glands and their transformation to preneoplastic and neoplastic counterparts. Several years ago it was demonstrated that both mouse and human mammary epithelial cells proliferate in response to progesterone (1-3). Despite this, among the two ovarian steroids, estrogens have been traditionally associated with an increased risk for breast cancer. In recent years, several studies, including our own, provide compelling evidence that progesterone/PR signaling may play a crucial role in the transformation of mammary epithelial cells. The identity of a particular cell is established during development and conversion of a normal cell to its malignant counterpart is accompanied by alterations in certain fundamental mechanisms that define its identity. Accordingly, to identify the mechanisms that trigger carcinogenesis, it is essential to understand the processes that ensure normal development and maintenance of this phenotype. This chapter summarizes our studies on PR during normal mammary development and their relevance to carcinogenesis.

Regulation of PR Expression and its Role During Mammary Development

Mammary development occurs mostly in the post natal female and is discontinuous. The initial phase of development accompanies puberty, commonly referred to as ductal morphogenesis, requires signaling through estrogen/ERα. The second phase of development accompanies pregnancy. It results in lobulo-alveolar development, and requires signaling through progesterone/PR. Similar to other estrogen target tissues, in the mammary gland, the expression of PR is regulated by estrogen (4), whereby it serves as a marker for both estrogen and progesterone action. Therefore, during various mammary gland physiological states, the PR expression patterns and their role need to be evaluated within the context of both estrogen and progesterone action.

In the mammary gland of pre-pubertal mice, PR gene expression is associated with the end bud cells (5), the putative undifferentiated progenitors which give rise to ductal cells. Studies on PR-null mutant mice reveal that PR is not essential for the ductal growth accompanying puberty (6). Thus, the expression of PR in the end bud cells does not reflect its requirement for ductal morphogenesis, but is indicative of the estrogenic sensitivity of these cells. In the mammary gland of adult females, the PR steady-state levels are modulated as a function of estrous/menstrual cycle. Estrogen regulates its expression positively which, in turn, is counteracted by progesterone. In mammary ducts of adult females, PR expression is heterogeneous, and the percentage of PR positive cells does not change either after ovariactomy or exposure to estradiol, despite changes in the steady state levels of PR (7), indicating that its expression is restricted to a subset of epithelial cells. Since epithelial cell proliferation is promoted by signaling through PR, which gives rise to alveolar cells, it is likely that in the adult female, PR positive cells represent the putative progenitors of alveolar cells (6).

A paradox that emerges from studies on the developmental regulation of PR is that while its expression is essential for lobulo-alveolar development, it is down regulated during pregnancy and lactation. This down-regulation is evident both at the level of mRNA and protein (4), and is accompanied by a significant decrease in the number of PR positive epithelial cells (8). It is well known that, for the most part, cellular proliferation and differentiation represent two incompatible phenomena, and lobulo-alveolar development is indicative of both morphological and functional differentiation. Thus, one of the keys for switching the epithelial cells from a proliferative to differentiated state may be the loss of PR expression. As such, in instances where the mechanisms that regulate PR expression are deregulated, cell proliferation can proceed in an unabated manner and thus, trigger transformation. Indeed, this appears to be the case, as revealed by our studies using one of the transgenic models created by our laboratory and discussed below.

Overexpression of the 'A' Form of PR (PR-A) Causes Transformation of Mammary Epithelial Cells

It is well established that PR exists in two molecular forms commonly referred to as the 'A' and 'B' forms whose ratio varies among target tissues (9). In adult rodents, the mammary gland ratio of the 'A':'B' forms is 3:1 (10). *In-vitro* studies have shown that the 'A' and 'B' forms can have different functions in the same cell. In addition, the PR activity of the individual form can vary among different types of cells (11). Furthermore, depending on the cell, the 'A' form can either inhibit or enhance the activity of the PR'B' form (11). Based on these observations, it is believed that appropriate cellular responsiveness to progesterone is dependent on regulated expression and/or activity of the two PR forms (12). Therefore, an aberration in normal mammary development may result from inappropriate progesterone signaling due to an imbalance in the expression and/or activities of the two PR forms. The imbalance, in turn, may have implications for mammary oncogenesis. To test this hypothesis, our laboratory created transgenic mice in which the native PR 'A':'B' ratio was altered by introducing additional PR 'A' or 'B' forms as transgene. A binary system was used for the expression of the transgene as shown in Figure 1. Detailed descriptions of the constructs have been published (13, 14).

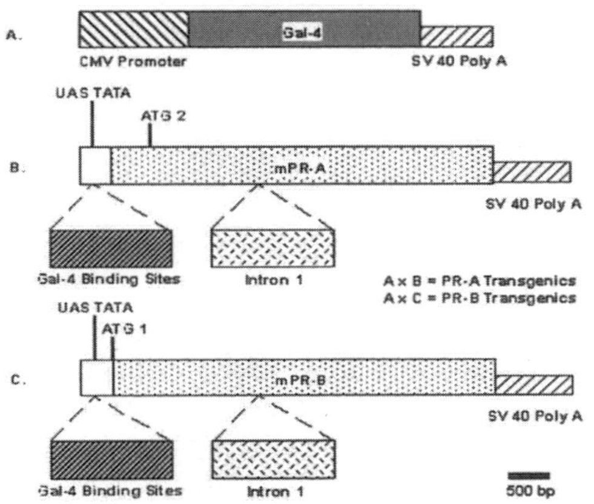

Figure 1. Schematic representation of plasmid construction for the binary system. (A) Insertion of the GAL-4 gene into the CMV promoter expression plasmid containing simian virus (SV) 40 splice and polyadenylylation sequences. (B) mPR cDNA (A form with only ATG 2) containing intron 1 and SV40 splice and polyadenylation sequences fused to UAS-TATA fragment containing four GAL-4 binding sites. E, *Eco*R1. (C) mPR cDNA (B form with only ATG1) containing intron 1 and SV40 splice and polyadenylation sequences fused to UAS-TATA fragment containing four GAL-4 binding sites.

Mammary development is abnormal in transgenic mice carrying an imbalance in the native ratio of 'A':'B' forms by overexpressing either the 'A' or 'B' form (referred to as PR-A and PR-B transgenics, respectively) (13, 14). In particular, PR-A transgenic mammary glands exhibit excessive ductal growth, and morphological and histological features associated with transformed cells, *i.e.,* loss in basement membrane integrity and cell-cell adhesion. The aberrant patterns of ductal growth were not restricted to the offspring of any particular founder mouse, establishing that the phenotype was due to the transgene expression, and not due to their integration into any unique site on the genome (15).

In a series of comprehensive studies, Medina, *et al.* identified certain molecular markers unique to mammary epithelial cells at various stages of transformation. In particular, they showed that in mouse mammary epithelial cells, a decrease in p21 expression, without an increase in cyclin D1, is indicative of immortalization, and precedes the onset of hyperplasia/preneoplasia (16). Among the characteristics that distinguish the presumptive immortalized epithelial cells from those in hyperplasia/preneoplasia are increases in cyclin D1 expression and cell proliferation, and a decrease in ERα expression (17).

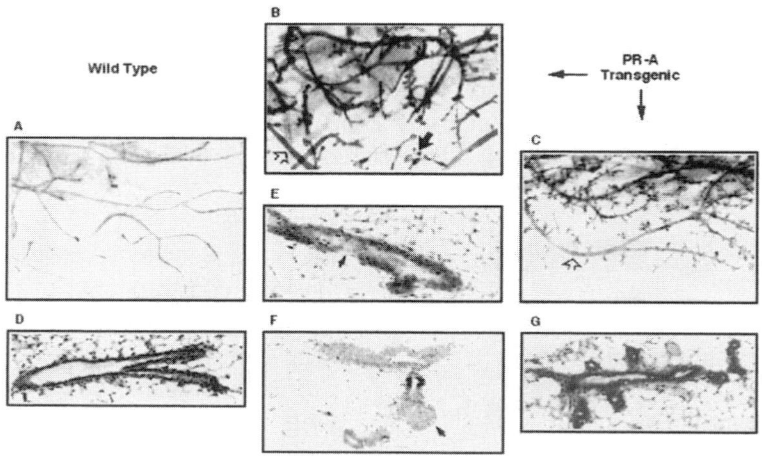

Figure 2. Morphological and histological characteristics of wild type and PR-A in adult transgenic mice. Whole-mounts (**A-C**) and histology (**D-G**) of mammary glands from adult (10-14 weeks old) control PR-A transgene-negative (**A and D**) and PR-A transgenic mice (**B, C, and E-G**) are shown. In B and C, open arrows point to thick ducts, while solid arrows point to clustered buds at the tip of ducts. Note that, in contrast to the monolayer associated with the duct from wild type mice (**D**), the ducts in PR-A transgenics are composed of multilayered cells (**E, F, and G**). In E, the arrow points to an indistinct epithelial-stromal boundary, and in F, the arrow points to disorganized masses of cells at the tip of a duct.

In mammary glands of PR-A transgenics, ducts with normal histology, displayed a decrease in p21 expression (Figure 3A) without an elevation in cyclin D1 (Figure 3B), or cell proliferation (Figure 3C). Similar to ducts with normal histology, in mammary dysplasias of PR-A transgenic, there was a decrease in p21 expression (Figure 3A), but was accompanied by an increase in cyclin D1 expression (Figure 3B) and cell proliferation (Figure 3C), and a decrease in ERα expression (Figure 3D). Therefore, these studies establish that mammary glands of PR-A transgenics contain at least two distinct populations of transformed epithelial cells: one, corresponding to immortalized cells, in early stages of transformation, present in the ducts with normal histology; and two, cells in later stages of transformation associated with dysplasias.

An imbalance in the native ratio of PR A:B isoforms also exists in PR-B transgenic mice, but the mammary phenotype of these mice is distinct from that of the PR-A transgenics (14). In fact, PR-B mammary glands have impaired ductal elongation. Analyses of these mammary glands did not reveal any detectable changes in either the number of BrdU-positive cells or the expression of cyclin D1, p21, or ERα (18).

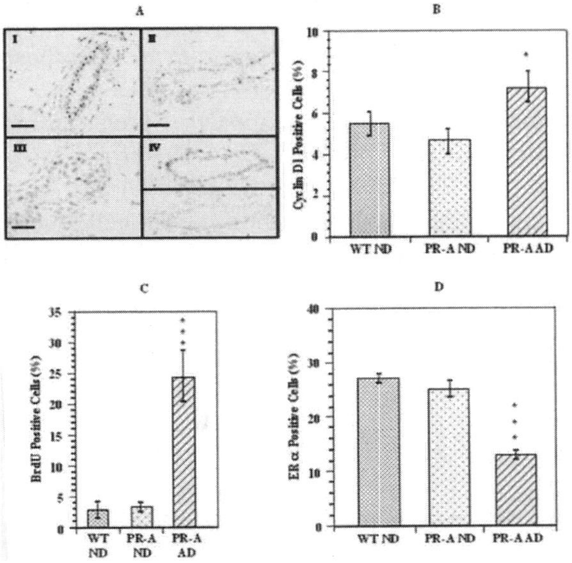

Figure 3. Cell proliferation, cyclin-D1, p21, and ERα expression in mammary glands of wild-type and PR-A transgenic mice. **Panel A** shows the expression profiles of p21 in a mammary duct of wild type mice (**I**), a mammary duct of PR-A transgenic with normal histology (**II**), an abnormal mammary duct of PR-A transgenic (**III**), and (**IV**) the absence of immunoreactivity with irrelevant mouse IgG (top), and deletion of primary antibody (bottom). Scale bar represents 20 μm. **Panel B** shows the number of cyclin D1-positive cells in normal (ND) and abnormal ducts (AD). **Panel C** shows the rate of cell proliferation, using BrdU-positive cells as index, in ND and AD. **Panel D** shows the number of ERα-positive epithelial cells in ND and AD.

Aberrant Features of Mammary Dysplasias in PR-A Transgenic Mice are Unaffected by Mifepristone

The presence of at least two distinct populations of transformed epithelial cells, in mammary glands of PR-A transgenics, was also apparent from their sensitivity to the antiprogestin, mifepristone. While mifepristone abolished BrdU immunostaining in the ducts displaying normal histology, and in those of wild type mice, it did not alter the number of BrdU-positive cells in dysplastic lesions (18). Similarly, in these lesions, mifepristone did not effect the expression of either cyclin D1 or ERα. Thus, it appears that deregulation of progesterone action, resulting from an overexpression of PR-A form, may not only trigger transformation of mammary epithelial cells, but may also dictate their phenotype in later stages of transformation.

Progesterone/PR Signaling in Mammary Carcinogenesis

Most of the classical studies on the developmental biology of the normal mammary gland and mammary oncogenesis have been done in rodents. Conclusions derived from these studies have been found to be applicable to the human condition (19). Furthermore, the pattern of PR localization in normal mouse mammary glands is similar to that reported previously for human breast, *i.e.,* (a) PR expression in the epithelial cells is heterogeneous, and (b) the connective tissue surrounding these epithelial cells are PR free (7, 20).

In normal mammary epithelial cells, the initial impact of progesterone signaling through PR results in an increase in proliferation. Therefore, in its capacity as a mammary gland mitogen, progesterone has the potential to trigger carcinogenesis. An excellent example of this P action is the animal model developed by Lanari *et al.* in which the progestin, medroxyprogesterone induces mammary tumors (21). Another example is the p53-null mutant mammary epithelium in which progesterone alone strongly enhances mammary tumorigenesis (22). Studies from our laboratory (as reviewed here) showed that a deregulation in progesterone action (as in PR-A transgenics) can alter the growth potential of epithelial cells, at least, in part, at the level of cell cycle, leading to transformation. Recent population based studies also demonstrated that, in uninterrupted combined hormone (estrogen + progestin) replacement therapy (CHRT), progesterone is the contributing factor for the increased risk for mammary carcinogenesis (23). Also, studies on normal mammary development established that the net outcome of signaling through PR is to drive the epithelial cells towards lobulo-alveolar development. Thus, it is noteworthy that women who are at increased risk for mammary oncogenesis due to CHRT develop lobular carcinomas (24).

Studies from our laboratory clearly established that an imbalance in the expression of the two PR isoforms, resulting from PR-A overexpression, may lead to transformation of mammary epithelial cells. An imbalance in the relative ratio of PR A:B isoforms has also been observed in certain human mammary tumors, and has often been associated with overexpression of the PR-A form (25-27).

Concluding Remarks

Our studies on the role of progesterone/PR in the mammary gland highlight their involvement in mammary epithelial cell proliferation during development. In addition, they establish that appropriate expression of the two PR isoforms is critical for normal mammary development. Finally, they demonstrate that an imbalance in the native ratio of the two isoforms, resulting in dominance of the PR-A form, can trigger transformation. Thus, PR-A transgenic mice can serve as an important experimental model for dissecting the mechanisms underlying progesterone-dependent mammary epithelial cell transformation, and progression *in vivo*. In addition, they may help to devise strategies for arresting the progression of transformation, and reversing the transformed phenotype.

Dedication

This article is dedicated to the memory of my dear friend Jane Toft, who lived a life of a scientist/teacher and wife/mother with courage, honesty, and devotion.

Acknowledgement

These studies were supported by NIH Grant CA 66541.

References

1. Bresciani F (1968) Topography of DNA synthesis in the mammary gland of the C3H mouse and its control by ovarian hormones: an autoradiographic study. Cell Tissue Kinetics 1:51.
2. Ferguson DJ, Anderson TJ (1981) Morphological evaluation of cell turnover in relation to the menstrual cycle in the "resting" human breast. Br J Cancer 44:177-181.
3. Going JJ, Anderson TJ, Battersby S, et al (1988) Proliferative and secretory activity in human breast during natural and artificial menstrual cycles. Amer J Pathol 130:193-204.
4. Shyamala G, Schneider W, Schott D (1990) Developmental regulation of murine mammary progesterone receptor gene expression. Endocrinology 126:2882-2889.
5. Silberstein GB, Van Horn K, Shyamala G, et al (1996) Progesterone receptors in the mouse mammary duct: distribution and developmental regulation. Cell Growth Differ 7:945-952.
6. Shyamala G (1999) Progesterone signaling and mammary gland morphogenesis. J Mammary Gland Biol Neoplasia 4:89-104.
7. Shyamala G, Barcellos-Hoff MH, Toft D, et al (1997) In situ localization of progesterone receptors in normal mouse mammary glands: absence of receptors in the connective and adipose stroma and a heterogeneous distribution in the epithelium. J Steroid Biochem Mol Biol 63:251-259.
8. Shyamala G, Chou YC, Louie SG, et al (2002) Cellular expression of estrogen and progesterone receptors in mammary glands: regulation by hormones, development and aging. J Steroid Biochem Mol Biol 80:137-148.
9. Tsai MJ, O'Malley BW (1994) Molecular mechanisms of action of steroid/thyroid receptor superfamily members. Ann Rev Biochem 63:451-486.
10. Schneider W, Ramachandran C, Satyaswaroop PG, et al (1991) Murine progesterone receptor exists predominantly as the 83-kilodalton 'A' form. J Steroid Biochem Mol Biol 38:285-291.
11. Vegeto E, Shahbaz MM, Wen DX, et al (1993) Human progesterone receptor A form is a cell- and promoter-specific repressor of human progesterone receptor B function. Mol Endocrinol 7:1244-1255.
12. McDonnell DP (1995) Unraveling the human progesterone receptor signal transduction pathway: Insights into antiprogestin action. Trends Endocrinol Metabol 6:133-138.

13. Shyamala G, Yang X, Silberstein G, et al (1998) Transgenic mice carrying an imbalance in the native ratio of A to B forms of progesterone receptor exhibit developmental abnormalities in mammary glands. Proc Natl Acad Sci USA 95:696-701.
14. Shyamala G, Yang X, Cardiff RD, et al (2000) Impact of progesterone receptor on cell-fate decisions during mammary gland development. Proc Natl Acad Sci USA 97:3044-3049.
15. Shyamala G, Louie SG, Camarillo IG, et al (1999) The progesterone receptor and its isoforms in mammary development. Mol Genet Metab 68:182-190.
16. Said TK, Moraes RC, Singh U, et al (2001) Cyclin-dependent kinase (cdk) inhibitors/cdk4/cdk2 complexes in early stages of mouse mammary preneoplasia. Cell Growth Differ 12:285-295.
17. Medina D (2002) Biological and molecular characteristics of the premalignant mouse mammary gland. Biochim Biophys Acta 1603:1-9.
18. Chou YC, Uehara N, Lowry JR, et al (2003) Mammary epithelial cells of PR-A transgenic mice exhibit distinct alterations in gene expression and growth potential associated with transformation. Carcinogenesis 24:403-409.
19. Cardiff RD, Wellings SR (1999) The comparative pathology of human and mouse mammary glands. J Mammary Gland Biol Neoplasia 4:105-122.
20. Press MF, Greene GL (1988) Localization of progesterone receptor with monoclonal antibodies to the human progestin receptor. Endocrinology 122:1165-1175.
21. Lanari C, Molinolo AA (2002) Progesterone receptors--animal models and cell signalling in breast cancer. Diverse activation pathways for the progesterone receptor: possible implications for breast biology and cancer. Breast Cancer Res 4:240-243.
22. Medina D, Kittrell FS, Shepard A, et al (2003) Hormone dependence in premalignant mammary progression. Cancer Res 63:1067-1072.
23. Weiss LK, Burkman RT, Cushing-Haugen KL, et al (2002) Hormone replacement therapy regimens and breast cancer risk(1). Obstet Gynecol 100:1148-1158.
24. Daling JR, Malone KE, Doody DR, et al (2002) Relation of regimens of combined hormone replacement therapy to lobular, ductal, and other histologic types of breast carcinoma. Cancer 95:2455-2464.
25. Graham JD, Yeates C, Balleine RL, et al (1995) Characterization of progesterone receptor A and B expression in human breast cancer. Cancer Res 55:5063-5068.
26. Mote PA, Johnston JF, Manninen T, et al (2001) Detection of progesterone receptor forms A and B by immunohistochemical analysis. J Clin Pathol 54:624-630.
27. Mote PA, Bartow S, Tran N, et al (2002) Loss of co-ordinate expression of progesterone receptors A and B is an early event in breast carcinogenesis. Breast Cancer Res Treat 72:163-172.

PART 2. BREAST CANCER II: MECHANISMS OF GENOMIC INSTABILITY

6

c-Myc Deregulation Promotes a Complex Network of Genomic Instability

Sabine Mai, Amanda Guffei, Thierry Fest, and J. Frederic Mushinski

Introduction

The oncoprotein c-Myc has been studied for decades. We have gathered important information about the role(s) c-Myc and other Myc family proteins play in neoplasia. PubMed, May 2003 (http://www.ncbi.nlm.nih.gov/) shows 334 citations for c-Myc and neoplasia; 53 for N-Myc and neoplasia; and 392 for Myc and neoplasia. This list is not complete because earlier studies dating from the 1920-40s are not included, but it gives an idea about how prolific the study of Myc oncoproteins in neoplasia has been.

Despite the intensive and extensive research carried out all over the world, we still have not identified a single pathway that suffices to explain how c-Myc initiates and promotes tumor formation, suggesting that multiple pathways must be involved. In one of these pathways, c-Myc appears to initiate a complex network of genomic instability that we are still deciphering. The analysis of this pathway is very challenging and complex, for c-Myc is a multifunctional protein that is able not only to alter the stability of the genome, but also to alter expression of genes profiles that have an impact on apoptosis, differentiation, proliferation, and angiogenesis, among others.

When referring to "effects on genomic stability" what do we actually refer to? Genomic instability is a widely used term, but the term is often vague. Is it increased mutations, aneuploidy, loss of tumor suppressors, specific genetic rearrangements - or all of the above? Before attempting to describe the effects of c-Myc on genomic instability, we will define the term genomic instability. Structural genomic instability encompasses point mutations, deletions, microsatellite instability, duplications, amplifications, translocations and inversions. Structural genomic instability does not include alterations in the number of chromosomes. Instead, numerical genomic instability, also called aneuploidy, is reflected in deviations from diploid, the normal state in which each cell contains two copies of each somatic chromosome. The terms nullisomy (absence of both copies of a chromosome), monosomy (presence of only one chromosome), trisomy (presence of three copies), tetrasomy (presence of four copies) or polysomy (presence of multiple copies of a chromosome) denote specific forms of numerical genomic instability. Finally, the term karyotypic instability, often also called 'chromosomal instability' combines structural and numerical genomic instability and indicates the simultaneous presence of both types of aberrations within one cell.

It is now appreciated that c-Myc induces a complex network of genomic instability that can include several components: locus-specific genomic instability, karyotypic or chromosomal instability, long-range illegitimate recombinations, point mutations, and DNA breakage. c-Myc also affects the overall nuclear structure, as documented by recent work (1). Additional effects of c-Myc on genomic stability are very likely to be discovered as we gain greater depth of understanding of cellular transformation in the years that lie ahead.

The History of *c-Myc*

c-*Myc* is the cellular counterpart of the viral oncogene *v-myc* that was identified as the transforming factor in avian retroviruses, MH-2, MC29 (2), CMII and OK10 (3). Subsequently, it was discovered that bursal B-cell lymphomas in retrovirus-infected fowl were caused by retroviral integrations in or near the c-*myc* proto-oncogene, which caused constitutive expression of c-*myc*. In some cases, these transcripts originated at the promoter in the long terminal repeat (LTR) of the retrovirus (4). In other cases, the nearby strong retroviral enhancer induced constitutive expression that originated in one of the c-*myc* promoters, if they were still attached to the coding region of the c-*myc* gene (5). Importantly, c-*myc* was recognized as a transformation-initiating oncogene in mammals when B-cell lymphomas in human beings [Burkitt lymphomas (6), rat immunocytomas in Louvain-strain rats (7), and mouse oil-induced BALB/c plasmacytomas (8, 9)] were found to be associated with constitutive expression of c-*myc* mRNA and protein. In these tumors, chromosomal translocations juxtaposed the *c-myc* gene to a gene that encoded either the *Ig* heavy chain or one of the *Ig* light chains. In these cases, the strong enhancers on the *Ig*-encoding genes were responsible for a constitutive activation of transcription originating at one of the c-*myc* promoters. This mechanism is analogous to the enhancer-insertion model of bursal lymphomas. Finally, evidence of a c-*myc*-activating extrachromosomal translocation has been found in extrachromosomal elements in a murine plasmacytoma (10). In this case, constitutive c-*myc* expression was initiated within the *c-myc/IgH*-carrying extrachromosomal elements.

c-Myc Deregulation in Cancer

c-Myc overexpression is associated with neoplasms of different tissues, including breast (11, 12), neuroblastoma (13), cervical carcinoma (14), malignant melanoma (15), prostate cancer (16), osteogenic sarcoma (17), and lymphoid cancer (6, 18, 19); (reviewed in 20). Thus, Myc protein overexpression is an important player in cellular transformation (21-25).

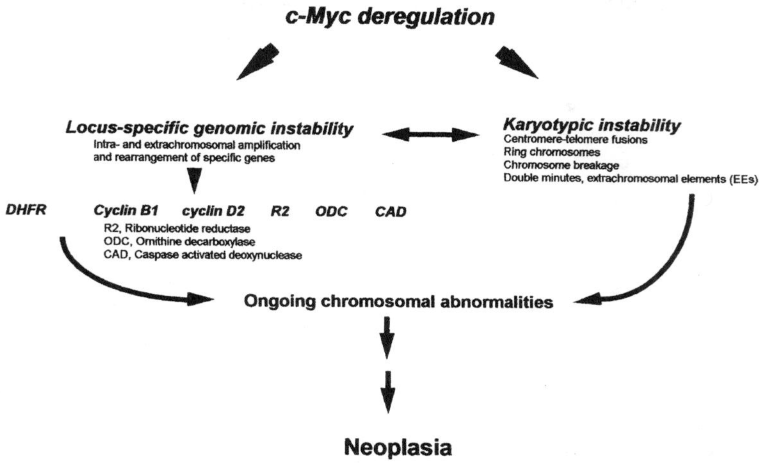

Figure 1. c-Myc-dependent locus-specific and karyotypic instability in neoplasia.

Locus-specific Genomic Instability. c-Myc induces the instability of some genes, but not of others. Therefore, we generated the term 'locus-specific' genomic instability (Figure 1). This type of instability was first described in the report that the *dihydrofolate reductase* (*DHFR*) gene was amplified as a result of inducible c-*myc* overexpression (26, 27). *DHFR* was amplified within 72 h of inducible c-Myc deregulation in a variety of cell lines of mouse, hamster, and human origin, and in several tissue types. The amplification of *DHFR* was reversible when c-Myc was induced transiently (27). In the same cells, however, other genes, such as *cyclin C*, *ribonucleotide reductase R1*, *syndecan-1,* and *glyceraldehyde phosphate dehydrogenase* (*GAPDH*) did not show genomic instability.

Subsequently, *in-vivo* studies showed that *DHFR* was amplified in primary mouse plasmacytomas, Myc-dependent B lymphocytic tumors that develop in susceptible mice, such as BALB/c (28). If *DHFR* amplification were an early event in c-Myc-induced tumorigenesis, one would expect to find evidence of *DHFR* instability early in tumorigenesis. Indeed, plasmacytoma induction studies showed that *DHFR* was amplified within the first week of c-*myc* deregulation (29) (Figure 2). The amplification of the *DHFR* gene occurred intra-and extra-chromosomally in mouse and human cells, however only intra-chromosomally in the hamster cell line CHO-9.

Subsequently, additional genes were shown to lose stability when c-*myc* expression was deregulated. Among them are: *ribonucleotide reductase R2* (*R2*) (30), *cyclin D2* (31), *ODC* (32) and *CAD* (33). All of these genes are amplified intra- and extra-chromosomally in the presence of constitutive c-*myc* expression. c-Myc-dependent induction of extra-chromosomally amplified genes is a novel product of c-Myc-dependent genomic instability. Several genes can be found in these extra-chromosomal elements. These genes carry histones and are able to replicate, thus behaving as functional genetic units (34). The generation of these extra-chromosomal elements may occur via illegitimate replication, DNA breakage, and/or DNA recombination (see additional discussion below).

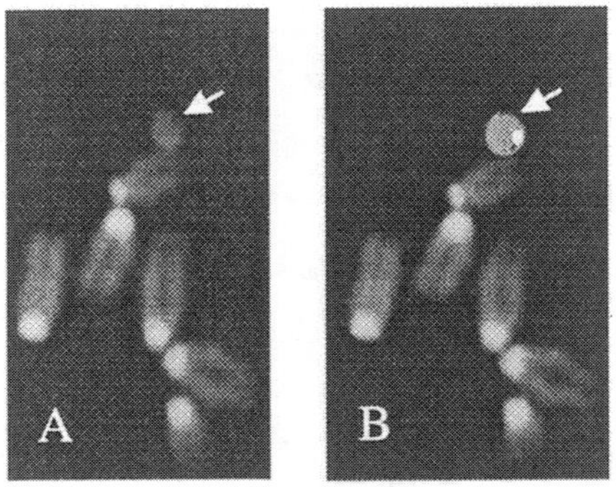

Figure 2. c-Myc-dependent amplification of *DHFR*. Fluorescent *in-situ* hybridization (FISH) demonstrates that the *DHFR* gene is amplified in a mouse plasmacytoma.

A. Part of a metaphase showing a large double minute chromosome (arrow).

B. Same as **A** showing *DHFR* and some pieces of *cyclin B1* on the double minute chromosome. *DHFR* is shown in red, *cyclin B1* in green. Chromosomes are counterstained with DAPI (blue). (Figure may be viewed in color at http://springeronline.com/0-387-23783-6)

Karyotypic Instability. c-*myc* deregulation also induces karyotypic instability (35, 36). This form of genomic instability was found in cell lines after prolonged c-*myc* deregulation (35). In addition, when injected into nude mice, cells that constitutively expressed c-*myc* developed karyotypic instability (36). The types of instability seen *in vivo* were identical to those seen *in vitro*: aneuploidy, telomere-centromere fusions, chromosome breakage, formation of unstable ring chromosomes, and generation of extra-chromosomal elements.

Long-range Illegitimate Recombinations. c-*myc* deregulation in murine B-cell lymphomas has been shown to induce recombination events that involved many different chromosomes (37). These illegitimate recombinations included translocations, deletions and inversions. DNA sequencing and spectral karyotyping showed that a wide variety of chromosomal regions had been affected and many different break points involved. Recent studies have also examined whether c-*myc* deregulation also increased the rate of point mutations or large-scale rearrangements. Davis, *et al.* (38) did not observe increased mutation rates in Myc-induced liver cancers, and Rockwood, *et al.* (37) reported no significant increase in point mutations in mouse Burkitt lymphoma with c-*myc* deregulation. However, Partlin, *et al.* (39) described a small increase in *HGPRT* mutations as a result of c-*myc* deregulation in Rat1a fibroblasts. The differences in these data may lie in the different cell systems used. Further studies are needed to clarify this point.

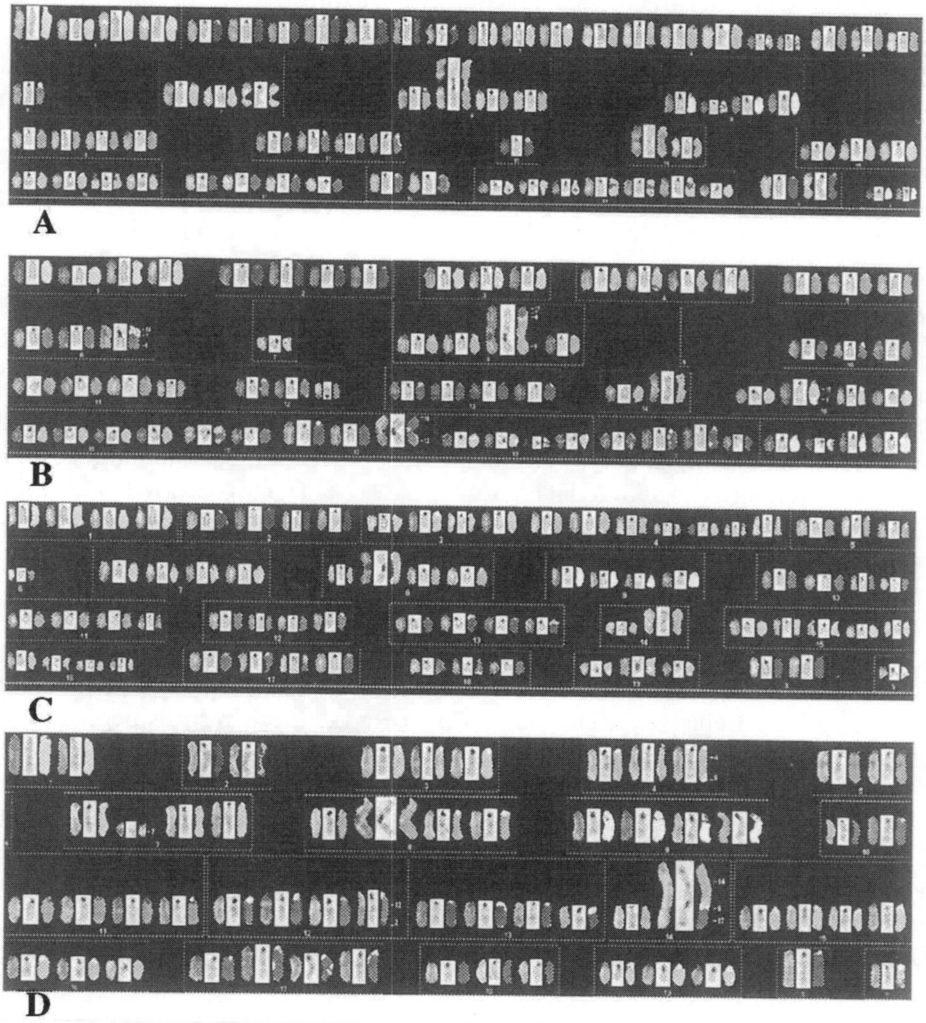

Figure 3. Wt MycER and mut MycER-dependent genomic instability as examined by spectral karyotyping. **A** and **B** show karyotypes before and 48 h after activation of wt MycER. **C** and **D** show karyotypes before and 72 h after mut MycER activation. (Figure in color at http://springeronline.com/0-387-23783-6)

Mutant Myc proteins also induce genomic instability, but with some unique features. The neoplastic B cells from human Burkitt lymphoma patients often express mutated forms of c-Myc protein. Up to 60% of the patients display mutations in the "Myc box II" motif (40). Recent analyses of both wild-type (wt) and mutant (mut) Myc proteins were aimed at determining whether both forms of Myc were potent inducers of genomic instability. BAF3 murine pro-B cells, stably transfected with constructs that constitutively expressed wt or mut MycER, unexpectedly showed that both forms of the Myc oncoprotein induced locus-specific gene amplification of *DHFR*. However, the Myc box II mutants protected the BAF3 cells from apoptosis while simultaneously triggering polyploidization. In contrast, wt Myc protein deregulation consistently induced apoptosis but no polyploidization (40). Thus, mutant Myc box II proteins promote instability in the absence of apoptosis and generate some aberrations that overlap with those induced by wt Myc deregulation. In addition, mut Myc mediates unique numerical genomic instability (Figure 3). A detailed analysis of the genetic changes induced by wild-type and mutant Myc proteins in this model system is currently being undertaken in our laboratories.

Reversible Tumorigenesis. c-Myc–dependent tumorigenesis is usually the product of long-term overexpression of constitutive expression of c-*myc*. Work by several groups has demonstrated very convincingly that c-Myc-initiated tumorigenesis is reversible as long as no additional irreversible genetic changes occurred (41, 11, 17, 19), proving that these neoplasms are critically dependent on the presence of c-Myc protein.

Mechanisms of c-Myc-Dependent Genomic Instability

The understanding of how c-Myc mediates the complex network of genomic instability is still elusive. Yet, some insight into the mechanism is beginning to emerge.

c-Myc-dependent Induction of DNA Breaks. Important new data have identified a hitherto unknown function of c-Myc, namely, the induction of DNA breakage. Vafa, *et al.* (42) described the c-Myc-mediated formation of reactive oxygen species, which in turn, led to DNA breakage as shown by the TUNEL assay and by DNA repair complex formation involving hMre11. This links c-Myc action directly to the initiation of genomic instability, since it is known that a single DNA break can be sufficient to induce gene amplifications, deletions, and rearrangements (43), leading to loss of heterozygosity (44) and translocations (45). Under the above conditions, c-Myc deregulation bypasses p53. What is more, c-Myc-induced extra-chromosomal elements (EEs) can be formed through DNA breaks and/or illegitimate replication and recombination (34).

c-Myc-dependent Induction of Illegitimate Rounds of Replication. Two basic mechanisms of gene amplification dominate discussion in the literature (46, 47). One implicates the replication-driven amplification of genes, while the other involves segregation-driven mechanisms. It is not yet clear which mechanism is more likely to be involved in c-Myc-dependent gene amplification. However, a combination of the two may be the most likely scenario.

The *replication-driven model* (also termed the 'onion skin' model) postulates that re-replication of a particular gene occurs within a single cell cycle, whereas, usually, each gene replicates only once in each cell division cycle. The advantage of this model is that it offers a mechanism to explain the formation of intra-chromosomal amplification. However, it can also account for the formation of extra-chromosomal DNA molecules that are released from their chromosomal location by recombination. A well-known example of re-replication is the amplification of chorion genes during *Drosophila* oogenesis (48).

The *segregation-driven models* include the 'deletion plus episome' model and the 'chromatid exchange' model (46, 47). These models require a series of events that usually include DNA breakage, crossing-over, recombination, and rejoining. During the segregation-driven process, genes may be deleted from their chromosomal location (49, 50) and then become EEs, episomes, or double minutes. This form of gene amplification is thought to be used in generating ribosomal DNA amplicons in *Tetrahymena*. Ribosomal DNA amplification in this organism involves specific breaks next to inverted repeats, which are known to facilitate illegitimate recombination events (51). For other experimental systems, it has also been shown that "breakage-bridge-fusion" cycles of amplification are instrumental in the generation of drug-induced amplicons (52, 53). Data by Pipiras, *et al.* (43) show that a double-strand break is sufficient to initiate gene amplification.

To describe the details of one appealing replication-driven mechanism of c-Myc-dependent gene amplification, we will focus on a c-Myc target gene in genomic instability, the *ribonucleotide reductase R2* (*R2*) gene on mouse chromosome 12 band A. As outlined above, the *R2* gene has been shown to be amplified as a result of c-*myc* deregulation in a mouse pre-B cell line (30). Using EMSA, two-dimensional gel electrophoresis, FISH, and BrdU-incorporation studies, our recent data showed that c-Myc induced the initiation of several rounds of replication of *R2* within a single cell cycle (54). These results strongly favor a replication-driven model of amplification of *R2* in the intra- and extra-chromosomal instability induced by c-*myc* overexpression. We are awaiting the results of similar studies for other c-Myc target genes in genomic instability and in other cell lines as well as from *in-vivo* studies.

c-Myc Overrides a Functional Wild-type p53. It is important to note that c-Myc triggers genomic instability while bypassing the pro-apoptotic effects of the tumor suppressor p53, possibly owing to Myc-induced downregulation of p53 expression. Previously, it has been shown for both c-Myc- and N-Myc-dependent instability that a functional wt p53 protein did not affect the oncogene-induced genomic instability of specific target genes (27, 35, 55). Recently, Vafa, *et al.* (42) demonstrated that the c-Myc effect on genomic instability could override p53 despite c-Myc's induction of DNA breakage, a well-known signal for up-regulation of p53 (56). This effect is critical for the persistence of cells bearing Myc-induced genomic aberrations, such as amplification. Normally such cells would die due to apoptotic events initiated by wt p53.

Conclusions

c-Myc is a versatile protein that has many different functions. With respect to genomic instability, it affects many endpoints that include gene amplification, translocation, deletions, insertions, long-range rearrangements, DNA breakage, and point mutations. This overall promotion of genomic instability suggests that c-Myc may have a central role in destabilizing the genome. We propose that c-Myc is a structural modifier of the genome and an important initiator and progressor molecule in neoplastic transformation.

References

1. Takaha N, Hawkins AL, Griffin CA, et al (2002) High mobility group protein I(Y): a candidate architectural protein for chromosomal rearrangements in prostate cancer cells. Cancer Res 62:647-651.
2. Duesberg PH, Vogt PK (1979) Avian acute leukemia viruses MC29 and MH2 share specific RNA sequences: evidence for a second class of transforming genes. Proc Natl Acad Sci USA 76:1633-1637.
3. Bister K, Duesberg PH (1980) Genetic structure of avian acute leukemia viruses. Cold Spring Harbor Symp Quant Biol 44:801-822.
4. Hayward WS, Neel BG, Astrin S (1981) Activation of a cellular onc gene by promoter insertion in ALV-induced lymphoid leukosis. Nature 290:475-480.
5. Payne GS, Bishop JM, Varmus HE (1981) Multiple rearrangements in viral DNA and an activated host oncogene (c-myc) in bursal lymphomas. Nature 295:209-214.
6. Della-Favera R, Bregni M, Erikson J, et al (1982) Human c-Myc oncogene is located on the region of chromosome 8 that is translocated in Burkitt's lymphoma cells. Proc Natl Acad Sci USA 79:7824-782.
7. Pear WS, Wahlstrom G, Nelson SF, et al (1988) 6;7 chromosomal translocation in spontaneously arising rat immunocytomas: evidence for c-myc breakpoint clustering and correlation between isotype expression and the c-myc target. Mol Cell Biol 8:441-451.
8. Ohno S, Babonits M, Wiener F, et al (1979) Nonrandom chromosome changes involving the Ig gene-carrying chromosomes 12 and 6 in pristane-induced mouse plasmacytomas. Cell 18:1001-1007.
9. Shen-Ong GL, Keath EJ, Piccoli SP, et al (1982) Novel myc oncogene RNA from abortive immunoglobulin gene recombination in mouse plasmacytomas. Cell 31:443-452.
10. Wiener, F, Kuschak, TI, Ohno, S, et al (1999) Deregulated expression of c-Myc in a translocation-negative plasmacytoma on extrachromosomal elements that carry IgH and myc genes. Proc Natl Acad Sci USA 96:13967-13972.

11. D'Cruz CM, Gunther EJ, Boxer RB, et al (2001) c-MYC induces mammary tumorigenesis by means of a preferred pathway involving spontaneous Kras2 mutations. Nat Med 7:235-239.
12. Stewart TA, Pattengale PK, Leder P (1984) Spontaneous mammary adenocarcinomas in transgenic mice that carry and express MTV/myc fusion genes. Cell 38:627-637.
13. Fan L, Iyer J, Zhu S, et al (2001) Inhibition of N-myc expression and induction of apoptosis by iron chelation in human neuroblastoma cells. Cancer Res 61:1073-1079.
14. Ngan HY, Cheung AN, Liu SS, et al (2001) Abnormal expression of pan-ras, c-myc and tp53 in squamous cell carcinoma of cervix: correlation with HPV and prognosis. Oncol Rep 8:557-561.
15. Kraehn GM, Utikal J, Udart M, et al (2001) Extra c-myc oncogene copies in high risk cutaneous malignant melanoma and melanoma metastases. Brit J Cancer 84:72-79.
16. Zhang X, Lee C, Ng PY, et al (2000) Prostatic neoplasia in transgenic mice with prostate-directed overexpression of the c-myc oncoprotein. Prostate 43:278-285.
17. Jain M, Arvanitis C, Chu K, et al (2002) Sustained loss of a neoplastic phenotype by brief inactivation of MYC. Science 297:102-104.
18. Adams JM, Harris AW, Pinkert CA, et al (1985) The c-myc oncogene driven by immunoglobulin enhancers induces lymphoid malignancy in transgenic mice. Nature 318:533-538.
19. Karlsson A, Giuriato S, Tang F, et al (2003) Genomically complex lymphomas undergo sustained tumor regression upon MYC inactivation unless they acquire novel chromosomal translocations. Blood 101:2797-803.
20. Nesbit CE, Tersak JM, Prochownik EV (1999) MYC oncogenes and human neoplastic disease. Oncogene 18:3004-3016.
21. Von Hoff DD, Needham-Van Devanter DR, Yucel Y, et al (1988) Amplified human MYC oncogenes localized to replicating submicroscopic circular DNA molecules. Proc Natl Acad Sci USA 85:4804-4808.
22. Von Hoff DD, McGill JR, Forseth BJ, et al (1992) Elimination of extrachromosomally amplified MYC genes from human tumor cells reduces their tumorigenicity. Proc Natl Acad Sci USA 89:8165-8169.
23. Eckhardt SG, Dai A, Davidson KK, et al (1994) Induction of differentiation in HL60 cells by the reduction of extrachromosomally amplified c-myc. Proc Natl Acad Sci USA 91:6674-6678.
24. Marcu KB, Bossone SA, Patel AJ (1992) myc function and regulation. Annual Rev Biochem 61:809-860.
25. Potter M, Marcu KB (1997) The c-myc story: where we've been, where we seem to be going. Curr Top Microbiol Immunol 224:1-17.
26. Mai S (1994) Overexpression of c-myc precedes amplification of the gene encoding dihydrofolate reductase, Gene 148:253-260.

27. Mai S, Hanley-Hyde J, Fluri M (1996) c-Myc overexpression associated DHFR gene amplification in hamster, rat, mouse and human cell lines. Oncogene 12:277-288.
28. Potter M, Wiener F (1992) Plasmacytomagenesis in mice: model of neoplastic development dependent upon chromosomal translocations. Carcinogenesis 13:1681-1697.
29. Taylor C, Mai S (1998) c-Myc-associated genomic instability of the dihydrofolate reductase locus in vivo. Cancer Detect Prev 22:350-356.
30. Kuschak TI, Taylor C, McMillan-Ward E et al (1999) The ribonucleotide reductase R2 gene is a non-transcribed target of c-Myc-induced genomic instability. Gene 238:351-365.
31. Mai S, Hanley-Hyde J, Rainey GJ, et al (1999) Chromosomal and extrachromosomal instability of the *cyclin D2* gene is induced by *Myc* overexpression. Neoplasia 1:241-252.
32. Smith GM, Dushniky LG, Symons SJ, et al (2001) c–Myc inducible chromosomal and extrachromosomal amplification of R2 and ODC. AACR Proceedings 42:902.
33. Fukasawa K, Wiener F, Vande Woude GF, et al (1997) Genomic instability and apoptosis are frequent in p53 deficient young mice. Oncogene 15:1295-1302.
34. Smith G, Taylor-Kashton C, Dushnicky L, et al (2003) c-Myc-Induced Extrachromosomal Elements Carry Active Chromatin. Neoplasia 5:110-120
35. Mai S, Fluri M, Siwarski D et al (1996) Genomic instability in MycER activated Rat1A-MycER cells. Chromosome Res 4:365-371.
36. Felsher DW, Bishop JM (1999) Transient excess of MYC activity can elicit genomic instability and tumorigenesis. Proc Natl Acad Sci USA 96:3940-3944.
37. Rockwood LD, Torrey TA, Kim JS, et al (2002) Genomic instability in mouse Burkitt lymphoma is dominated by illegitimate genetic recombinations, not point mutations. Oncogene 21:7235-7240.
38. Davis CD, Dacquel EJ, Schut HA, et al (1996) In vivo mutagenicity and DNA adduct levels of heterocyclic amines in Muta mice and c-myc/lacZ double transgenic mice. Mutat Res 356:287-296.
39. Partlin MM, Homer E, Robinson H, et al (2003) Interactions of the DNA mismatch repair proteins MLH1 and MSH2 with c-MYC and MAX. Oncogene 22:819-825.
40. Fest T, Mougey V, Dalstein V, et al (2002) c-MYC overexpression in Ba/F3 cells simultaneously elicits genomic instability and apoptosis. Oncogene 21:2981-2990.
41. Felsher DW, Bishop JM (1999) Reversible tumorigenesis by MYC in hematopoietic lineages. Mol Cell 4:199-207.
42. Vafa O, Wade M, Kern S, et al (2002) c-Myc can induce DNA damage, increase reactive oxygen species, and mitigate p53 function: a mechanism for oncogene-induced genetic instability. Mol Cell 9:1031-1044.

43. Pipiras E, Coquelle A, Bieth A, et al (1998) Interstitial deletions and intrachromosomal amplification initiated from a double-strand break targeted to a mammalian chromosome. EMBO J 17:325-333.
44. Moynahan ME, Jasin M (1997) Loss of heterozygosity induced by a chromosomal double-strand break. Proc Natl Acad Sci USA 94:8988-8993.
45. Richardson C, Jasin M (2000) Frequent chromosomal translocations induced by DNA double-strand breaks. Nature 405:697-700.
46. Hamlin JL, Ma C (1990) The mammalian dihydrofolate reductase locus. Biochim Biophys Acta 1087:107-125.
47. Stark GR (1993) Regulation and mechanisms of mammalian gene amplification. Adv Cancer Res 61:87-113.
48. Spradling AC (1981) The organization and amplification of two chromosomal domains containing Drosophila chorion genes. Cell 27:193-201.
49. Carroll SM, DeRose ML, Gaudray P, et al (1988) Double minute chromosomes can be produced from precursors derived from a chromosomal deletion. Mol Cell Biol 8:1525-1533.
50. Morris T, Thacker J (1993) Formation of large deletions by illegitimate recombination in the HPRT gene of primary human fibroblasts. Proc Natl Acad Sci USA 90:1392-1396.
51. Butler DK, Yasuda LE, Yao MC (1996) Induction of large DNA palindrome formation in yeast: implications for gene amplification and genome stability in eukaryotes. Cell 87:1115-1122.
52. Ma C, Martin S, Trask B, et al (1993) Sister chromatid fusion initiates amplification of the dihydrofolate reductase gene in Chinese hamster cells. Genes Dev 7:605-620.
53. Poupon M-F, Smith KA, Chernova OB, et al (1996) Inefficient growth arrest in response to dNTP starvation stimulates gene amplification through bridge-breakage-fusion cycles. Mol Biol Cell 7:345-354.
54. Kuschak TI, Kuschak BC, Taylor CL, et al (2002) c-Myc initiates illegitimate replication of the ribonucleotide reductase R2 gene. Oncogene 21:909-920.
55. Chernova OB, Chernov MV, Ishizaka Y, et al (1998) MYC abrogates p53-mediated cell cycle arrest in N-(phosphonacetyl)-L-aspartate-treated cells, permitting CAD gene amplification. Mol Cell Biol 18:536-545.
56. Lane DP (1992) Cancer. p53, guardian of the genome. Nature 358:15-16.

7

Deregulation of Cyclin E and Genomic Instability

Charles H. Spruck, Adrian P. L. Smith, Susanna Ekholm Reed, Olle Sangfelt, Jaimie Keck, Heimo Strohmaier, Juan Méndez, Martin Widschwendter, Bruce Stillman, Anders Zetterberg, and Steven I. Reed

Introduction

Cyclin E is a positive regulator of cyclin-dependent kinase 2 (Cdk2) in metazoan species (1). In somatic mammalian cells, cyclin E is expressed periodically during an interval extending from the latter part of G1 until mid-S phase (2-4). Consistent with this time frame, cyclin E associated Cdk2 kinase activity is thought to mediate S phase events, including initiation of DNA replication and histone biosynthesis. However, elevated levels of cyclin E have been correlated with a broad spectrum of human malignancies (5-10). A causal relationship between cyclin E deregulation and carcinogenesis is supported by a transgenic mouse model, where cyclin E was expressed under hormonal control in the mammary epithelium during pregnancy and lactation and a small but significant number of animals developed mammary adenocarinomas (11). These findings raise two significant questions. How does cyclin E get deregulated and how does such deregulation lead to carcinogenesis?

Regulation of Cyclin E Expression

Cyclin E is transcribed in a periodic burst late in G1. Based on both western blot analysis of synchronized cell cultures and immunofluorescence analysis of single cells, cyclin E protein levels peak at the G1/S boundary and decline as cells proceed through S phase (2-4). The decline of cyclin E levels, once transcription is terminated, is linked to the short half-life of cyclin E in S phase cells (12, 13). Cyclin E is degraded via ubiquitin-mediated proteolysis (12, 14). During S phase, cyclin E becomes phosphorylated on T380, which targets it for poly-ubiquitination by the protein-ubiquitin ligase SCF^{Cdc4} and subsequent turnover (12, 14).

Mechanisms of Cyclin E Deregulation in Cancer

The first report of cyclin E gene amplification as tumor-linked mechanism for elevation of cyclin E expression was based on a breast cancer (BC) derived cell line (15). Subsequently cyclin E gene amplification has been reported at varying frequencies in a variety of different tumor types, particularly gastrointestinal and ovarian cancer (16-18). In BC, however, cyclin E gene amplification is rare (16). We therefore sought to determine if deregulation of cyclin E could be mediated by defects in the cyclin E degradation pathway. First, we analyzed tumor-derived DNA for mutations in cyclin E itself that might impair phosphorylation-dependent ubiquitination. Single stranded conformational polymorphism (SSCP) analysis of DNA from more than 100 tumors yielded no mutations that could account for cyclin E stabilization. Therefore, we turned to analysis of the cyclin E ubiquitation machinery. SCF protein-ubiquitin ligases are characterized by specificity factors, known as F-box proteins that target specific sub-groups of proteins (19, 20). We analyzed the F-box protein that targets cyclin E, hCdc4 (13, 21, 22) for mutations in tumor-derived cell lines and in primary tumors. One out of 8 BC-derived cell lines tested was found to have mutated *hCDC4* and loss of heterozygosity (LOH) (13). This correlated with stable cyclin E. Next, primary ovarian and endometrial tumors were analyzed by SSCP for *hCDC4* mutation. Whereas 0/40 ovarian tumors had mutations in *hCDC4* based on SSCP analysis, mutations were found in 16% (8/51) of endometrial tumors (23). Where appropriate linked markers were available, LOH could be demonstrated in the *hCDC4*-mutated tumors. These data strongly suggest that hCdc4 is a classical tumor suppressor. Consistent with this, *hCDC4* maps to chromosomal region 4q32, a site of frequent allelic loss in a broad spectrum of human cancers. Finally, *hCDC4* mutation status may have prognostic value. Of the endometrial tumors analyzed, *hCDC4* mutation correlated with higher grade, more advanced stage, and metastasis (23). The latter parameter is significant, since metastasis is not frequent in endometrial cancer.

hCDC4 Mutation Leads to Deregulation of Cyclin E Relative to the Cell Cycle

Analysis of cyclin E in the breast tumor-derived cell line found to be mutated for *hCDC4* (SUM149PT) indicated that the protein was overexpressed relative to non-transformed human mammary epithelial cells and most other BC derived cell lines (13). However, more interestingly, when the SUM149PT cell line was analyzed by immunofluorescence microscopy all cells in an asynchronous population were cyclin E-positive. This is in contrast to other tumor-derived cell lines and non-transformed cells where only 30-60% of cells stain positively for cyclin E. Therefore, loss of Cdc4 function, presumably leading to loss of SCF^{Cdc4} protein-ubiquitin ligase activity, results in deregulation of cyclin E relative to cell cycle progression.

Cyclin E status was also determined in endometrial tumors. Similarly to the SUM149PT cell line, endometrial tumors that had *hCDC4* mutations showed a deregulation of cyclin E relative to the cell cycle. Sections from *hCDC4*-mutated tumors typically resulted in well over 60% cyclin E positive nuclei, whereas similar analysis of non-mutated tumors always resulted in levels significantly lower than 60%. Even tumors that had extremely high cyclin E levels, but were wild-type at the *hCDC4* locus, exhibited a normal distribution of cyclin E positive and negative nuclei. Surprisingly, cyclin E protein levels in endometrial tumors did not correlate well with *hCDC4* mutational status (23). Most tumors that had high cyclin E levels, based on western blotting, did not have *hCDC4* mutations. Conversely, most tumors that had *hCDC4* mutations had only moderate levels of cyclin E. These results suggest that when the SCF^{hCDC4} pathway is inactivated, other modes of cyclin E turnover can compensate. However, these alternative pathways cannot restore cell cycle regulation of cyclin E.

Cyclin E Deregulation Leads to Chromosome Instability

Malignant cells are often characterized by genomic instability. It is thought that such instability allows cells to evade surveillance systems that have evolved to prevent cancer and allows cancer cells to adapt to the environmental pressures characteristic of malignant growth. In order to determine whether deregulation of cyclin E confers genomic instability on non-transformed cells, we created cell lines where cyclin E could be expressed ectopically for long periods of time. Both rodent fibroblasts (rat-1 cells) and immortalized human mammary epithelial cells carrying tetracycline-repressible cyclin E cDNAs were grown for four weeks in the presence and absence of tetracycline (24). At the end of the experiment, cells were scored for gene amplification (by screening for PALA resistant clones) and for chromosome instability. It was found that whereas constitutive ectopic expression of cyclin E did not lead to gene amplification, there was strong reproducible evidence of chromosome instability (24). On a genome-wide basis, both chromosome losses and gains were detected, although losses appeared to be more frequent. In addition, elevated levels of polyploidy were observed. With respect to chromosome loss and gain, events appeared to be random, rather than chromosome specific, arguing against a selection for particular karyotypic alterations (24). The fact that both chromosome losses and gains occur suggests that sister chromatid non-disjunction may be occurring at elevated levels in cells with deregulated cyclin E. Although chromosome instability may generally confer a hyper-adaptive phenotype on malignant cells, a more specific mechanistic link may be an acceleration of LOH at tumor suppressor loci (Figure 1). Specifically, elevated levels of chromosome loss are expected to synergize with mutation of tumor suppressor genes to promote malignancy.

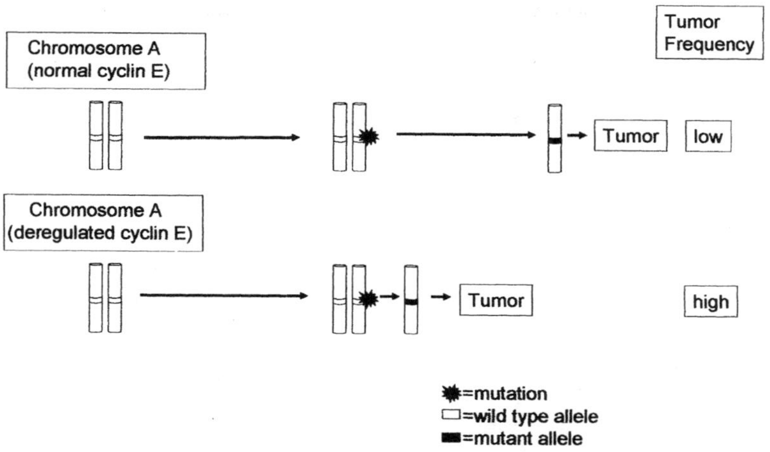

Figure 1. Cyclin E deregulation promotes carcinogenesis by accelerating LOH. A hypothetical chromosome (Chromosome A) is first mutated at a tumor suppressor locus, and then eventually experiences LOH resulting in tumor formation (upper sequence). Since both events are rare, tumor frequency is low. In cyclin E deregulated cell (lower sequence), tumor suppressor mutation rate is unaffected but LOH is accelerated, leading to increase in tumor frequency.

Cyclin E Deregulation Synergizes With a Tumor Suppressor Mutation in a Mouse Model

In order to test whether cyclin E-mediated chromosome loss could indeed synergize with tumor suppressor mutations, we utilized a mouse model. Trangenic mice were constructed that were heterozygously mutated at either the *Rb* or *p53* locus and that carried a hormone-responsive cyclin E transgene expressed during pregnancy and lactation in the mammary epithelium. In this instance, an allele of cyclin E refractory to ubiquitin-mediated proteolysis was used to maximally deregulate cyclin E through the cell cycle. Female mice were subjected to two pregnancies, aged, and monitored for mammary tumorigenesis. While the cyclin E transgene and the heterozygous *p53* mutation alone each conferred mammary tumorigenesis at a rate of 10% or less, approximately 60% of mice carrying both the *p53* mutation and the cyclin E transgene developed mammary tumors. This strong significant genetic interaction is consistent with the hypothesis that cyclin E-mediated chromosome instability causes increased levels of LOH. Indeed, when tumors were analyzed for *p53* gene status, allelic loss had occurred in all cases.

No such interaction was observed between cyclin E deregulation and *Rb* heterozygosity. However, there is a caveat related to this experiment in that *Rb* heterozygous mice have a limited lifespan, and must be euthanized at year of age because of a highly penetrant pituitary tumor phenotype. On the other hand, mammary tumorigenesis induced by cyclin E deregulation alone or in concert with *p53* heterozygosity usually occurs between 12 and 18 mo. Consequently, a potential interaction between cyclin E deregulation and *Rb* heterozygosity may have been inaccessible due to experimental design limitations.

How Does Cyclin E Deregulation Mediate Chromosome Instability?

Cell lines deregulated for cyclin E expression exhibit several cell cycle perturbations. Flow cytometric analysis reveals a shortened G1 phase, consistent with cyclin E driving cells into S phase prematurely, and lengthened S and G2/M phases, respectively. Analysis of BrdU pulse-labeled cells by immunofluorescence microscopy demonstrated directly that cyclin E deregulation reduces the rate of DNA replication. A hypothesis that could potentially link cyclin E deregulation to impairment of DNA replication is that inappropriate cyclin E/Cdk2 activity at the M/G1 boundary could interfere with pre-replication complex assembly. It has been demonstrated in yeast and in Xenopus egg extracts that cyclin-dependent kinase activity reduction is a prerequisite for assembly of pre-replication complexes (25). In somatic mammalian cells, pre-replication complexes begin to assemble during telophase (26). Therefore cells ectopically expressing cyclin E and controls were synchronized to enrich for telophase cells, which were then analyzed for pre-replication complex status both biochemically and by immunofluorescne microscopy. Pre-replication complexes contain a multi-protein complex known as Orc, to which Cdc6, Cdt1 and a hetero-hexomer consisting of Mcm2-7 are loaded in a sequential fashion (Figure 2). For immunofluorescence analysis, detergent extracted cells were fixed and analyzed for chromatin bound Cdc6 and Mcm proteins. For biochemical analysis, cells were fractionated to obtain a chromatin pellet, which was then analyzed for bound proteins by SDS-PAGE and western blotting. Both analyses revealed that all proteins were loaded onto chromatin equivalently in cyclin E-deregulated and control cells except for Mcm4, which was underrepresented on the chromatin of cyclin E-deregulated cells. Treating cells with the Cdk inhibitor roscovitine during mitotic exit reversed this effect, confirming that impairment of Mcm4 loading is a consequence of cyclin E/Cdk2 activity. Therefore, it is likely that the inefficient S phase observed in response to cyclin E deregulation is a consequence of a reduced number of competent pre-replication complexes resulting from an inhibition of Mcm4 loading.

How might impaired DNA replication lead to chromosome instability? We speculate that occasional failure of checkpoints that monitor the completion of DNA replication will allow cells with incompletely replicated genomes to enter mitosis, leading to chromatid non-disjunction events (Figure 3). The resolution of these non-disjunction events in progeny cells is likely to result in chromosome gains and losses as well as other form of aneuploidy.

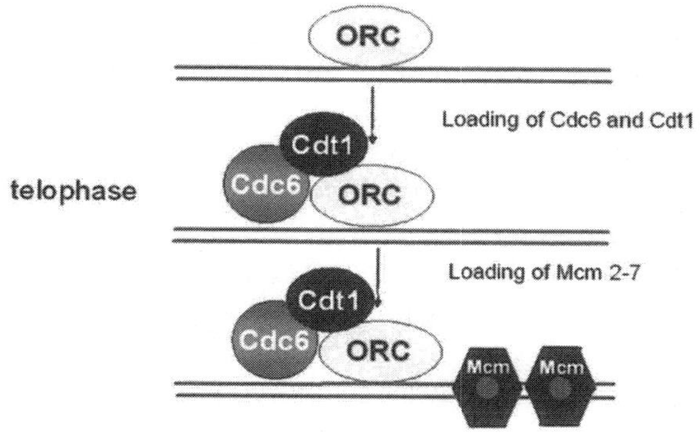

Figure 2. Sequential assembly of pre-replication complexes. The ORC complex first binds defining origins of replication. Cdc6 and Cdt1 are then loaded in an ORC-dependent manner in telophase. This potentiates the loading of the Mcm heterohexamer.

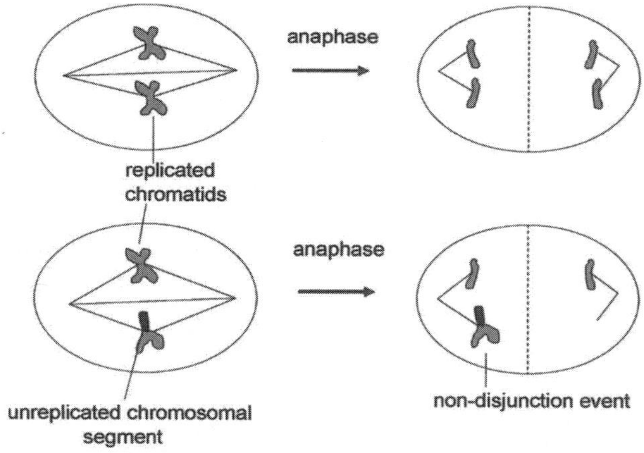

Figure 3. Incomplete DNA replication can lead to chromatid non-disjunction evens and aneuploidy. Normally, replicated sister chromatids align at metaphase and then are segregated to opposite poles at anaphase. However, an unreplicated segment (indicated in red) will not allow chromatid separation and segregation. Instead non-disjunction will occur, with both of the incompletely replicated chromatids segregating to the same pole and going to the same daughter cell. This will lead to chromosome loss and other forms of aneuploidy.

Cyclin E deregulation also leads to an accumulation of G2 and/or M phase cells. Comparing the distribution of mitotic phases in cyclin E-deregulated and control cells, there is an excess of metaphases and fewer anaphases. At high levels of cyclin E expression, some cells appear to be blocked at metaphase. This is consistent with cyclin E deregulation interfering with the metaphase-anaphase transition. We speculate that the increase in polyploidy observed when cyclin E is deregulated is a consequence of cells failing at the metaphase-anaphase transition and eventually exiting mitosis without division.

Concluding Remarks

Cyclin E deregulation has been associated with an array of human malignancies. Evidence suggests that the most critical parameter may be deregulation relative to the cell cycle, which can occur if the normal pathway of cyclin E turnover is inactivated. The resulting expression of cyclin E at inappropriate times of the cell cycle can then interfere with efficient progression through both S phase and M phase. Although not yet proven, we propose that these cyclin E-mediated cell cycle perturbations ultimately result in chromosome instability.

References

1. Ekholm SV, Reed SI (2000) Regulation of G(1) cyclin-dependent kinases in the mammalian cell cycle. Curr Opin Cell Biol 12:676-684.
2. Ekholm SV, Zickert P, Reed SI, et al (2001) Accumulation of cyclin E is not a prerequisite for passage through the restriction point. Mol Cell Biol 21:3256-3265.
3. Dulic V, Lees E, Reed SI (1992) Association of human cyclin E with a periodic G1-S phase protein kinase. Science 257:1958-1961.
4. Koff A, Giordano A, Desai D, et al (1992) Formation and activation of a cyclin E-cdk2 complex during the G1 phase of the human cell cycle. Science 257:1689-1694.
5. Keyomarsi K, Conte D, Jr., Toyofuku W, et al (1995) Deregulation of cyclin E in breast cancer. Oncogene 11:941-950.
6. Donnellan R, Chetty R (1999) Cyclin E in human cancers. FASEB J 13:773-780.
7. Erlanson M, Landberg G (2001) Prognostic implications of p27 and cyclin E protein contents in malignant lymphomas. Leuk Lymphoma 40:461-470.
8. Li JQ, Miki H, Ohmori M, et al (2001) Expression of cyclin E and cyclin-dependent kinase 2 correlates with metastasis and prognosis in colorectal carcinoma. Hum Pathol 32:945-953.
9. Porter PL, Malone KE, Heagerty PJ, et al (1997) Expression of cell-cycle regulators p27Kip1 and cyclin E, alone and in combination, correlate with survival in young breast cancer patients [see comments]. Nat Med 3:222-225.
10. Sandhu C, Slingerland J (2000) Deregulation of the cell cycle in cancer. Cancer Detect Prev 24:107-118.
11. Bortner DM, Rosenberg MP (1997) Induction of mammary gland hyperplasia and carcinomas in transgenic mice expressing human cyclin E. Mol Cell Biol 17:453-459.
12. Won KA, Reed SI (1996) Activation of cyclin E/CDK2 is coupled to site-specific autophosphorylation and ubiquitin-dependent degradation of cyclin E. Embo J 15:4182-93.

13. Strohmaier H, Spruck CH, Kaiser P, et al (2001) Human F-box protein hCdc4 targets cyclin E for proteolysis and is mutated in a breast cancer cell line. Nature 413:316-322.
14. Clurman BE, Sheaff RJ, Thress K, et al (1996) Turnover of cyclin E by the ubiquitin-proteasome pathway is regulated by cdk2 binding and cyclin phosphorylation. Genes Dev 10:1979-1990.
15. Keyomarsi K, Pardee AB (1993) Redundant cyclin overexpression and gene amplification in breast cancer cells. Proc Natl Acad Sci USA 90:1112-6.
16. Courjal F, Louason G, Speiser P, et al (1996) Cyclin gene amplification and overexpression in breast and ovarian cancers: evidence for the selection of cyclin D1 in breast and cyclin E in ovarian tumors. Int J Cancer 69:247-253.
17. Becker KF, Keller G, Hoefler H (2000) The use of molecular biology in diagnosis and prognosis of gastric cancer. Surg Oncol 9:5-11.
18. Tahara E (1995) Genetic alterations in human gastrointestinal cancers. The application to molecular diagnosis. Cancer 75:1410-1417.
19. Jackson PK, Eldridge AG, Freed E, et al (2000) The lore of the RINGs: substrate recognition and catalysis by ubiquitin ligases. Trends Cell Biol 10:429-439.
20. Kipreos ET, Pagano M (2000) The F-box protein family. Genome Biol 1:REVIEWS3002.
21. Koepp DM, Schaefer LK, Ye X, et al (2001) Phosphorylation-dependent ubiquitination of cyclin E by the SCFFbw7 ubiquitin ligase. Science 294:173-177.
22. Moberg KH, Bell DW, Wahrer DC, et al (2001) Archipelago regulates Cyclin E levels in Drosophila and is mutated in human cancer cell lines. Nature 413:311-316.
23. Spruck C, Strohmaier H, Sangfelt O, et al (2002) hCDC4 Gene mutations in endometrial cancer. Cancer Res 62:4535-4539.
24. Spruck CH, Won KA, Reed SI (1999) Deregulated cyclin E induces chromosome instability. Nature 401:297-300.
25. Diffley JF (2001) DNA replication: building the perfect switch. Curr Biol 11:R367-370.
26. Mendez J, Stillman B (2000) Chromatin association of human origin recognition complex, cdc6, and minichromosome maintenance proteins during the cell cycle: assembly of prereplication complexes in late mitosis. Mol Cell Biol 20:8602-8612.

8

Centrosome Amplification and the Origin of Chromosomal Instability in Breast Cancer

Jeffrey L. Salisbury

Introduction

Aneuploidy and chromosomal instability (CIN) are defining features of most aggressive breast cancers (BC). One consequence of CIN is a constantly changing genetic makeup of cancer cells - this in turn is a major driving force behind cancer cell heterogeneity, tumor progression, and acquisition of resistance to chemotherapeutics. How CIN arises in cancer and the mechanisms underlying this process have become a topical focus of cancer research. Yet it was nearly a century ago that Theodor Boveri first recognized that aneuploidy in cancer cells could arise through defects in the machinery for chromosomal segregation (1). Based on observations of abnormal chromosomal segregation in early sea urchin embryo development following dispermic fertilization and similarities to chromosomal anomalies seen in cancer, Boveri proposed that malignant tumors arise through centrosome defects that result in improper cell division (1). At about this same time, Galeotti came to a similar conclusion from his studies on tumors (2). Despite these compelling arguments and a strident call to the medical research community in *JAMA* by Maynard Metcalf a decade later (3) imploring that "*Boveri's work should be the starting point for any studies of causes, inheritance or cure of cancer*" it was not until near the end of the last century that investigations on human tumors and mouse models began to corroborate Boveri's astute prescience (4-8). In this article, a review of centrosome structure and function, and the regulation of centrosome duplication in normal cells will be presented. Using recent studies on BC as an exemplary model, a discussion will follow on how deregulation of centrosome behavior can arise, result in centrosome amplification, and lead to CIN in cancer.

Centrosome Structure and Behavior in Normal Cells

The centrosome resides near the cell center (hence its name) and consists of a pair of centrioles and a surrounding matrix of pericentriolar material (PCM) that anchor microtubule nucleation sites and consequently determines the number and organization of microtubules in interphase cells. Like chromosomes, centrosomes double in number once in each cell cycle in a process that is initiated with centriole duplication. The centriole pair embodies an intrinsic counting mechanism that establishes the number of centrosome equivalents in the cell such that a pair of centrioles equals one, and two pair of centrioles equals two centrosome equivalents (9). Centrosomes increase in size through the recruitment of PCM and centrosomes of G2/M cells show a dramatic increase in microtubule nucleating activity. At the time of cell division the two centrosomes (one residing at each spindle pole) organize microtubules of the bipolar mitotic spindle. In normal cells, spindle architecture is such that the two oppositely oriented half-spindle microtubule arrays (each arising from one of the two spindle poles) cast microtubules outward to engage and orient chromosomes so that sister chromatids face and can engage microtubules originating from opposite spindle poles. When all chromosomes are appropriately oriented and attached to microtubules originating from both spindle poles, the sister chromatids separate and move toward opposite poles by a molecular motor-driven process and dynamic shortening of their attached microtubules. Two new daughter cells form by cytokinesis and each inherit a complete complement of chromosomes along with one of the spindle poles that acts as the centrosome in the next cell cycle.

Centrosome Amplification in Cancer

Recent studies implicate centrosome abnormalities in the pathogenesis of cancer (4, 10-14). The term "centrosome amplification" refers to centrosomes that appear larger than normal, centrosomes that contain more than four centrioles, and/or when more than two centrosomes are present within a cell. In addition to these structural abnormalities, amplified centrosomes also show protein hyperphosphorylation and altered functional properties such as an increased microtubule nucleating capacity (4, 8, 15-17). Electron microscope studies revealed supernumerary centrioles in centrosomes of humans and animal model tumors, including leiomyosarcoma, neuroblastoma, glioma, and thymic carcinoid tumors (18-23). Systematic analyses of centrosomes in human breast carcinomas and a mouse model for prostate cancer revealed a range of abnormalities in centrosome structure including: excess number of centrioles, increased pericentriolar material, abnormal centriole orientation, and inverted polarity of centrosome location (5, 24). These structural centrosome abnormalities have been implicated as a potential cause of loss of cell and tissue architecture seen in cancer (*i.e.,* anaplasia) through altered centrosome function in microtubule nucleation and organization, and to result in chromosome missegregation during mitosis as a consequence of multipolar spindle formation.

Correlation of Centrosome Amplification, Aneuploidy, and Chromosomal Instability

A key question is whether or not centrosome amplification leads to CIN and aneuploidy or is a consequence of them; the proverbial chicken and egg riddle. Aneuploidy is characterized as the *state* of an abnormal karyotype, having gains and/or losses of whole chromosomes. Aneuploidy occurs early in the development of many tumor types, suggesting that it may play a role in both tumorigenesis and tumor progression. Indeed, aneuploidy is present in the great majority of malignant tumors, in contrast to benign tumors, which are most often diploid. Aneuploidy can be distinguished from the persistent generation of chromosomal variations, termed "CIN", which reflects the rate of change in karyotype (25). Quantitative analysis of CIN can be determined as the percent of cells with a chromosome number different from the modal chromosome number. Thus, tumors may show either "stable aneuploidy" (low CIN) or "unstable aneuploidy." Unstable karyotypes may lead to phenotypic heterogeneity in cancer, reflecting the persistent generation of new chromosomal variations (26, 27).

The development of aneuploidy may be a consequence of centrosome amplification, which can lead to the formation of multipolar spindles and miss-segregate sister chromatids during mitosis, and as a result to high CIN. CIN occur exclusively in aneuploid tumors and tumor-derived cell lines in contrast to diploid tumors, which contain centrosomes that are functionally and structurally normal (4, 26, 28). The degree of genomic instability in aneuploid tumors parallels the degree of centrosome abnormalities in cell lines from breast (29), pancreas (13), prostate (30), colon (28), and cervix tumors (31), from short-term culture of mouse mammary tumors (32), and from SV40 ST over-expressing fibroblasts (33). When tissues were examined, centrosome abnormalities were higher in high-grade prostate tumors (30) and high-grade cervical tumors (31) than in low-grade tumors. In prostate cancer, centrosome amplification has been implicated in the development of abnormal mitoses and CIN facilitating progression to advanced stages of the disease (30, 34, 35). Strong support for a direct mechanistic link between centrosome amplification and CIN is suggested by the significant linear correlation between centrosome amplification and the rate of change in karyotype (CIN) seen in human breast tumors (26). Although such correlation alone does not necessarily imply cause and effect, these observations have led many authors to propose the hypothesis that centrosome amplification is the primary cause of genomic instability observed in most tumors (13, 26, 31, 33). As discussed above, Boveri first recognized these features of cancer cells nearly a century ago and proposed that centrosome defects could lead to mitotic and subsequent chromosomal abnormalities (1). An alternative hypothesis has been proposed that CIN seen in cancer cells is caused by aneuploidy, that is that aneuploidy itself destabilizes the karyotype and thus initiates CIN leading to widespread heterogeneity in tumor cell phenotypes (36-39).

Several independent lines of evidence support the proposition that centrosome abnormalities drive genomic instability. In a recent study of human breast tumors, all specimens of ductal carcinoma *in-situ* examined showed significant centrosome amplification, while aneuploidy is present, on average, suggesting that centrosome amplification is an early event that occurs prior to invasion in breast tumors (26). Furthermore, cells transfected to express the HPV E7 oncoprotein undergo centrosome amplification prior to developing nuclear morphology associated with aneuploidy (40, 41). Finally, in a xenograft model of pancreatic cancer, metastatic foci showed a higher incidence of centrosome amplification than did the primary xenograft, and abnormal centrosome numbers were accompanied by a higher frequency of abnormal mitoses (42). Taken together, these studies suggest that centrosome amplification may be an early event in tumorigenesis that can drive CIN and lead to genotypic and phenotypic diversity of cells within a tumor.

Coordination of the DNA, Cell, and Centrosome Cycles in Normal Cells

Because the fidelity of equal segregation of sister chromatids into daughter cells depends on the bipolar nature of the mitotic spindle it is essential that cells maintain a strict linkage between the DNA, cell, and centrosome cycles. Cell cycle progression is governed by the location, activation and inactivation of the serine/threonine cyclin-dependent protein kinases (Cdks) (43). A direct role for the Cdks in regulating the mitotic activity of centrosomes was first suggested by the localization of cyclin B and Cdk1 ($p34^{cdc2}$) at the centrosome during mitosis, and from experiments implicating cyclin A and B in the control of microtubule dynamics (44-47). Additional evidence pointed to a role for Cdk2 activity in linking centrosome duplication and the DNA cycle. Both processes are dependent on Cdk2 activation and are blocked by the Cdk2 inhibitors butyrolactone-I or roscovitine (48, 49). In addition, centrosome duplication was arrested by protein inhibitors of Cdk2 (p21/waf1 or p27), or by immuno-depletion of Cdk2 or cyclin E, and centrosome duplication could be restored by excess purified cdk2/cyclin E (49-51). Importantly, separation of the centriole pair, an early event in the centrosome duplication cycle, is dependent on Cdk2/cyclin E activity, suggesting that a Cdk-mediated phosphorylation event regulates centriole pair cohesion (51-54). Finally, DNA replication and centrosome duplication also depend on the phosphorylation status of retinoblastoma tumor suppressor Rb, which in turn governs the availability of the E2F transcription factor to promote S phase progression (55). Taken together these findings establish the mechanism by which the DNA and centrosome cycles are coordinated: both DNA replication and centrosome duplication are controlled by the Rb pathway, both processes depend on downstream transcriptional consequences of E2F activity, and both processes require Cdk2/cyclin activation.

Centrosome behavior and function during the cell cycle are also regulated by protein phosphorylation events in addition to those directly mediated by the Cdk-cyclins (56). At the onset of mitosis centrosome protein phosphorylation increases dramatically and then falls precipitously at the metaphase/anaphase transition (57-60). Several protein kinases and protein phosphatases have been identified that localize at centrosomes and affect the phosphorylation status of centrosome targets. Importantly, certain of these centrosome-associated kinases and their target substrates may become altered during the development of centrosome amplification in cancer. For example, over-expression of the breast tumor amplified kinase, BTAK/STK15 (also known as Aurora A) can lead to centrosome amplification and CIN in breast epithelial cell lines and in mouse models for mammary tumorigenesis (8, 16, 61, 62). Studies on human tissues have also shown that inappropriate phosphorylation of centrosome proteins can serve as a sensitive marker for centrosome amplification in tumors (4).

Deregulation of the Centrosome Cycle in Cancer

G1/S and G2/M checkpoints enforce the orderly completion of cell cycle events, and when triggered, they inhibit the formation and/or activation of Cdks and thereby induce cell cycle arrest (63, 64). Cell cycle checkpoints operate through the action tumor suppressor proteins p53 and Rb, and their downstream activation target Cdk2. Interestingly both p53 and Cdk2 may also physically reside at the centrosome, albeit only transiently (65, 66). The physical presence of key proteins involved in checkpoint control at the centrosome has led to the suggestion that the centrosome itself may provide an important structural context for coordinating cell cycle regulation (56, 65-70).

Centrosome abnormalities in cancer are correlated with loss of p53 function in carcinomas of the breast, head and neck, and prostate, and in neuroectodermal tumors (14, 34, 71). In tumors that retained wild-type p53, amplified centrosomes were frequently associated with overexpression of MDM2, which abrogates p53 function by promoting its degradation (71). Furthermore, gain-of-function p53 mutations and p53 null mice can result in deregulation of centrosome duplication leading to the generation of functionally amplified centrosomes and aberrant mitoses (72-74). Interestingly, in some cancers p53 mutations and cyclin E overexpression may act synergistically since together they increased the frequency of centrosome defects in cultured cells and in mouse models (75).

Centrosome homeostasis is controlled at the G1/S and G2/M checkpoints (Figure 1) through transcriptional regulation by p53 of several downstream targets including the Cdk inhibitor p21/waf1 (76, 77). As discussed earlier, p21/waf1 blocks centrosome duplication through inhibition of Cdk2/cyclin E activity. This conclusion is supported by experiments in which anti-sense targeting of p21/waf1 in human cell lines resulted in endoreduplication and centrosome amplification (78).

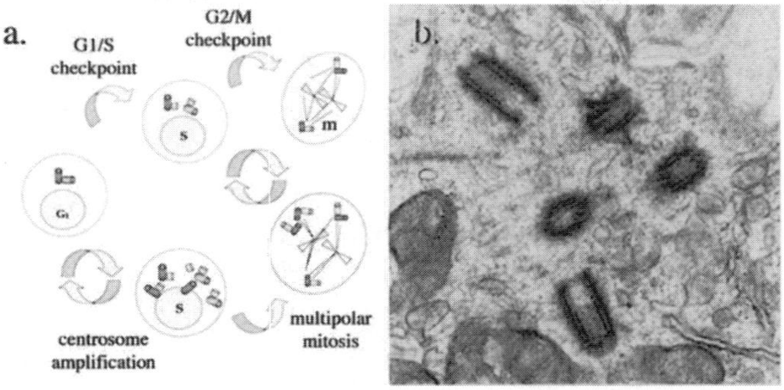

Figure 1. (**a**) Model illustrating the centrosome and DNA cycles in cells with normal and with defective checkpoint controls as discussed in the text. (**b**) Electron micrograph of an amplified centrosome with five centrioles from a human mammary tumor (courtesy of Wilma Lingle, Mayo Clinic).

Interestingly, while introduction of wild-type p53 into p53$^{-/-}$ mouse embryonic fibroblasts re-established centrosome homeostasis, overexpression of p21/waf1 only partially restored control of centrosome duplication in p53-null fibroblasts, suggesting that alternative downstream p53 targets may also be involved in the regulation of centrosome homeostasis (79, 80). In this regard, GADD45, another downstream product of the p53 pathway, has been implicated in activation of the G2/M checkpoint and regulation of the centrosome cycle (81-83). Alternative mechanisms, independent of p53 function, may also lead to deregulation of centrosome homeostasis (25, 26, 84, 85). For example high-risk human papillomavirus (HPV) v-oncogenes, E6 and E7, have been implicated in the induction of centrosome amplification in human cell lines (86). HPV E6 and E7 interfere with centrosome homeostasis by targeting different pathways. Whereas E6 may operate through inactivation of p53 function resulting in accumulation of excess centrosomes by failure of the G2/M checkpoint leading to defects in cytokinesis, E7 may lead to centrosome amplification through inactivation of the Rb and G1/S checkpoint resulting in abnormal centrosome duplication (86). In addition, as discussed above, centrosome amplification can be induced by over-expression of the centrosome kinase BTAK/STK15 (8), mutations in the BRCA1 and BRCA2 tumor suppressor genes (87-92) or by over-expression of the PCM structural protein pericentrin (30). Thus, centrosome defects and consequent genomic instability may result from inactivation of cell cycle checkpoints, inappropriate activation of key centrosome kinases, or alterations in structural proteins of the centrosome itself. These observations suggest several regulatory pathways operate in parallel to ensure linkage between the DNA, cell, and centrosome cycles. In the development of cancer, centrosome defects may result from an imbalance between negative and positive regulators that converge on G1/S and G2/M checkpoints, or directly on components of the centrosome itself.

Centrosome Amplification as Potential Indicator of Tumor Aggressiveness

Can centrosome amplification be utilized as an indicator of tumor progression, the potential to develop aggressive tumor phenotypes, and to serve as a prognostic indicator of clinical outcome? Centrosome amplification is not only characteristic of aneuploid tumors in general, but also is more pronounced in advanced stage malignancies, in recurrent tumors, and in cell lines that show more aggressive malignant phenotypes in xenograph animal models (10, 11, 16, 29). These observations suggest that centrosome amplification might be useful in monitoring tumor progression and phenotypic diversity in cancer. In association with other established prognostic factors, centrosome amplification may be helpful in predicting outcomes and survival of patients with cancer. And finally, because the centrosome may serve as a structural context for the action of key cell cycle regulators, it may also represent a critical target for therapeutic intervention.

Acknowledgements

I have been fortunate to have many creative and talented individuals work in my laboratory on centrosome defects in cancer. I would especially like to acknowledge, Susan Barrett, Robert Busby, Lynn Cordes, Vivian Negron, Mark Sanders, Kelly Suino, and doctors Antonino D'Assoro and Wilma Lingle.

References

1. Boveri T (1914) Zur Frage der Entstehung Maligner Tumoren. Baltimore: Jena: Fischer Verlag (1929 English translation by M. Boveri reprinted as "The origin of malignant tumors", The Williams and Wilkins Co.), 119.
2. Ewing J (1919) Neoplastic Diseases: a text-book on tumors. 1st ed. Philadelphia: W.B. Saunders Co, 1027.
3. Metcalf MM (1925) Boveri's work on cancer. JAMA 84:1140.
4. Lingle WL, Lutz WH, Ingle JN, et al (1998) Centrosome hypertrophy in human breast tumors: implications for genomic stability and cell polarity. Proc Natl Acad Sci USA 95:2950-2955.
5. Lingle WL, Salisbury JL (1999) Altered centrosome structure is associated with abnormal mitoses in human breast tumors. Am J Pathol 155:1941-1951.
6. Fukasawa K, Wiener F, Vande Woude GF, et al (1997) Genomic instability and apoptosis are frequent in p53 deficient young mice. Oncogene 15:1295-1302.
7. Pihan GA, Purohit A, Wallace J, et al (1998) Centrosome defects and genetic instability in malignant tumors. Cancer Res 58:3974-3985.

8. Zhou H, Kuang J, Zhong L, et al (1998) Tumour amplified kinase STK15/BTAK induces centrosome amplification, aneuploidy and transformation. Nat Genet 20:189-193.
9. Mazia D (1987) The chromosome cycle and the centrosome cycle in the mitotic cycle. Int Rev Cytol 100:49-92.
10. Gustafson LM, Gleich LL, Fukasawa K, et al (2000) Centrosome hyperamplification in head and neck squamous cell carcinoma: a potential phenotypic marker of tumor aggressiveness. Laryngoscope 110:1798-1801.
11. Kuo KK, Sato N, Mizumoto K, et al (2000) Centrosome abnormalities in human carcinomas of the gallbladder and intrahepatic and extrahepatic bile ducts. Hepatology 31:59-64.
12. Pihan GA, Wallace J, Zhou Y, et al (2003) Centrosome abnormalities and chromosome instability occur together in pre-invasive carcinomas. Cancer Res 63:1398-1404.
13. Sato N, Mizumoto K, Nakamura M, et al (2001) Correlation between centrosome abnormalities and chromosomal instability in human pancreatic cancer cells. Cancer Genet Cytogenet 126:13-19.
14. Weber RG, Bridger JM, Benner A, et al (1998) Centrosome amplification as a possible mechanism for numerical chromosome aberrations in cerebral primitive neuroectodermal tumors with TP53 mutations. Cytogenet Cell Genet 83:266-269.
15. Meraldi P, Nigg EA (2002) The centrosome cycle. FEBS Lett 521:9-13.
16. Miyoshi Y, Iwao K, Egawa C et al (2001) Association of centrosomal kinase STK15/BTAK mRNA expression with chromosomal instability in human breast cancers. Int J Cancer 92:370-373.
17. Salisbury JL, Lingle WL, White RA, et al (1999) Microtubule nucleating capacity of centrosomes in tissue sections. J Histochem Cytochem 47:1265-1274.
18. Friedlander M (1982) Centrioles and centrospheres in giant cells of human gliomas. J Submicrosc Cytol 14:401-406.
19. Kaneko H, Ishikawa S, Sumida T, et al (1980) Ultrastructural studies of a thymic carcinoid tumor. Acta Pathol Jpn 30:651-658.
20. Seifert HW (1978) Electron microscopic investigation on cutaneous leiomyosarcoma. Arch Dermatol Res 263:159-169.
21. Sharp GA, Osborn M, Weber K (1981) Ultrastructure of multiple microtubule initiation sites in mouse neuroblastoma cells. J Cell Sci 47:1-24.
22. Sharp GA, Weber K, Osborn M (1982) Centriole number and process formation in established neuroblastoma cells and primary dorsal root ganglion neurones. Eur J Cell Biol 29:97-103.
23. Ring D, Hubble R, Kirschner M (1982) Mitosis in a cell with multiple centrioles. J Cell Biol 94:549-56.

24. Schatten H, Hueser CN, Chakrabarti A (2000) From fertilization to cancer: the role of centrosomes in the union and separation of genomic material. Microsc Res Tech 49:420-427.
25. Lengauer C, Kinzler K, Vogelstein B (1997) Genetic instability in colorectal cancers. Nature 386:623-627.
26. Lingle WL, Barrett SL, Negron VC, et al (1978) Centrosome amplification drives chromosomal instability in breast tumor development. Proc Natl Acad Sci USA 99:1978-1983.
27. Mitelman F (1994) Chromosomes, genes, and cancer. CA Cancer J Clin 44:133-135.
28. Ghadimi BM, Sackett DL, Difilippantonio MJ, et al (2000) Centrosome amplification and instability occurs exclusively in aneuploid, but not in diploid colorectal cancer cell lines, and correlates with numerical chromosomal aberrations. Genes Chromosomes Cancer 27:183-190.
29. D'Assoro AB, Barrett SL, Folk C, et al (2002) Amplified centrosomes in breast cancer: a potential indicator of tumor aggressiveness. Breast Cancer Res Treat 75:25-34.
30. Pihan GA, Purohit A, Wallace J, et al (2001) Centrosome defects can account for cellular and genetic changes that characterize prostate cancer progression. Cancer Res 61:2212-2219.
31. Skyldberg B, Fujioka K, Hellstrom AC, et al (2001) Human papillomavirus infection, centrosome aberration, and genetic stability in cervical lesions. Mod Pathol 14:279-284.
32. Montagna C, Andrechek ER, Padilla-Nash H, et al (2002) Centrosome abnormalities, recurring deletions of chromosome 4, and genomic amplification of HER2/neu define mouse mammary gland adenocarcinomas induced by mutant HER2/neu. Oncogene 21:890-898.
33. Gaillard S, Fahrbach KM, Parkati R, et al (2001) Overexpression of simian virus 40 small-T antigen blocks centrosome function and mitotic progression in human fibroblasts. J Virol 75:9799-9807.
34. Ouyang X, Wang X, Xu K, et al (2001) Effect of p53 on centrosome amplification in prostate cancer cells. Biochim Biophys Acta 1541:212-220.
35. Schatten H, Ripple M, Balczon R, et al (2000) Androgen and taxol cause cell type-specific alterations of centrosome and DNA organization in androgen-responsive LNCaP and androgen- independent DU145 prostate cancer cells. J Cell Biochem 76:463-477.
36. Duesberg P (1999) Are centrosomes or aneuploidy the key to cancer? Science 284:2091-2092.
37. Duesberg P, Rasnick D (2000) Aneuploidy, the somatic mutation that makes cancer a species of its own. Cell Motil Cytoskeleton 47:81-107.
38. Duesberg P, Rausch C, Rasnick D, et al (1998) Genetic instability of cancer cells is proportional to their degree of aneuploidy. Proc Natl Acad Sci USA 95:13692-13697.

39. Li R, Sonik A, Stindl, R, et al (2000) Aneuploidy vs. gene mutation hypothesis of cancer: Recent study claims mutation but is found to support aneuploidy. PNAS 97:3236-3241.
40. Duensing S, Duensing A, Crum CP, et al (2001) Human papillomavirus type 16 E7 oncoprotein-induced abnormal centrosome synthesis is an early event in the evolving malignant phenotype. Cancer Res 61:2356-2360.
41. Duensing S, Lee LY, Duensing A, et al (2000) The human papillomavirus type 16 E6 and E7 oncoproteins cooperate to induce mitotic defects and genomic instability by uncoupling centrosome duplication from the cell division cycle. Proc Natl Acad Sci USA 97:10002-10007.
42. Shono M, Sato N, Mizumoto K, et al (2001) Stepwise progression of centrosome defects associated with local tumor growth and metastatic process of human pancreatic carcinoma cells transplanted orthotopically into nude mice. Lab Invest 81:945-952.
43. Pines J (1999) Four-dimensional control of the cell cycle. Nat Cell Biol 1:E73-79.
44. Bailly E, Bordes N, Bornens M, et al (1992) A high molecular weight centrosomal protein of mammalian cells is antigenically related to myosin II. Cell Motil Cytoskeleton 23:122-132.
45. Debec A, Montmory C (1992) Cyclin B is associated with centrosomes in Drosophila mitotic cells. Biol Cell 75:121-126.
46. Verde F, Dogterom M, Stelzer E, et al (1992) Control of microtubule dynamics and length by cyclin A- and cyclin B- dependent kinases in Xenopus egg extracts. J Cell Biol 118:1097-1098.
47. Verde F, Labbe JC, Doree M, et al (1990) Regulation of microtubule dynamics by cdc2 protein kinase in cell-free extracts of Xenopus eggs. Nature 343:233-238.
48. Keezer SM, Gilbert DM (2002) Sensitivity of the origin decision point to specific inhibitors of cellular signaling and metabolism. Exp Cell Res 273:54-64.
49. Matsumoto Y, Hayashi K, Nishida E (1999) Cyclin-dependent kinase 2 (Cdk2) is required for centrosome duplication in mammalian cells. Curr Biol 9:429-432.
50. Hinchcliffe EH, Li C, Thompson EA, et al (1999) Requirement of Cdk2-cyclin E activity for repeated centrosome reproduction in Xenopus egg extracts. Science 283:851-854.
51. Lacey KR, Jackson PK, Stearns T (1999) Cyclin-dependent kinase control of centrosome duplication. Proc Natl Acad Sci USA 96:2817-2822.
52. Okuda M, Horn HF, Tarapore P, et al (2000) Nucleophosmin/B23 is a target of CDK2/cyclin E in centrosome duplication. Cell 103:127-140.
53. Peter M, Nakagawa J, Doree M et al (1999) Identification of major nucleolar proteins as candidate mitotic substrates of cdc2 kinase. Cell 60:791-801.

54. Tokuyama Y, Horn HF, Kawamura K, et al (2001) Specific phosphorylation of nucleophosmin on Thr(199) by cyclin- dependent kinase 2-cyclin E and its role in centrosome duplication. J Biol Chem 276:21529-21537.
55. Meraldi P, Lukas J, Fry A, et al (1999) Centrosome duplication in mammalian somatic cells requires E2F and Cdk2-cyclin A. Nature Cell Biol 1:88-93.
56. Fry AM, Mayor T, Nigg EA (2000) Regulating centrosomes by protein phosphorylation. In: Palazzo RE, Schatten GP (eds) Centrosomes in Cell Replication and Early Development, San Diego:Academic Press, 291-312.
57. Lutz W, Lingle WL, McCormick D (2001) Phosphorylation of centrin during the cell cycle and its role in centriole separation preceding centrosome duplication. J Biol Chem 276:20774-20780.
58. Rao PN, Zhao JY, Ganju RK, et al (1989) Monoclonal antibody against the centrosome. J Cell Sci 93:63-69.
59. Vandre DD, Borisy GG (1989) Anaphase onset and dephosphorylation of mitotic phosphoproteins occur concomitantly. J Cell Sci 94:245-258.
60. Vandre DD, Feng Y, Ding M (2000) Cell cycle-dependent phosphorylation of centrosomes: Localization of phosphopeptide specific antibodies to the centrosome. Microscopy Res Tech 49:458-466.
61. Katayama H, Zhou H, Li Q, et al (2001) Interaction and feedback regulation between STK15/BTAK/Aurora-A kinase and protein phosphatase 1 through mitotic cell division cycle. J Biol Chem 276:46219-46224.
62. Goepfert TM, Adigun YE, Zhong L, et al (2002) Centrosome amplification and overexpression of aurora A are early events in rat mammary carcinogenesis. Cancer Res 62:4115-4122.
63. Hartwell LH, Weinert TA (1989) Checkpoints: controls that ensure the order of cell cycle events. Science 246:629-634.
64. Murray AW (1992) Creative blocks: cell-cycle checkpoints and feedback controls. Nature 359:599-604.
65. Giannakakou P, Sackett DL, Ward Y, et al (2000) p53 is associated with cellular microtubules and is transported to the nucleus by dynein. Nat Cell Biol 2:709-717.
66. Pockwinse SM, Krockmalnic G, Doxsey SJ, et al (1997) Cell cycle independent interaction of CDC2 with the centrosome, which is associated with the nuclear matrix-intermediate filament scaffold. Proc Natl Acad Sci USA 94:3022-3027.
67. Doxsey SJ (2001) Centrosomes as command centres for cellular control. Nat Cell Biol 3:E105-108.
68. Sluder G, Hinchcliffe EH (1998) The apparent linkage between centriole replication and the S phase of the cell cycle. Cell Biol Int 22:3-5.
69. Brown CR, Doxsey SJ, White E, et al (1994) Both viral (adenovirus E1B) and cellular (hsp 70, p53) components interact with centrosomes. J Cell Physiol 160:47-60.

70. Morris VB, Brammall J, Noble J, et al (2000) p53 localizes to the centrosomes and spindles of mitotic cells in the embryonic chick epiblast, human cell lines, and a human primary culture: An immunofluorescence study. Exp Cell Res 256:122-130.
71. Carroll PE, Okuda M, Horn HF, et al (1999) Centrosome hyperamplification in human cancer: chromosome instability induced by p53 mutation and/or Mdm2 overexpression. Oncogene 18:1935-1944.
72. Fukasawa K, Choi T, Kuriyama R, et al (1996) Abnormal centrosome amplification in the absence of p53. Science 271:1744-1747.
73. Murphy KL, Rosen JM (2000) Mutant p53 and genomic instability in a transgenic mouse model of breast cancer. Oncogene 19:1045-1051.
74. Tarapore P, Fukasawa K (2000) p53 mutation and mitotic infidelity. Cancer Invest 18:148-155.
75. Mussman JG, Horn HF, Carroll PE, et al (2000) Synergistic induction of centrosome hyperamplification by loss of p53 and cyclin E overexpression. Oncogene 19:1635-1646.
76. el-Deiry WS, Tokino T, Velculescu VE, et al (1993) WAF1, a potential mediator of p53 tumor suppression. Cell 75:817-825.
77. Harper JW, Adami GR, Wei N, et al (1993) The p21 Cdk-interacting protein Cip1 is a potent inhibitor of G1 cyclin- dependent kinases. Cell 75:805-816.
78. Mantel C, Braun SE, Reid S, et al (1999) p21(cip-1/waf-1) deficiency causes deformed nuclear architecture, centriole overduplication, polyploidy, and relaxed microtubule damage checkpoints in human hematopoietic cells. Blood 93:1390-1398.
79. Tarapore P, Fukasawa K (2002) Loss of p53 and centrosome hyperamplification. Oncogene 21:6234-6240.
80. Tarapore P, Horn HF, Tokuyama Y, et al (2001) Direct regulation of the centrosome duplication cycle by the p53- p21Waf1/Cip1 pathway. Oncogene 20:3173-3184.
81. Kastan MB, Zhan Q, el-Deiry WS, et al (1992) A mammalian cell cycle checkpoint pathway utilizing p53 and GADD45 is defective in ataxia-telangiectasia. Cell 71:587-597.
82. Wang XW, Zhan Q, Coursen JD, et al (1999) GADD45 induction of a G2/M cell cycle checkpoint. Proc Natl Acad Sci USA 96:3706-3711.
83. Hollander MC, Sheikh MS, Bulavin DV et al (1999) Genomic instability in Gadd45a-deficient mice. Nat Genet 23:176-184.
84. Donehower LA, Harvey M, Slagle BL, et al (1992) Mice deficient for p53 are developmentally normal but susceptible to spontaneous tumours. Nature 356:215-221.
85. Eshleman JR, Casey G, Kochera ME, et al (1998) Chromosome number and structure both are markedly stable in RER colorectal cancers and are not destabilized by mutation of p53. Oncogene 17:719-725.

86. Duensing S, Munger K (2003) Centrosomes, genomic instability, and cervical carcinogenesis. Crit Rev Eukaryot Gene Expr 13:9-23.
87. Deng CX, Brodie SG (2000) Roles of BRCA1 and its interacting proteins. Bioessays 22:728-737.
88. Hsu LC, White RL (1998) BRCA1 is associated with the centrosome during mitosis. Proc Natl Acad Sci USA 95:12983-12988.
89. Maul GG, Jensen DE, Ishov AM, et al (1998) Nuclear redistribution of BRCA1 during viral infection. Cell Growth Differ 9:743-755.
90. Okada S, Ouchi T (2003) Cell cycle differences in DNA damage-induced BRCA1 phosphorylation affect its subcellular localization. J Biol Chem 278:2015-2020.
91. Tutt A, Gabriel A, Bertwistle D, et al (1999) Absence of Brca2 causes genome instability by chromosome breakage and loss associated with centrosome amplification. Curr Biol 9:1107-1110.
92. Xu X, Wagner KU, Larson D, et al (1999) Conditional mutation of Brca1 in mammary epithelial cells results in blunted ductal morphogenesis and tumour formation. Nat Genet 22:37-43.

PART 3. BREAST AND PROSTATE: EARLY *IN-SITU* LESIONS

9

Premalignant Breast Disease: Anatomic Lesions and Hormonal Associations

David L. Page

Introduction

Breast cancer (BC) risk associations are many, and specific patterns of hyperplastic lesions, as revealed in biopsy tissue, are among the most widely accepted. A review of these lesions is presented in this chapter, along with their magnitude of BC risk and regional implications in women. In this chapter, the term "premalignant' is used broadly, and indicate by the term 'disease' that the risk implications are of a magnitude that may have possible practical importance for an individual woman. Many of the risk associations that are well-known are just above 1.0. An age-corrected relative risk (RR) of 50% elevation in a period of 10-15 years after biopsy, has been adopted as a reference point to make suggested quantitative references (1) (*i.e.,* slight = 1.5-2.0 x; moderate = 4.0-5.0 x). Particularly important, and not recognized in the initial, formal epidemiologic studies of hyperplastic lesions in the breast, is the varied implication for site(s) and type of later BC presentation.

Available information concerning tissue and estrogen receptor (ER) expression in breast and various lesions will be reviewed. Finally, epidemiologic associations of BC risk will be presented that offer insight into possible interactions between hormonal milieu and hyperplastic lesions.

Estrogen Receptor and Normal Cell Cycle in Breast Tissue

ER protein, particularly the more widely studied ERα has varied associations in the specific hyperplasic and atypical *in-situ* lesions of the breast (2). Although these have yet to be linked in the way of predicting later BC risk in the decade after biopsy. However, this association of ERα and hyperplastic epithelial lesions promises to add to the estimation of both BC risk, and perhaps in indicating responsiveness to hormonal elements in the prevention setting (3). Estrogen, mediated through the ERα, has a central role in normal growth and development of the breast (4) as well as a well-developed and proven role in the evaluation of established neoplastic disease (5). Cell proliferation involving E is also central to the expression of other genes, including most prominently, progesterone receptor

(PR) (6). PR mediates the mitogenic effect of progesterone which further stimulates proliferation (4). This latter effect has now been demonstrated epidemiologically in that hormone replacement therapy (HRT) after menopause has only a mild effect on BC risk if it involves estrogen only (7), whereas the addition of progestins adds significantly to the likelihood of later BC development in the post-menopausal woman (8-10)

Most of the studies of ERα expression in normal ducts and lobules, as well as *in-situ* neoplasia, are immunohistochemical studies focused on the ERα, although the potential cross reactivity for the much more recently discovered ERβ is not currently clear (11). Normal terminal duct-lobular units express ERα approximately 90% of the time, and in perhaps a third of these the expression is quite limited and only seen in a minority of cells. During the childbearing years (12), the number of ERα positive cells varies with the menstrual cycle and is much higher during the follicular phase. Proliferation however peaks during the luteal phase (13), suggesting that the mitogenic effect of estrogen is either delayed or indirect, and mediated by other interactions, probably mostly with progesterone (14, 15) and prolactin. In post-menopausal women, the expression of ERα is relatively high (2) and stable in the absence of HRT.

Hyperplastic and Premalignant Lesions: Precursors and Markers of Increased Risk of Breast Cancer

The various identified breast lesions of interest are listed in Table 1 and 2 along with their major known associations. The precise separation of these lesions in practical histopathology has evolved over time, and as a practical matter in clinical medicine, there remain some differences in application in the clinical arena. Basically, most of the studies presented in this review have been accepted by major pathology associations in North America and Britain.

The ductal carcinomas *in-situ* (DCIS) are well accepted as precursors lesions. They probably have a magnitude of risk in the range of 50% over a ten year period with some variation in both size at time of diagnosis and risk differing between the low and high grade varieties. Presentation of DCIS raises the very important consideration not frequently discussed, the nature of the subsequently developing invasive lesion. The majority of the invasive carcinomas developing at least in the short period of several years after initial identification of an inadequately removed low grade DCIS are low grade invasive cancers, and many of these lesions may have prolonged periods without invasion (16, 17).

Whilst, the high grade DCIS are associated with subsequent high grade invasive tumors with metastatic and death dealing capacity (18), the subsequent invasive carcinomas after low grade DCIS are largely local. The invasive tumors developing after identification atypical ductal hyperplasia (ADH) after biopsy may occur anywhere in either breast (19, 20), and their type is probably randomly distributed amongst those ordinarily seen.

Table 1. Histologically Defined Lesions of Female Breast with Proven Predictiveness for Later Carcinoma.

ELUCA **Enlarged Lobular Units with Columnar Alteration** (Columnar Alteration with Secretion or Hyperplastic Unfolded Lobules)	Quite common, usually co-exist with hyperplastic lesions RR ~ 1.2-1.4
PDWA **Proliferative Disease Without Atypia** most are usual pattern hyperplasia of EHLA (Epithelial Hyperplasia Lacking Atypia)	Defined by their association with a RR of 1.5-2.0 x Changes present in about 25% of biopsies
ADH **Atypical Ductal Hyperplasia** Similar histology to low grade DCIS, **but** of limited extent	General distribution of cancer risk in each breast, Magnitude = 4.0-5.0 x
ALH **Atypical Lobular Hyperplasia** Major lesion of lobular neoplasia series	Regional distribution of risk. Cancer in same breast as ALH 70% of the time, and possibly favoring same region in same breast. RR ~ 4.0 x, decreasing with menopause.

Lobular Neoplasia: Atypical Lobular Hyperplasia/Lobular Carcinoma *In-situ*

First described in 1941, this series of characteristic lesions have had perceptions of their implications changed frequently. The lobular neoplasia series includes lesions identified as atypical lobular hyperplasia (ALH) and lobular carcinoma *in-situ* (LCIS) with some variety of individual authors using different terms and concepts. However, in studies rigorously performed with specific criteria, the ALH lesion (21) is both dominant and best studied using specific criteria. With more extensive density of local disease, LCIS has a slightly greater increased risk that may not be of great clinical importance (22).

While ALH has been long considered a risk indicator along with LCIS, it is now quite evident that subsequently developing carcinomas tend to be regional, favoring the breast in which the ALHs was originally diagnosed (23) rather than the model for clinical-decision making accepted for many years which was one of equal risk of later cancer in each breast. Studies by Rosen, *et al.* published in 1981 (24), showed a very low risk for later invasive carcinoma in the contralateral risk of women with LCIS, especially of biopsy samples taken from the contralateral breast had shown no lobular neoplasia. Also, Page, *et al.* demonstrated that after ALH, 70% of subsequent invasive carcinomas develop in the same breast with only 30% in the contralateral breast (23). There is a favored association of these later carcinomas having lobular and tubular features, but this association is not terribly strong (22). However, the association of the type of invasive carcinoma having some lobular features further support that the ALH lesions and LCIS are true precursors with a model certainly intermediate between general increased risk of ADH and most risk indicators, and local increased risk precursor model evidenced by DCIS.

Table 2. Relative Risk of Invasive BC After Biopsy. Follow–up 10 Years After Initial Biopsy, Compared to Similar Women of Similar Age.*

Nashville Cohort Study	**Proliferative Breast Disease (PBD) Without Atypia** **Atypical Hyperplasia** (both ADH and ALH combined)	1.6 to 2.3 3.1 to 8.8
Harvard Nurses' Study Case Control Design	**PBD Without Atypia** **Atypical Hyperplasia**	1.2 to 2.2 2.6 to 5.9
BCDDP Case Control Design Specific Histologic Review	**PBD Without Atypia** **Atypical Hyperplasia**	0.77 to 2.2 1.7 to 11.0
BCDDP Review of Original Pathology reports		

*95% confidence interval

Precursors and increased BC risk markers are usually considered as quite separate elements. However, it is quite evident that if one has many individual lesions with low but greater risk than normal tissue that, as a practical matter, multiple increased risk lesions may actually serve more as markers of increased risk as a practical matter than precursors. Thus, the magnitude of increased subsequent risk of individual lesions as well as their numbers and distribution in the breast become important considerations (23). The various lesions with risk implications also have important variations in incidence related to age and reproductive factors (25).

Estrogen Receptor Evidence in Breast Tissue with Risk Associations

The DCISs have been relatively well-characterized with regard to ERα. Its presence in varying degrees in up to 75% of cases, with the higher grade lesions with extensive necrosis being regularly negative (26). It is likely that estrogen exposure is important during the prolonged period of BC development where this proliferation stimulus may allow random genetic alterations to accumulate.

Other molecular biomarkers, which have a role in well-developed invasive lesions and high grade DCIS, include erbB2 (also known as HER-2/neu) which is regularly present in high grade DCIS, (certainly over 70%) (27), and is regularly absent in lower grade lesions. This is also true for the tumor suppressor gene p53 (28, 29), and other molecular markers that have more similarity between histologic grade *in-situ* and invasive cancers than they have differences between invasive and *in-situ* lesions in aggregate (30). Thus, DCIS is proven to be a local precursor of later invasive BC, unique and most important among the precursor lesions of invasive BC (16). The great majority of so called precursor lesions with the acceptance and exception of high grade DCIS express ERα often in seeming increased density. These observations support the possibility that increased susceptibility to estrogen hormonal stimuli to these precursor lesions is an important element in the evolution of BC. This must be understood against the background of complexities in the biologic variability and age-dependence of breast epithelial proliferation (31) that includes variations among individual lobular units in breast (12).

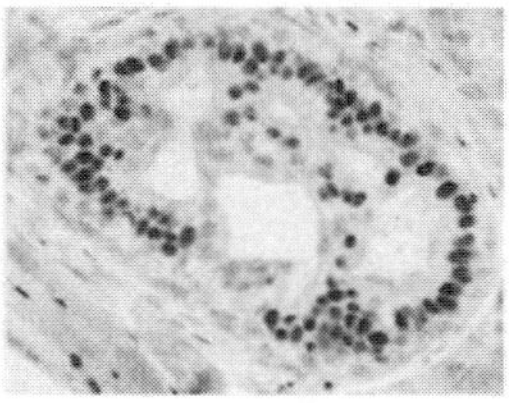

Figure 1. Moderate hyperplasia without atypia decorated for ERα in nuclei. Note that some of the hyperplastic cells in the center of the lumen are ERα positive. Also, some of the luminal cells elsewhere are positive, but the myoepithelial cells adjacent to the deliminating basement membrane are ERα negative.

However, preliminary studies quantitating ERα in individual lesions (Figure 1) of usual pattern hyperplasia [epithelial hyperplasia lacking atypia (EHLA)] were not successful in demonstrating that higher ERα levels were associated with a higher BC risk in a follow-up study design (32). Within this set of lesions that indicate a later increased RR of BC development of 1.5 to 2.0 x, cancer risk was not further stratified by varied ERα levels. In the same cohort of patients (33), the absence of TGFβ in these same EHLA lesions indicated a further increased RR over that of women having the lesions alone.

The Nature of Subsequent Invasive Carcinoma

Although not widely accepted or utilized in clinical decision making, it is quite evident that there are some important associations between hormonal manipulation and the nature of the subsequently developing invasive BC (34). Simply, many BCs, particularly in the post-menopausal age range (35), are of low grade, low stage, and high ERα content indicating not excellent survival in the absence of treatment. Furthermore, excellent survival is further fostered by treatment with tamoxifen and other anti-estrogenic compounds.

Several studies have recently indicated that the carcinomas developing or diagnosed in women on HRT are of increased incidence of these particularly good prognosis lesions. Exactly the mechanism involved is unknown, but it is likely that growth of these lesions is fostered, and they are able to be in greater number than in women not on HRT, and diagnosed before they might become more advanced. Indeed, the increased number of genetic alterations associated with poor prognosis carcinomas may well develop during the progression of these lower grade proliferative lesions.

A corollary bit of evidence is the type of carcinomas developing in women on tamoxifen in the prevention setting. The P-1 prevention trial of the National Surgical Adjuvant Breast Project demonstrated quite clearly that tamoxifen treatment selectively decreased the incidence of BC, strikingly so, in women who recalled that their previous breast biopsy (IES) had either atypical hyperplasia, or LCIS, or possibly other atypias. However, in the relatively short period of this study, it was remarkable that the percentage and incidence of higher grade, ERα negative carcinomas was similar in both the treated and untreated group (36). Therefore, this study did not demonstrate a survival advantage, both because of the relatively short nature of the study as well as because of the great difference between good and bad prognosis of subsequent lesions. Thus, it becomes important to consider the nature of the hormonal milieu as well as the lesional sub-strata upon which it is acting (37).

The relation between the presence of increased risk indicators in tissue biopsies and modulation of later BC risk by classic epidemiologic milestones such as parity have been reported in a large cohort of women from Nashville, TN (38, 39) (Dupont+). The interactions were striking and correlated with magnitude of each factor's influence on cancer risk. Thus, nulliparous women with various risk-associated lesions experienced a further elevation in BC risk in each category above that of parous women (38). Most striking, the protective effect of pregnancy prior to the age of 20 interacted with the usually moderately high magnitude of risk associated with atypical hyperplasia to a risk for women with early pregnancy and atypical hyperplasia closer to unity (38).

Conclusions

This review emphasized the role of individual varieties of cellular and tissue changes in the human breast and their relation to BC risk. Further, any identifiable associations with hormonal relationships are emphasized. Therefore, much of this evidence from tissue based studies and epidemiologic studies of subsequent BC risk, as well as concurrent associations of ERα and other hormone substances in various lesions. There are very few follow up studies, therefore, most of these relationships must remain as implications derived from concurrent association in breast tissue (40) associated with developed carcinomas (41). The cancer implications and associations of hormone expression in premalignant lesions are few but promising because of the obvious importance of the lines of evidence for each element. However, the integration of hormone expression with varied lesions in the prospective evaluation of BC development and progression awaits further experience.

References

1. Fitzgibbons PL, Henson DE, Hutter RV (1998) Benign breast changes and the risk for subsequent breast cancer: an update of the 1985 consensus statement. Cancer Committee of the College of American Pathologists. Arch Pathol Lab Med 122:1053-1055.
2. Allred DC, Mohsin SK, Fuqua SA (2001) Histological and biological evolution of human premalignant breast disease. Endocr Relat Cancer 8:47-61.
3. Pappas SG, Jordan VC (2002) Chemoprevention of breast cancer: current and future prospects. Cancer Metastasis Rev 21(3-4):311-21.
4. Pike MC, Spicer DV, Dahmoush L, et al (1993) Estrogens, progestogens, normal breast cell proliferation, and breast cancer risk. Epidemiol Rev 15:17-35.
5. Harvey JM, Clark GM, Osborne CK, et al (1999) Estrogen receptor status by immunohistochemistry is superior to the ligand-binding assay for predicting response to adjuvant endocrine therapy in breast cancer. J Clin Oncol 17:1474-1481.
6. Barnes DM, Hanby AM (2001) Oestrogen and progesterone receptors in breast cancer: past, present and future. Histopathology 38:271-274.
7. Li CI, Malone KE, Porter PL, et al (2003) Relationship between long durations and different regimens of hormone therapy and risk of breast cancer. JAMA 289:3254-3263.
8. Chlebowski RT, Hendrix SL, Langer RD, et al (2003) Influence of estrogen plus progestin on breast cancer and mammography in healthy postmenopausal women: the Women's Health Initiative Randomized Trial. JAMA 289:3243-3253.
9. Olsson HL, Ingvar C, Bladstrom A (2003) Hormone replacement therapy containing progestins and given continuously increases breast carcinoma risk in Sweden. Cancer 97:1387-1392.
10. Schairer C, Lubin J, Troisi R, et al (2000) Menopausal estrogen and estrogen-progestin replacement therapy and breast cancer risk. JAMA 283:485-491.
11. Roger P, Sahla ME, Makela S, et al (2001) Decreased expression of estrogen receptor beta protein in proliferative preinvasive mammary tumors. Cancer Res 61:2537-2541.

12. Anderson TJ (1986) Effects on breast tissue of exogenous oestrogens and progestogens. Acta Obstet Gynecol Scand Suppl 134:9-13.
13. Anderson TJ, Ferguson DJ, Raab GM (1982) Cell turnover in the "resting" human breast: influence of parity, contraceptive pill, age and laterality. Br J Cancer 46:376-382.
14. Going JJ, Anderson TJ, Battersby S, et al (1988) Proliferative and secretory activity in human breast during natural and artificial menstrual cycles. Am J Pathol 130:193-204.
15. Anderson TJ, Howell A, King RJ (1987) Comment on progesterone effects in breast tissue. Breast Cancer Res Treat 10:65-66.
16. Page DL, Simpson JF (1999) Ductal carcinoma in situ--the focus for prevention, screening, and breast conservation in breast cancer. N Engl J Med 340:1499-1500.
17. Page DL, Dupont WD, Rogers LW, et al (1995) Continued local recurrence of carcinoma 15-25 years after a diagnosis of low grade ductal carcinoma in situ of the breast treated only by biopsy. Cancer 76:1197-1200.
18. Bijker N, Peterse JL, Duchateau L, et al (2001) Risk factors for recurrence and metastasis after breast-conserving therapy for ductal carcinoma-in-situ: analysis of European Organization for Research and Treatment of Cancer Trial 10853. J Clin Oncol 19:2263-2271.
19. Page DL, Dupont WD, Rogers LW, et al (1985) Atypical hyperplastic lesions of the female breast. A long-term follow-up study. Cancer 55:2698-2708.
20. Marshall LM, Hunter DJ, Connolly JL, et al (1997) Risk of breast cancer associated with atypical hyperplasia of lobular and ductal types. Cancer Epidemiol Biomarkers Prev 6:297-301.
21. Schnitt SJ (2003) Benign breast disease and breast cancer risk: morphology and beyond. Am J Surg Pathol 27:836-841.
22. Page DL, Kidd TE, Jr., Dupont WD, et al (1991) Lobular neoplasia of the breast: higher risk for subsequent invasive cancer predicted by more extensive disease. Hum Pathol 22:1232-1239.
23. Page DL, Schuyler PA, Dupont WD, et al (2003) Atypical lobular hyperplasia as a unilateral predictor of breast cancer risk: a retrospective cohort study. Lancet 361:125-129.
24. Rosen PP, Braun DW, Jr., Lyngholm B, et al (1981) Lobular carcinoma in situ of the breast: preliminary results of treatment by ipsilateral mastectomy and contralateral breast biopsy. Cancer 47:813-819.
25. Gertig DM, Stillman IE, Byrne C, et al (1999) Association of age and reproductive factors with benign breast tissue composition. Cancer Epidemiol Biomarkers Prev 8:873-879.
26. Jensen RA, Page DL (2003) Ductal carcinoma in situ of the breast: impact of pathology on therapeutic decisions. Am J Surg Pathol 27:828-831.

27. Fitzgibbons PL, Page DL, Weaver D, et al (2000) Prognostic factors in breast cancer. College of American Pathologists Consensus Statement 1999. Arch Pathol Lab Med 124:966-978.
28. Simpson JF, Page DL, Edgerton ME (2001) p53 in mammary ductal carcinoma in situ, mutations in high-grade lesions only? J Natl Cancer Inst 93:666-667.
29. Lukas J, Niu N, Press MF (2000) p53 mutations and expression in breast carcinoma in situ. Am J Pathol 156:183-191.
30. Gobbi H, Arteaga CL, Jensen RA, et al (2000) Loss of expression of transforming growth factor beta type II receptor correlates with high tumour grade in human breast in-situ and invasive carcinomas. Histopathology 36:168-177.
31. Visscher DW, Gingrich DS, Buckley J, et al (1996) Cell cycle analysis of normal, atrophic, and hyperplastic breast epithelium using two-color multiparametric flow cytometry. Anal Cell Pathol 12:115-124.
32. Gobbi H, Dupont WD, Parl FF, et al (2003) Estrogen receptor expression in epithelial hyperplasia without atypia and breast cancer risk: A nested case control study. Modern Pathology 16:29A.
33. Gobbi H, Dupont WD, Simpson JF, et al (1999) Transforming growth factor-beta and breast cancer risk in women with mammary epithelial hyperplasia. J Natl Cancer Inst 91:2096-2101.
34. Kenemans P, Bosman A (2003) Breast cancer and post-menopausal hormone therapy. Best Pract Res Clin Endocrinol Metab 17:123-137.
35. Fisher CJ, Egan MK, Smith P, Wicks K, Millis RR, Fentiman IS. Histopathology of breast cancer in relation to age. Br J Cancer 75:593-596.
36. Fisher B, Costantino JP, Wickerham DL, et al (1998) Tamoxifen for prevention of breast cancer: report of the National Surgical Adjuvant Breast and Bowel Project P-1 Study. J Natl Cancer Inst 90:1371-1388.
37. Byrne C, Connolly JL, Colditz GA, et al (2000) Biopsy confirmed benign breast disease, postmenopausal use of exogenous female hormones, and breast carcinoma risk. Cancer 89:2046-2052.
38. Dupont WD, Page DL (1987) Breast cancer risk associated with proliferative disease, age at first birth, and a family history of breast cancer. Am J Epidemiol 125:769-779.
39. Dupont WD, Page DL, Parl FF, et al (1999) Estrogen replacement therapy in women with a history of proliferative breast disease. Cancer 85:1277-1283.
40. Allred DC, Hilsenbeck SG (1998) Biomarkers in benign breast disease: risk factors for breast cancer. J Natl Cancer Inst 90:1247-1248.
41. O'Connell P, Pekkel V, Fuqua SA, et al (1998) Analysis of loss of heterozygosity in 399 premalignant breast lesions at 15 genetic loci. J Natl Cancer Inst 90:697-703.

10

Aromatase Overexpression: Effect of Tissue Estrogen on Phenotypic and Biochemical Changes in Aromatase Transgenic Mice

Rajeshwar Rao Tekmal, Nameer Kirma, Usha Mandava, and Roopa Luthra

Introduction

The etiology of breast cancer (BC) is thought to involve a complex interplay of genetic, hormonal, and environmental factors that influence the physiological status of the host. Epidemiological studies indicate estrogen (E)-dependent physiological changes can increase a women's risk of developing BC (1). These findings implicate the involvement of steroid hormones, especially E, in BC development. E exerts its mitogenic effect on mammary tissue through its interaction with estrogen receptors (ERα/β). Several studies have shown that the proliferative effect of E on epithelial cells is indirect and may involve a paracrine regulatory mechanism (3, 4). The observation that E-dependent tumors eventually lose this E-dependence suggests the presence of other factors such as growth factors and others are involved in cellular proliferation and tumor maintenance. The role of some of these factors and their E regulation in normal mammary gland (MG) development as well as tumorigenesis has been the focus of several studies (5-8).

The source of sex steroids differ between pre- and postmenopausal women. Es is synthesized by the ovary or by extra-glandular tissues, and delivered, via the endocrine route, to breast tissue. E synthesis is catalyzed by an enzyme complex known as aromatase (ARO), whose activity results in aromatization of the A ring of androgens, to form the phenolic ring characteristic of Es (9-11). The capacity of certain BCs to synthesize Es because of the presence of intratumoral ARO has been established for almost 20 years (9). The potential clinical importance of this intracrine mechanism has led to widespread clinical and laboratory investigations with aromatase inhibitors (AIs) have been used to treat women with hormone dependent BC (11). This chapter will focus on the possibility that Es may initiate, as well as promote BC, and that these processes occur in breast tissues that overexpress ARO.

Aromatase Transgenic Model and Aromatase Expression

To address the importance of tissue E, we developed a mouse transgenic model overexpressing ARO (12). Transgenic mice overexpressing ARO under the control of mouse mammary tumor virus long terminal repeat (MMTV-LTR) were generated using standard transgenic methodology. To date (over 40 generations), all females have had normal pregnancies, and are able to successfully nurse their young. However, homozygous males are infertile. In general, ARO expression is very low or undetectable in non-transgenic virgin MG tissue. Figure 1 shows the expression of ARO transgenic MG tissue. It is expressed both in epithelial and stromal cells. Although, ARO is overexpressed in MGs, and tissue E concentration is higher than in non-transgenic littermates, the E_2 circulating levels in female transgenic mice are not significantly different.

Transgenic virgin female MGs, even at very young age, show extensive terminal branching with localized alveolar growth. No such changes are present in virgin female non-transgenic littermates. Histological observations show the presence of hyperplastic and dysplastic lesions. Some hyperplastic legions are surrounded by excess stromal tissue like fibroadenomas. This suggests that early estrogenic activity due to ARO overexpression is sufficient for these preneoplastic changes to occur in MG epithelial cells of these transgenic mice.

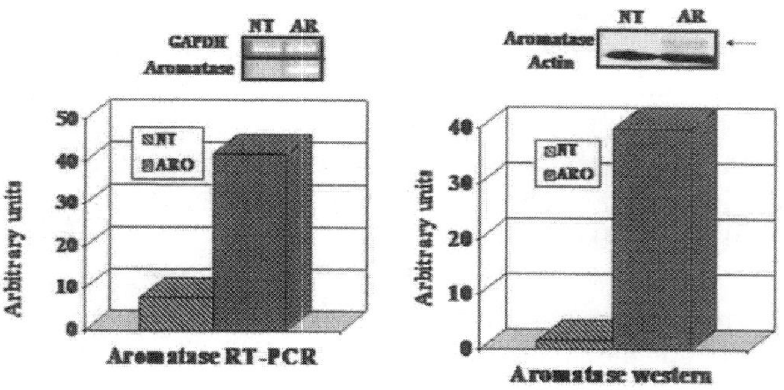

Figure 1. Aromatase expression in transgenic (ARO) and nontransgenic (NT) MG. Densitometric examination of aromatase expression by RT-PCR and Western blot analysis.

Compared to involuted MGs from non-transgenic females, those of transgenic females show persistent ductal and alveolar hyperplasia throughout the MGs; in addition, persistence of ductal hyperplasia is evident even after several months of MG involution. Hyperplasia increases with age and appears similar to ductal carcinoma like changes; however, no tumors appear even after two years of age (Figure 2). Furthermore, the development of hyperplasia and other changes are persistent even without circulating ovarian Es in ovariectomized transgenic mice, suggesting that the E present in the MG has direct effects on the induction and maintenance of these changes in the MGs.

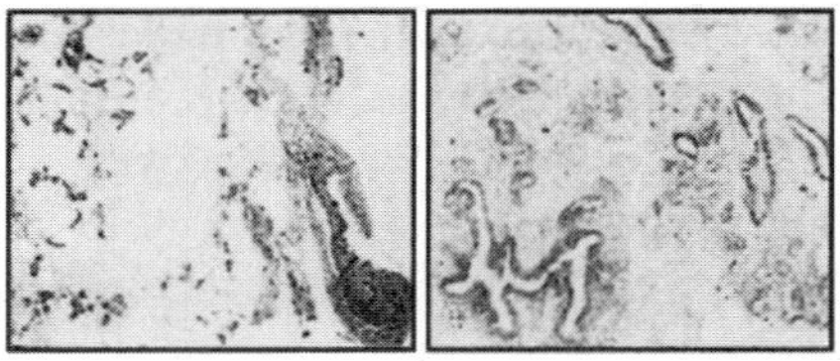

Figure 2. Representative histological MG sections from aromatase transgenic female mice. **A.** Ductal hyperplasia and dysplasia, and **B.** Ductal carcinoma-like changes in aged aromatase transgenic female MGs.

Biochemical Changes in Aromatase Transgenic Females

Previous studies (12) have shown that ARO overexpression, leads to hyperplastic, dysplastic, and other premalignant changes in ARO transgenic mice MGs. Increased levels of expression of ERα and β in transgenic MG tissue suggest that both ER isoforms may play an important role in the mediation of the estrogenic response resulting in MG preneoplastic development in this transgenic animal model. In addition, we have reported an increase in progesterone receptor (PR) levels in the transgenic MGs (13). The expression of both ERα and β and PR in the absence of ovarian E, and their upregulation underscore the relevance of ARO overexpression in increased estrogenic activity, and the regulation of these receptors in MG tissue.

We have identified several growth factor genes with altered expression in the ARO transgenic MG (Figure 3). Our results support the notion that estrogenic action results in the overexpression of cellular factors responsible for the suppression of apoptosis and promotion of cellular proliferation during tumorigenesis, including cyclin D1 and E1, and PCNA (13). Interestingly, the levels of EGFR expression and its ligand, EGF, were lower in the transgenic MG tissue. Yarden, *et al.* (6) have shown that estrogenic action is bimodal in respect to the regulation of EGFR expression. The first phase is a transient induction that is not dependent on protein synthesis, while the second phase is repression dependent on protein synthesis. The latter observation suggests that the synthesis of an additional factor is required for E repression of EGFR (6). Surprisingly, TGFβ1 showed increased overexpression in the transgenic MG as compared to the non-transgenic control. TGFβ1 is known to play a negative role in cellular proliferation by promoting cell cycle arrest at G1 (14, 15). However, recent studies (16) suggest that TGFβ1 is a pleiotropic protein that exhibits positive effects on cellular growth and may, under normal conditions, be involved in maintaining homeostasis. Recent evidence has shown that the cyclin-dependent kinase inhibitor (CDKI), p27, induced by TGFβ1, prevents the progression of the cell cycle past G1 by interfering with the activity of cyclin/CDK complexes (17). Consistent with the previously observed induction of p27 by TGFβ1, our results also show an increase in p27 expression levels in the MGs of ARO transgenic animals. Despite the increase in the CDKI p27 expression, the observed net effect exerted by ARO overexpression in our model results in abnormal cellular proliferation in the transgenic MG tissue.

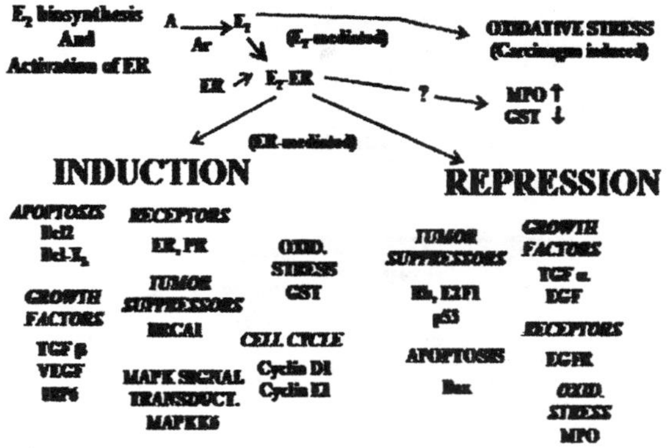

Figure 3. Schematic representation of various biochemical changes in ARO transgenic females. Increased estrogenic activity mediated by ARO overexpression in MGs of transgenic females results in a change in the regulation of certain E-dependent genes.

Estrogenic stimulation of BC cell lines results in increased cellular proliferation, and expression of cyclin D1 and CDKI p21 as well as sequestration of p27 (and/or p21, depending on the cellular context) away from cyclin E/CDK2 complexes, giving rise to the release of the inhibitory effects of cyclin E/CDK2, phosphorylation of Rb, and its inactivation (17, 18). Our data demonstrate that the ratio of p27 bound to cyclin D1 to that bound to cyclin E is several folds higher in the transgenic MG as compared to that of the non-transgenic tissue. Taken together, our data are consistent with the evidence (13) suggesting that increased cyclin D1 levels sequester more p27 from cyclinE/CDK2, leading to increased cellular proliferation. In addition to the possible regulation of Rb phosphorylation, the expression of Rb mRNA is downregulated in the ARO transgenic mice MG. This may contribute to the reduction of Rb activity and the inhibition of cell cycle progression. Also, the expression of p53 is decreased in the ARO transgenic MG. However, increase in BRCA1 expression in ARO mice is consistent with recent studies (19, 20).

Aromatase Transgenic Mice are Susceptible to Carcinogens

To test the hypothesis that tissue E-induced preneoplastic changes induced may be susceptible to carcinogens like DMBA, and that exposure to these carcinogens may result in acceleration and/or increase in the incidence of BC (21). Exposure to a single sub-threshold concentrations of DMBA (0.5 µg/mouse) resulted in development of MG tumors in 25-1 of the transgenic animals and non in wild type mice. All the DMBA-exposed transgenic females had microscopic evidence of tumor formation/neoplastic progression. When the ARO transgenic female animals were exposed to a higher DMBA dose (1.0 µg/mouse/week/4 weeks), more than 50% developed palpable MG tumors within 4.0 mo, while 100% show microscopic evidence of tumor formation within the same time period, compared to non-transgenic littermates (Table 1). Increased EGFR expression and its ligands, most notably TGFα, in DMBA-induced MG tumors, suggest a synergistic role of these growth factors in tumor progression and consistent with other studies showing the involvement of increased levels of EGFR and its ligands in BC (23, 24).

Table 1. Tumor Incidence in Aromatase Transgenic Mice After DMBA Exposure.

Genotype	Palpable Tumors[1,2]	Micro-scopic Tumors[1,2]	Adeno-carcinoma[1,2]	Papilloma/ Adeno-carcinoma[1,2]	DCIS/ Hyper-plasia[1,2]
Wild type (n = 25)	0	0	0	0	17
Aromatase (n = 16)	60	100	45	30	100

[1] 4.0 mo after the last dose of DMBA exposure.
[2] % number of mice with tumors and tumor types

Previous data (6, 13, 22) suggest that the ERα has a negative regulatory effect on EGFR expression and its ligands. These results are consistent with the clinical observations that loss of ERα in BC results in the upregulation of growth factors like TGFα, leading to a more aggressive progression of the disease. We e observed changes in ERα expression between DMBA-treated and untreated ARO transgenic mice. These findings exclude the possibility that the increased TGFα is due to the downregulation of ERα. One possible explanation is the alteration of other factors involved in ERα-mediated regulation, leading to the loss of ERα negative control on TGFα expression. Another explanation to consider is that the levels of E_2, and consequently activated E_2-bound ER, may be possibly decreased due to the metabolic conversion of E_2 to catechols and OH-Es as part of the carcinogenic process.

High-doses of exogenous Es are associated with experimental BC (25). Exogenous Es significantly enhance or promote the carcinogenic effects of other chemical carcinogens, such as DMBA, and consequently, they may enhance MG carcinogenicity. This synergy between E and DMBA in the development of MG tumors was also observed in these studies. Tumors were observed only in DMBA-treated ARO-overexpressing mice, but none in similarly treated non-transgenic or in untreated ARO transgenic mice. Only preneoplastic changes, *i.e.,* ductal hyperplasia, were observed in the untreated ARO transgenic mice (12, 13). These observations point to the involvement of ARO overexpression, and the resulting increase in E levels, in the development of MG hyperplasia. DMBA may act at this stage, with a larger population of epithelial cells susceptible to mutagenesis, further enhancing the observed estrogenic effects in the DMBA-treated ARO-overexpressing mice. On the other hand, DMBA may act as the initiator in the MGs of the treated ARO-overexpressing mice, and that the continuously elevated E levels promote tumor development in these mice.

Recently, several lines of evidence (26-28) suggested a role for E_2 (non ER-mediated) and its metabolites in genotoxic effects leading to carcinogenesis, which includes oxidative stress and adduct formation. Since no changes were observed in ERα expression in the DMBA-treated ARO transgenic mice, it is possible that the observed changes were mediated by the metabolic conversion of E_2 to its quinone derivatives, resulting in mutagenic DNA adduct formation (26-28). The increase in tissue E_2 levels alone in the untreated ARO transgenic mice does not seem to have carcinogenic effects. However, through these mechanisms, E metabolism may result in increased DMBA carcinogenicity (29) in the treated ARO transgenic mice. DMBA is a ligand for the aryl hydrocarbon receptor (AhR), a transcription factor that induces the expression of the oxidative P450 enzymes, CYP1-A1 and -B1 (30). DMBA treatment enhances MG AhR and CYP1B1 expression in a rat model system (30). Conversely, AhR is drastically increased in the DMBA-treated ARO transgenic mice as compared to the other mice groups. These studies suggest that E may act as a mitogenic, thus providing an expanded target population for subsequent initiating events.

Aromatase Overexpression Results in Gynecomastia and Leydig Tumors in Male Transgenic Mice

Overexpression of ARO in male transgenic mice results in increased MG epithelial growth as early as 3 mo of age with well-organized ductal structures and lobulo-alveolar growth. We did not observe further progression of hyperplastic and dysplastic changes in the ARO transgenic mice with age (31). In addition, our data demonstrated that the overexpression of ARO plays a significant role in the formation of Leydig cell tumors, and that these cells are targets for estrogenic action and are involved in E-mediated tumorigenesis (32). The presence of ERs in human testicular Leydig cell tumors indicate that these cells are the source of E in both rodents and human testis (33-36). Our studies show that the level of ER expression and ARO are higher in the testicular tissue of these mice compared to that of non-transgenic mice, and that the serum E_2 levels are significantly higher in transgenic mice vs non-transgenic littermates. Therefore, we suggest that an enhanced ER phenotype of Leydig cells and increased peripheral aromatization may contribute to Leydig cell tumorigenesis. The studies described above demonstrate that tissue Es play a direct role inducing gynecomastia and testicular cancer in male ARO transgenic mice and are consistent with recent observations with ARO knock out mice (37).

Effect of Continuos Estrogen on TGFα Expression in MMTV-TGFα x MMTV-Aromatase Transgenic Cross

Whether synergistic action of E with growth factors leads to MG tumors. To test this hypothesis, we crossed the ARO transgenic mice with a TGFα-overexpressing transgenic strain (22). We expected to observe synergistic interactions in the ARO x TGFα cross strain leading to enhanced changes that would lead to increased MG tumor formation in these mice. However, in the ARO x TGFα double transgenic mice, we did not observe any MG tumor formation or histological changes beyond those we have observed in the parental ARO strain, even at age > 2 years. Our previous results have shown a decrease of EGFR and TGFα expression in the MG of ARO transgenic mice as compared to the nontransgenic MG (13). Consistently, other studies have suggested that EGFR expression, the receptor for TGFα, is negatively regulated by Es via the ligand-activated ERα, and that the loss of ERα correlates with an increase in EGFR expression in BC (6). The decrease in EGFR and TGFα expression in the ARO x TGFα cross strain as compared to the TGFα parental strain is in agreement with our previous data (13), suggesting a negative regulation of EGFR expression as a result of ARO overexpression. Our previous results have shown that the increase in ERα expression in the ARO MG corresponds to an increase in the expression of cyclin D1 and PCNA, cell cycle progression and proliferation markers, respectively. The observation that cyclin D1 protein levels are drastically decreased while the levels of the inhibitor p27 are unchanged in the double transgenic animals, as compared to the parental strains, suggests that more p27 molecules can bind and inactivate the cyclin E1-CDK2 complexes (13, 14, 38, 39). Our results demonstrate that continuous MG E exposure, due to ARO overexpression, counteracts the mitogenic effects of MG TGFα overexpression of ARO x TGFα double transgenic mice, presumably due to the negative regulation of EGFR. Interestingly, ERα, cyclin D1, and PCNA levels are also decreased in the double transgenic strain. This is probably due to an antagonistic relationship between E and TGFα. The decrease expression of these cellular proliferation genes may account for the absence of MG tumor formation of the ARO x TGFα cross.

Use of an Aromatase Inhibitor as a Chemopreventive Agent

AIs are effective blockers of this enzyme, and may serve as therapeutic agents for hormone-dependent BC (10, 11). Our studies on the efficacy of AIs to abrogate ARO-induced preneoplastic/neoplastic changes in transgenic mice indicated that AIs could completely abrogate or reduce ARO-induced hyperplastic and other changes in breast tissue (41-43). We have also shown that low doses of letrozole (0.25 or 0.5 μg/day/animal), completely abrogate ARO-induced preneoplastic changes. These doses did not affect the uterus, the ovary, or E_2 and FSH circulating levels. No difference was found in the expression of ERα and PR in the uterus of the treated animals. The data suggest that very low doses of letrozole may eliminate E-mediated BC in transgenic mice without affecting normal uterine and ovarian functions. Subsequent studies (44) have demonstrated that letrozole, at low doses, is effective in abrogating ARO-induced breast hyperplasia. Moreover, mice were physiologically normal after treatment as indicated by normal pregnancy and lactation, and lack of birth defects in the pups born to letrozole-treated mothers.

Although, letrozole inhibited ARO-induced MG hyperplasia in the ARO transgenic mice, the presence of some MG proliferation in aged animals, previously treated with letrozole, suggests that inhibition of MG growth is not a permanent. The slight increase in ERα expression, and other proliferation markers in aged animals were in agreement with the morphological changes expected. Furthermore, the decreased ductal branching in letrozole-treated mice and continued low levels of PR is consistent with the importance of ERα/PR in MG ductal branching (45).

Concluding Remarks

Studies summarized herein demonstrate that overexpression of ARO leads to in increased tissue estrogenic activity, induction of MG hyperplastic and dysplastic lesions, gynecomastia, and testicular cancer in male ARO transgenic mice. The preneoplastic lesions induced due to increased tissue estrogenic activity are susceptible to carcinogens. The ARO overexpression-induced MG changes were inhibited with very low letrazole doses, an ARO inhibitor without any effect on normal physiology. The potential clinical importance of this intracrine growth support may provide future clinical and laboratory investigations.

References

1. Harris JR, Lippman ME, Vernonesi U, et al (1992) Breast Cancer. N Engl J Med 327:319-328; 390-398; and 473-480.
2. Santen RJ, Santner SJ, Pauley RJ, et al (1997) Estrogen production via the aromatase enzyme in breast cancer carcinoma: which cell type is responsible? J Steroid Biochem Mol Biol 61:267-271.
3. Anderson E, Clarke RB, Howell A (1998) Estrogen responsiveness and control in normal human breast proliferation. J Mammary Gland Biol Neoplasia 3:23-35.
4. Russo IH, Russo J (1998) Role of hormones in mammary cancer initiation and progression. J Mammary Gland Biol Neoplasia 3:49-60.
5. Schroeder JA, Lee DC (1997) Transgenic mice reveal roles for TGFα and EGF receptor in mammary gland development and neoplasia. J Mammary Gland Biol Neoplasia 2:119-129.
6. Yarden RI, Lauber AH, El-Ashry D, et al (1996) Bimodal regulation of epidermal growth factor receptor by estrogen in breast cancer cells. Endocrinology 137:2739-2747.
7. Nakamura J, Savinov A, Lu Q, et al (1996) Estrogen regulated vascular endothelial growth/permeability factor expression in 7,12-dimethylbenz(a)anthracene-induced rat mammary tumors. Endocrinology 137:5589-5596.

8. Sicinski P, Donaher JL, Parker SB, et al (1995) Cyclin D1 provides a link between development and oncogenesis in the retina and breast. Cell 82:621-630.
9. Siiteri PK (1982) Review of studies on estrogen biosynthesis in the human. Cancer Res 42:3269s-3275s.
10. Brodie AMH, Santen RJ (1994) Aromatase and its inhibitors in breast cancer--overview and perspective. Breast Cancer Res Treat 30:1-6.
11. Miller WR, O'Neill J (1976) Oestradiol synthesis from C19 steroids by human breast cancers. Br J Cancer 33:116-118.
12. Tekmal RR, Ramachandra N, Gubba S, et al (1996) Overexpression of int-5/aromatase in mammary glands of transgenic mice results in the induction of hyperplasia and nuclear abnormalities. Cancer Res 56:3180-3185.
13. Kirma N, Gill K, Mandava U, et al (2001) Overexpression of aromatase leads to hyperplasia and changes in the expression of genes involved in apoptosis, cell cycle, growth and tumor suppressor functions in the mammary glands of transgenic. Cancer Res 61:1910-1918.
14. Polyak K, Kato J, Solomon MJ, et al (1994) $p27^{kip1}$, a cyclin-CdK inhibitor, links transforming growth factor-c and contact inhibition to cell cycle arrest. Genes Develop 8:9-22.
15. Toyoshima H, Hunter T (1994) p27, a novel inhibitor of G1 cyclin-CdK protein kinase activity, is related to p21. Cell 78:67-74.
16. Hartsough MT, Mulder KM (1997) Transforming growth factor-β signaling in epithelial cells. Pharmacol Ther 75:21-41.
17. Vidal A, Koff A (2000) Cell-cycle inhibitors: three families united by a common cause. Gene 247:1-15.
18. Prall OW, Sarcevic P, Musgrove EA, et al (1997) Estrogen-induced activation of Cdk4 and Cdk2 during G1-S phase progression is accompanied by increased Cyclin D1 expression and decreased Cyclin-dependent kinase inhibitor association with Cyclin E-Cdk2. J Biol Chem 272:10882-10894.
19. Romagnolo D, Annab AA, Thompson TE, et al (1998) Estrogen upregulation of BRCA1 expression with no effect of localization. Mol Carcinog 22:102-109.
20. Bennet LM, McAllister KA, Malphurs J, et al (2000) Mice heterozygous for a BRCA1 and BRCA2 mutation display distinct mammary gland and ovarian phenotypes in response to Diethylstilbestrol. Cancer Res 60:3461-3469.
21. Keshava N, Mandava U, Kirma N, et al (2001) Acceleration of mammary neoplasia in aromatase transgenic mice by 7, 12-dimethylbenz(a)anthracene. Cancer Lett 167:125-133.
22. Kirma N, Mandava U, Gill K, et al (2001) Synergistic action of growth factors with breast tissue estrogen: Down regulation TFG, EGF, and EGFR in aromatase x TFG cross. J Steroid Biochem Mol Biol 78:419-426.

23. Sandgren EP, LuettekeNC, Palmiter RD, et al (1990) Overexpression of TGFα in transgenic mice: induction of epithelial hyperplasia, pancreatic metaplasia, and carcinoma of the breast. Cell 61:1121-1135.
24. Fox SB, Harris AL (1997) The epidermal growth factor receptor in breast cancer. J Mammary Gland Biol Neopl 2:131-141.
25. Lupulescu A (1983) Hormones and Carcinogenesis. New York: Praeger Scientific, 89-144.
26. Yager JD (2000) Endogenous estrogens as carcinogens through metabolic conversion. J Natl Cancer Inst Monogr 27:67-73.
27. Cavlieri E, Frenkel K, Liehr JG, et al (2000) Estrogens as endogenous genotoxic agents-DNA adducts and mutations. J Natl Cancer Inst Monogr 27:75-94.
28. Jefcoate CR, Liehr JG, Santen R, et al (2000) Tissue-specific synthesis and oxidative metabolism of estrogen. J Natl Cancer Inst Monogr 27:95-112.
29. Todorovic R, Ariese F, Devanesan P, et al (1997) Determination of benzo[a]anthracene-DNA adducts formed in rat mammary glands. Chem Res Toxicol 10:941-947.
30. Trombino AF, Near RI, Matulka RA, et al (2000) Expression of aryl hydrocarbon receptor/transcription factor (AhR) and AhR-regulated *CYP*1 gene transcripts in a rat model of mammary tumorigenesis. Breast Cancer Res Treat 63:117-131.
31. Gill K, Kirma N, Tekmal RR (2001) Overexpression of aromatase in transgenic male mice results in gynecomastia and other biochemical changes in mammary glands. J Steroid Biochem Mol Biol 77:13-18.
32. Fowler K, Gill K, Kirma N, et al (2000) Overexpression of aromatase leads to development of testicular Leydig cell tumors: An animal model for hormone-mediated testicular cancer. Am J Pathol 156:347-353.
33. Hawkins A, Inkster S (1993) Aromatase in the human testis. J Steroid Biochem Mol Biol 44:549-555.
34. Payne AH, Perkins LM, Georgiou M, et al (1987) Intratesticular site of aromatase activity and possible function of testicular estradiol. Steroids 50:435-448.
35. Berensztein E, Belgorosky A, de Davila MT, et al (1995) Testicular steroid biosynthesis in a boy with large cell calcifying sertoli cell tumor producing prepubertal gynecomastia. Steroids 60:220-225.
36. Vakensi P, Coussieu C, Pauwles A, et al (1987) Feminizing Leydig cell tumor: endocrine and incubation studies. J Endocrinol Invest 10:187-193.
37. Fisher CR, Graves KH, Parlow AF, et al (1998) Characterization of mice deficient in aromatase (ArKO) because of targeted disruption of the cyp19 gene. Proc Natl Acad Sci USA 95:6965-6970.
38. Pestell RG, Albanese C, Reutens AT, et al (1999) The cyclins and cyclin-dependent kinase inhibitors in hormonal regulation of proliferation and differentiation. Endocrine Reviews 20:501-534.

39. Prall OWJ, Sarcevic P, Musgrove EA, et al (1997) Estrogen-induced activation of Cdk4 and Cdk2 during G1-S phase progression is accompanied by increased Cyclin D1 expression and decreased Cyclin-dependent kinase inhibitor association with Cyclin E-CDK2. J Biol Chem 272:10882-10894.
40. Tekmal RR, Gill K, Kirma N, et al (1999) Aromatase overexpression and breast hyperplasia, an in vivo model: Continued overexpression of aromatase is sufficient to hyperplasia maintain without circulating estrogens, and use of aromatase inhibitors abrogate these preneoplastic changes in mammary glands. Endocrine-Related Caner 6:307-314.
41. Yue W, Wang J, Savinov A, et al (1995) Effect of aromatase inhibitors on growth of mammary tumors in nude mouse model. Cancer Res 55:3073-3077.
42. Yue W, Zhou D, Chen S, et al (1994) A new nude mouse model for postmenopausal breast cancer using with MCF-7 cells transfected with the human aromatase gene, Cancer Res 54:5092-5095.
43. Mandava U, Kirma N, Tekmal RR (2001) Aromatase overexpression transgenic mice model: Cell type specific expression and use of letrozole to abrogate mammary hyperplasia without affecting normal physiology. J Steroid Biochem Mol Biol 79:27-34.
44. Luthra R, Kirma N, Jones J, et al (2003) Use of letrozole as a chemopreventive agent in aromatase overexpressing transgenic mice. J Steroid Biochem Mol Biol 86:461-467.
45. Shyamala G, Chou YC, Louie SG, et al (2002) Cellular expression of estrogen and progesterone receptors in mammary glands: regulation by hormones, development and aging. J Steroid Biochem Mol Biol 80:137-148.

11

Prostate Epithelial Carcinogenesis: Putative and Controversial Precursor Lesions

Cristina Magi-Galluzzi and Angelo M. De Marzo

Introduction

Prostate cancer (PCA) is the most common non-skin malignancy diagnosed in men in the USA and the second leading cause of cancer deaths among North American and Western European men. The incidence of PCA and the rate of death due to the disease increase with age. Less than 1% cases of PCA are diagnosed under the age of 40, although small PCAs have been detected in up to 29% men 30 to 40 years of age in autopsy series (1). The incidence and mortality rates are highest among African American men, being 2.0-fold higher than in Caucasian Americans, whereas lower rates are characteristic of the Asian population (2).

Etiologic factors associated with PCA are varied and comprise both genetic and environmental influences. Among the genetic factors aging, family clustering, race, and hormonal influences seem to play a major role. A diet high in animal fat and red meat and poor in fruits and vegetables, and the preventive use of antioxidants associated with a high intake of tomatoes are also critical factors (3, 4).

Accumulating data suggests that normal and neoplastic prostate cells may be subjected to multiple genome-damaging stresses, and that both diet and male sex steroids may modulate the level of threatening insults. A model of carcinogenesis and progression similar to colon cancer has been proposed for PCA (Figure 1) (5). This model predicts multiple steps in the process from normal prostatic epithelium to invasive cancer through the intermediate step of prostatic intraepithelial neoplasia (PIN) and/or proliferating intermediate luminal cells. The most important feature of such a model is the gradual accumulation of changes in critical genes rather than their order of occurrence.

PIN, in particular the high-grade form, has been extensively studied and considered as the major morphological identifiable putative precursor of PCA (6, 7). However, it is likely that PCA might also arise from precursor lesions other than PIN, some not yet identifiable with certainty at the present time. PCA develops in two different regions of the gland, with the majority of the lesions (approximately 80%) found in the peripheral zone (PZ), and most of the remaining cancers in the region surround the peri-urethral region, termed the transition zone (TZ).

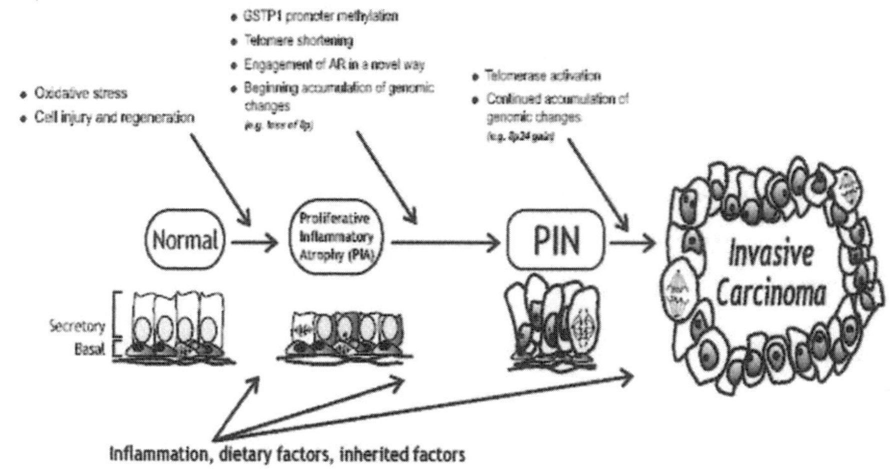

Figure 1. The molecular pathogenesis of prostate cancer.

Prostatic Intraepithelial Neoplasia (PIN)

PIN represents the neoplastic transformation of the epithelial lining of the prostatic ducts and acini without invasion of the basal layer (6). In high-grade PIN (HGPIN), the glandular units of the prostate are lined by cytologically malignant, enlarged epithelial cells with a high nuclear cytoplasmic ratio. HGPIN has been extensively studied and is thought to be the precursor lesion of PCA, particularly for the lesions occurring in the PZ.

PIN and PCA foci have similar genetic changes. In early carcinoma lesions, McNeal has demonstrated that foci of HGPIN are often directly merging with cancer (8). This supports the hypothesis that PIN is a likely precursor lesion to PCA. The concordance of DNA content between PIN and PCA (9); the over-expression of the oncoprotein Bcl-2 (10), the expression of PCNA, MIB1, AgNOR (11, 12) and androgen receptor (AR) (13, 14) similarly to PCA are all supportive of the above hypothesis.

α-Methylacyl-CoA Racemase (AMACR) or protein P504S is an enzyme involved in β-oxidation of fatty acids. It is over-expressed in PCA (15), and in its putative precursor lesion PIN (15-17), further supporting the link between the two entities. Glutathione S-transferase (GSTP1) is an inducible phase II detoxifying enzyme for reactive oxygen species and organic electrophiles. It is inactivated by promoter hyper-methylation in human (18-20). GSTP1 promoter hyper-methylation occurs in at least 70% of HGPIN (18, 21), and in 90% of PCA cases. GSTP1 is involved in protecting cells against carcinogens and is the most somatic DNA alteration in PCA and PIN. It could predispose PIN cells to further mutations.

The most common genetic alterations found in HGPIN are the loss of sequences from chromosome 8p and other chromosome arms (22), with less frequent gain of sequences from chromosome 8q. Inactivation of tumor suppressor genes and over-expression of oncogenes in those regions may be important for the initiation and progression of PCA.

In 1941, McClintock suggested that telomere dysfunction is one potential mechanism of chromosomal instability (23). Telomeres are specialized structures that cap the ends of linear chromosomes, essential for maintaining chromosomal stability. Natural shortening with each cell division serves as a mitotic clock. Shortening has been hypothesized as a protective mechanism against proliferation. As compared with normal prostate tissue, telomere shortening had previously been reported in PCA (24-26). More recently telomere shortening was shown to occur very commonly in HGPIN (27, 28). Significant telomere shortening is associated with increased levels of chromosomal aberration. These findings support the possibility that telomere erosion may be an important feature in PCA oncogenesis.

Proliferative Inflammatory Atrophy (PIA)

Since chronic inflammation of long-standing duration has been linked to the development of carcinomas in several organ systems, such as liver, stomach and esophagus (29, 30), inflammatory cell-mediated oxidant stress may be a key pathogenetic mechanism driving PCA (31). In prostate, chronic inflammation is associated with focal atrophy.

Atrophy of the prostate is identified as a reduction in the volume of pre-existing glands and stroma and can be divided in two major patterns: diffuse and focal. Diffuse atrophy results from a decrease in circulating androgens and involves the entire prostate in a relatively uniform manner (32). In contrast, focal atrophy is not related to decreasing circulating androgens, and it occurs as focal areas of atrophic epithelium within a background of surrounding normal appearing non-atrophic epithelium. Franks (33) indicated that focal prostatic atrophy lesions occur predominantly in the 'outer portion of the prostate' referred also as PZ by McNeal (34). These lesions increase in frequency with advancing age. Others confirmed these findings (35-38). Prior to the widespread study of PIN, various atrophic lesions were described as potential PCA precursors. Franks in 1954, and Liavag in 1968 also suggested an association between PCA and atrophy (33, 37). However, other authors did not find an association between PCA and atrophy (1, 4, 5, 44). While most focal prostatic atrophy lesions have been considered to be quiescent (46), cells in some of these lesions appear proliferative (37, 41, 42). Ruska, *et al.* recently demonstrated that although there is no increase in the apoptotic index, between benign non-atrophic epithelium and focal prostatic atrophy, atrophy lesions exhibit a markedly increased immunohistochemical staining index for the proliferation marker Ki-67 (42). These findings support the hypothesis that focal atrophy represents either a *de novo* proliferative lesion or a regenerative lesion resulting from replacement of cellular loss (43).

Since atrophic lesions are proliferative and usually associated with an inflammatory component, predominantly chronic, they have been termed proliferative inflammatory atrophy (PIA) (43). The term designates discrete foci of proliferative glandular epithelium with the morphologic appearance of simple atrophy or post-atrophic hyperplasia occurring in association with inflammation. The key features of these lesions are the presence of two distinct cell layers, mononuclear and/or polymorphonuclear inflammatory cells in both the epithelial and stromal compartments, and stromal atrophy with variable amounts of fibrosis. Atrophic lesions occur predominantly in the PZ, but can also be present in the TZ, but rarely in the central zone. Morphologic transitions within the same acinar/ductal unit between HGPIN and PIA occur quite frequently. PIA is often found near small cancers, and at times have genetic alterations similar to those detected in high grade PIN and adenocarcinomas (5, 44).

Most cell divisions in the normal human prostate epithelium occur in the basal cell compartment (45-47). Yet HGPIN, and adenocarcinoma cells possess phenotypic and morphological features of secretory cells. The cell proliferation has been shifted up from the basal into the secretory compartment in high-grade PIN and in PCA (45, 47). Based on these findings, as well as patterns of cytokeratin expression from two different groups, it has been postulated that the prostatic cell type that is the target of neoplastic transformation is an intermediate cell, with some features of basal cells and some of secretory cells (43, 48, 49). To better elucidate the cell types and to explore the feasibility that focal atrophy may be related to PCA and HGPIN, the expression of both basal cell-specific and secretory cell-specific markers together with few molecular markers implicated in PCA have been examined (43, 48).

$p27^{KIP1}$ is a cyclin-dependent kinase inhibitor whose expression is reduced in the majority of PCA and in HGPIN (50, 51). $p27^{KIP1}$ is down-regulated in prostatic atrophy, consistent with its proposed role as a repressor of prostatic epithelial cell proliferation (52). The increased expression of Bcl-2 in atrophy as compared to the adjacent normal epithelium, is also consistent with the observed very low level of apoptosis (42). A striking increase in the expression of GSTP1 was found in many of the atrophic cells, suggesting a stress-induced response. The atrophic luminal cells consistently stained intensely with the monoclonal antibody CAM5.2 that recognizes cytokeratin 8 and 18. These proteins are expressed in normal secretory cells, with somewhat weaker expression in basal cells. Many of the atrophic luminal cells also stain positively for keratin 5, which is distinct from normal secretory luminal cells which are negative for this marker (48). Many of the luminal atrophic cells also stain positively for nuclear AR, with the majority of the cells staining weakly, however some the cells show intense staining. As further evidence of partial secretory cell differentiation in PIA, positive staining for PSA and prostate specific acid phosphatase (PSAP) was found in the luminal layer of cells. although the staining of these cells was generally weak in intensity as compared to surrounding normal gland cells (43). The atrophic lesions showed a variable, yet marked, increase in staining for proliferation markers as compared to normal appearing epithelium. In normal, non-atrophic prostatic epithelium, GSTP1 was expressed predominantly in basal cells. The data related to the elevated levels of GSTP1 in proliferative atrophy is highly suggestive of a stress-induced response in these cells. Some of the proliferative secretory type cells that lack expression of GSTP1 may be targets for genetic alterations and neoplastic transformation. In view of these data, it has been suggested that PIA may indeed give rise to PCA directly or indirectly via PIN development. Different findings provide supportive evidence for the hypothesis that PIA may represent a PIN precursor. The shift in proliferation occurring in atrophy, the phenotype of many cells in PIA being consistent with that of an immature secretory type cells similar to HGPIN and PCA, and the fact that PIA, HGPIN, and PCA all occur with high prevalence in the PZ, but low in the central zone of the prostate support this hypothesis. Moreover, in randomly sampled PIA lesion from radical prostatectomy patients, approximately 30% of the cases show areas of atrophy merging directly with areas of HGPIN within the same glands (44). In view of these findings we have proposed to integrate the atrophy and PIN hypothesis into PCA. The interpretation of atrophy together with additional studies using animal models of inflammation and atrophy may help to further understand the progression of PCA.

Atypical Adenomatous Hyperplasia (AAH) or Adenosis

The data that exist on the prevalence of atypical adenomatous hyperplasia (AAH) also known as adenosis (53, 54) is controversial. The vast majority of adenosis foci are found in the TZ. They are usually associated with benign prostatic hyperplasia (BPH) nodules and may share some morphologic features with well-differentiated carcinoma (55). The prevalence of AAH in the literature is highly variable (from 1.6% to 19.6%) (56). A study of a series of needle biopsy specimens reported a lower prevalence, approximately 0.8% (57). Data reporting higher numbers of TZ cancers in African Americans than in Caucasians may imply a higher prevalence of adenosis in the former (58). The relationship between adenosis and PCA remains somewhat disputed and ill defined, although adenosis lesions may at times express elevated levels of racemase (59) and show genetic alterations characteristic of PCA (60, 61).

Benign Prostatic Hyperplasia (BPH)

Interestingly, the virtually ubiquitous process of BPH originates in the TZ of the prostate (62), and approximately 20% of all PCA occur in the TZ. Adenocarcinomas originating in the TZ are usually well-differentiated tumors with Gleason score generally lower than tumors from the PZ (63), and with architectural features that closely resemble foci of BPH. Over the years, the findings of BPH and

PCA together in standard transurethral resection of the prostate (TURP) specimens has led to speculations that carcinomas may arise from hyperplastic lesions (64-66). Leav, *et al.* recently reported the possibility to define areas of transition from hyperplasia to carcinoma in six BPH nodules using AMACR (P504S), p63, or 34βE12 antibodies (67). They reported enhanced AMACR expression in benign glands within cancer-containing nodules, as well as in BPH lesions adjacent to carcinomas and suggested that the up-regulation of the enzyme may precede morphological evidence of neoplastic transformation. Further studies are needed to confirm the possible relationship between BPH and PCA.

Conclusions

Rapid advances in technology provide a good platform for the systemic cataloging and characterization of the normal and cancerous phenotype and underlying genotype of PCA. This should set the stage for the development of new prognostic and therapeutic strategies. The understanding of genetic events which occur during the progression of PIN and other preneoplastic lesions to PCA would be useful for prevention, early detection, and treatment of this common disease.

References

1. Sakr WA, Grignon DJ, Crissman JD, et al (1994) High grade prostatic intraepithelial neoplasia (HGPIN) and prostatic adenocarcinoma between the ages of 20-69: an autopsy study of 249 cases. In Vivo 8:439-443.
2. Hsing AW, Deng J, Sesterhenn IA, et al (2000) Body size and prostate cancer: a population-based case-control study in China. Cancer Epidemiol Biomarkers Prev 9:1335-1341.
3. Gann PH, Ma J, Giovannucci E, et al (1999) Lower prostate cancer risk in men with elevated plasma lycopene levels: results of a prospective analysis. Cancer Res 59:1225-1230.
4. Giovannucci E, Rimm EB, Colditz GA, et al (1993) A prospective study of dietary fat and risk of prostate cancer. J Natl Cancer Inst 85:1571-1579.
5. De Marzo AM, Putzi MJ, Nelson WG (2001) New concepts in the pathology of prostatic epithelial carcinogenesis. Urology 57:103-114.
6. Bostwick DG (1989) Prostatic intraepithelial neoplasia (PIN). Urology 34:16-22.
7. Bostwick DG (2000) Prostatic intraepithelial neoplasia. Curr Urol Rep 1:65-70.
8. McNeal JE, Villers A, Redwine EA, et al (1991) Microcarcinoma in the prostate: its association with duct-acinar dysplasia. Hum Pathol 22:644-652.

9. Weinberg DS, Weidner N (1993) Concordance of DNA content between prostatic intraepithelial neoplasia and concomitant invasive carcinoma. Evidence that prostatic intraepithelial neoplasia is a precursor of invasive prostatic carcinoma. Arch Pathol Lab Med 117:1132-1137.
10. Colombel M, Symmans F, Gil S, et al (1993) Detection of the apoptosis-suppressing oncoprotein bcl-2 in hormone-refractory human prostate cancers. Am J Pathol 143:390-400.
11. Helpap B, Riede C (1995) Nucleolar and AgNOR-analysis of prostatic intraepithelial neoplasia (PIN), atypical adenomatous hyperplasia (AAH) and prostatic carcinoma. Pathol Res Pract 191:381-390.
12. Magi-Galluzzi C, Murphy M, Cangi MG, et al (1998) Proliferation, apoptosis and cell cycle regulation in prostatic carcinogenesis. Anal Quant Cytol Histol 20:343-350.
13. Magi-Galluzzi C, Montironi R, Prete E, et al (1996) Effect of microwave oven heating times on androgen receptor antigen retrieval from paraffin-embedded prostatic adenocarcinoma. Anticancer Res 16:2931-2936.
14. Magi-Galluzzi C, Xu X, Hlatky L, et al (1997) Heterogeneity of androgen receptor content in advanced prostate cancer. Mod Pathol 10:839-845.
15. Luo J, Zha S, Gage W, et al (2002) Alpha-methylacyl-CoA racemase: A new molecular marker for prostate cancer. Cancer Res 62:2220-2226.
16. Magi-Galluzzi C, Luo J, Isaacs WB, et al (2003) Alpha-methylacyl-coA racemase: a variably sensitive immunohistochemical marker for the diagnosis of small prostate cancer foci on needle biopsy. Am J Surg Pathol 27:1128-1133.
17. Rubin MA, Zhou M, Dhanasekaran SM, et al (2002) Alpha-methyl-acyl coenzyme A racemase as a tissue biomarker for prostate cancer. JAMA 287:1662-1670.
18. Brooks JD, Weinstein M, Lin X, et al (1998) CG island methylation changes near the GSTP1 gene in prostatic intraepithelial neoplasia. Cancer Epidemiol Biomarkers Prev 7:531-536.
19. Lee WH, Isaacs WB, Bova GS, et al (1997) CG island methylation changes near the GSTP1 gene in prostatic carcinoma cells detected using the polymerase chain reaction: a new prostate cancer biomarker. Cancer Epidemiol Biomarkers Prev 6:443-450.
20. Lee WH, Morton RA, Epstein JI, et al (1994) Cytidine methylation of regulatory sequences near the pi-class glutathione S-transferase gene accompanies human prostatic carcinogenesis. Proc Natl Acad Sci USA 91:11733-11737.
21. Nakayama M (2003) Hypermethylation of the human GSTP1 CpG island is present in a subset of proliferative inflammatory atrophy lesions but not normal or hyperplastic epithelium of the prostate: a detailed study using laser capture microdissection. Am J Pathol 163:923-933.

22. Qian J, Jenkins RB, Bostwick DG (1999) Genetic and chromosomal alterations in prostatic intraepithelial neoplasia and carcinoma detected by fluorescence in situ hybridization. Eur Urol 35:479-483.
23. McClintock B (1941) The stability of broken ends of chromosomes in Zea mays. Genetics 26:234-282.
24. Meeker AK, Garcia-Moreno B, Shortle D (1996) Contributions of the ionizable amino acids to the stability of staphylococcal nuclease. Biochemistry 35:6443-6449.
25. Meeker AK, Sommerfeld HJ, Coffey DS (1996) Telomerase is activated in the prostate and seminal vesicles of the castrated rat. Endocrinology 137:5743-5746.
26. Sommerfeld HJ, Meeker AK, Piatyszek MA, et al (1996) Telomerase activity: a prevalent marker of malignant human prostate tissue. Cancer Res 56:218-222.
27. Meeker AK, Hicks JL, Platz EA, et al (2002) Telomere shortening is an early somatic DNA alteration in human prostate tumorigenesis. Cancer Res 62:6405-6409.
28. Vukovic B, Park PC, Al-Maghrabi J, et al (2003) Evidence of multifocality of telomere erosion in high-grade prostatic intraepithelial neoplasia (HPIN) and concurrent carcinoma. Oncogene 22:1978-1987.
29. Bartsch H, Frank N (1996) Blocking the endogenous formation of N-nitroso compounds and related carcinogens. IARC Sci Publ, 189-201.
30. Weitzman SA, Gordon LI (1990) Inflammation and cancer: role of phagocyte-generated oxidants in carcinogenesis. Blood 76:655-663.
31. Nelson WG, DeMarzo AM, DeWeese TL (2001) The molecular pathogenesis of prostate cancer: focus on the earliest steps. Eur Urol 39 Suppl 4:8-11.
32. McNeal JE (1968) Regional morphology and pathology of the prostate. Am J Clin Pathol 49:347-357.
33. Franks LM (1954) Atrophy and hyperplasia in the prostate proper. J Pathol Bacteriol 68:617-621.
34. McNeal JE, Redwine EA, Freiha FS, et al (1988) Zonal distribution of prostatic adenocarcinoma. Correlation with histologic pattern and direction of spread. Am J Surg Pathol 12:897-906.
35. Billis A. 1998. Prostatic atrophy: An autopsy study of a histologic mimic of adenocarcinoma. Mod Pathol 11:47-54.
36. Gardner WA, Jr., Culberson DE (1987) Atrophy and proliferation in the young adult prostate. J Urol 137:53-56.
37. Liavag I (1968) Atrophy and regeneration in the pathogenesis of prostate carcinoma. Acta Pathol Microbiol Scan 73:338-350.
38. McNeal JE (1997) Prostate cancer volume. Am J Surg Pathol 21:1392-1393.

39. Anton RC, Kattan MW, Chakraborty S, et al (1999) Post atrophic hyperplasia of the prostate: lack of association with prostate cancer. Am J Surg Pathol 23:932-936.
40. Billis A, Luis AM (2003) Inflammatory Atrophy of the prostate. Arch Pathol Lab Med 127:840-844.
41. Feneley MR, Young MP, Chinyama C, et al (1996) Ki-67 expression in early prostate cancer and associated pathological lesions. J Clin Pathol 49:741-748.
42. Ruska KM, Sauvageot J, Epstein JI (1998) Histology and cellular kinetics of prostatic atrophy. Am J Surg Pathol 22:1073-1077.
43. De Marzo AM, Marchi VL, Epstein JI, et al (1999) Proliferative inflammatory atrophy of the prostate: implications for prostatic carcinogenesis. Am J Pathol 155:1985-1992.
44. Putzi MJ, De Marzo A (2000) Morphological transitions between proliferative inflammatory atrophy and high-grade prostatic intraepithelial neoplasia. Urology 56:828-832.
45. Bonkhoff H, Stein U, Remberger K (1994) The proliferative function of basal cells in the normal and hyperplastic human prostate. Prostate 22:114-118.
46. Giannulis I, Montironi R, Magi-Galluzzi C, et al (1993) Frequency and location of mitoses in prostatic intraepithelial neoplasia (PIN). Anticancer Res 13:2447-2452.
47. McNeal JE, Haillot O, Yemoto C (1995) Cell proliferation in dysplasia of the prostate: analysis by PCNA immunostaining. Prostate 27:258-268.
48. van Leenders GJ, Gage WR, Hicks JL et al (2003) Intermediate cells in human prostate epithelium are enriched in proliferative inflammatory atrophy. Am J Pathol 162:1529-1537.
49. Verhagen AP, Ramaekers FC, Aalders TW, et al (1992) Colocalization of basal and luminal cell-type cytokeratins in human prostate cancer. Cancer Res 52:6182-6187.
50. Cordon-Cardo C, Koff A, Drobnjak M, et al (1998) Distinct altered patterns of p27KIP1 gene expression in benign prostatic hyperplasia and prostatic carcinoma. J Natl Cancer Inst 90:1284-1291.
51. De Marzo AM, Meeker AK, Epstein JI, et al (1998) Prostate stem cell compartments: expression of the cell cycle inhibitor p27Kip1 in normal, hyperplastic, and neoplastic cells. Am J Pathol 153:911-919.
52. Lloyd RV, Erickson LA, Jin L, et al (1999) p27kip1: a multifunctional cyclin-dependent kinase inhibitor with prognostic significance in human cancers. Am J Pathol 154:313-323.
53. Brawn PN (1982) Adenosis of the prostate: a dysplastic lesion that can be confused with prostate adenocarcinoma. Cancer 49:826-833.
54. McNeal JE (1978) Origin and evolution of benign prostatic enlargement. Invest Urol 15:340-345.

55. Bostwick DG, Qian J (1995) Atypical adenomatous hyperplasia of the prostate. Relationship with carcinoma in 217 whole-mount radical prostatectomies. Am J Surg Pathol 19:506-518.
56. Gaudin PB, Epstein JI (1994) Adenosis of the prostate. Histologic features in transurethral resection specimens. Am J Surg Pathol 18:863-870.
57. Gaudin PB, Epstein JI (1995) Adenosis of the prostate. Histologic features in needle biopsy specimens. Am J Surg Pathol 19:737-747.
58. Pinto JT, Qiao C, Xing J, et al (2000) Alterations of prostate biomarker expression and testosterone utilization in human LNCaP prostatic carcinoma cells by garlic-derived S-allylmercaptocysteine. Prostate 45:304-314.
59. Yang XJ, Wu CL, Woda BA, et al (2002) Expression of alpha-methylacyl-CoA racemase (P504S) in atypical adenomatous hyperplasia of the prostate. Am J Surg Pathol 26:921-925.
60. Cheng L, Shan A, Cheville JC, et al (1998) Atypical adenomatous hyperplasia of the prostate: a premalignant lesion? Cancer Res 58:389-391.
61. Doll JA, Zhu X, Furman J, et al (1999) Genetic analysis of prostatic atypical adenomatous hyperplasia (adenosis). Am J Pathol 155:967-971.
62. McNeal JE (1965) Morphogenesis of prostatic carcinoma. Cancer 18:1659-1666.
63. Erbersdobler A (2002) Tumour grade, proliferation, apoptosis, microvessel density, p53, and bcl-2 in prostate cancers: differences between tumours located in the transition zone and in the peripheral zone. Eur Urol 41:40-46.
64. Armenian HK, Lilienfeld AM, Diamond EL, et al (1974) Relation between benign prostatic hyperplasia and cancer of the prostate. A prospective and retrospective study. Lancet 2:115-117.
65. Greenwald P, Kirmss V, Polan AK, et al (1974) Cancer of the prostate among men with benign prostatic hyperplasia. J Natl Cancer Inst 53:335-340.
66. Guess HA (2001) Benign prostatic hyperplasia and prostate cancer. Epidemiol Rev 23:152-158.
67. Leav I, McNeal JE, Ho SM, et al (2003) Alpha-methylacyl-CoA racemase (P504S) expression in evolving carcinomas within benign prostatic hyperplasia and in cancers of the transition zone. Hum Pathol 34:228-233.

PART 4. PROMOTION/PREVENTION OF HORMONE CANCERS

12

Estrogen Can Prevent Breast Cancer by Mimicking the Protective Effect of Pregnancy

Satyabrata Nandi, Raphael C. Guzman, Gudmundur Thordarson, and Lakshmanaswamy Rajkumar

Introduction

Breast cancer (BC) is the most common cancer in women worldwide. The estimated incidence of BC in the USA, in the year 2003 is approximately 211,300 with 40,200 deaths in women (1). In spite of better screening, earlier diagnosis, and improved therapeutic procedures, there is as yet no cure for metastatic BCs. BC incidence is very high in developed countries and is increasing rapidly in developing countries.

Attempts at prevention of BC are important areas of clinical and experimental investigations, and many different approaches are being used for such studies (2). Currently, there are only three clinical procedures for BC prevention. Women at high risk for BC can opt for risk reduction by bilateral mastectomy and/or oophorectomy, as well as chemoprevention using long-term treatment with the anti-estrogen, Tamoxifen (2). Neither of these procedures is universally acceptable, all have serious side effects, causing physical and psychosocial problems. These are severely anguishing choices for women.

There is a fourth choice for BC prevention in women. It involves undergoing multiple pregnancies along with prolonged nursing beginning at a relatively early age (3). Thus, pregnancy- and nursing-related protection from BC is a universal phenomenon, common to women of all ethnicities, worldwide. Although pregnancy and nursing are the only natural phenomena with few adverse side effects, it appears unlikely that many of the 21st Century women would choose this option. An acceptable alternative would be to determine the reasons for this protective effect, and to introduce it to nulliparous women, preferably by a short-term, safe, non-invasive procedure. Therefore, this universal protective effect of pregnancy is clearly of major consideration in devising experimental strategies for BC prevention.

Rats and mice (4-9) that undergo a full-term pregnancy also have a greatly reduced susceptibility to chemical carcinogen-induced mammary carcinogenesis (MC) compared to nulliparous rats and mice. Our current research goals are to develop safe, short-term hormonal prevention procedures that will mimic the protective effect of pregnancy against MC in nulliparous rats exposed to potent chemical carcinogens. In order to accomplish this goal, it is absolutely essential to characterize the parous phenotype of women, mice, and rats in relation to their refractoriness to MC. This chapter is divided into two parts. First, we summarize the current status of our knowledge of the parous phenotype in these three species. Second, we will describe studies of our published and unpublished results in attempts to develop a non-invasive, safe, and short-term hormonal treatment in nulliparous rats, mimicking the protective effect of full-term pregnancy. Surprisingly, in nulliparous rats, short-term exposure to high pregnancy levels of 17β-estradiol (E_2) mimics the protective effect of pregnancy against mammary cancer (10, 11). Recent comparisons of pregnancy levels of hormones between Chinese and Caucasian women with low and high BC risks, respectively, have observed significantly higher levels of E_2 in the low risk Chinese women (18).

Parous Phenotypes

Human. International studies by MacMahon, *et al.* (12, 13) suggested that women undergoing a full-term pregnancy before the age of 18, with or without lactation, have about one-third the risk for developing BC compared to women undergoing a first pregnancy after the age of 35. Recently, meta-analysis of BC risk reduction related to breast feeding has been performed based on data from 47 epidemiological studies from 30 countries (3). Their analysis shows that breast-feeding is associated with a substantial risk reduction. The risk for developing BC decreases by 4.3% for each 12 mo of nursing in addition to a decrease of 7% for each birth.

In a series of reports, Musey, *et al.* (14-16) studied the long-term effects of a first pregnancy on the hormonal environment. They observed a persistent reduction in serum levels of prolactin (PRL), but not of luteinizing hormone, follicle stimulating hormone, estrone, or E_2. Wang, *et al.* reported that increasing numbers of pregnancies were associated with a step-wise reduction in plasma PRL levels (17). A step-wise decrease in circulatory PRL levels with increasing parity fits well with the current finding that increasing parity causes a step-wise decrease in risk for developing BC. Therefore, all these data are supportive of Musey's (14-16) suggestion that parity protection against BC is likely due to a permanent reduction in the promotional hormone environment.

Mice. Marchant (4, 5) reported that virgin mice of the IF strain and their F1 hybrids, highly susceptible to chemical carcinogen-induced MC, became almost completely refractory to the same agent following full-term pregnancy and lactation. Medina and Smith (9) reported a significant reduction in mammary cancer incidence in parous BD2fF and C3H/Sm mice compared to virgin mice treated with 7,12 dimethylbenz[a]anthracene (DMBA). Our studies (19) demonstrated that parous BALB/c mice are highly refractory to N-methyl-N-nitrosourea (MNU)-induced MC. However, they become highly susceptible to MC by the same carcinogen when exposed to high levels of PRL during initiation and promotion. To our knowledge, this is the first conclusive demonstration that refractoriness to MC in parous mice can be reversed by increasing the endogenous levels of PRL by isografting pituitaries before carcinogen exposure. Additionally, these studies also suggested that mammary glands containing secretory lobules and lacking terminal end buds (TEBs) are fully susceptible to carcinogenesis (19, 20). Thus, refractoriness of parous mice to carcinogenesis is likely due to the pregnancy-induced permanent reduction in secretion of the mammogenic hormone, PRL, a situation similar to that observed in parous women (14-17).

Rats. In the rat, Dao, *et al.* (6, 7) reported that treatment with 3-methylcholanthrene failed to induce mammary cancers in pregnant or post partum lactating Sprague Dawley (SD) rats. Moon (8) reported that SD rats that had undergone a full-term pregnancy, followed by involution of their mammary glands, were more resistant to DMBA-induced MC than their age-matched nulliparous controls. He observed that mammary glands of parous rats required twice as much estrogen compared to the glands of nulliparous rats to induce the equivalent amount of growth. Chakravarty (21) found that the splenocytes from parous rats had cytolytic activity against mammary cancer cells weeks after pregnancy, and the effect decreased with time.

In the 1970s Russo, *et al.* have been the premier investigators of studies dealing with the characterization of the rat parous phenotype and the reason for their refractoriness to MC. Their morphological, biochemical, and endocrinological studies have resulted in some of the most widely accepted conclusions regarding the refractoriness phenomenon. They are of the opinion that quantitative differences in TEBs, especially in TEBs in the process of differentiation to alveolar buds account for the differences in susceptibility between nulliparous and parous rats. According to this view, TEBs are the site of origin of MCs in rats (22), they undergo full-differentiation during pregnancy, become secretory during lactation, and are totally absent in the mammary gland following post-lactation involution. They concluded that structural changes in the mammary gland, and not the transitional hormonal milieu during pregnancy and lactation, are the reason for the protective effect seen in parous rats. They have also reported that parity in the rat results in a phenotype where the mammary epithelial cells (MEC) show decreased proliferation, increased DNA repair capability, decreased carcinogen binding, activation of genes controlling apoptosis, and increased inhibin expression (23). All of these parity-associated characteristics insure refractoriness of these rats to carcinogen-induced MC.

In contrast, studies by us and our colleagues (24, 25) have led us to the conclusion that parous rats are not protected from carcinogen-induced initiation, but are protected from promotion. This protection results from a pregnancy-induced persistent reduction in circulatory levels of mammogenic hormones (26). As in mice (19), the refractoriness of parous rats to chemical carcinogen-induced MC can be overcome by treatment with exogenous ovarian hormones, E_2 with or without progesterone (P) at initiation and promotion (25). Again, as in mice, the mammary glands of parous rats with full lobulo-alveolar development can give rise to a high incidence of MCs following exposure to MNU.

Following is a summary of the rat parous phenotype: 1) Parous rats, lacking TEBs, exposed to MNU, give rise to a very low incidence of palpable MCs, but a high incidence of microscopic cancers (latent MCs) (24). 2) The refractoriness of parous rats to the development of palpable MCs, following exposure to MNU, can be overcome by treating rats during initiation and promotion with $E_2 \pm P$ (25). These studies have also shown that the mammary glands of parous rats with full lobulo-alveolar development are highly susceptible to carcinogen-induced MC (25). 3) Parous rats, compared to nulliparous rats, have reduced circulatory levels of growth hormone (GH) and PRL, and a decrease in the expression of estrogen receptor (ERα) and epidermal growth factor receptor (EGFR) in the mammary gland (26, 27).

Thus, studies in parous rodents, by us and our colleagues seem to indicate that pregnancy-induced protection from MC in rats and mice, exposed to carcinogens, is likely due to persistently reduced levels of mammogenic hormones in their circulation. As mentioned above, human studies have also indicated a step-wise decrease in the secretion of PRL with each successive pregnancy. Recent prospective studies by Hankinson, *et al.* (28) have shown a correlation of high plasma PRL levels with high BC risk in post-menopausal women.

Hormonal Prevention of Breast Cancer: Mimicking the Protective Effect of Pregnancy

Hormonal prevention strategies have used exogenous hormonal treatment to mimic the protective effect of pregnancy against BC. Huggins, *et al.* (29) reported that high levels of E_2 and P given for 30 days beginning 15 days after DMBA administration inhibited MC in SD rats. They proposed that treatment with high levels of these hormones destroyed the potential cancer cells. Grubbs, *et al.* (30, 31) and McCormick and Moon (32) demonstrated that treatment of SD rats with high levels of E_2 and P, or E_2 alone following treatment with MNU, a direct acting chemical carcinogen, was as effective as ovariectomy in preventing MC. Grubbs (31) suggested that the primary action of the hormones was to cause differentiation of the pre-neoplastic cells. Recent studies have reported on persistently altered expression of genes in the mammary glands of parous rats (33). They reported that certain genes involved in growth promotion are persistently down regulated in mammary glands. They also reported that TGF-β3 and several of its transcriptional targets were up regulated. Other investigators studying protection induced by E_2 + P treatment have found persistent alterations in the expression of several known genes. Specifically, they have identified persistent changes in ERα, p53, and retinoblastoma binding protein (RbAp46), a gene implicated in the regulation of cell proliferation and differentiation following hormone treatment (34, 35).

Russo, *et al.* (36-39) reported that human chorionic gonadotrophin (hCG) administration before or after DMBA treatment resulted in protection from mammary cancer in SD rats. In 1990 (36), they reported that pregnancy or hCG (100 IU/day) treatment (50-71 days of age), followed by DMBA exposure significantly reduced the incidence of carcinomas compared to controls. Srivastava, *et al.* (23) reported that rats exposed to DMBA on day 45 of age followed by no treatment or hCG treatment from days 65 to 105 resulted in the occurrence of carcinomas in 100% of the controls (3.5 cancers/rat), and in 81% of the hCG treated (0.9 cancers/rat). This hCG treatment also significantly lowered the incidence of microscopic mammary lesions, intraductal proliferations, and ductal carcinomas *in-situ* (DCIS). Russo, *et al.* (36) concluded that hCG treatment, like pregnancy, caused lobular differentiation and that both procedures can efficiently prevent chemically-induced MC.

In an attempt to mimic the protective effect of pregnancy against MC, we have used a modification of the E_2 + P treatment, begun earlier by Huggins (29) and confirmed by others (30-32, 40). We were attracted to this mode of treatment because very high levels of E_2 are a common feature of pregnancy in all three species. Our modifications included the use of steroids in silastic capsules resulting in sustained constant levels of steroids in the circulation for a short period (7-21 days; gestation period in the rat is 21 days), and assays of the blood levels of these hormones at the completion of the treatment. We also determined the incidence of palpable overt carcinomas, and the incidence of latent mammary cancers by microscopic examination of mammary gland wholemounts.

On the basis of our and our colleagues' studies, we have arrived at an entirely unexpected set of results which have allowed us to generate the following hypothesis regarding the nature of parity protection, and the reason for this protection against mammary cancer in parous rats.

First, our results show that the protective effect of pregnancy is unlikely due to terminal differentiation of TEBs, considered to be the target cells for mammary cancer (10, 11). We now hypothesize that parous females are protected only against promotion-progression and not against initiation. Parous females, treated with carcinogens develop latent microscopic cancers that normally do not undergo promotion-progression to frank carcinomas. A full-term pregnancy results in a persistently reduced promotional environment in parous females which prevents latent cancers from progressing to frank cancers.

Causative Factors for Refractoriness of Parous Rats to Mammary Carcinogenesis

Parous Rats Treated with MNU Have a High Incidence of Latent Mammary Cancers. In collaboration with Dr. Airo Tsubura, the role of parity before and after MNU treatment was studied (24). Pregnancy and lactation before (22%) or after (25%) MNU treatment reduced the incidence of frank MCs compared to age-matched nulliparous females (72%) and (94%), respectively. At termination, rats that underwent pregnancy prior to MNU (67%) or after MNU (50%) both had high incidences of microscopic latent mammary cancers. These findings suggest that MCs are initiated at a fairly high incidence in parous rats, but many of these cancers do not progress to frank MCs during the 12 mo after MNU treatment. Thus, these studies indicate that the MEC are highly susceptible to initiation of carcinogenesis and are not completely refractory to the carcinogenic process.

Refractoriness to Frank Mammary Carcinogenesis in Parous Rats can be Overcome by Hormone Treatment. In a separate study, MNU was given to parous rats or parous rats treated with 20 μg E_2 or 20 mg P or E_2 + P in silastic capsules for 7 days followed by continuous promotional treatment with E_2, P, or E_2 + P in silastic capsules (25). 10% of parous rats not treated with promotional hormone developed MCs. 67% of those treated with E_2, 50% of the rats treated with P, and 92% of those treated with E_2 + P developed MCs. These results show that the parity refractoriness to MNU can be reversed by treatment with ovarian steroids that increase proliferation in the mammary gland.

The Promotional Environment for Mammary Carcinogenesis is Decreased in the Parous Rats Compared to that of Nulliparous Rats

Transplantation Studies. The tumorigenicity of MNU-treated MEC transplanted to virgin or parous Lewis rats was tested (41). The objective of these studies was to determine whether the systemic environment in the parous rat is supportive of tumorigenesis. Young nulliparous Lewis rats (50-60 days of age) were treated with MNU. One month after treatment, the treated mammary glands were removed and enzymatically dissociated. The purified epithelial cells were transplanted (10^6 per site) to the subscapular fat pads of parous or age-matched virgins (AMV) hosts. MCs were found in 9/16 virgin hosts and in only 1/7 parous hosts within 6 mo of transplantation. These preliminary findings suggest that transformed MEC are capable of producing carcinomas when transplanted into virgin hosts, and that the systemic factors in the parous host have a decreased ability to support the promotion of transformed MEC to frank carcinomas.

Parous vs. Nulliparous Rats: Mammogenic Hormones and Mammary Epithelial Cells. Administration of MNU to 50-60-day old nulliparous rats, 120-day-old nulliparous rats, and 120-day-old parous rats, resulted in a high incidence of mammary cancers in the nulliparous rats (97 % in 50-60 day old rats; 75 % in 120 day old rats), no MCs developed in the parous rats (26). The concentrations in the serum of mammotropic hormones were measured at the time of MNU treatment. GH concentration was reduced in parous rats (16.1 ng/ml) compared to young nulliparous (59.3 ng/ml), and age matched nulliparous (38.6 ng/ml). PRL levels were 14.6 ng/ml in parous, 25.7 ng/ml in young nulliparous, and 25.5 ng/ml in age-matched nulliparous. Levels of E_2, P, corticosterone, and thyroxine were not different in the three groups. The concentrations of ERα and EGFR were decreased in the mammary glands of parous rats (3.5 fmol/mg protein, 3,500 cpm/mg) compared to young nulliparous (12.1 fmol/mg, 5000 cpm/mg), and age-matched nulliparous (7.9 fmol/mg, 5000 cpm/mg). Concentrations of GH receptor were not significantly different among the 3 groups. These studies led us to hypothesize that protection from mammary cancer induced by pregnancy is not due to the loss of target cells for carcinogenesis but is due to a decrease in the systemic promotional environment.

Development of Short-Term Hormone Treatments for Prevention of MNU-induced Mammary Carcinogenesis in Rats

The Protective Effect of Parity can be Mimicked by Short-term Treatment with E_2 or E_2 + P. Pregnant rats are exposed to very high circulatory levels of E_2 and P that induce lobulo-alveolar differentiation in the mammary gland. To mimic this protective effect of pregnancy, we treated 7-week-old virgin rats with MNU followed by treatment with E_2 + P or perphenazine (PPZ, dopamine receptor inhibitor) for 3 weeks beginning at 9 weeks of age (10). Both of these treatments caused late pregnancy-like lobule development in mammary glands. 90% of the control rats, 73% of the PPZ-treated, and only 9% of the rats treated with sustained exposure to E_2 + P in silastic capsules developed mammary cancer during the 9 mo-period of observation following MNU treatment. The E_2 + P-treated rats developed 93% fewer MCs compared to controls not receiving hormonal treatment. Both PPZ and E_2 + P treatments induced, similar to pregnancy, lobular growth, secretory differentiation, and involution after cessation of treatment of the mammary gland. Assays of blood levels of E_2 and P indicate that after PPZ treatment, only the P levels (P = 101.5 ng/ml, E_2 = 16.6 pg/ml,) were increased to pregnancy levels. There was no increase in circulating E_2 levels compared to untreated control virgin rats (E_2 = 18.3 pg/ml, P = 12.7 ng/ml). Treatment with E_2 + P resulted in pregnancy levels of E_2 (168.8 pg/ml), and lower than pregnancy levels of P (25.8 ng/ml). These circulating steroid levels are highly protective from MC. These studies suggested that pregnancy levels of E_2 are the likely reason for the protective effect of pregnancy against mammary cancer, and that lobular differentiation alone is insufficient to confer protection.

E_2 Alone but Not P Induced Protection from Mammary Carcinogenesis. We tested the effect of E_2 or P administered singly for their ability to induce protection from MC (10). Rats were treated with MNU at 7 weeks of age. A silastic capsule containing 30 mg of E_2 or 30 mg P was implanted at 9 weeks of age and removed 3 weeks later. 100 % of the control rats developed mammary cancer by 9 mo (2.9

MCs/rat) after MNU treatment. E_2 treatment alone provided protection from MC, 38% of E_2-treated rats developed mammary cancer (0.5 cancers/rat). In contrast, P treatment enhanced MC in terms of cancer load compared to controls. 100% of the rats treated with P developed mammary cancer (3.6 cancers/rat). Treatment with a combination of 30 mg of E_2 and 30 mg of P had the lowest incidence of mammary cancer (11%, 0.1 cancers/rat).

Pregnancy Levels of E_2 Alone Induce Protection from Mammary Carcinogenesis. We have demonstrated that E_2 alone or in combination with P is effective in preventing mammary cancer, while P alone is not. We determined the lowest dosage of E_2 given alone that would be effective in preventing mammary cancer (11). Rats were injected with MNU at 7 weeks of age. Two weeks later, the rats were treated with 30 mg, 200 μg, 100 μg or 20 μg of E_2 in silastic capsules for three weeks. Blood levels of E_2 immediately after treatments were 144, 95, 68, and 50 pg/ml, respectively. Untreated control rats had blood levels of 8-14 pg/ml of E_2. The E_2 blood levels from the 30 mg-, 200 μg-, and 100 μg-treated rats were in the range found during pregnancy. Control rats had 100% mammary cancer incidence, 9.0 mo after treatment, while rats treated with 30 mg, 200 μg, or 100 μg of E_2 had a cancer incidence ranging from 15 to 20%. The cancer incidence was not different from controls at the lowest E_2 dose, but the multiplicity was reduced. The multiplicity of mammary cancers was 0.2 (30 mg), 0.3 (200 μg), 0.6 (100 μg), 1.2 (20 μg), and 3.0 (controls). These results demonstrate that short-term treatments with doses of E_2 equivalent to levels during second half of pregnancy (100 μg or higher) are highly effective in preventing mammary cancer. Treatment with E_2 equivalent to low pregnancy levels (100-200 μg) does not induce full lobulo-alveolar differentiation, yet it is highly effective in conferring protection. Studies have also shown that 1 or 2 weeks treatment with pregnancy levels of E_2 (200 μg) and P (30 mg) in silastic capsule is highly effective in reducing the mammary cancer incidence in MNU-treated rats (11).

Short Term Protective Hormone Treatment has No Adverse Effects on the Health and Reproductive Physiology of Treated Rats. We examined whether short term E_2 + P treatment had any significant effects on the health and reproductive capabilities of treated rats. Nulliparous rats were treated with 200 μg E_2 + 30 mg P in silastic capsules from 9 to 12 weeks of age, and compared to age-matched controls treated with empty silastic capsules. This treatment is highly effective in preventing carcinogen-induced BC. All the hormone treated rats had regular estrous cycles 4-6 weeks after removal of the hormone treatment. Treated and control rats were mated at 16 weeks of age. Control rats had an average of 8 pups at birth and hormone-treated rats had 9. The weight of the pups was 5.9 and 6.0 g for control and treated rats, respectively. The weight at weaning was 35 g and 33 g for pups from control and treated rats, respectively. There were no significant differences in the monthly body weights of control or hormone-treated rats. The animals were necropsied at 600 days of age showing no difference in body weight, or pathology between control vs. the E_2 + P treated rats.

Comparisons of Parous and Hormone Protected Rats: Similar or Different?

Parous rats are susceptible to carcinogen-induced initiation and develop a high incidence of latent mammary cancers. However, these rats are protected from promotion and develop a very low incidence of overt mammary cancers. We determine whether the phenotypes of rats receiving protective hormone treatment mimic the protective effect of parous rats. That is, whether or not protective hormone treatments result in phenotypes susceptible to carcinogen-induced initiation, and develop latent mammary cancers, but remain protected from promotion/progression by failing to develop overt mammary cancers.

Previously, we described two protective treatments for the prevention of mammary cancers in carcinogen-exposed rats. Both treatments used E_2 at pregnancy levels for 7-21 days, either alone or in combination with P. Acronyms for these treatments are short-term E_2 treatment (STET) and short-term E_2 + P treatment (STEPT). Our data indicate that as in parous rats, STET/STEPT rats, exposed to MNU, develop a high incidence of latent mammary cancers and a low incidence of overt cancers. In both situations, exposure to prolonged hormonal promotion results in a high incidence of overt mammary cancers.

In a typical experiment, nulliparous control rats exposed to MNU at 7 weeks of age had a 100% incidence of both latent and overt mammary cancers. Similar rats receiving STET with three different doses of E_2 (100 μg, 200 μg, 30 mg in silastic capsules) for 21 days developed less than 20% overt mammary cancer, and a 100% incidence of latent mammary cancers (42). These results indicate that, as in parous rats, STET rats are protected from promotion, but fully susceptible to initiation.

In a separate study, parous rats and rats receiving STEPT before or after MNU exposure developed a low incidence of overt mammary cancers and a high incidence of latent cancers. However, the rats receiving promotional hormonal (E_2 ± P) treatment for a prolonged period developed a high incidence of overt mammary cancers (25, 43). Thus, as in parous rats, the refractoriness of STEPT rats to chemical carcinogen-induced overt MC is completely reversible with increased promotional environment.

A decreased promotional environment appears to be the major reason for the protective effect seen in parous and STET/STEPT rats against carcinogen-induced MC. We found that all three phenotypes (parous, STET, STEPT) have persistently reduced circulatory levels of GH and PRL (26, 44) as well as reduced ERα and PR expression in the MEC. These results provide conclusive evidence that STET/STEPT rats, like parous rats, are highly susceptible to carcinogen-induced initiation. The mammary glands of all three phenotypes, completely lacking TEBs, contain target cells fully capable of giving rise to latent mammary cancers. Our recent unpublished studies show that latent mammary cancers in MNU-exposed parous and STET/STEPT rats have significantly reduced immunohistochemically detectable cyclin D1 positive cells compared to those in nulliparous rats (42). Analogous phenomena have been observed between Japanese men in Japan with a low incidence of prostate cancer (PCA) and Caucasians in USA with a high incidence (45). Men from both countries have a similar incidence of latent PCA. The lower incidence of overt PCA in Japanese men is likely due to an inadequate promotional hormonal environment.

Summary

Our data, based on the MNU-exposed parous studies and STET/STEPT rats, indicate that these three phenotypes are alike in terms of their susceptibility to MC. All are: susceptible to initiation and development of latent cancers; have a drastically reduced incidence of overt mammary cancers; have a decreased promotional environment with persistently reduced circulatory levels of GH and PRL, as well as a decreased expression of ERα and PRs in the MEC. Finally, all protocols can develop a high incidence of overt mammary cancer with increased promotion with $E_2 \pm P$.

Together, all these data provide strong evidence in support of the notion that STET/STEPT phenotypes can mimic the protective effect of pregnancy, and that high pregnancy levels of E_2 may be the chief reason for the protective effect against MC. Sustained exposure to high pregnancy levels of E_2 resets the hypothalamic-pituitary axis resulting in a persistent reduction in the secretion of circulatory PRL and GH, involved in the promotion and progression of the initiated latent MC cells to overt carcinomas. Based on this hypothesis, we have succeeded in developing two hormonal intervention strategies, using pregnancy levels of E_2 for the prevention of carcinogen-induced MC in nulliparous rats. These treatments (STET/STEPT) require sustained exposure to hormones for only 7-21 days. These treatments are as effective as ovariectomy or full-term pregnancy and lactation without any loss of ovarian function, reproductive ability, longevity, or any other parameters, compared to normal controls. Understanding the biology and the mechanism of the E_2-induced persistent reduction in circulatory mammogenic hormones is likely to lead to better ways for BC prevention in women who choose to remain nulliparous.

References

1. Jemal AJ, Murray T, Samuels A, et al (2003) Cancer Statistics 2003. CA Cancer J Clin 53:5-26.
2. Stefanek M, Hartmann L, Nelson W (2001) Risk-reduction mastectomy: clinical issues and research needs. J Natl Cancer Inst 93:1297-1306.

3. Collaborative Group on Hormonal Factors in Breast Cancer (2002) Breast cancer and breast feeding: collaborative reanalysis of individual data from 47 epidemiological studies in 30 countries including 50,302 women with breast cancer and 96,973 women without the disease. Lancet 360:187-195.
4. Marchant J (1955) Influence of pregnancy and lactation on the incidence of mammary carcinoma induced with methylchoanthrene in female mice of the IF strain. J Pathol Bact 70:415-418.
5. Marchant J (1958) The inhibitory effect of continued lactation on the incidence of chemically-induced breast tumours in mice of the IF strain. Br J Cancer 12:55-61.
6. Dao TL, Bock FG, Greiner MJ (1960) Mammary carcinogenesis by 3-methylchoanthrene. II. Inhibitory effect of pregnancy and lactation on tumor induction. J Natl Cancer Inst 25:991-1003.
7. Sinha DK, Pazik JE, Dao TL (1988) Prevention of mammary carcinogenesis in rats by pregnancy: Effect of full-term and interrupted pregnancy. Br J Cancer 57:390-394.
8. Moon RC (1969) Relationship between previous reproductive history and chemically induced mammary cancer in rats. Int J Cancer 4:312-317.
9. Medina D, Smith GH (1999) Chemical carcinogen-induced tumorigenesis in parous, involuted mouse mammary glands. J Natl Cancer Inst 91:967-969.
10. Guzman RC, Yang J, Rajkumar L, et al (1999) Hormonal prevention of breast cancer: Mimicking the protective effect of pregnancy. Proc Natl Acad Sci USA 96:2520-2525.
11. Rajkumar L, Guzman RC, Yang J, et al (2001) Short-term exposure to pregnancy levels of estrogen prevents mammary carcinogenesis. Proc Natl Acad Sci USA 98:11755-11759.
12. MacMahon B, Cole P, Lin M, et al (1970) Age at first birth and breast cancer risk. Bull Wld Hlth Org 43:209-221.
13. MacMahon B, Cole P, Brown J (1973) Etiology of human breast cancer: A review. J Natl Cancer Inst 50:21-42.
14. Musey V, Collins DC, Musey PI, et al (1987) Long-term effect of a first pregnancy on the secretion of prolactin. N Engl J Med 316:229-234.
15. Musey V, Collins DC, Musey, PI, et al (1987) Age-related changes in the female hormonal environment during reproductive life. Am J Obstet Gynecol 157:312-317.
16. Musey V, Collins DC, Brogan DR, et al (1987) Long term effects of a first pregnancy on the hormonal environment: estrogens and androgens. J Clin Endocrinol Metab 64:111-118.
17. Wang DY, De Stavola BL, Bulbrook RD, et al (1988) The permanent effect of reproductive events on blood prolactin levels and its relation to breast cancer risk: a population study of postmenopausal women. Eur J Cancer Clin Oncol 21:1225-1231.

18. Lipworth L, Hsieh CC, Wide L, et al (1999) Maternal pregnancy hormone levels in an area with a high incidence (Boston, USA) and in an area with a low incidence (Shanghai, China) of breast cancer. Br J Cancer 79:7-12.
19. Swanson SM, Guzman RC, Collins G, et al (1995) Refractoriness to mammary carcinogenesis in the parous mouse is reversible by hormonal stimulation induced by pituitary isografts. Cancer Lett 90:171-181.
20. Swanson SM, Guzman RC, Christov K, et al (1994) Pituitary-isografted mice are highly susceptible to MNU-induced mammary carcinogenesis irrespective of the level of alveolar differentiation. Carcinogenesis 15:1341-1346.
21. Chakravarty PK, Sinha D (2000) Pregnancy induced mammary tumor specific effector cells are present long after parturition in a breast cancer model in rats. Cancer Letters 154:1-7.
22. Russo J, Russo IH (1978) DNA labeling index and structure of the rat mammary gland as determinants of its susceptibility to carcinogenesis. J Natl Cancer Inst 61:1451-1459.
23. Srivastava P, Russo J, Russo IH (1999) Inhibition of rat mammary tumorigenesis by human chorionic gonadotropin associated with increased expression of inhibin. Mol Carcinogenesis 26:10-19.
24. Yang J, Yoshizawa K, Nandi S, et al (1999) Protective effects of pregnancy and lactation against N-methyl-N-nitrososurea-induced mammary carcinomas in female Lewis rats. Carcinogenesis 20:623-628.
25. Thordarson G, Van Horn K, Guzman RC, et al (2001) Parous rats regain high susceptibility to chemically induced mammary cancer after treatments with various mammotropic hormones. Carcinogenesis 22:1027-1033.
26. Thordarson G, Jin E, Guzman RC, et al (1995) Refractoriness to mammary tumorigenesis in parous rats: is it caused by persistent changes in the hormonal environment or permanent biochemical alterations in the mammary epithelia? Carcinogenesis 16:2847-2853.
27. Bridges RS, Felicio LF, Pellerin LJ, et al (1993) Prior parity reduces post-coital diurnal and nocturnal prolactin surges in rats. Life Sci 53:439-445.
28. Hankinson SE, Willet WC, Michaud DS, et al (1999) Plasma prolactin levels and subsequent risk of breast cancer in postmenopausal women. J Natl Cancer Inst 91:629-634.
29. Huggins C, Moon RC, Morii S (1962) Extinction of experimental mammary cancer, I. Estradiol 17β and progesterone. Proc Natl Acad Sci USA 48:379-386.
30. Grubbs CJ, Peckham JC, McDonough KD (1983) Effect of ovarian hormones on the induction of 1-methyl-1-nitrosourea-induced mammary cancer. Carcinogenesis 4:495-497.
31. Grubbs CJ, Hill DL, McDonough KC, et al (1983) N-nitroso-N-methylurea-induced mammary carcinogenesis: effect of pregnancy on preneoplastic cells. J Natl Cancer Inst 71:625-628.

32. McCormick GM, Moon RC (1973) Effect of increasing doses of estrogen and progesterone on mammary carcinogenesis in the rat. Eur J Cancer 9:483-486.
33. D'Cruz CM, Moody SE, Master SR, et al (2002) Persistent parity-induced changes in growth factors, TGF-beta3, and differentiation in the rodent mammary gland. Mol Endocrinol 16:2034-2051.
34. Ginger MR Gonzalez-Rimbau MF Gay JP, et al (2001) Persistent changes in gene expression induced by estrogen and progesterone in the rat mammary gland. Mol Endocrinol 15:1993-2009.
35. Sivaraman L, Conneely OM, Medina D, et al (2001) P53 is a potential mediator of pregnancy and hormone-induced resistance to mammary carcinogenesis. Proc Natl Acad Sci USA 98:12379-12384.
36. Russo IH, Koszalka M, Russo J (1990) Human chorionic gonadotropin and rat mammary cancer prevention. J Natl Cancer Inst 82:1286-1289.
37. Russo J, Russo IH (1994) Toward a physiological approach to breast cancer prevention. Cancer Epidemiol Biomarkers Prev 3:353-364.
38. Russo IH, Russo J (1988) Hormone prevention of mammary carcinogenesis: A new approach in anticancer research. Anticancer Res 8:1247-1264.
39. Russo IH Russo J (2000) Hormonal approaches to breast cancer prevention. J Cell Biochem Suppl 34:1-6.
40. Sivaraman L, Stephens LC, Markaverich BM, et al (1998) Hormone-induced refractoriness to mammary carcinogenesis in Wistar-Furth rats. Carcinogenesis 19:1573-1581.
41. Abrams TJ, Guzman RC, Swanson SM, et al (1998) Changes in the parous rat mammary gland environment are involved in parity-associated protection against mammary carcinogenesis. Anticancer Res 18:4115-4122.
42. Reddy M, Nguyen S, Farjamrad F, et al (2002) Short-term hormone treatment with pregnancy levels of estradiol prevents mammary carcinogenesis by preventing promotion of carcinogen initiated cells. Proceedings Amer Assoc Cancer Res Annual Meeting, 43:824.
43. Guzman RC, Rajkumar L, Leung J, et al (2000) Refractoriness to breast cancer induced by treatment with estradiol (E) and progesterone (P) in nulliparous rats is similar in phenotype to that in parous rats. Proceedings the Amer Assoc Cancer Res Annual Meeting, 48:1964.
44. Rajkumar L, Guzman RC, Thordarson G, et al (2002) Estradiol prevents mammary cancer in rats: mimicking the protective effect of pregnancy. Proceedings Frontiers Cancer Prev Res Annual meeting, A208.
45. Shiraishi T, Watanabe M, Matsuura H, et al (1994) The frequency of latent prostatic carcinoma in young males: the Japanese experience. In vivo 8:445-448.

13

Aromatase Inhibition and Breast Cancer

William R. Miller

Introduction

There is compelling evidence that estrogens (Es) both increase risk of developing breast cancer (BC) and maintain the growth and progression of established tumors (1). Recently, drugs have been developed which specifically and potently block E biosynthesis (aromatase inhibitors). These agents now occupy a central role in the treatment of postmenopausal patients with BC, and are being considered as preventative agents in women at high risk of the disease. The success of these drugs is founded on: (A) The peripheral production of Es in postmenopausal women. (B) The key role of aromatase in the pathway of E synthesis. (C) The unique endocrinology of postmenopausal BCs. (D) The evolution of different types of aromatase inhibitors. (E) The mechanisms whereby Es initiate and promote BC. Therefore, the objectives of this chapter are to review (i) the pivotal role of aromatase in maintaining E within the breast and its tumors, and agents which have been developed as inhibitors of this enzyme, (ii) the anti-tumor responses which may be observed in hormone sensitive BCs following treatment with aromatase inhibitors, and (iii) the biological rationale for using aromatase inhibitors to prevent BC.

Sites of Estrogen Production

The major sites of E production differ in pre- and postmenopausal women. Before the menopause, the ovary is mainly responsible for circulating levels of E. However, peripheral synthesis of E is of greater importance in postmenopausal women when ovarian E biosynthesis ceases. The postmenopausal ovary along with the adrenal cortex produces substantial amounts of androgen (2, 3) that can be used as substrate for synthesis of E at peripheral sites such as fat (4), skin (5), muscle (6), liver (7), and BC (8), all of which possess aromatase activity.

Aromatase

The pathway of E biosynthesis is summarized in Figure 1. The first step of cholesterol side-chain cleavage is rate-limiting but the last step catalyzed by aromatase is unique to E biosynthesis. Aromatase removes the methyl group between the steroid A and B rings, and the A ring becomes aromatic. NADPH is required as a cofactor, being generated by a NADP reductase and a transfer of electrons via a specific CYP 19 p450 arom (9).

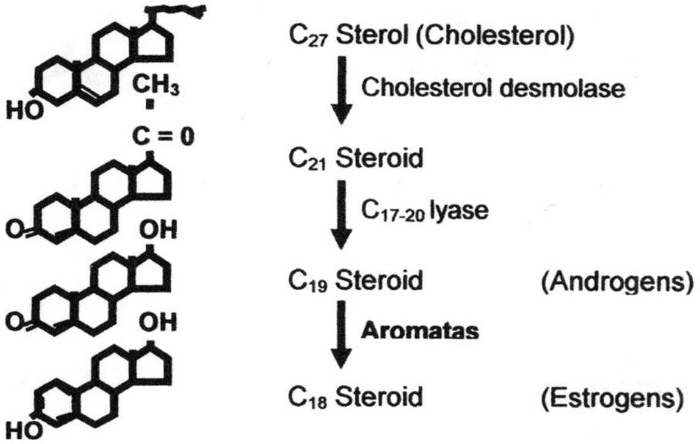

Figure 1. Classical pathway of E biosynthesis from cholesterol

Aromatase Inhibitors

E production may be reduced by inhibiting any step in the biosynthetic pathway but blockade of aromatase (the last step in the sequence) is less likely to affect other steroid classes. Consequently, there has been an impetus to develop selective and potent inhibitors of the aromatase enzyme.

Aromatase inhibitors can be subdivided into steroidal (Type I) and non-steroidal (Type II) agents (10) (Figure 2). Type I agents are androgen analogues that bind to the enzyme, competing with the natural substrate. Inhibitors such as formestane (11) and exemestane (12) are thought to be metabolized by the aromatase enzyme into reactive intermediates resulting in tight or irreversible binding. Because the enzyme is inhibited as an outcome of its own metabolism, the drugs have been termed "suicide inhibitors" (13) or "inactivators". A consequence of such "inactivation" is that the inhibition is usually specific and long-term. Unlike other reversible agents, continued presence of drug is not required for inhibition (14), and duration of inhibitory effects is dependent upon de novo synthesis of aromatase.

The development of Type II inhibitors followed the recognition that the anti-epileptic drug, aminoglutethimide, inhibited aromatase (15). Aminoglutethimide lacks specificity but newer second and third generation non-steroidal aromatase inhibitors have greater selectivity and efficacy and less toxicity. The lead third generation type II inhibitors are anastrozole (16) and letrozole (17). Both are triazoles that bind reversibly to the aromatase enzyme, interacting with the heme iron in the cytochrome p450 and occluding the substrate binding site (18).

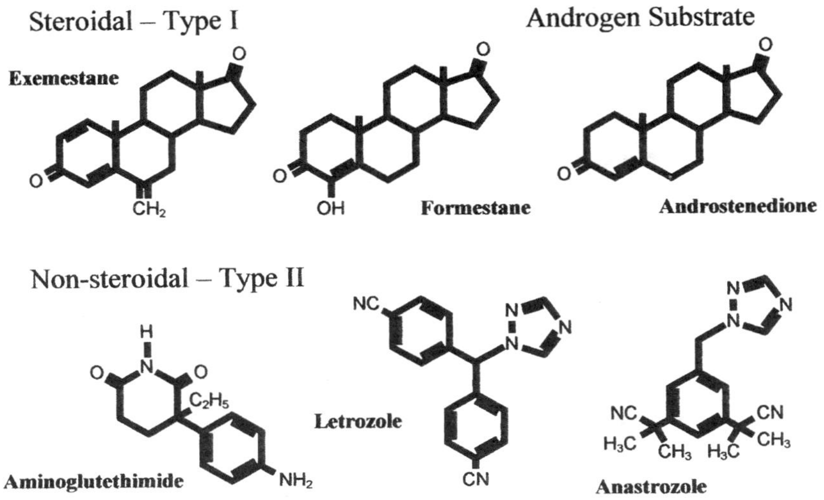

Figure 2. Different classes of aromatase inhibitors. Type I, steroidal inhibitors are androgen analogues. Type II, non-steroidal inhibitors, *i.e.,* aminoglutethimide, letrozole, and anastrozole are azoles.

Potency has been tested in a variety of model systems (19) including placental microsomes, particulate fractions of BC and whole-cell systems such as cultured fibroblasts. The effects of anti-aromatase agents in these three types of test systems are summarized in Table 1. Aminoglutethimide, anastrozole, letrozole, formestane, and exemestane all inhibit aromatase activity in each of the systems, but the newer agents are orders of magnitude more potent than aminoglutethimide. Thus, whereas μmolar concentrations are required for aminoglutethimide, only nanomolar concentrations are needed with letrozole, anastrozole, exemestane, and formestane. In general, the newer Type II inhibitors (anastrozole and letrozole) are more potent than Type I inhibitors (formestane and exemestane). Amongst Type II agents, letrozole appears most effective. Letrozole and exemestane are more active in whole-cell cultures than in disrupted-cell preparations (20). The newer anti-aromatase agents also have great specificity. For example, exemestane only affects other steroid hydroxylases at concentrations that are at least 2,500 times higher than those affecting aromatase.

Table 1. Inhibition of Aromatase Activity in Whole- and Disrupted-cell Preparations.

	Placental Microsomes		Breast Cancer Homogenates		Mammary Fibroblast Cultures	
	IC_{50} (nM)	Relative Potency	IC_{50} (nM)	Relative Potency	IC_{50} (nM)	Relative Potency
Aminoglutethimide	3,000	1	4500	1	8,000	1
Anastrozole	12	250	10	450	14	570
Letrozole	12	250	2.5	1,800	0.8	10,000
Formestane	50	60	30	150	45	180
Exemestane	50	60	15	300	5	1,600

Endocrinology of Aromatase Inhibitors

Effects of aromatase inhibitors have been assessed on *in-vivo* peripheral aromatase activity and circulating Es. To measure peripheral aromatase levels, radioactive androgen precursors are given to women, and conversion to Es assayed by measuring the radioactivity in Es purified from either urine or plasma (21-23). Such studies have shown that aminoglutethimide, 4-hydroxyandrostenedione, and fadrozole achieve 90-92% inhibition of aromatization (23-25), whereas anastrozole, letrozole, and exemestane suppress all *in-vivo* aromatization by >96% (21-23, 26-28).

These effects on peripheral aromatization translate into decreased levels of Es in the circulation. Thus, daily anastrozole (1 mg), letrozole (2.5 mg), and exemestane (25 mg) reduce circulating E in postmenopausal women to levels often below the detection amount of current assays (21, 28). Additionally, differences in potency between inhibitors that are apparent on whole body aromatase can also be detected at the level of circulating Es. Thus, exemestane will reduce E levels in patients relapsing on the first-generation inhibitor aminoglutethimide (29).

These dramatic effects produced by the third-generation aromatase inhibitors on Es are invariably achieved without measurable effects on the synthesis and levels of other circulating steroid hormones (16, 17, 35).

Breast Endocrinology

The postmenopausal breast is unusual in that (i) steroid hormone levels, in particular those of estradiol, may be substantially higher than those in circulation (30), (ii) the breast is able to accumulate E from the circulation against gradient (31), and (iii) both mammary adipose tissue (8, 9) and BCs (8) are capable of E biosynthesis. The relative contribution of uptake into the breast and local biosynthesis of endogenous Es within breast tumors can be assessed from infusion studies. These suggest substantial inter-subject variation; in some breasts, E levels are largely accounted for by uptake, whereas in others by synthesis (31, 32). However, tumors with highest levels of E are invariably those with the greatest *in-situ* synthesis (32, 33). It has therefore been important to monitor the efficacy of novel aromatase inhibitors on aromatase activity within breast tumors themselves.

Accurate assessment of the inhibitory potential of aromatase inhibitors requires *in-situ* methodology. These protocols provide evidence of potent effects within the breast (34). Thus, marked decreases in tumor *in-situ* aromatase activity are observed in the majority of BCs, although occasional cancers appear resistant to inhibitors (Figure 3). Inhibitory effects are also associated with changes in tumor Es which decrease markedly with treatment (34).

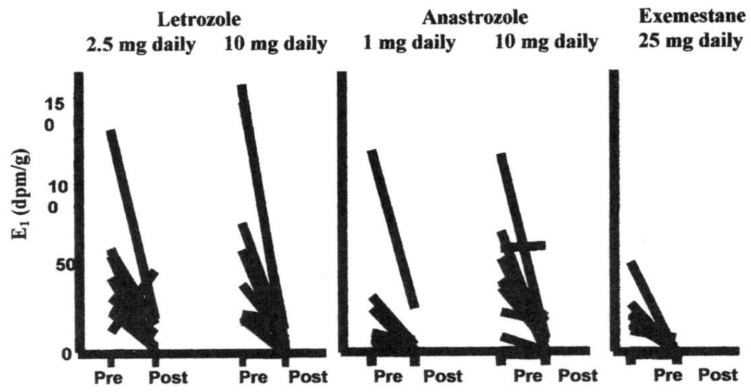

Figure 3. The effect of neoadjuvant treatment with aromatase inhibitors on *in-situ* tumor E synthesis. One patient in the letrozole (2.5 mg), two in the letrozole (10 mg), two in the anastrozole (1 mg), three in the anastrozole (10 mg), and three in the exemestane groups had no evidence for tumor *in-situ* synthesis either before or after treatment. Pre: Measurements performed immediately before treatment. Post: Measurements performed after 3 mo treatment.

Aromatase Inhibitors and Treatment of Breast Cancer

Neoadjuvant protocols in which therapy is given with the primary cancer remaining within the breast allow tumor responses to be assessed in individual patients by monitoring changes in tumor volume during treatment. Impressive anti-tumor effects have been observed in selected groups of patients with estrogen receptor (ER)-α rich cancers. (19, 35) (Table 2). Marked reduction in tumor size provides clinical benefits, and many patients who initially required mastectomy or were inoperable can be treated by more conservative breast surgery (19, 35) (Table 3).

Table 2. Median Tumor Volume Reduction in Series of Patients with Locally Advanced BC who Received Neoadjuvant Endocrine Therapy in the Edinburgh Breast Unit[1]

Agent	Number of Patients	Patients with > 50% Reduction, n (%)	Patients with < 50% Reduction or <25% Increase, n (%)	Patients with >25% Increase, n (%)
Letrozole	36	32 (89)	3 (8)	1 (3)
Anastrozole	23	18 (78)	5 (13)	0
Exemestane	12	10 (83)	2 (17)	0

[1] Tumor volume changes assessed by ultrasound during the 3.0-mo treatment period.

Table 3. Patients with Locally Advanced BC Requiring Mastectomy Before and After Neoadjuvant Endocrine Therapy in the Edinburgh Breast Unit

Agent	Number of Patients	Number Initially Requiring Mastectomy	Number Requiring Mastectomy After Treatment	Conversion Rate, %[1]
Letrozole	36	24	2	93
Anastrozole	24[2]	19	2	89
Exemestane	12	10	2	80

[1] Percentage of patients initially considered only for mastectomy that underwent breast-conserving surgery following treatment.
[2] Includes one patient who did not complete full treatment.

Effects on tumor morphology and histopathology may also be monitored by examining sequential biopsies before and during neoadjuvant therapy (36, 37). The latest generation of aromatase inhibitors is capable of producing marked changes in tumor morphology; clear reductions in cellularity/increases in fibrosis are observed in most tumors following treatment (38, 39).

Therapy with aromatase inhibitors also produce a marked and consistent reduction in the proportion of tumor cells staining positively with the Mib1 antibody, a useful surrogate marker for proliferative activity (40). Similarly, following treatment, staining for progesterone receptors (PR), a marker of E response, are significantly reduced (often to zero) in over 90% of the tumors studied (40). These results are consistent with anti-estrogen and anti-tumor effects of aromatase inhibitors.

Estrogens, Aromatase and Risk of Breast Cancer

Many of the etiological factors associated with BC have a hormonal component (1). Additionally, high BC risk has been related to increased E exposure from either endogenous (41-46) or exogenous sources (47, 48). Polymorphisms in the aromatase gene in women at high BC risk (49, 50), and enhanced aromatase activity in the breasts of women with BC (51, 52) have been reported. Conversely, E deprivation may protect against the disease (53, 54). For example, ovarian ablation before the age of 35 years is calculated to reduce subsequent BCs by two thirds (53).

These observations provide the rationale for the use of endocrine therapy as preventative measures against BC in women with high risk of the disease. Four trials using tamoxifen have been published (55-57), as has a fifth trial using raloxifene (58). Although there are differences among studies, a recent meta-analysis of the tamoxifen trials indicated that results were compatible with a 42% reduction in short term incidence of breast cancer with tamoxifen use (59).

These studies provide the impetus to use aromatase inhibitors as preventative agents and determine whether they may be more effective than selective ER modulators (SERMs).

Differences Between SERMs and Aromatase Inhibitors

SERMs and aromatase inhibitors are often thought of as being equivalent; however differences in mechanism of action may lead to variations in biological and clinical effects. For example, aromatase inhibitors reduce endogenously synthesized Es; in contrast, SERMs generally do not inhibit synthesis, and E levels remained unaltered (60), or, in the case of pre-menopausal women, may increase (61, 62). This difference may be important because metabolites of natural Es can have influences independent of ER (63). Although aromatase inhibitors reduce levels of Es synthesized endogenously, they will not diminish the activity of exogenous Es and adrenal androgens that interact with the ER (64-66). In contrast, SERMs will interfere with ER signaling irrespective of ligand. However, as the third generation aromatase inhibitors appear more effective than tamoxifen (67-71), it is unlikely that exogenous Es are responsible for the induction and growth of most BCs (64).

Finally, in contrast to tamoxifen, aromatase inhibitors are without E agonist activity. This is illustrated by the effects of treatment on the expression of markers of estrogenic activity such as PRs. Whilst aromatase inhibitors reduce the tumor expression, tamoxifen's most common effect is to increase expression (39) (Table 4). The phenotype of an aromatase inhibitor-treated tumor is ER+/PR-, whereas that of tamoxifen-treated is ER-poor/PR–rich.

Table 4. Comparative Effects on PR Expression.

	Decrease	No Change	Increase
Aromatase Inhibitors	43	3	1
Tamoxifen	12	13	27

Role of Estrogens in the Development of Breast Cancer

The mechanism by which Es increase risk of overt BC is still the subject of debate (Figure 4). Es can stimulate the proliferation of breast epithelial cells, leading to genetic mistakes and a transformed cellular phenotype (72). Additionally, Es can accelerate the growth of occult cancers resulting in increased incidence of overt disease (73). These promotional properties of Es appear to be largely mediated through ER signaling system. Consequently, both SERMs and aromatase inhibitors should attenuate such promotion.

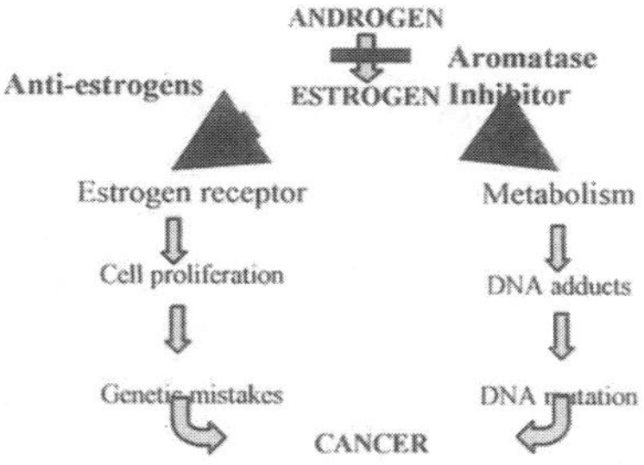

Figure 4. Mechanisms whereby E may cause breast cancer: (1) via the ER, (2) via metabolism of E. Note that anti-estrogens block only ER-mediated events, whereas aromatase inhibitors block both ER and metabolic pathways

However, evidence is accumulating that E metabolites are genotoxic (74). In particular, metabolism of Es via catechols (2-/4-hydroxyestradiol or hydroxyestrone) may produce reactive quinones which can directly interact with and mutate DNA, initiating carcinogenesis (75-77). Furthermore, these reactions also generate reactive oxygen metabolites that may also damage DNA. The enzymes in this metabolic pathway may be raised in women at risk of BC (72, 75, 76, 78). These genotoxic processes appear independent of ER signaling. SERMs are therefore unlikely to reduce genotoxic damage (Figure 4): in contrast aromatase inhibitors which reduce E levels would. Indeed, it has been shown that BC cells metabolize androgens into E-guanine adducts, and the aromatase inhibitor, letrozole, is capable of inhibiting their formation (78).

Aromatase Inhibitors and Prevention of Breast Cancer

The above rationale and positive results with SERMs underpin the case for trials of aromatase inhibitors in the preventative setting. However, the potent endocrine effects of third generation inhibitors make long-term monitoring of bone and lipid profiles mandatory. Ongoing adjuvant trials with aromatase inhibitors provide information on the incidence of new contralateral BCs and side-effect profiles. The large ATAC trial shows a significant reduction in contralateral cancer in the aromatase inhibitor (anastrozole) group compared with tamoxifen (35 vs 20, $p = 0.007$, compared with an expected 85 cases with no treatment,) (79). Aromatase inhibitors also have not been successful in premenopausal women (80-82) because of the high levels of aromatase activity in the premenopausal ovary and compensatory feedback loops (13). However, the new aromatase inhibitors may be sufficiently powerful to produce effective blockade, but they are not yet used routinely in premenopausal women

Pilot chemoprevention studies which target high risk women, using third generation aromatase inhibitors are being planned or underway. One study has identified a small cohort of women with either DCIS or benign breast disease who were treated for 3 months with letrozole, the primary objective being to measure surrogate biomarkers of BC risk, and bone and lipid metabolism (83). The WISE trial aims to recruit women at risk on account of high circulating Es and randomize to either letrozole (1 year) or placebo. The National Institute of Canada is conducting double blind, multicentrer trials to evaluate the effect of letrozole, exemestane, or placebo on postmenopausal women with high breast density (84).

There are plans for two large Phase III trials. One will compare anastrozole with placebo in 6,000 postmenopausal women (IBIS II), the other will randomize 5,100 women to receive either exemestane or placebo or exemestane plus a COX-II inhibitor (NCIC CTL MAP3).

Summary

The latest generation of aromatase inhibitors comprises highly potent and specific endocrine agents. They have a central role in the management of hormone-dependent BC, and their potential as preventative agents in women at high risk of the disease is currently being explored in clinical trials.

References

1. Miller WR (1996) Estrogen and Breast Cancer. R G Landes & Co.
2. Judd HL, Judd GE, Lucas WE, et al (1974) Endocrine function of the postmenopausal ovary: concentration of androgen and estrogens in ovarian and peripheral vein blood. J Clin Endocrinol Metab 39:1020-1024.
3. Baird DT, Uno A, Melby JC (1969) Adrenal secretion of androgens and estrogens. J Endocrinol 45:135-136.
4. Perel E, Killinger DW (1979) The interconversion and aromitization of androgens by human adipose tissue. J Steroid Biochem 10:623-627.

5. Schweikert HU, Milewich L, Wilson JD (1976) Aromatization of androstenedione by cultured human fibroblasts. J Clin Endocrinol Metab 43:785-795.
6. Larionov AA, Vasyliev DA, Mason JI, et al (2003) Aromatase in skeletal muscle. J Steroid Biochem & Mol Biol 84:485-492.
7. Smuk M, Schwers J (1977) Aromatization of androstenedione by human adult liver *in vitro*. J Clin Endocrinol Metab 45:1009-1012.
8. Miller WR, Mullen P, Sourdaine P, et al (1997) Regulation of aromatase activity within the breast. J Steroid Biochem 61:193-202.
9. Bulun SE, Simpson ER (1994) Regulation of aromatase expression in human tissues. Breast Canc Res Treat 30:19-29.
10. Miller WR (1989) Aromatase inhibitors in the treatment of advanced breast cancer. Cancer Treatment Rev 16:83-93.
11. Brodie AMH, Brodie HJ, Romanoff L, et al (1982) Inhibition of estrogen biosynthsis and regression of mammary tumors by aromatase inhibitors. Hormonal Cancer Advances Exp Med Biol 138:179-190.
12. Lonning PE (2000) Pharmacology and clinical experience with exemestane. Expert Opin Invest Drugs 9:1897-1905.
13. Johnston JO, Metcalf BW (1984) Aromatase: a target enzyme in breast cancer. In: Novel Approaches to Cancer Chemotherapy, London: Academic Press, 307-328.
14. Miller WR, Vidya R, Mullen P, et al (2001) Induction and suppression of aromatase by inhibitors. In: Miller WR, Santen RJ (eds) Aromatase Inhibition and Breast Cancer. New York:Marcel Dekker Inc, 213-225.
15. Santen RJ, Santner S, Davis B, et al (1978) Aminoglutethimide inhibits extraglandular estrogen production in postmenopausal women with breast cancer. J Clin Endocrinol Metab 47:1257-1265.
16. Plourde PV, Dyroff M, Dukes M (1994) Arimidex: a potent and selective fourth-generation aromatase inhibitor. Breast Canc Res Treatment 30:103-111.
17. Iveson TJ, Smith IE, Ahern J, et al (1993) Phase I study of the oral nonsteroidal aromatase inhibitor CGS 20267 in postmenopausal patients with advanced breast cancer. Cancer Res 53:266-270.
18. Graham-Lorence S, Peterson JA (1996) P450s: structural similarities and functional differences. FASEB J 10:206-214.
19. Miller WR, Jackson J (2003) The therapeutic potential of aromatase inhibitors. Expert Opin Invest Drugs 12:1-15.
20. Bhatnagar AS, Miller WR (1999) Pharmacology of inhibitors of estrogen biosynthesis. In: Oettel M, Schillinger E (eds) Handbook of Experimental Pharmacology: Estrogens and Antiestrogens. Berlin: Springer-Verlag, 223-230.

21. Lonning PE (1996) Pharmacology of new aromatase inhibitors. The Breast 5:202-208.
22. Geisler J, King N, Anker G, et al (1998) In vivo inhibition of aromatization by exemestane, a novel irreversible aromatase inhibitor, in postmenopausal breast cancer patients. Clin Canc Res 4:2089-2093.
23. Miller WR, Dixon JM (2001) Local endocrine effects of aromatase inhibitors within the breast. J Steroid Biochem Mol Biol 79:93-102.
24. MacNeill FA, Jones AL, Jacobs S, et al (1992) The influence of aminoglutethimide and its analogue rogletimide on peripheral aromatisation in breast cancer. Brit J Cancer 66:692-697.
25. Jones AL, MacNeill F, Jacobs S, et al (1992) The influence of intramuscular 4-hydroxyandrostenedione on peripheral aromatisation in breast cancer patients. Euro J Cancer 10:1712-1716.
26. Lonning PE, Jacobs S, Jones A, et al (1991) The influence of CGS 16969A on peripheral aromatisation in breast cancer patients. Brit J Cancer 63:789-793.
27. Dowsett M, Jones A, Johnston SR, et al (1995) In vivo measurement of aromatase inhibition by letrozole (CGS 20267) in postmenopausal patients with breast cancer. Clin Canc Res 1:1511-1515.
28. Geisler J, King N, Dowsett M, et al (1996) Influence of anastrozole (Arimidex), a selective non-steroidal aromatase inhibitor, on in vivo aromatisation and plasma estrogen levels in postmenopausal women with breast cancer. Brit J Cancer 74:1286-1291.
29. Thurlimann B, Paridaens R, Serin D, et al (1997) Third-line hormonal treatment with exemestane in postmenopausal patients with advanced breast cancer progressing on aminoglutethimde: a phase II multicentre multi-national trial. Euro J Cancer 33:1767-1773.
30. Vermeulen A (1986) Human mammary cancers as a site of sex steroid metabolism. Cancer Surveys 5:585-595.
31. Miller WR (1997) Uptake and synthesis of steroid hormones by the breast. Endocrine Related Cancer 4:307-311.
32. Reed AJ, Aherne GW, Ghilchik MW, et al (1991) Concentrations of oestrone and 4-hydroxyandrostenedione in malignant and normal breast tissues. Int J Cancer 49:562-565.
33. James VHT, Reed MJ, Adams EF, et al (1989) Estrogen uptake and metabolism in vivo. In: Beck JS, editor. Estrogen and the Human Breast. Edinburgh:The Royal Society of Edinburgh, Vol 95B:185-193.
34. Miller WR, Dixon JM (2001) Local endocrine effects of aromatase inhibitors within the breast. J Steroid Biochem Mol Biol 79:93-102.
35. Dixon JM (2002) Neoadjuvant therapy: surgical perspectives. In: WR Miller, Ingle JN (eds) Endocrine Therapy in Breast Cancer. New York:Marcel Dekker, 197-212

36. Dixon JM, Love CDB, Renshaw L, et al (1999) Lessons from the use of aromatase inhibitors in the neoadjuvant setting. Endocrine Related Canc 6:227-230.
37. Sasano H, Suzuki T, Moriya T (2002) Pathology of breast cancer following neoadjuvant therapy. In: Miller WR, Ingle JN (eds) Endocrine Therapy in Breast Cancer. New York: Marcel Dekker 213-222.
38. Miller WR, Dixon JM, Macfarlane L, et al (2002) Pathological features of breast cancer response following neoadjuvant treatment with either letrozole or tamoxifen. Euro J Cancer 39:462-468.
39. Anderson TJ, Dixon JM, Stuart M, et al (2002) Effect of neoadjuvant treatment with anastrozole on tumour histology in postmenopausal women with large operable breast cancer. Brit J Cancer 87:334-338.
40. Miller WR, Dixon JM, Cameron DA, et al (2001) Biological and clinical effects of aromatase inhibitors in neoadjuvant therapy. J Steroid Biochem Mol Biol 79:103-107.
41. Dorgan JF, Longcope C, Stephenson HE Jr, et al (1997) Serum sex hormone levels are related to breast cancer risk in postmenopausal women. Environ Hlth Perspect 105(S3):583-585.
42. Toniolo PG, Levitz M, Zeleniuch-Jacquotte A, et al (1995) A prospective study of endogenous estrogens and breast cancer in postmenopausal women. J Natl Canc Inst 87:190-197.
43. Berrino F, Muti P, Micheli A, et al (1996) Serum sex hormone levels after menopause and subsequent breast cancer. J Natl Canc Inst 88:291-296.
44. Thomas HV, Key TJ, Allen DS, et al (1997) A prospective study of endogenous serum hormone concentrations and breast cancer risk in premenopausal women on the island of Guernsey. Brit J Cancer 76:401-405.
45. Thomas HV, Key TJ, Allen DS, et al (1997) A prospective study of endogenous serum hormone concentrations and breast cancer risk in postmenopausal women on the island of Guernsey. Brit J Cancer 75:1075-1079.
46. Thomas HV, Reeves, GK, Key TJA (1997) Endogenous estrogen and postmenopausal breast cancer: a quantitative review. Canc Causes Control 8:922-928.
47. Collaborative Group on Hormonal Factors in Breast Cancer (1997) Breast cancer and hormone replacement: Collaborative reanalysis of data from 51 epidemiological studies of 52,705 women with breast cancer and 108,411 women without breast cancer. Lancet 350:1047-1059
48. Million Women Study Collaborators. (2003) Breast cancer and hormone-replacement therapy in the Million Women Study. Lancet 362:419-427
49. Kristensen VN, Andersen TI, Lindblom A, et al (1998) A rare CYP19 (aromatase) variant may increase the risk of breast cancer. Pharmacogenetics 8:43-48.

50. Haiman CA, Hankinson SE, Speizer FE, et al (1999) A tetranucleotide repeat polymorphism in CYP19 and breast cancer risk. Proc Am Assoc Cancer Res 40:A1290.
51. O'Neill JS, Miller WR (1987) Aromatase activity in adipose tissue from women with benign and malignant breast disease. Brit J Cancer 56:601-604.
52. O'Neill JS, Elton RA, Miller WR (1988) Aromatase activity in adipose tissue from breast quadrants: a link with tumour site. Brit Med J 296:741-743.
53. Feinleib M (1968) Breast cancer and artificial menopause: a cohort study. J Natl Canc Inst 41:315-329.
54. Trichopoulos D, MacMahon B, Cole P (1972) Menopause and breast cancer risk. J Natl Canc Inst 48:605-613.
55. Fisher B, Costantino JP, Wickerham DL, et al (1998) Tamoxifen for prevention of breast cancer: report of the National Surgical Adjuvant Breast and Bowel Project P-1 Study. J Natl Cancer Inst 90:1371-1388.
56. Early Breast Cancer Trialist's Collaborative Group (EBCTCG) (1998) Tamoxifen for early breast cancer: an overview of the randomised trials. Lancet 351:1451-1467.
57. Powles T, Eeles R, Ashley S, et al (1998) Interim analysis of the incidence of breast cancer in the Royal Marsden Hospital tamoxifen randomised chemoprevention trial. Lancet 352:98-101.
58. Cummings SR, Eckert S, Krueger KA, et al (1999) The effect of raloxifene on risk of breast cancer in postmenopausal women: results from the MORE randomized trial. JAMA 281:2189-2197.
59. Cuzick J (2002) A brief review of the international breast cancer intervention study (IBIS), the other current breast cancer prevention trials, and proposals for future trials. Ann NY Acad Sci 2002:123-133.
60. Santen RJ, Manni A, Harvey H, et al (1990) Endocrine treatment of breast cancer in women. Endocrine Rev 11:1-45.
61. Sherman BM, Chapler FK, Crickard K, et al (1979) Endocrine consequences of continuous antiestrogen therapy with tamoxifen in premenopausal women. J Clin Invest 64:398-404.
62. Groom GV, Griffiths K (1976) Effect of the anti-estrogen tamoxifen on plasma levels of lutenizing hormone, follicle stimulating hormone, prolactin, oestradiol and progesterone in normal premenopausal women. J Endocrinol 70:421-428.
63. Liehr JG (2002) Breast carcinogensis and its prevention by inhibition of estrogen genotoxicity. In: Miller WR, Ingle JN (eds) Endocrine Therapy in Breast Cancer. New York:Marcel Dekker, 287-301.
64. Miller WR, Sharpe RM (1998) Environmental estrogens and human reproductive cancers. Endocrine Related Cancer 5:69-96.

65. Setchell KDP, Borriello SP, Hulme P, et al (1984) Nonsteroidal estrogens of dietary origin: possible roles in hormone-dependent disease. Amer J Clin Nutr 40:569-578.
66. Adams JB, Garcia M, Rochefort H (1981) Estrogenic effects of physiological concentrations of 5-androstene 3ß17ß diol and its metabolism in MCF-7 human breast cancer cells. Cancer Res 42:470-476.
67. Mourisden H, Gershanovich M, Sun Y, et al (2001) Superior efficacy of letrozole versus tamoxifen as first-line therapy for postmenopausal women with advanced breast cancer: results of a Phase III study of the International Letrozole Breast Cancer Group. J Clin Oncol 19:2596-2606.
68. Nabholtz JM, Buzdar A, Pollak M, et al (2000) Anastrozole is superior to tamoxifen as first-line therapy for advanced breast cancer in postmenopausal women: results of a North American multicenter randomized trial. Arimidex Study Group. J Clin Oncol 18:3758-3767.
69. Dixon JM, Love CD, Bellamy CO, et al (2001) Letrozole as primary medical therapy for locally advanced and large operable breast cancer. Breast Canc Res Treat 66:191-199.
70. Eiermann W, Paepke S, Appfelstaedt J (2001) Letrozole Neoadjuvant Breast Cancer Study Group. Preoperative treatment of postmenopausal breast cancer patients with letrozole. A randomized double-blind multicenter study. Ann Oncology 12:1527-1532.
71. The ATAC Trialist's Group (2002) Anastrozole alone or in combination with tamoxifen versus tamoxifen alone for adjuvant treatment of postmenopausal women with early breast cancer: first results of the ATAC randomised trial. Lancet 359:2131-2139.
72. Fiegelson HS, Ross RK, Yu MC, et al (1996) Genetic susceptibility to cancer from exogenous and endogenous exposures. J Cell Biochem 25S:15-22.
73. Miller WR (1996) Estrogens and endocrine therapy for breast cancer. In: Estrogen and Breast Cancer. Austin, Texas:RG Landes Company, 125-150.
74. Yager JD, Liehr JG (1996) Molecular mechanisms of estrogen carcinogenesis. Annual Rev Pharmacol Toxicol 36:203-232.
75. Santen RJ, Yue, W, Naftolin F, et al (1999) The potential of aromatase inhibitors in breast cancer prevention. Endocrine Related Cancer 6:235-243.
76. Raftogianis R, Creveling C, Weinshilboum R, et al (2000) Estrogen metabolism by conjugation. J Natl Canc Inst Monograph 27:131-124.
77. Roy D, Liehr JG (1999) Estrogen, DNA damage and mutations. Mutation Res 424:107-115.
78. Yue W, Santen RJ, Wang JP, et al (2003) Genotoxic metabolites of estradiol in breast: potential mechanism of estradiol induced carcinogenesis. J Steroid Biochem& Mol Biol 86:477-486.

79. The ATAC Trialist's Group (2002) Anastrozole alone or in combination with tamoxifen versus tamoxifen alone for adjuvant treatment of postmenopausal women with early breast cancer: first results of the ATAC randomised trial. Lancet 359:2131-2139.
80. Santen RJ, Samojlik E, Wells SA (1980) Resistance of the ovary to blockade of aromatization with aminoglutethimide. J Clin Endocrinol Metab 51:473-477.
81. Harris AL, Dowsett M, Jeffcoate SL, et al (1982) Endocrine and therapeutic effects of aminoglutethimide in premenopausal patients with breast cancer. J Clin Endocrinol Metab 55:718-720.
82. Wander HE, Blossey HC, Nagel GA (1986) Aminoglutethimide in the treatment of premenopausal patients with metastatic breast cancer. Euro J Canc Clin Oncol 22:1371-1374.
83. Harper-Wynne C, Ross G, Sacks N, et al (2002) Effects of the aromatase inhibitor letrozole on normal breast epithelial cell proliferation and metabolic indices in postmenopausal women: a pilot study for breast cancer prevention. Cancer Epidemiol Biomarkers Prev 11:614-621.
84. Goss PE (2002) Aromatase inhibitors and chemoprevention of breast cancer. In: Miller WR, Ingle JN (eds): Endocrine Therapy in Breast Cancer, Basel, CH:Marcel Dekker, 309-317.

PART 5. PROSTATE I: ANDROGEN/ESTROGEN ACTION

14

Androgens and Prostate Cancer Etiology: Sorting Through the Evidence

Ronald K. Ross, Leigh Pearce, Juergen K.V. Reichardt, and Gerhard A. Coetzee

Introduction

Prostate cancer (PCA) is by far the most common cancer in USA men. 221,000 cases were diagnosed in 2003, far exceeding lung cancer, the second most common cancer, for which 92,000 cases occurred (1). In fact, PCA is the most common cancer in either gender, exceeding even female breast cancer for which approximately 211,000 cases were diagnosed. Although some PCA is curable when detected at an early stage, and other PCAs follow a somewhat indolent clinical course, PCA is, nonetheless, also the second leading cause of cancer deaths in USA men, with nearly 29,000 in 2003.

The epidemiology of PCA is complex and effective treatment of PCA, at least for advanced stage disease, has remained elusive. While there remain many important unanswered questions regarding both prevention and therapeutic strategies for PCA, a complete understanding of three important and possibly interrelated issues would likely pave the way to substantial control of this ever growing epidemic. 1) PCA is among the most internationally and racial ethnically variant of all cancers. African-Americans have by far the highest rates of PCA in the world, whereas Asian populations in their native countries (*i.e.,* Japanese, Chinese, and Koreans) have by far the lowest. Historically, there has been as much as a 100.0-fold difference in PCA risk between these extremes (2). The reasons for this remarkable difference remain poorly understood. 2) While the difference in incidence of clinically detectable PCA among different countries and among various racial-ethnic groups is extraordinary, differences in the underlying prevalence of "occult" or clinically non-detectable disease across these same populations is substantially less (3). Moreover, the prevalence of occult disease, as determined by autopsy, approaches 30% for men in their early 50's, which far exceeds the lifetime incidence of clinical PCA even among very high risk groups such as African-Americans (3). It is likely, though unproven, that clinically detectable disease evolves from these occult lesions. If so, it is clear that only a fraction of these lesions progress, and at highly variable rates among different individuals and across populations. The reasons for this variability are unknown. 3) Nearly all advanced PCA responds initially to androgen ablation therapy, but after a variable period of remission, every PCA becomes refractory to such treatment, and then progresses incessantly to eventual death in nearly all patients. (4). Why PCA becomes hormone refractory remains a mystery.

We believe that understanding why there exists such extraordinary variation in incidence across racial-ethnic groups, what controls the varying rates of progression from occult to clinically relevant disease between individuals and across populations, and what causes PCA to escape androgen ablation therapy, are the three keys to effectively controlling the disease.

The pioneering work of Huggins in the 1940's showing that androgen ablation therapy was an effective, albeit temporary, treatment for early metastatic PCA (5), combined with the longstanding knowledge that normal prostate growth is under androgen control (6), has led to a large body of research devoted to understanding fully the role of androgens in PCA etiology, progression and therapy. In fact, our research program at the Keck School of Medicine at USC has focused on androgen signaling pathways in trying to further our understanding of the three crucial questions posed above. *We focus here primarily on the etiologic question.*

Major Risk Factors

PCA is probably the least understood epidemiologically of all major cancers. In fact there are still only three fully established risk factors: 1) **Age.** PCA is the most age-related of all epithelial cancers. Historically, it has been extremely rare prior to age 40 (more cases are now being diagnosed with the introduction of prostate specific antigen (PSA) screening to detect pre-clinical disease), but thereafter the increase in incidence with age is greater than for any other cancer (incidence increasing at approximately the 8th-9th power of age) (7). 2) **Family History.** The strong familiality of PCA has been fully recognized only during the past decade or so (8). Men with a first degree relative (brother or father) with PCA have approximately a 2.0-3.0-fold increase in risk compared to men with no such family history. Risk is greater if the relative had PCA at a relatively young age or if multiple relatives have been affected (8). 3) **Race-ethnicity/International Variation.** As noted above, there exists extraordinary racial-ethnic and across country variation in PCA risk, with African-American men having the highest risk worldwide and Asian populations the lowest. Asian-American men have PCA rates that are substantially higher than those in their homelands, but even among highly acculturated Asian-Americans, such as Japanese in Los Angeles (LA), rates never approach those of whites much less African-Americans. Unfortunately, we are unaware of any truly comparable data on PCA incidence rates among blacks in Africa with which to compare the rates in African-Americans. PCA rates in most areas of the USA showed marked secular trends following the implementation of large scale use of PSA testing in the late 1980s. In LA, this led to rapid increases in incidence which began to taper and decline in the mid 1990s, but rates remain well above those of the pre-PSA era. Noteworthy, these increases and subsequent declines occurred somewhat in parallel among different racial-ethnic groups with the peak among African-Americans and Latinos offset about two years compared to whites (Figure 1).

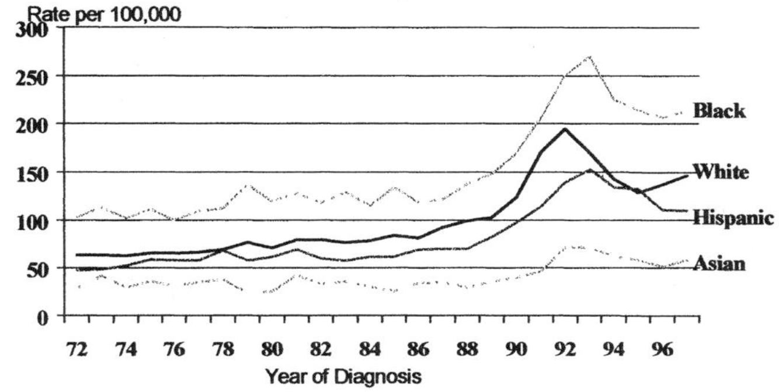

Figure 1. Time trends of prostate cancer incidence rates by race/ethnicity, LA Co., CA, 1972-1997. All rates are age-adjusted to the 1970 USA population.

Overview: Androgens and Prostate Cancer

Our initial interest in the role of androgens in PCA pathogenesis and progression was stimulated by our more general interest in the role of cell division in carcinogenic transformation (9), and the recognition that androgens are essential for such division to occur in prostate tissue (6). Other work which supported the possible importance of androgens further stimulated that interest, in particular that all existing animal models of adenocarcinoma of the prostate development have an absolute requirement for androgens (10). In some instances, androgens alone are sufficient to cause the disease (11).

In the late 1980s and early 1990s, we initiated a series of studies to determine whether there was any evidence of underlying differences in the androgen milieu among populations at widely varying risk of PCA, *i.e.,* African-Americans, USA whites, and rural Japanese, that might suggest an androgen role in this variability. We found that African-American men have substantially higher testosterone levels at a young age than young men in the other two groups (12), and that Japanese men, while not having lower testosterone than whites as anticipated, have substantially lower circulating levels of androstanediol glucuronide (13). This hormone is a whole body index of 5α-reductase, the enzyme required to bioactivate testosterone in prostate cells through irreversible conversion to dihydrotestosterone (DHT) (6). Our research, while focused entirely on healthy populations, nonetheless suggested the possibility that there are differences in testosterone biosynthesis and metabolism across racial-ethnic groups that might help explain their varying rates of PCA. Of course, demonstrating that testosterone or testosterone metabolite levels in healthy men predict their subsequent risk of PCA would provide more direct evidence for a role of androgens in PCA etiology. Although several such prospective studies have been published with inconsistent results (10), the best-designed study to test this hypothesis in terms of laboratory rigor and adequate statistical power, the Physicians Health Follow-up Study, found that men in the highest quartile of circulating testosterone had 2.6 times the likelihood of developing PCA compared to men in the lowest quartile ($p = 0.004$ for trend) (14).

We have been exploring whether these observed racial-ethnic differences in androgenic milieu might have a genetic basis; in particular, whether there might exist genetic variants in androgen metabolism, transport or activity genes that impact PCA risk within and between racial-ethnic groups. Initially, we have explored two genes – the androgen receptor (AR) and steroid 5α-reductase type II genes – and these investigations have proven helpful in understanding the possible connection of candidate genes to PCA generally, and have also helped us lay out a more global strategy for future investigations.

The Androgen Receptor Gene

The AR is a member of the superfamily of steroid receptors and like most receptors in this family has a DNA binding domain, a hormone binding domain and a transcription modulatory domain which for the AR is completely encoded by the large exon 1 of the AR gene (15). The AR gene maps to the long arm of the X chromosome (15). Within exon 1 are two highly polymorphic tri-nucleotide repeats [a $(CAG)_n$ and a $(GGC)_n$]. We became interested initially in the CAG repeat with the recognition that an expansion of that repeat from the normal size range of approximately 8-33 to 36 or greater, was the cause of an uncommon adult onset motor neuron disease, spinal and bulbar muscular atrophy, or Kennedy's disease (16). Men with Kennedy's disease have evidence of hypoandrogenization, including gynecomastia, low sperm counts, and sub-fertility (17). *In-vitro* studies have demonstrated that ARs with an expanded glutamine tract encoded by this repeat, bind androgens normally but transactivate androgen responsive genes substantially sub-optimally (16). As the expansion of this repeat is the sole cause of Kennedy's disease and the apparent sole cause of the hypoandrogenization in such men, we hypothesized that there might be altered androgen activity related to the size of the CAG repeat within the normal range of repeats, *i.e.*, that androgen transactivation activity is reduced with increasing number of repeats (18). We further hypothesized that shorter repeats, if they are associated with higher androgen transactivation activity, might also be associated with PCA risk. Consistent with this hypothesis, African-American men have the shortest average repeat length, Asians the longest, with whites intermediate, as expected from their respective PCA rates (19).

This hypothesis gained additional credibility when it was shown in transfection assays that transactivation activity is negatively correlated with repeat length within the normal range (20), just as it was in the expanded repeat range seen in Kennedy's disease. We proceeded to explore this hypothesis directly in a small population based case-control study in LA County, in which the case group was comprised of white men under age 65 at PCA diagnosis and the control group was comprised of a comparably aged group of healthy white men from the neighborhoods in which the cases lived at the time of diagnosis. We showed in this study that men with CAG repeat length under 20 had a 2.0-fold increase in PCA risk compared to men with a repeat length of 20 or more. This effect was particularly pronounced for advanced disease (OR = 2.36, 95% CI = 1.02, 5.49) (21). This hypothesis has been among the most investigated of all molecular epidemiologic hypotheses of PCA to date (22). Many of these investigations are summarized in Table 1. Not all of these studies are of the same quality, some suffering from low statistical power, others from cases chosen from convenience samples or from controls drawn from non-representative populations. Nonetheless, while there is considerable inconsistency in findings across studies, most find some inverse relationship – either overall, limited to advanced disease or for early onset PCA only – between PCA risk and CAG repeat length.

It is interesting to think about how changes in androgen activity due to incremental changes in CAG repeat length might be expected to alter PCA risk. Cancer rates increase as a logarithmic function of age; for PCA, the most age-related of all cancers, incidence increases by age raised to approximately the 8th power. For most epithelial cancers, including PCA, the relationship between incidence, I, with age, t, can be represented by the equation [$I(t) = axt^k$], which produces a straight line of slope k when the logarithm of incidence is plotted against the logarithm of age (39). The fundamental idea is that "aging" of a tissue relates to its cell kinetics, *i.e.,* its effective mitotic rate. When the tissue is not undergoing cell division, the rate of aging is zero, whereas aging is maximal when the mitotic rate is maximal.

Under this model if a single repeat increment results in a 10% increase in androgen activity and, therefore, in prostate tissue "aging," this would translate into potentially as much as a 2.4-fold lifetime increase in PCA risk; a 2% increase would translate into a 20% increase in risk, whereas a 1% increase would translate into a 9% increase, and a 0.5% into a 5% lifetime increase (Table 2).

We have done additional work to understand better mechanistically how CAG repeat length affects androgen activity, and also have begun preliminary studies of how this genetic effect might interact with other genetic variants in the androgen signaling pathway to modify PCA risk. Coetzee, *et al.* have shown, for example, the importance of co-activator proteins in realizing the full impact of CAG repeat length on transactivation activity. His lab has shown that members of the p160 family of co-activator proteins (A1B1 and GRIP1) bind to the region of the poly-glutamine tract encoded by the $(CAG)_n$ repeat and that, *in-vitro*, ligand alone (*i.e.,* DHT) is insufficient for any impact of CAG length on transactivation activity (40). Adding A1B1 or GRIP1 together with DHT not only substantially increases transactivation activity in this system, but also clarifies the substantial impact of the length of the poly-glutamine tract on such activity (40). On average, there is approximately a 0.5%-2% decline in activity with each additional poly-glutamine in these *in-vitro* systems. Based on Table 2, under reasonable assumptions regarding PCA incidence, this would result in a 4% or greater change in lifetime PCA incidence.

Table 1. Summary of Studies Evaluating the Roles of the AR CAG Microsatellite in PCA Risk, Progression, and Age at Onset (22).

Study	Subjects	Risk	Stage/ Grade	Age at Onset
Pilot studies				
Irvine, *et al.* (15)	USA White	yes	N/A[1]	N/A
Hardy, *et al.* (23)	USA White	N/A	no	yes
Ingles, *et al.* (21)	USA White	yes	yes	N/A
Hakimi, *et al.* (24)	USA White	yes	yes	no
Matched case-control studies				
Giovannucci, *et al.* (25)	USA White	yes	yes	no
Stanford, *et al.* (26)	USA White	yes	no	yes
Hsing, *et al.* (27)	Chinese	yes	no	no
Beilin, *et al.* (28)	Australian White	no	no	yes
Other studies				
Ekman, *et al.* (29)	Swedish White	yes	N/A	N/A
Edwards, *et al.* (30)	British White	no	no	N/A
Correa-Cerro, *et al.* (31)	French/German White	no	no	no
Bratt, *et al.* (32)	Swedish White	no	yes	yes
Lange, *et al.* (33)	USA White High Risk	no	no	no
Nam, *et al.* (34)	Canadian	N/A	yes	N/A
Latil, *et al.* (35)	French White	no	no	yes
Modugno, *et al.* (36)	USA White	yes	N/A	N/A
Miller, *et al.* (37)	USA White	no	N/A	N/A
Panz, *et al.* (38)	South Africans Black & White	yes	yes	N/A

[1] N/A = Not applicable or not assessed

Table 2. Hypothesized Relationship Between Changes in Androgen Activity and Lifetime PCA Risk Resulting from an Incremental Change in CAG repeat Length[1].

Δ in % Androgen Activity	Lifetime Increase in % Incidence
10	114
2	13
1	8
0.5	4

[1] Based on the following equation: $[I(t) = axt^k]$ where I = incidence of PCA at age (t); k = 8

Although the full set of genes under various conditions that are activated by the AR are still poorly understood, the PSA gene is one whose expression is consistently tied to androgen receptor activity. We have found some evidence that PSA levels are correlated with $(CAG)_n$ size in the AR gene. A non-silent SNP (A→G) has been described in the androgen response element of the PSA gene (41). We explored preliminarily the relationship of that SNP with PCA risk in conjunction with CAG size in the AR gene. The GG PSA genotype was associated with advanced PCA regardless of AR CAG genotype and CAG genotype was associated with risk regardless of PSA genotype. However, the strongest relationship with PCA risk was for the combined PSA GG, AR CAG short genotypes (OR = 9.6, 95% CI = 2.0, 45.5) (Table 3) (42).

Table 3. PCA Risk, Cross-classified by PSA and AR Genotypes[+].

Genotype PSA	AR	No. of Controls (%)	No. of Cases (%)	95% CI Odds Ratio	No. of Advanced Cases (%)	95% CI Odds Ratio
Not GG	Long[1]	72 (52)	21 (37)	1.0 (referent)	5 (19)	1.0 (referent)
GG	Long[1]	30 (22)	12 (21)	1.4 (0.6-3.1)	9 (35)	4.3 (1.3-14.0)
Not GG	Short[1]	31 (22)	16 (28)	1.8 (0.8-3.8)	8 (31)	3.7 (1.1-12.3)
GG	Short[1]	6 (4)	8 (14)	4.6 (1.4-14.7)	4 (15)	9.6 (2.0-45.5)

[1] Long = > 20 CAG repeats [2] Short = < 20 CAG repeats

Our research activity related to the AR gene has been useful from several perspectives. It has illustrated the potential importance of focusing on polygenic origins, the critical importance of combining functional studies with epidemiologic analyses in understanding genotype/disease relationships, and the potential for transcribing knowledge gained in understanding the genetic origin of one disease to understanding others.

The Steroid 5α-Reductase Type 2 Gene

The other gene that our research program has focused on is the steroid 5α reductase type 2 (SRD5A2) gene. There are at least two 5α reductase isozymes encoded by 2 genes on different chromosomes (43). Although early work suggested that only the type II enzyme was expressed in prostate tissue, more recent work suggests that the more ubiquitous type I enzyme is as well (44). We began our work with the SRD5A2 gene by systematically sequencing the gene in a group of men with very high or very low circulating levels of androstanediol glucuronide, an index of whole body 5α reductase activity. We identified initially 25 single nucleotide polymorphisms (SNPs) in the protein coding region – 10 missense and 15 silent (*i.e.,* third base pair substitutions) (45). We have subsequently identified 11 additional SNPs in the putative promoter region and 5' and 3' UTRs. We have reconstituted each of the missense coding region SNPs by site directed mutagenesis in the SRD5A2 cDNA, and then overexpressed these and assayed them biochemically and pharmacogenetically (Table 4). We found an apparent 200-fold range in enzyme activity, measured as the apparent V_{max}, and a 60-fold pharmacogenetic variation as measured by finasteride inhibition (46). Several of these SNPs looked particularly interesting based on these *in-vitro* studies, in particular the A49T (alanine replaced by threonine at codon 49) polymorphism which was associated with an apparent substantial increase in enzyme activity, and the R227Q (arginine at codon 227 replaced by glutamine) polymorphism which was associated with almost non-detectable enzyme activity. Subsequently, we showed that both African-American and Latino carriers of the T allele of A49T had substantial and statistically significant increases (approximately 5.0-6.0-fold) in advanced PCA compared to men with wild type genotype (47). We have found among such carriers to date in either whites or Japanese-Americans. Hsing, *et al.* found an increase in risk in Chinese men carrying the T allele (48), but other investigations of A49T have yielded inconsistent results (49). The R227Q polymorphism has been found only in Asian, especially Chinese men, but never as a homozygous variant. In fact, the R227Q mutation has been reported in Chinese with male psudo-hermaphroditism consistent with both the observed low *in-vitro* activity, and to our inability to find this mutation in a homozygous state among either healthy individuals or PCA cases (50).

As with our work on the AR, the SRD5A2 work has also proven valuable in developing strategies for understanding genotype/disease relationships. In particular, this work has provided a useful paradigm of proceeding from targeted biochemical studies to detailed gene studies to SNP-based functional studies to epidemiologic association studies. The biochemical and molecular epidemiologic studies of 5α reductase and PCA led directly to the ongoing USA PCA chemoprevention trial of finasteride, a specific 5α reductase type II inhibitor (51). There exists the distinct possibility that trial efficacy might depend strongly on underlying SRD5A2 genotypes, as we have demonstrated strong pharmacogenetic variation *in-vitro* (Table 4).

Table 4. Biochemical Characterization of SRD5A2 Mutant Enzymes *In vitro* (46)

Enzyme	Km NADPH (μM)	Km V_{max} (μM)	Finasteride (nmoles/min/mg)	K_i Protein (nM)	(T) (%)
WT (normal)	0.9	8	1.9	60	100
C5R	0.9	8	1.8	-	103
P30L	2.1	21	0.5	420	102
A49T	2.7	7	9.9	180	98
V89L	0.6	8	1.1	113	114
T187M	1.1	47	0.8	-	125
R227Q	4.6	38	0.06	260	91
F234L	1.6	21	1.4		87

Future Research Agenda

Androgen signaling pathways in the prostate are, of course, exceedingly complex and include roles for products of genes involved in androgen biosynthesis, transport, bioactivation and metabolism, as well as various coactivator proteins and the products of various downstream genes. Additional complexity relates to the role of intersecting pathways [*i.e.,* obesity-insulin-insulin like growth factor signaling pathway (52)], and to the recent recognition that AR transactivation can occur both through alternative ligands, as well as through non-classical ligand pathways (53). While androgen-related genes other than the AR and SRD5A2 have been preliminarily investigated in relationship to PCA, none have been extensively studied (49). It is clear that a full understanding of the contribution of genetic variants to this etiologic pathway overall will require a fuller understanding of the complexity of the pathway, and of variation and functional relevance of this variation in the genes themselves.

To date, the general approach to candidate gene association studies for PCA, as well as most other chronic diseases, has been primarily a SNP genotype-based approach with a somewhat arbitrary, in many cases, selection of SNPs.

Although utilization of biochemical correlate or *in-vitro* studies has reduced the randomness of this process in some instances (as with the AR and SRD5A2 work described above) there is a clear need for a more systematic pathway driven approach. Investigators in our Epidemiology Program at USC have been collaborating with molecular geneticists at the Whitehead Institute Center for Genomic Research at MIT to fully characterize SNP variations in a large series of candidate genes including many known to be involved in androgen signaling pathways. The goal of this research is to conduct disease association studies utilizing a SNP haplotype-based, rather than the traditional SNP genotype-based approach. The pattern of linkage disequilibrium (LD) varies widely across the human genome, but discrete regions of high LD (blocks) exist across populations (54). Within these blocks haplotype diversity is limited. Due to the high degree of LD within these blocks, a subset of haplotype tag SNPs (htSNPs) can be selected to efficiently identify the common haplotypes (55). While empirical data that this SNP haplotype-based approach is superior to SNP genotype-based methodology in disease association studies are limited, this approach provides the huge potential advantage of characterizing variation across an entire genomic locus in a highly cost efficient manner. Importantly for PCA, our collaborative effort with the Whitehead is conducting these detailed characterizations across racial-ethnic groups.

References

1. Jemal A, Murray T, Samuels A, et al (2003) Cancer statistics 2003. CA Cancer J Clin 53:5-26, 2003.
2. Ross RK, Schottenfeld D (1996) Prostate cancer. In: Schottenfeld D, Fraumeni JF (eds) Cancer Epidemiology and Prevention, 2nd Edition. New York:Oxford University Press, 1180-1206.
3. Sakr WA, Grignon DJ, Crissman JD, et al (1994) High grade prostatic intraepithelial neoplasia (HGPIN) and prostatic adenocarcinoma between the ages of 20-69: an autopsy study of 249 cases. In Vivo 8:439-443.
4. Hellerstedt BA, Pienta KJ (2002) The current state of hormonal therapy for prostate cancer. CA Cancer J Clin 52:154-179.
5. Huggins C, Hodges CV (1941) Studies on prostatic cancer: effect of castration, of estrogen, and of androgen injection on serum phosphatases in metastatic carcinoma of the prostate. Cancer Res 1:293-297.
6. Cheng E, Lee C, Grayhack J (1993) Endocrinology of the prostate. In: Lepor H, Lawson RK (eds) Prostate Diseases. Philadelphia:W.B. Saunders, 57-71.
7. Henderson BE, Ross RK, Pike MC, et al (1982) Endogenous hormones as a major factor in human cancer. Cancer Res 42:3232-3239.
8. Carter BS, Beaty TH, Steinberg GD, et al (1992) Mendelian inheritance of familial prostate cancer. Proc Natl Acad Sci USA 89:3367-3371.

9. Ross RK, Pike MC, Coetzee GA, et al (1998) Androgen metabolism and prostate cancer: establishing a model of genetic susceptibility. Cancer Res 58:4497-4504.
10. Hsing AW, Reichardt JK, Stanczyk FZ (2002) Hormones and prostate cancer: current perspectives and future directions. Prostate 52:213-235.
11. Noble RL (1977) The development of prostatic adenocarcinoma in Nb rats following prolonged sex hormone administration. Cancer Res 37:1929-1933.
12. Ross RK, Bernstein L, Judd H, et al (1986) Serum testosterone levels in healthy young black and white men. J Natl Cancer Inst 76:45-48.
13. Ross RK, Bernstein L, Lobo RA, et al (1992) 5-Alpha reductase activity and risk of prostate cancer among Japanese and US White and Black males. Lancet 339:887-889.
14. Gann PH, Hennekens CH, Ma J, et al (1996) Prospective study of sex hormone levels and risk of prostate cancer. J Natl Cancer Inst 88:1118-1126.
15. Irvine RA, Yu MC, Ross RK, et al (1995) The CAG and GGC microsatellites of the androgen receptor gene are in linkage disequilibrium in men with prostate cancer. Cancer Res 55:1937-1940.
16. Mhatre AN, Trifiro MA, Kaufman M, et al (1993) Reduced transcriptional regulatory competence of the androgen receptor in X-linked spinal and bulbar muscular atrophy. Nat Genet 5:184-188.
17. Tut T, Ghadessy FJ, Trifiro MA, et al (1997) Long polyglutamine tracts in the androgen receptor are associated with reduced trans-activation, impaired sperm production, and male infertility. J Clin Endocrinol Metab 82:3777-3782.
18. Coetzee GA, Ross RK (1994) Prostate cancer and the androgen receptor. J Natl Cancer Inst 86:872-873.
19. Edwards A, Hammond HA, Jin L, et al (1992) Genetic variation at five trimeric and tetrameric tandem repeat loci in four human population groups. Genomics 12:241-253.
20. Chamberlain NL, Driver ED, Miesfeld RL (1994) The length and location of CAG trinucleotide repeats in the androgen receptor N-terminal domain affect transactivation function. Nucleic Acids Res 22:3181-3186.
21. Ingles SA, Ross RK, Yu MC, et al (1997) Association of prostate cancer risk with genetic polymorphisms in vitamin D receptor and androgen receptor. J Natl Cancer Inst 89:166-170.
22. Tilley WD, Buchanan G, Coetzee GA (2003) Androgen receptor signaling in prostate cancer. In: Henderson BE, Ponder B, Ross RK (eds) Hormones, Genes and Cancer. Oxford:Oxford University Press, 288-315.

23. Hardy DO, Scher HI, Bogenreider T, et al (1996) Androgen receptor CAG repeat lengths in prostate cancer: correlation with age of onset. J Clin Endocrinol Metab 81:4400-4405.
24. Hakimi JM, Schoenberg MP, Rondinelli RH, et al (1997) Androgen receptor variants with short glutamine or glycine repeats may identify unique subpopulations of men with prostate cancer. Clin Cancer Res 3:1599-1608.
25. Giovannucci E, Stampfer MJ, Krithivas K, et al (1997) The CAG repeat within the androgen receptor gene and its relationship to prostate cancer. Proc Natl Acad Sci USA 94:3320-3323.
26. Stanford JL, Just JJ, Gibbs M, et al (1997) Polymorphic repeats in the androgen receptor gene: molecular markers of prostate cancer risk. Cancer Res 57:1194-1198.
27. Hsing AW, Gao YT, Wu G, et al (2000) Polymorphic CAG and GGN repeat lengths in the androgen receptor gene and prostate cancer risk: a population-based case-control study in China. Cancer Res 60:5111-5116.
28. Beilin J, Harewood L, Frydenberg M, et al (2001) A case-control study of the androgen receptor gene CAG repeat polymorphism in Australian prostate carcinoma subjects. Cancer 92:941-949.
29. Ekman P, Gronberg H, Matsuyama H, et al (1999) Links between genetic and environmental factors and prostate cancer risk. Prostate 39:262-268.
30. Edwards SM, Badzioch MD, Minter R, et al (1999) Androgen receptor polymorphisms: association with prostate cancer risk, relapse and overall survival. Int J Cancer 84:458-465.
31. Correa-Cerro L, Wohr G, Haussler J, et al (1999) (CAG)nCAA and GGN repeats in the human androgen receptor gene are not associated with prostate cancer in a French-German population. Eur J Hum Genet 7:357-362.
32. Bratt O, Borg A, Kristoffersson UH, et al (1999) CAG repeat length in the androgen receptor gene is related to age at diagnosis of prostate cancer and response to endocrine therapy, but not to prostate cancer risk. Br J Cancer 81:672-676.
33. Lange EM, Chen H, Brierley K, et al (2000) The polymorphic exon 1 androgen receptor CAG repeat in men with a potential inherited predisposition to prostate cancer. Cancer Epidemiol Biomarkers Prev 9:439-442.
34. Nam RK, Elhaji Y, Krahn MD, et al (2000) Significance of the CAG repeat polymorphism of the androgen receptor gene in prostate cancer progression. J Urol 164:567-572.
35. Latil AG, Azzouzi R, Cancel GS, et al (2001) Prostate carcinoma risk and allelic variants of genes involved in androgen biosynthesis and metabolism pathways. Cancer 92:1130-1137.

36. Modugno F, Weissfeld JL, Trump DL, et al (2001) Allelic variants of aromatase and the androgen and estrogen receptors: toward a multigenic model of prostate cancer risk. Clin Cancer Res 7:3092-3096.
37. Miller EA, Stanford JL, Hsu L, et al (2001) Polymorphic repeats in the androgen receptor gene in high-risk sibships. Prostate 48:200-205.
38. Panz VR, Joffe BI, Spitz I, et al (2001) Tandem CAG repeats of the androgen receptor gene and prostate cancer risk in black and white men. Endocrine 15:213-216.
39. Pike MC (1987) Age-related factors in cancers of the breast, ovary, and endometrium. J Chronic Dis 40:59S-69S.
40. Irvine RA, Ma H, Yu MC, et al (2000) Inhibition of p160-mediated coactivation with increasing androgen receptor polyglutamine length. Hum Mol Genet 9:267-274.
41. Rao A, Cramer SD (1999) Identification of a polymorphism in the ARE I region of the PSA promoter. Proc Am Assoc Cancer Res 40:65.
42. Xue W, Irvine RA, Yu MC, et al (2000) Susceptibility to prostate cancer: interaction between genotypes at the androgen receptor and prostate-specific antigen loci. Cancer Res 60:839-841.
43. Russell DW, Wilson JD (1994) Steroid 5 alpha-reductase: two genes/two enzymes. Annu Rev Biochem 63:25-61.
44. Thigpen AE, Silver RI, Guileyardo JM, et al (1993) Tissue distribution and ontogeny of steroid 5 alpha-reductase isozyme expression. J Clin Invest 92:903-910.
45. Reichardt JKV, Makridakis N, Henderson BE, et al (1995) Genetic variability of the human SRD5A2 gene: implications for prostate cancer risk. Cancer Res 55:3973-3975.
46. Makridakis NM, di Salle E, Reichardt JK (2000) Biochemical and pharmacogenetic dissection of human steroid 5 alpha-reductase type II. Pharmacogenetics 10:407-413.
47. Makridakis NM, Ross RK, Pike MC (1999) Association of mis-sense substitution in SRD5A2 gene with prostate cancer in African-American and Hispanic men in Los Angeles, USA. Lancet 354:975-978.
48. Hsing AW, Chen C, Chokkalingam AP, et al (2001) Polymorphic markers in the SRD5A2 gene and prostate cancer risk: a population-based case-control study. Cancer Epidemiol Biomarkers Prev 10:1077-1082.
49. Coughlin SS, Hall IJ (2002) A review of genetic polymorphisms and prostate cancer risk. Ann Epidemiol 12:182-196.
50. Guang W, Zhou L, Zeng R, et al (1999) Detection of R277Q mutation of SRD5alpha2 gene by amplification refractory mutation system. Zhonghua Yi Xue Yi Chuan Xue Za Zhi 16:390-391.
51. Stoner E (1996) 5 alpha-reductase inhibitors/finasteride. Prostate Suppl 6:82-87.

52. Chan JM, Stampfer MJ, Giovannucci E (1998) Plasma insulin-like growth factor-I and prostate cancer risk: a prospective study. Science 279:563-566.
53. Buchanan G, Irvine RA, Coetzee GA, et al (2001) Contribution of the androgen receptor to prostate cancer predisposition and progression. Cancer Metastasis Rev 20:207-223.
54. Gabriel SB, Schaffner SF, Nguyen H, et al (2002) The structure of haplotype blocks in the human genome. Science 296:2225-2229.
55. Stram DO, Haiman CA, Hirschhorn JN, et al (2003) Choosing haplotype-tagging SNPs based on unphased genotype data using a preliminary sample of unrelated subjects with an example from the Multiethnic Cohort Study. Hum Hered 55:27-36.
56. Veldscholte J, Ris-Stalpers C, Kuiper GG, et al (1990) A mutation in the ligand binding domain of the androgen receptor of human LNCaP cells affects steroid binding characteristics and response to anti-androgens. Biochem Biophys Res Commun 173:534-540.
57. Buchanan G, Greenberg NM, Scher HI (2001) Collocation of androgen receptor gene mutations in prostate cancer. Clin Cancer Res 7:1273-1281.
58. Koivisto P, Visakorpi T, Kallioniemi OP (1996) Androgen receptor gene amplification: a novel molecular mechanism for endocrine therapy resistance in human prostate cancer. Scand J Clin Lab Invest Suppl 226:57-63.
59. Gregory CW, He B, Johnson RT, et al (2001) A mechanism for androgen receptor-mediated prostate cancer recurrence after androgen deprivation therapy. Cancer Res 61:4315-4319.
60. Wen Y, Hu MC, Makino K, et al (2000) HER-2/neu promotes androgen-independent survival and growth of prostate cancer cells through the Akt pathway. Cancer Res 60:6841-6845.
61. Akalu A, Dlmajian DA, Highshaw RA, et al (1999) Somatic mutations at the SRD5A2 locus encoding prostatic steroid 5 alpha-reductase during prostate cancer progression. J Urol 161:1355-1358.

15

Id-1 Protein as a New Marker for PCA

YC Wong, MT Ling, XH Wang, XS Ouyang, ALM Cheung, Franky L Chan

Introduction

Prostate cancer (PCA) is one of the most common cancers and is the second killer among men in the USA. Despite extensive research in the last two decades, the mechanism of prostate carcinogenesis remains not well understood. In order to study the process of prostate carcinogenesis in a systematic fashion, we employed an approach modified from Noble's original design by implanting a combination of testosterone and estrogens capsules subcutaneously for a period of up to12 mo (1). The hormone capsules were replaced every 3 mo to ensure that a high level of serum hormone was maintained. In this model, hyperplasia/dysplasia was detected as early as 2-3 mo, and fully developed prostate adenocarcinomas first appeared ~ 5 mo post-treatment. After 10-12 mo, almost all rats developed PCA. During the process of carcinogenesis, there were changes in expression of a number of growth factors, including IGF-1, VEGF, TGFβ, HGF, and their receptors together with a change in secretory protein profiles (1-4). Concurrently, there was also a change in stromal structural organization involving derangement as well as dedifferentiation of smooth muscle to fibroblastic cells (2, 5).

Detection of Id-1 Over-expression in PCA in the Rat Model

Using this model, we studied the gene expression profile during prostate carcinogenesis using a cDNA array method. The results were confirmed by RT-PCR, western blotting, and immunohistochemical (IHC) analyses. Seventeen genes were differentially expressed, and three of them with the highest level of over-expression were selected for further analysis. They included: Testosterone-repressed prostatic message-2 (TRPM-2), matrix metalloproteinase-7, (MMP-7), and inhibitor of differentiation or DNA binding (Id-1) (6). Increased expression of TRPM-2 and MMP-7 was observed in both pre- and malignant samples after sex hormone treatment, indicating their role in the early stages of hormone response and PCA development. In contrast, Id-1 was expressed at relatively low levels in all pre- malignant samples, but its level of expression increased in malignant cells, suggesting its potential role as a PCA biomarker. More significantly, we found that the level of Id-1 expression was higher in poorly differentiated lesions than in well-differentiated carcinomas (Figure 1), suggesting that the levels of Id-1 expression may be correlated with tumor malignancy (6).

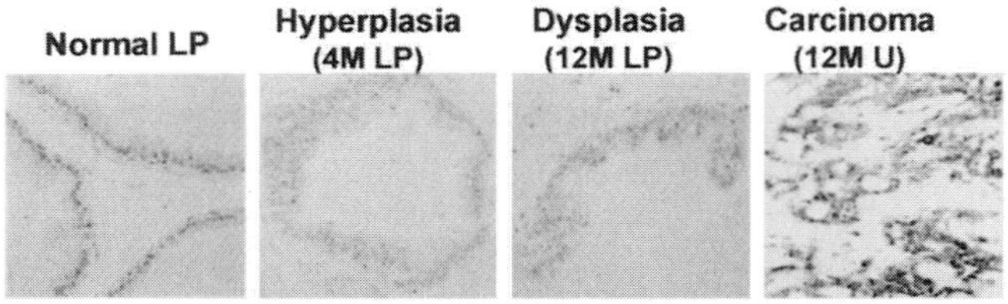

Figure 1. IMH study of Id-1 protein expression. Sections from normal LPs, hyperplasia, dysplasia, and carcinoma regions were stained with specific Id-1 antibodies. Note absent to weak Id-1 expression detected in normal LPs and in non-malignant lesions, while PCA cells exhibit high Id-1 immunoreactivity (400 x).

Nature of Id-1 Protein

Id proteins belong to a protein family with molecular weights ranging from 13-20 kDa and helix-loop-helix (HLH) domains. There are four family members; Id-1 to Id-4. Their function is to inhibit basic HLH transcription factors from binding to DNA by heterodimerization with basic HLH proteins, thus, inhibiting the transcription of differentiation associated genes (9, 10). Id-1 was first discovered as a negative regulator of DNA binding and an inhibitor of cell differentiation in myoblasts (7). It is expressed in a variety of cell types including epithelial cells. Recently, Id-1 expression has been reported in many tumor types, *i.e.,* breast and pancreatic cancers. Increased expression of Id-1 has been associated with proliferation potential and invasion ability (11-13). In addition, Id-1 expression does not only restore the ability of DNA synthesis in senescent human fibroblast cells, but, in addition, results in the immortalization of human keratinocytes (9, 14), indicating its positive role in cell proliferation and carcinogenic process. Moreover, Id-1 appears to be an essential factor promoting G1/S cell cycle transition in certain cell types (14-16). In human breast cancer (BC), Id-1 is more commonly expressed in invasive carcinomas compared to ductal carcinomas *in-situ* (DCIS) (11).

Role of Id-1 in Human PCA

Since in the Noble rat model, the levels of Id-1 expression correlate to those of PCA, we hypothesize that Id-1 over-expression occur in human PCA. In addition, if Id-1 levels are positively correlated with tumor malignancy, Id-1 may serve as a useful PCA prognostic marker. To test our hypothesis, we examined the expression of Id-1 protein and mRNA by IMH and *in-situ* hybridization analysis in human normal prostate, benign prostatic hyperplasia (BPH), and cancer biopsies (Figure 2) (17). In addition, we correlated the difference in Id-1 expression with PCA grade using the Gleason classification (17).

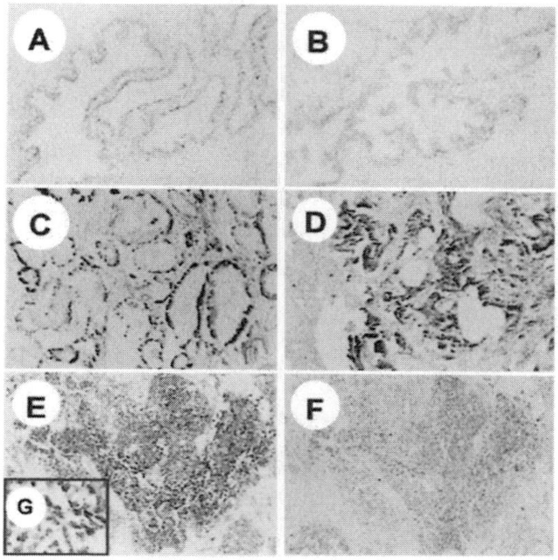

Figure 2. Id-1 *in-situ* hybridization in normal prostate, BPH, and malignant prostate specimens. **(A).** Normal prostate. Note the absence of Id-1 mRNA expression (200 x). **(B).** BPH. Note the very weak to undetectable Id-1 mRNA expression (200 x). **(C).** Well-differentiated PCA showing moderate levels of Id-1 mRNA expression (200 x). **(D).** Moderately well- differentiated PCA. Note the relatively strong Id-1 mRNA expression (200 x). **(E)** and **(F).** Poorly differentiated PCA. Note strong Id-1 mRNA expression (100 & 400 x). **(G).** Negative control using sense probe to determine non nonspecific staining (100 x).

In normal and BPH tissue samples, we observed negative to weak Id-1 expression. In contrast, all PCA biopsies displayed substantial Id-1 expression in tumor cells at both the mRNA and protein levels. Furthermore, Id-1 expression was highly prominent in poorly differentiated (high Gleason grade) than in well-differentiated (low Gleason grade) carcinomas, suggesting that its level of expression may be associated with tumor malignancy (17).

Functional Studies of Id-1

In order to determine the effect of Id-1 expression on human PCA cell growth, we transfected an Id-1 expression vector into a PCA cell line, LNCaP, which exhibited undetectable levels of Id-1 in the absence of fetal calf serum (FCS).

Ectopic Id-1 Expression and its Effect on PCA Growth. Id-1 expression in LNCaP cells was high in the presence of 10% FCS; it decreased with lower FCS concentrations, and became undetectable after culturing in serum free media (SFM) for 48 h. To determine the effect of ectopic Id-1 expression, a retroviral vector containing the full-length human *Id-1* cDNA (pBabe-Id-1) was transfected into LNCaP cells, and 10 stable clones were selected (18). In the absence of FCS, 7/10 clones expressed Id-1 at different levels in the absence of FCS (Figure 3). The remaining 3 clones remained *Id-1* negative. The effect of *Id-1* in PCA cell proliferation was determined in the transfectant clones. The results showed that the introduction of *Id-1* increased LNCaP cell proliferation and Id-1 expression (Figure 4). These data indicate that ectopic expression of *Id-1* gene activity enhances the growth of transfectant clones in the absence of FCS.

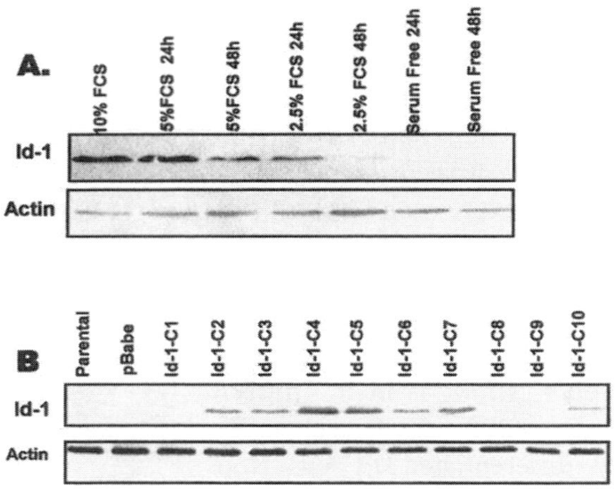

Figure 3. Id-1 expression in parental LNCaP and stable transfectants. **(A)** Serum-dependent-Id-1 expression in LNCaP cells. Cells were cultured in RPMI 1640 medium containing different FCS concentrations for 24 to 48 h before Western blot analysis. Note that Id-1 expression declines as the level of FCS decreases. **(B)** Id-1 expression levels in stable transfectant clones (Id-1-C1-10), vector control (pBabe), and parental LNCaP cells cultured in SFM for 48 h. Note that 7/10 clones express different levels of Id-1 protein, while Id-1 expression is undetectable in the controls. Expression of actin was used as an internal loading control.

Effect of Id-1 Expression on DNA Synthesis and Cell Cycle Distribution. Next, we determine whether Id-1-induced cell growth was due to its ability to initiate DNA synthesis in PCA cells in SFM (18). Cell cycle analyses showed an increased number of S phase cells in Id-1-expressing cells (Figure 5). In addition, the increase in S phase cells was correlated with the levels of *Id-1* expression; *i.e.,* higher *Id-1* expression levels, higher S phase number, thus higher cell proliferation rate. This finding was confirmed by the analysis of DNA synthesis by BrdU incorporation (Figure 6). The DNA synthesis rate in clones with higher Id-1 levels was similar to those of control LNCaP-pBabe cultured in 10% FCS. These data indicate that the increased cell growth involves shortening of the cell cycle with a concurrent increase in DNA synthesis (18).

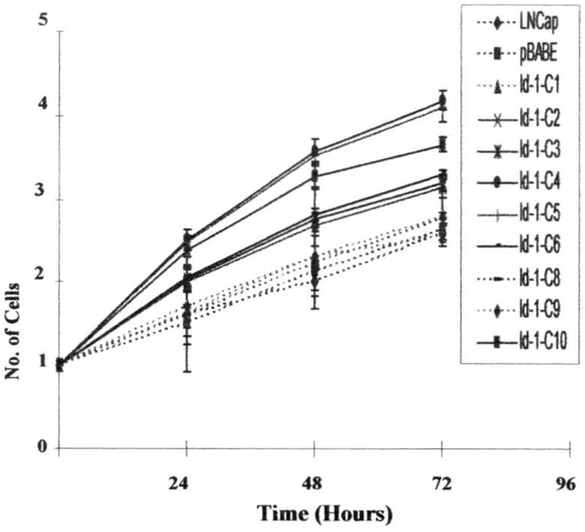

Figure 4. Growth rate of LNCaP and Id-1 transfectants cells. 10^3 cells were cultured in 24-well plates in SFM. Every 24 h, cell number was determined using Trypan-blue. Values represent the mean ± SD of three experiments. Note that the cell growth rate is associated with increased Id-1 expression.

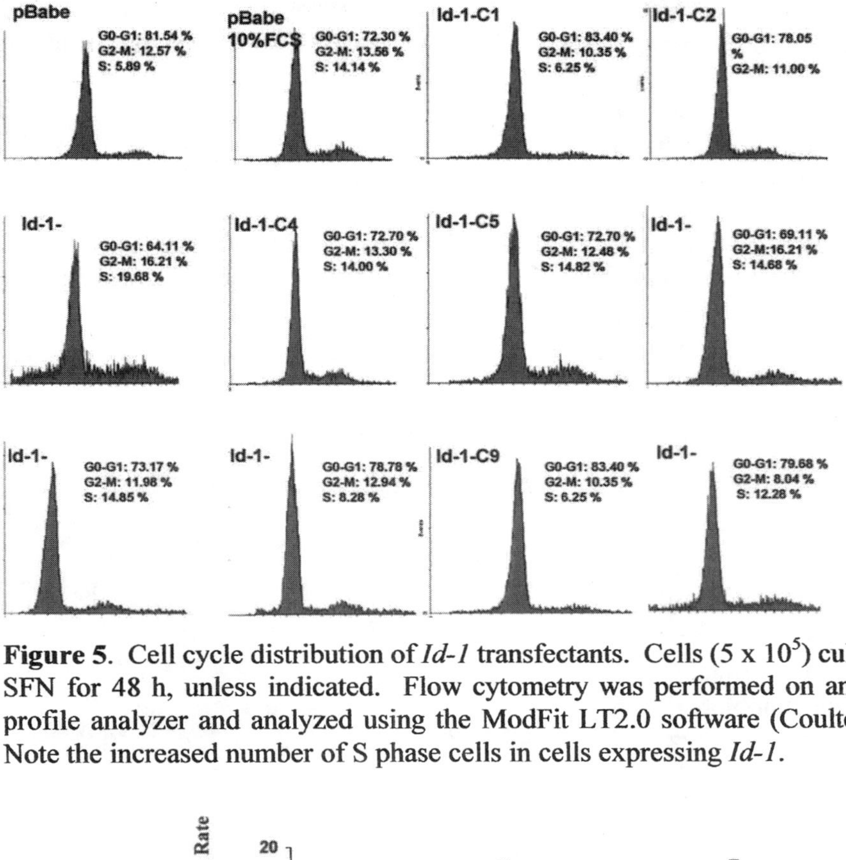

Figure 5. Cell cycle distribution of *Id-1* transfectants. Cells (5 x 10^5) cultured in SFN for 48 h, unless indicated. Flow cytometry was performed on an EPICS profile analyzer and analyzed using the ModFit LT2.0 software (Coulter) (18). Note the increased number of S phase cells in cells expressing *Id-1*.

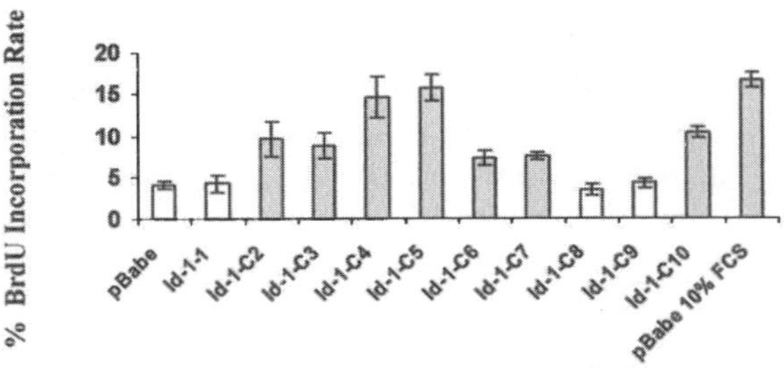

Figure 6. BrdU incorporation. Cells cultured in SFM for 48 h before testing. At least 500 cells/experiment were counted. Values represent the mean ± SD of three separate experiments (18).

Effect of Id-1 Expression on Rb/p16^{INK4a} Pathway. To investigate the mechanisms involved in Id-1-induced LNCaP cell proliferation, we determined the expression levels of p16^{INK4a},CDK4, p21^{Waf1}, p27^{Kip1}, CDK2, and Rb in Id-1 expressing clones (18). The results showed that the expression of p16^{INK4a} was very low or undetectable in all of the *Id-1* expressing clones, while a 2.0-3.0-fold increase in p16^{INK4a} expression levels was observed in the controls and the *Id-1* negative clones (Figure 7). These data demonstrate that *Id-1* expression reduced the levels of p16^{INK4a} protein expression in LNCaP cells. In addition, we found that the phosphorylated forms of CDK4 and 2 were apparent in all of the *Id-1* expressing clones, but not in the controls or Id-1 negative clones (Figure 7). No significant changes were found in the levels of expression of p21^{Waf1} or p27^{Kip1} in the *Id-1* expressing clones. Further phosphorylated Rb was found in all of the clones; however there was no evidence of Rb phosphorylation in the controls or *Id-1* negative clones (Figure 7). Our results indicate that *Id-1* induced inactivation of the p16^{INK4a}/Rb pathway may be responsible for the increased cell proliferation observed in PCA cells.

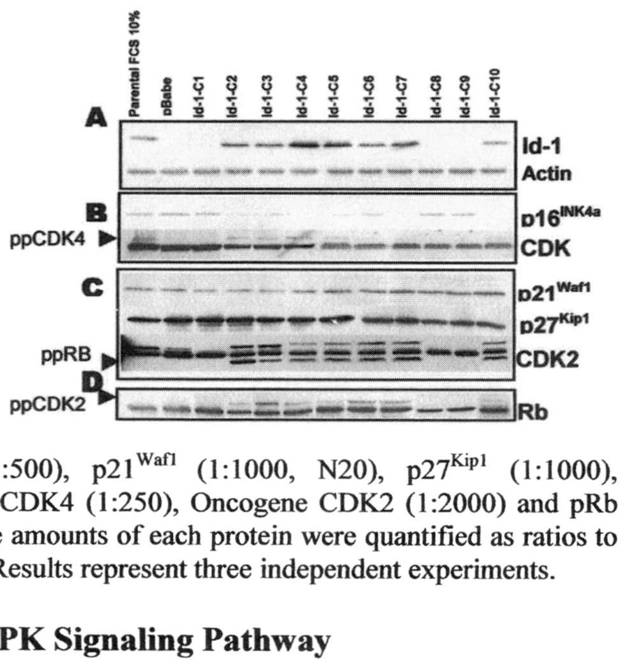

Figure 7. Western blotting analysis of p16^{INK4a}, CDK4, p21^{Waf1}, p27^{Kip1}, CDK2, and Rb expression in *Id-1* transfectants and controls. Cells were cultured in SFM for 48 h before harvesting. Aliquots (50 μg protein) were immunoblotted with Santa Cruz Biotechnology antibodies Id-1 (C20, 1:200), p16^{INK4a} (N20, 1:500), p21^{Waf1} (1:1000, N20), p27^{Kip1} (1:1000), Transduction Laboratories CDK4 (1:250), Oncogene CDK2 (1:2000) and pRb (1:500, Ab-1). The relative amounts of each protein were quantified as ratios to actin (1:500, Amersham). Results represent three independent experiments.

Id-1 Activity and MAPK Signaling Pathway

Earlier, we have shown that expression of *Id-1* may be associated with advanced stages of PCA (17). This implicates that *Id-1* may function as an oncogene. Our earlier studies further suggest that one of the oncogenic functions of *Id-1* may be through inactivation of p16^{INK4a}, leading to the dysfunction of p16^{INK4a}/Rb tumor suppressor pathway (18). Recently, *Id-1* has been shown to facilitate human fibroblasts escaping senescence through the Ras-Raf-MEK signaling pathway (19). In addition, one of the downstream effectors of the MAPK pathway, early growth response-1(Egr-1), has been reported to interact with Id-1 (20). Since activation of the MAPK pathway and the over-expression of one of its downstream effectors, Egr-1, are associated with advanced cancer (21, 22), we investigated the involvement of MAPK in *Id-1*-induced PCA cell proliferation.

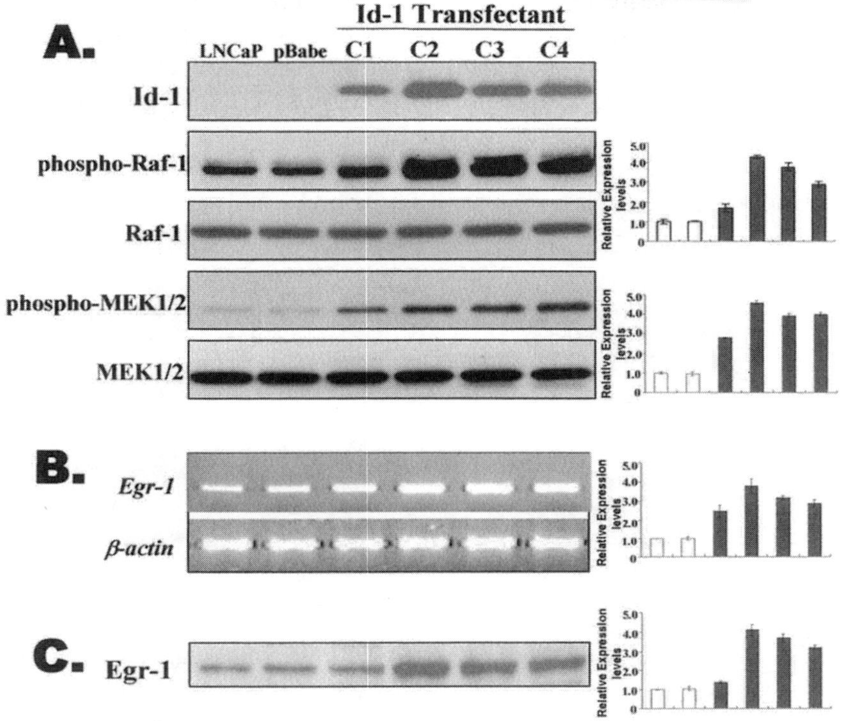

Figure 8. Effect of ectopic Id-1 expression on Raf-1, MEK1/2, and Egr-1 expression in LNCaP cells. Stable Id-1 tansfectants and the vector control pBabe were generated as previously described (18). Cultured cells in SFM RPMI 1640 medium for 48 hr. **A and C**: Protein aliquots (30 μg) immunobloted with phospho-Raf-1 (Ser259), phospho-MEK-1/2 (Ser 217/221), MEK-1/2; Id-1, Raf-1, Egr-1, and β-actin. **B**: RT-PCR analysis of *Egr-1* mRNA expression (25, 26).

Effect of Ectopic Expression of Id-1 on Raf-1, MEK1/2 and Egr-1 Expression. Using four stable transfectant clones (C1-C4), we determined the possible contribution of MAPK in *Id-1*-induced serum independent cell proliferation in LNCaP cells (23). Western blotting data showed, in the absence of FCS for 48 h, there were no detectable levels of Id-1 expression in the control, pBabe, and LNCaP cells, while the Id-1 protein was expressed in all four transfectant clones. In addition, there was an increase in expression of the phosphorylated form of Raf-1 and MEK1/2 proteins in all the *Id-1* clones (18). The level of expression of the phosphorylated Raf-1 and MEK1/2 was much lower in the controls. However, there was no increase in total levels of each protein among the controls and the *Id-1*-expressing clones (Figure 8).

Since activation of Raf-1 and MEK1/2 signaling is through phosphorylation, these data indicate that exogenous *Id-1* expression in LNCaP cells resulted in a direct activation of Raf/MEK1/2 kinases. The expression of Egr-1, one of the downstream effectors, was also increased both at the protein and mRNA levels (Figure 8). These results strengthen our views on the involvement of MAPK signaling pathway in *Id-1*-induced serum independent LNCaP cell growth (23).

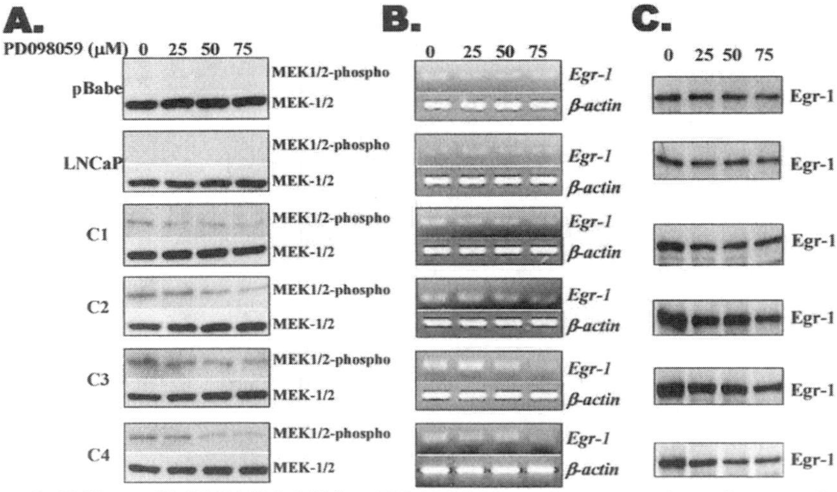

Figure 9. Effect of MEK1/2 inhibitor PD098059 on the expression of MEK1/2 and Egr-1 in Id-1 transfectants and their controls. Culture cells in SFM for 48 h were treated with different concentrations of PD098059 for 24 h (24). **A** and **C**: Western blotting analysis of MEK1/2 and Egr-1 expression before and after treatment with PD098059. **B**: RT-PCR analysis of *Egr-1* expression. Note that in Id-1-expressing clones treated with PD098059 there is a decrease in the expression of MEK1/2 phosphorylation and Egr-1.

Effect of MEK1/2 Inhibitor PD098059 on the Expression of MEK1/2 and Egr-1 in Id-1 Transfectant Clones. In order to determine whether activation of the MAPK signaling pathway played a role in *Id-1*-induced serum-independent cell proliferation, all *Id-1*-expressing clones and the controls were treated with PD098059, a selective inhibitor of MEK1/2 (24). The data showed a decrease in phosphorylated MEK1/2 protein levels after the inhibitor treatment. This decrease in expression concentration dependent, and was not detect in the control pBabe and parental LNCaP cells (Figure 9). On the other hand, the basal levels of MEK1/2 protein (both the none and the phosphorylated forms) remained unchanged under the same conditions. When the treated cells were analyzed for Egr-1 expression, both the protein and mRNA levels were also affected. The decreased levels were again dependent to the concentration of the inhibitor (Figure 9). These data indicate that inhibition of MEK1/2 phosphorylation by PD098059 leads to inactivation of MEK1/2 kinase, thus resulting in a decreased MAPK signaling transduction activity (23).

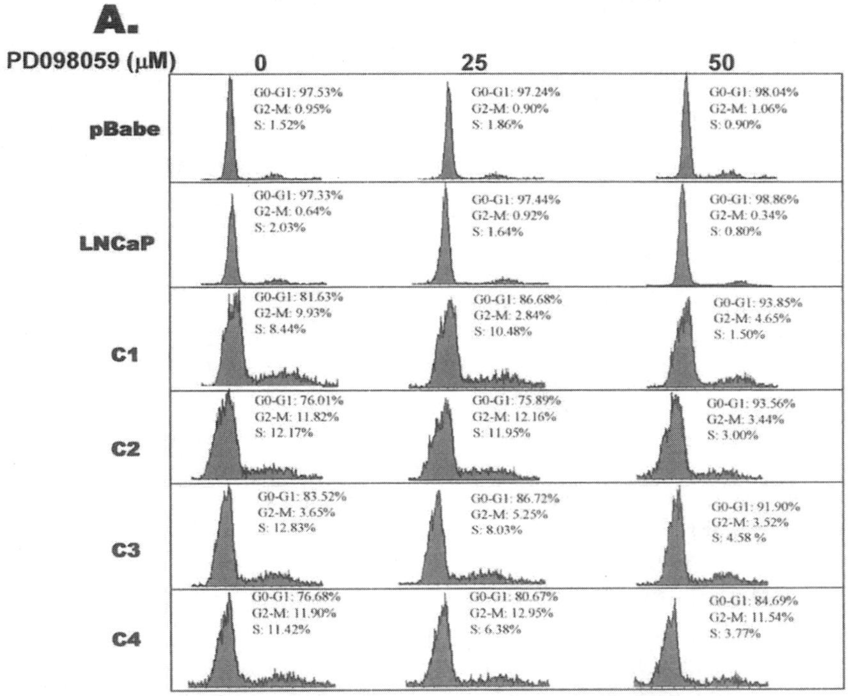

Figure 10A. Effect of PD098059 on cell cycle distribution. Cells (5×10^5) were plated in 5% FCS culture medium. After 24 h, the medium was replaced by SFM for 48 h, and PD098059 or DMSO (solvent control) was added. Cells were harvested 24 h later by trypsinization, and fixed in ice cold 70% ethanol for at least 1 h. Flow cytometric analysis was performed on an EPICS profile analyzer and analyzed using the ModFit LT2.0 software. Note the decreased % of S phase cells shown in the Id-1 transfectants after exposure to PD098059.

Effect of PD098059 on Id-1 Induced Serum Independent PCA Cell Growth. In order to investigate the effect of the inhibition of MAPK signaling pathway on Id-1-induced cell proliferation on LNCaP cells, we examined the cell cycle distribution and growth rate after exposure to PD098059 and compared to the untreated controls (23). The results showed that in untreated cells, the % of S phase cells in pBabe and parental LNCaP controls was small (1-2%), while in the four transfectant clones increased 8-13% (Figure 10). Upon PD098059treatment, a MAPK inhibitor, at 50 µM for 48 h, the % S phase cells of the four transfectant clones was reduced to 1.5-4.5%, similar to the controls (Figure 10A).

In addition, the data on cell growth were compared to the cell cycle studies. In the absence of inhibitor, the *Id-1*-expressing clones showed an increased growth rate compared to the controls. After PD098059 treatment, the growth rate of *Id-1*-expressing cells was reduced with increasing concentrations of the inhibitor.

At a concentration of 75 μM, the growth rate was similar to that observed in the controls (Figure 10B). These results indicate that inhibition of Raf/MAPK signaling pathway leads to inhibition of *Id-1*-induced cell proliferation in LNCaP cells (23).

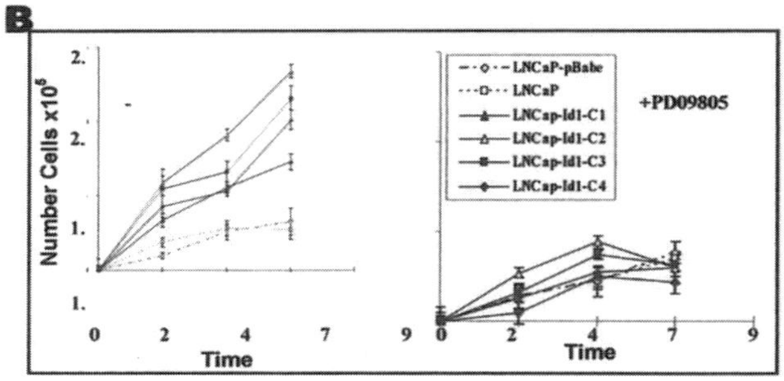

Figure 10B. Effect of PD098059 (75 μM) on Id-1-induced PCA cell growth. Cells (1.5 x 10^5) were plated in SFM 48 h, and PD098059 (75 μM) or DMSO (solvent control) was added. Values represent the mean ± SD of triplicate wells/experiment, and each experiment was repeated at least three times. Note that PD098059 treatment in the Id-1-transfectants results in decreased cell growth.

Effect of Treatment of Id-1 Anti-sense on Expression Id-1 and Egr-1 Proteins in PCA Cell Lines Constitutively Expressing Id-1. In order to confirm further the association between *Id-1* and MAPK signaling pathway, we treated PC3 and DU145 cell lines that express relatively high levels of Id-1 protein, with anti-sense oligonucleotides complementary to *Id-1* mRNA and a control oligomer (14). No change was detected in the controls. However, in the *Id-1* anti-sense-treated cells, the expression of both Id-1 and Egr-1 proteins was decreased (Figure 11). The decrease in expression of Id-1 and Egr-1 proteins was associated with a decrease in cell proliferation rate. These data further strengthen the view that MAPK signaling pathway activation is essential for Id-1-induced serum-independent cell growth in PCA cells (23).

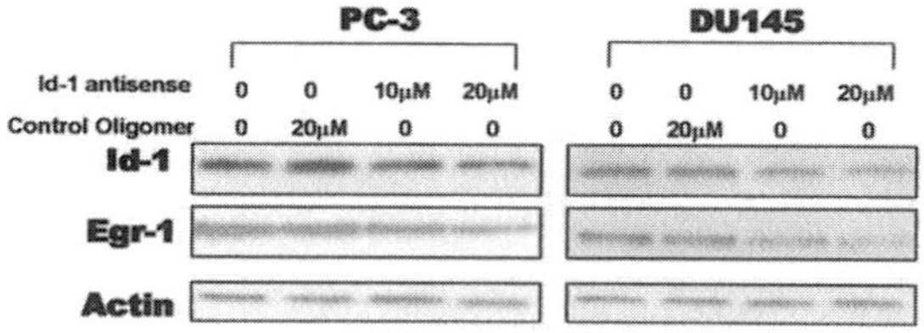

Figure 11. Reduction of Id-1 and Egr-1 protein expression by antisense oligomers complementary to Id-1 mRNA in PC-3 and DU145 cells. Western blotting analysis of Egr-1 from cells cultured in SFM for 48 h and treated with control oligomer (20 μM) or *Id-1* antisense oligonucleotide (10 or 20 μM) for 48 h (14). Note that inactivation of Id-1 by antisense oligonucleotides leads to decreased Egr-1 expression in both cell lines.

Summary and Conclusions

Several conclusions can be reached from the data presented. **(1).** Id-1 protein is an important biomarker for PCA. **(2).** The level of its expression is related to the malignancy of the PCA. **(3).** Id-1 expression levels in PCA tissue samples has the potential to be used as a prognostic marker since high Id-1 expression is related to poor prognosis; and low expression to good prognosis. **(4).** One of the possible pathways for *Id-1* gene action is through the inactivation of the p16/Rb tumor suppressor pathway. **(5).** The activation of the MAPK signaling pathway is a cell proliferation pathways induced by *Id-1*. **(6).** Activation of the *Id-1* gene may be an important step in the transition stage of PCA cells from androgen-dependent to androgen-independent state, the most malignant form of PCA.

Acknowledgements

This work was supported by RGC grants to YCW (HKU 7186/99M and HKU7314/01M) and Area of Excellent Scheme (Project No. AoE/P-10/01).

References

1. Wang YZ, Wong YC (1998) Sex hormone induced prostatic carcinogenesis in the Noble rat: the role of insulin-like growth factor-1 (IGF-1) and vascular endothelial growth factor (VEGF) in the development of PCA Prostate 35:165-177.
2. Wong YC, Xie W, Tsao SW (2000) Structural changes and alteration in expression of TGF-beta1 and its receptors in prostatic intraepithelial neoplasia (PIN) in the ventral prostate of Noble rats. Prostate 45:289-298.
3. Tam NNC, Chung SSM, Lee DTW, et al (2000) Aberrant expression of hepatocyte growth factor and its receptor, c-Met, during sex hormone-induced prostatic carcinogenesis in Noble rat. Carcinogenesis 21:2183-2191.
4. Xie W, Wong YC, Tsao SW, et al (2000) Expression of a kallikrein-like protein in prostatic intraepithelial neoplasia in ventral prostate of the Noble rat. Prostate 42:8-17.
5. Wong YC, Tam NNC (2002) Dedifferentiation of stromal smooth muscle as a factor in prostate carcinogenesis. Differentiation 70:633-645.
6. Ouyang XS, Wang X, Lee DTW, et al (2001) Up-regulation of TRPM-2, MMP-7 and Id-1 during sex hormone-induced prostate carcinogenesis in the Noble rat. Carcinogenesis 22:965-973.
7. Benezra R, Davis RL, Lockshon D, et al (1990) The protein Id: a negative regulator of helix-loop-helix DNA binding proteins. Cell 61:49-59.
8. Desprez PY, Hara E, Bissell MJ, et al (1995) Suppression of mammary epithelial cell differentiation by helix-loop-helix protein Id-1. Mol Cell Biol 15:3398-3404.
9. Alani RM, Young AZ, Shifflett CB (2001) Id-1 regulation of cellular senescence through transcriptional repression of p16INK4a. Proc Natl Acad Sci USA 91:7812-7816.
10. Maruyama H, Kleeff J, Wildi S, et al (1999) Id-1 and Id-2 are overexpressed in pancreatic cancer and in dysplastic lesions in chronic pancreatitis. Am J Path 155:815-822.

11. Lin CQ, Singh J, Murata K, et al (2000) A role for Id-1 in the aggressive phenotype and steroid hormone response of human BC cells. Cancer Res 60:1332-1340.
12. Desprez PY, Lin CQ, Thomasset N, et al (1998) A novel pathway for mammary epithelial cell invasion induced by the helix-loop-helix protein Id-1. Mol Cell Biol 18:4577-4588.
13. Langlands K, Down GA, Kealey T (2000) Id proteins are dynamically expressed in normal epidermis and dysregulated in squamous cell carcinoma. Cancer Res 60:5929-5933.
14. Hara E, Yamaguchi T, Nojima H, et al (1994) Id-related genes encoding helix-loop-helix proteins are required for G1 progression and are repressed in senescent human fibroblasts. J Biol Chem 269:2139-2145.
15. Barone MV, Pepperkok R, Peverali FA, et al (1994) Id proteins control growth induction in mammalian cells. Proc Natl Acad Sci USA 91:4985-4988.
16. Peverali FA, Ramqvist T, Saffrick R, et al (1994) Regulation of G1 progression by E2A and Id helix-loop-helix proteins. EMBO 13:4291-4301.
17. Ouyang XS, Wang X, Lee DTW, et al (2002) Overexpression of Id-1 in PCA. J Urol 167:2598-2602.
18. Ouyang XS, Wang X, Ling MT, et al (2002) Id-1 stimulates serum independent PCA cell proliferation through inactivation of $p16^{INK4a}$/pRB pathway. Carcinogenesis 23:721-725.
19. Ohtani N, Zebedee Z, Huot TJ, et al (2001) Opposing effects of Ets and Id proteins on p16INK4a expression during cellular senescence. Nature 409:1067-1070.
20. Tournay O, Benazra R (1996) Transcription of the dominant-negative helix-loop-helix protein Id1 is regulated by a protein complex containing the immediate early response gene Egr-1. Mol Cell Biol 16:2418-2430.
21. Eid MA, Kumar MV, Iczkowski KA, et al (1998) Expression of early growth response genes in human PCA. Cancer Res 58: 2461-2468.
22. Gioeli D, Mandell JW, Petroni GR, et al (1999) Activation of mitogen-activated protein kinase associated with PCA progression. Cancer Res 59:279-281.
23. Ling MT, Wang X, Ouyang XS, et al (2002) Activation of MAPK signaling pathway is essential for Id-1 induced serum independent PCA cell growth. Oncogene 21:8498-8505.
24. Dudley DT, Pang L, Decker SJ, et al (1995) A synthetic inhibitor of the mitogen-activated protein kinase cascade. PNAS USA 92:7686-7689.
25. Kawashita Y, Ohtsuru A, Kaneda Y, et al (1999) Regression of hepatocellular carcinoma in vitro and in vivo by radiosensitizing suicide gene therapy under the inducible and spatial control of radiation. Hum Gene Ther 10:1509-1519.
26. Ling MT, Chan KW, Choo CK (2001) Androgen induces differentiation of a human papillomavirus16 E6/E7 immortalized prostate epithelial cell line. J. Endocrinol. 170:287-296.

PART 6. PROSTATE II: ANDROGEN RECEPTOR: DEPENDENCE/RESISTANCE

16

Four Stages of Prostate Cancer: Suppression and Eradication by Androgen and Green Tea Epigallocatechin Gallate

Shutsung Liao, John M. Kokontis, Chih-pin Chuu, Stephen Hsu, Junichi Fukuchi, Mai Dang, and Richard A. Hiipakka

Introduction

We have shown that prostate cancer (PCA) cells can exist in four stages of progression in culture media or in experimental animals based on their androgen (A) dependency or sensitivity. This is in contrast with the common characterization of prostate tumors as occurring in two forms, simply based on whether the tumors are A-dependent or A-independent. Recognition of four different forms of PCA strongly suggests that the current design of hormonal therapy by anti-androgenic agents requires revision; otherwise the practice may be very harmful to patients. In this chapter, we summarize our effort in understanding the four forms of PCA cells and show that their growth or proliferation can be suppressed or eradicated in culture or in athymic mice by selectively utilizing anti-androgen, A, or green tea (-)epigallocatechin-3-gallate (EGCG).

Androgens and Green Tea Catechins are Two Groups of Ancient Medicines

As and green tea are ancient medicines that have been in use for many thousands years. More than 2,000 years ago, in China, androgenic crystals were prepared from urine and organs by sublimation and used to treat individuals lacking 'maleness activity'. In oriental culture, tea beverage was used for prevention and treatment of many diseases for over 3,000 years, although scientific and medical evaluation of tea started only very recently (1).

In this chapter, we summarize our effort in establishing models of human PCA progression in culture and in athymic mice to understand how the process of the PCA progression may occur in patients. We have used these model systems to explore novel methods for control and eradication of PCA by As and the green tea catechin, epigallocatechingallate (EGCG).

Control of Androgen Action and Medicinal Applications

In the early 1960s, we found that As can rapidly enhance RNA synthesis in target organs, such as the ventral prostate of rats, suggesting that As act by modulating gene expression in target cell nuclei (2-4). Subsequently, we (5, 6) and Bruchovsky and Wilson (7) showed that, in many target organs, testosterone, the major A produced by testis and circulating in blood, is converted by 5α-reductase to 5α-dihydrotestosterone (5α-DHT). 5α-DHT is the active A that binds to a specific nuclear androgen receptor (AR) (8-12). The 5α-DHT-AR complex, apparently in conjunction with other chromosomal proteins (13), then regulates specific transcription of genes and production of specific proteins that modulate cellular activities and organ functions. Cloning and sequence determination of the genes for AR (10, 11) and 5α-reductase (14) have shown that mutations of these genes are responsible for A-insensitivity syndromes, including pseudo-hermaphroditism.

The molecular steps required for A action in target cells provide two effective methods for control of testosterone-regulated responses: (a) the use of a 5α-reductase inhibitor to suppress 5α-DHT production, and (b) the use of anti-androgens to block the interaction of 5α-DHT with AR (12). Both methods are now being utilized as therapies for A-related disorders including PCA, prostate enlargement, as well as male pattern baldness. As will be described below, both 5α-reductase inhibitors and anti-androgens are useful in the study of PCA progression and treatment.

5α-Reductase Inhibitors

Since testosterone activation is dependent on 5α-reductase, synthetic inhibitors of the reductase have been prepared by pharmaceutical companies. The synthetic 4-aza-steroid, finasteride, is now prescribed as Proscar for benign prostate hyperplasia (BPH), and as Propecia for male pattern baldness.

Many natural compounds that inhibit 5α-reductase have been described. In 1992 (15), we demonstrated that γ-linolenic acid [C18:3 (cis-9,12,15)] (GLA) (Figure 1), an essential fatty acid in many plant oils was an inhibitor of 5α-reductase. We also found that EGCG (Figure 1) was a 5α-reductase inhibitor (16, 17). Other gallated catechins are also active, but non-gallated catechins are inactive. The gallate group is important for the inhibitory activity. However, gallic acid and the methyl ester of gallic acid are inactive. Curcumin and alizarin (Figure 1) are also 5α-reductase inhibitors (17). The biological activity of these inhibitors has been tested *in-vivo* using flank organs of male hamsters as an animal model. Topical application of these inhibitors suppressed testosterone-dependent growth of the flank organ (18, 19).

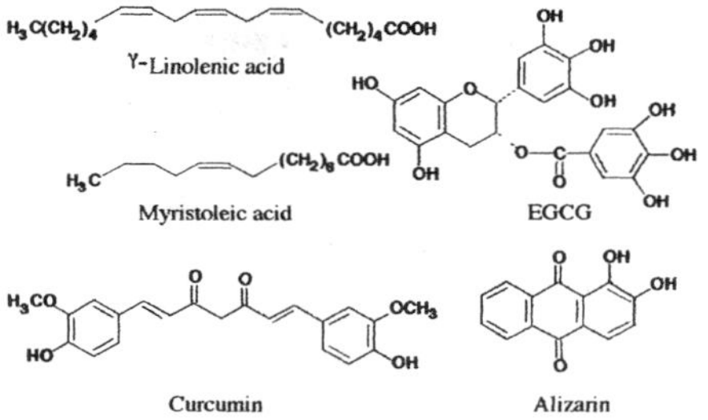

Figure 1. Structures of natural inhibitors of 5α-reductases.

Two isozymes of 5α-reductase have been identified (14). The specific roles of the individual isozymes are not well understood. Finasteride is a selective inhibitor of the type 2 isozyme of 5α-reductase whereas GLA, curcumin, alizarin, can inhibit both the type 1 and type 2 isozymes. EGCG was a better inhibitor of the type 1 isozyme than of type 2. The biomedical significance of this difference is unclear. While oral Proscar has been shown to be effective in treating BPH, the effectiveness of natural inhibitors for benign or cancerous prostate growth has not been demonstrated.

Establishment of Models for Studying Prostate Cancer Cell Progression

Because PCA is initially dependent on As for its growth, hormonal therapy (A deprivation), pioneered by Charles Huggins 60 years ago (20) using castration or more recently anti-androgens, has been a standard PCA therapy. Although over 70% of patients may benefit from this therapy, PCA recurs in most of these patients in one to three years as tumors that do not need A for growth (A-independent). For lack of effective therapy, patients die from this A-independent PCA.

We believe that a better understanding of the process of PCA cell progression is very important for establishing better methods for PCA treatment. To investigate PCA progression, we established a model system (Figure 2) using a clone derived from the human PCA cell line, LNCaP, whose growth was dependent on the presence of nanomolar concentrations of testosterone, 5α-DHT or 17β-hydroxy-17-methyl-estra-4,9,11-trien-3-one (R1881) (21-23). The original cancer cell (named LNCaP 104-S) population was cultured through weekly passages and after about 40-70 passages in A-depleted culture medium, these cells progressed to A-independent cancer cells that we named LNCaP 104-R1 cells. These cells can grow well in culture without A. After continuous culture of 104-R1 cells in A-depleted medium for 60-120 additional passages, these 104-R1 cells were transformed into faster-growing cells, named 104-R2 cells. This transition of A-dependent 104-S cells to A-independent 104-R1 and 104-R2 cells is accompanied by dramatically increased AR expression apparently without new mutations in the ligand binding domain of AR gene. Despite an increase in cellular levels of AR, we found that the growth of both 104-R1 and 104-R2 cells in culture are suppressed by physiological concentrations of A. In the presence of a high A concentration (>20 nM), some 104-R1, but not 104-R2 cells, can revert back to A-dependent cells (104-R1Ad) that behave like 104-S cells.

Recently, we found an alternative method to generate A-repressed cells from 104-S cells by using the anti-androgen, Casodex (bicalutamide), to suppress the growth of 104-S cells and isolate cells that can grow in the presence of anti-androgen. These cells, called CDXR cells, behave like 104-R2 cells and can not revert back to A-dependent cells. These A-independent cells also contain high levels of AR and their growth is suppressed by low A concentrations (<1 nM). When CDXR cells were cultured further in the presence of A, we were able to isolate A-insensitive cells, named IS cells. IS cells express very low levels of AR and are not stimulated or suppressed by A in the culture medium (Figure 2). These cells resemble human prostate PC-3 cells, since they do not have AR and can grow in the absence or presence of A.

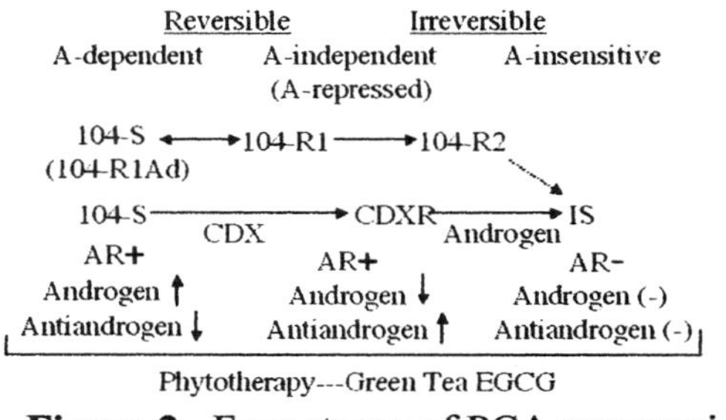

Figure 2. Four stages of PCA progression and treatment.

Growth Suppression of Prostate Cancer Cells in Culture by Androgen

The cellular level of AR mRNA was 2-3 fold higher in the LNCaP 104-R1 and 104-R2 cells than in LNCaP 104-S cells. AR protein level increased 10-20 fold during this transition from A-dependent 104-S cells to A-independent 104-R1 or 104-R2 cells. The growth of both 104-R1 and 104-R2 cells, as well as CDXR cells in culture was suppressed by physiological concentrations (<1 nM) of testosterone, 5α-DHT, or R-1881. Non-androgenic steroids, such as 5β-DHT, 17β-estradiol, medroxyprogesterone, and cortisol, did not suppress 104-R tumor growth (24). AR in 104-R (R1 and R2) cells was functional, since A induction of prostate-specific antigen (PSA) mRNA increased up to 20 times in these 104-R cells.

A suppression of 104-R cells is apparently due to a G1 arrest during cell cycling. In 104-R cells, R1881 at 0.1-1 nM repressed cell growth and induced the cyclin-dependent kinase (cdk) inhibitor, $p27^{kip1}$ (23). The same concentrations of R1881 that promote the growth of 104-S cells, reduced the cellular level of $p27^{kip1}$ in 104-S cells. CDXR cells also behave like 104-R cells; A increased $p27^{kip1}$ level in CDXR cells, and caused G1 arrest and suppress CDXR cell proliferation.

The effect of testosterone on c-*myc* gene expression correlated well with the proliferative activity of both 104-S and 104-R cells. R1881, at 0.1 nM, induced c-*myc* mRNA level in 104-S cells but repressed the mRNA level to less than 20% of the control value in 104-R cells. At high concentrations (~20 nM), R1881 inhibited both proliferation and c-*myc* expression in these cells. The retroviral overexpression of c-*myc* could block the A repression of LNCaP cells (21). A also suppresses c-*myc* in CDXR cells.

Androgen Specific Suppression and Eradication of Androgen-independent Prostate Tumors in Athymic Mice

A-dependent 104-S tumors grew very well in normal but not in castrated athymic mice. In contrast, A-independent LNCaP 104-R cells (24) or CDXR cells grew as tumors in castrated athymic mice but not in normal athymic male mice. Administration of testosterone, 5α-DHT, or R1881 to castrated mice prevented the growth of these A-independent tumors, and suppresses 104-R prostate tumors already present in these animals. As in cell culture, non-androgenic steroid hormones (progesterone, cortisol and 17β-estradiol) did not suppress the growth of these tumors in athymic mice. In many animals, testosterone caused regression of 104-R1, 104-R2, and CDXR tumors to less than 10% of the original tumor size within 30 days. In fact, in some mice, testosterone administration actually eradicated 104-R or CDXR tumors, and no tumor re-growth was observed thereafter for more than five mo. Testosterone did not affect the growth of A-insensitive human PCA cell PC-3 tumors or IS tumors in athymic mice.

Testosterone treatment of mice bearing 104-R tumors reduced c-*myc* mRNA in the tumors, but increased PSA mRNA in tumors and PSA level in serum before tumor regression (24). The 5α-reductase inhibitor, finasteride (Proscar) or anti-androgens, such as Casodex, blocked the repressive effect of testosterone on these xenografts and stimulated the tumor growth, suggesting that the growth suppression required conversion of testosterone to 5α-DHT and binding of 5α-DHT to AR (24). This observation suggested that if, in patients, testosterone can suppress PCA growth, the use of these drugs may enhance the growth of A-independent PCAs.

As in the cell culture system, A-independent 104-R1 cells can adapt to the presence of A and become A-dependent cells in normal mice (24). Adapted tumors behaved like A-dependent tumors. These cells in culture and tumors grown in mice not castrated can then be again controlled by anti-androgens or 5α-reductase inhibitors, such as Proscar. Adaptation to A was not observed with 104-R2 or CDXR tumors in mice.

EGCG Suppression of Prostate and Breast Tumors

Green tea consumption has been linked to lower incidence of some cancers in humans and animals. Epidemiological studies, however, have not provided consistent evidence about the anti-tumorigenic effect of green tea in humans.

For a better understanding of the ability of green tea to control different forms of prostate tumors, we produced tumors in athymic mice by subcutaneously inoculating athymic mice with AR positive and A-dependent LNCaP 104-S cells, AR positive LNCaP 104-R2, 104-R1 or CDXR cells whose growth is repressed by A, or AR negative PC-3 or IS cells whose growth is neither stimulated or repressed by A. We found that green tea EGCG (>98% pure, 1 mg/20g body weight daily), injected intraperitoneally (ip), significantly inhibited the growth and rapidly (in 1-2 week) reduced the size of all types of human prostate tumors in athymic mice. Structurally-related catechins, such as epicatechin gallate (ECG) that lacks only one of the eight hydroxyl groups in EGCG, were inactive. Epicatechin (EC) and epigallocatechin (EGC) were also inactive (25).

Since both A-dependent and A-independent prostate tumors respond to tumor suppression by EGCG. EGCG action was not related to modulation of A activity or due to 5α-reductase inhibition. In addition, the growth of human breast tumors in athymic mice produced by human breast cancer MCF-7 cells, were also clearly inhibited by ip injection of EGCG.

It is possible that the low clinical incidence of prostate and breast cancer in some Asian countries is, in part, related to high green tea consumption. The frequency of the latent, localized PCA does not vary significantly among geographically different populations, but the clinical incidence of metastatic PCAr varies considerably among countries (low in Japan and high in the USA). If consumption of green tea beverage is related to this difference, EGCG may play an important role in preventing the progression or metastasis of PCA cells.

EGCG Modulation of Food Intake and Endocrine Systems

The mechanism by which EGCG suppresses prostate tumor growth may be very complex. Many *in-vitro* effects of EGCG, including inhibition of cancer cell mobility, inhibition of key enzymes and protein factors, induction of apoptosis, and inhibition of angiogenesis, have been shown (1). It is very difficult to assess whether these *in-vitro* observations are related to *in-vivo* effects because EGCG and other catechins can interact non-specifically with enzymes or other macromolecules as well as cellular membranes.

We have studied the effects of EGCG on endocrine systems (1, 26, 27) We found that EGCG, given to rats by ip injection, could within one week reduce body weight by about 20%. Other structurally related catechins, such as EC, EGC, or ECG were not effective at the same dose. Reduction of body weight appeared to be due to EGCG-induced reduction in food intake. EGCG, therefore, may influence neuropeptides and cause the loss of appetite.

After 7 days of daily ip treatment with EGCG, circulating levels of testosterone are reduced by about 75% in male rats and 17β-estradiol levels by 34% in female rats. The weights of A-sensitive organs, such as ventral prostate and seminal vesicles and estrogen-sensitive organs, such as the uterus and ovary were reduced by about 50%. Other catechins were not as effective as EGCG. We also found that the serum level of LH is reduced by 40-50%, suggesting that low LH production led to the reduced blood levels of sex hormones. In both male and female rats, we observed significant reduction in blood levels of leptin, IGF-I, and insulin (1, 26). Some of these peptide hormones may modulate the levels of sex hormone and indirectly alter tumor growth in the animals.

In male rats treated with EGCG for one week, the serum level of protein, fatty acids, and glycerol were not altered, but significant reductions in serum glucose (-32%), lipids (-15%), triglycerides (-46%) and cholesterol (-20%) were observed. Based on proximate composition analysis, there was no change in % water and protein content, a moderate decrease in carbohydrate content, but a very large reduction in fat content, decreasing from 4.1% in control to 1.4% in EGCG-treated group. EGCG treatment also decreases subcutaneous fat by 40 to 70%, and abdominal fat by 20 to 35% in male rats (1, 26). Reduction of body fat may also influence the hormonal levels in animals and influence tumor growth.

Although orally administered EGCG is not as effective as ip injected EGCG (26), probably due to poor intestinal EGCG absorption, long-term oral use of green tea beverage or EGCG-containing drinks may mimic the effects of ip injected EGCG. This was clearly shown by Gupta, *et. al.* (28) who used a transgenic adenocarcinoma of the mouse prostate (TRAMP) model that mimics progressive forms of human PCA. When these mice were orally infused with green tea polyphenols, in an amount equivalent to six cups of green tea per day in humans, PCA development was significantly inhibited and mice survival was increased. The green tea polyphenol infusion almost completely inhibited distant site metastasis.

Concluding Remarks

Our studies revealed that our current understanding of PCA progression is very inadequate, especially for designing clinical treatment of PCA after the initial hormonal therapy has failed and A-independent tumors reappeared. The four stages we have identified take a long period of time, over years, for progression, which resembles the PCA patient situation. Based on our findings, at the initial stage after recurrence of A-independent tumors (R1-form), As at physiological concentration can suppress the growth of R1-tumors or promote some of these cells to revert back to A-dependent tumors (S-form).

These tumors may then be treated again with anti-androgenic agents including natural or synthetic 5α-reductase inhibitors and anti-androgens. Repetitive cyclic treatment by an A and an anti-androgen may be appropriate at this stage and may delay and reduce the number of cancer cells progressing to cancer cells (R2-form) that cannot revert back to the S or R-1 form. R1 and R2-form tumors can be treated with As, but 5α-reductase inhibitors or anti-androgens should not be employed for treatment of the R1 or R2-form tumors since they can interfere with the A suppression of these tumors. R2-tumors after A treatment eventually lose the ability to produce AR and become IS-form tumors that cannot be stimulated or suppressed by As. Fortunately, green tea EGCG is an effective treatment for the A-insensitive tumors and may be used at this final stage of PCA progression. In fact EGCG is effective for suppression of all forms of prostate tumors regardless of their A sensitivity. Therefore, infusion of EGCG-rich green tea beverage or polyphenol products may be advisable for patients with prostate tumors at any stage. Since we have shown that rodents may gradually adapt to the continuous use of EGCG (1), possibly due to induction of enzymes or proteins that can degrade EGCG or increase its excretion, an intermittent use of EGCG may be advisable for PCA treatment.

Our studies suggest that the use of 5α-reductase inhibitors and anti-androgens for treatment of PCA patients needs careful evaluation of individual patients. These drugs may stimulate tumor growth if the patient's prostate tumors are at the A-repressed stage and behave like 104-R or CDXR tumors. This might be the reason that, in a recent PCA prevention trial, finasteride was shown to reduce the risk of PCA by about 25%, but increased the chance of acquiring more aggressive tumors (29).

It is also important to note that our studies also suggest that, only physiological doses of A are needed for suppression of the growth of R1, R2, or CDXR-like tumors in patients. The amount of green tea EGCG required may also be achievable. Many chemotherapeutic agents for PCA treatment are very cytotoxic and have not been very effective. The use of two groups of ancient natural medicines, A and green tea EGCG, may provide a novel approach for PCA control and cure.

Acknowledgement

Studies described in this article were supported by US NIH Grants AT000850 and CA-58073.

References

1. Liao S, Kao YH, Hiipakka RA (2001) Green tea: biochemical and biological basis for health benefits. Vitamins and Hormones 62:1-94.

2. Liao S, Leininger KR, Sagher D, et al (1965) Rapid effect of testosterone on ribonucleic acid polymerase activity of rat ventral prostate. Endocrinology 77:763-765.
3. Liao S, Barton RW, Lin AH (1966) Differential synthesis of ribonucleic acid in prostatic nuclei: Evidence for selective gene transcription induced by androgen. Proc Natl Acad Sci USA 55:1593-1600.
4. Liao S, Lin AH, Barton RW (1966) Selective stimulation of ribonucleic acid synthesis in prostatic nuclei by testosterone. J Biol Chem 241:38963871.
5. Anderson KM, Liao S (1968) Selective retention of dihydrotestosterone by prostatic nuclei. Nature 219:277-279.
6. Fang S, Anderson KM, Liao S (1969) Receptor proteins for androgens: On the role of specific proteins in selective retention of 17ß-hydroxy-5α-androstan-3-one by ventral prostate in vivo and in vitro. J Biol Chem 244:6584-6595.
7. Bruchovsky N, Wilson JD (1968) The conversion of testosterone to 5α-androstan-17β-ol-3-one by rat prostate in vivo and in vitro. J Biol Chem 243:2012-2021
8. Liao S, Fang S (1969) Receptor proteins for androgens and the mode of action of androgens on gene transcription in ventral prostate. Vitamins and Hormones 27:17-90.
9. Fang S, Liao S (1971) Androgen receptor: Steroid-and tissue-specific retention of a 17ß-hydroxy-5α-androstan-3-one protein complex by cell nuclei of ventral prostate. J Biol Chem 246:16-24.
10. Liao S, Kokontis JM, Sai T, et al (1989) Androgen receptors: Structures, mutations, antibodies and cellular dynamics. J. Steroid Biochem 34:41-51.
11. Hiipakka RA, Liao S (1998) Molecular mechanism of androgen action. Trend in Endocrinol Metab 9:317-324.
12. Fan S, Liao S (1969) Antagonistic action of anti-androgens on the formation of a specific dihydrotestosterone-receptor protein complex in rat ventral prostate. Mol. Pharmacol. 5:428-431.
13. McKenna NJ, O'Malley BW (2002) Minireview: Nuclear receptor coactivator—An Update. Endocrinology 143:24612465.
14. Russell DW, Wilson JD (1994) Steroid 5α-reductase: Two genes/two enzymes. Annu Rev Biochem 63:25-61.
15. Liang T, Liao S (1992) Inhibition of steroid 5α-reductase by specific aliphatic unsaturated fatty acids. Biochem J 285:557-562.
16. Liao S, Hiipakka RA (1995) Selective inhibition of steroid 5α-reductase isozymes by tea epicatechin-3-gallate and epigallocatechin-3-gallate Biochem Biophys Res Commun 214:833-838.

17. Hiipakka RA, Zhang H-Z, Dai W, et al (2002) Structure-activity relationship for inhibition of human 5α-reductases by polyphenols. Biochem Pharmacol 63:1165-1176.
18. Liang T, Liao S (1997) Growth suppression of hamster flank organs by topical application of γ-linolenic and other fatty acid inhibitors of 5α-reductase. J Invest Dermatol 109:152-157.
19. Liao S, Lin J, Dang MT, et al (2001) Growth suppression of hamster organs by topical application of catechins, alizarin, curcumin, and myristoleic acid. Arch Dermatol Res 293:200-205.
20. Huggins C, Steven, RE, Hodges CV (1941) Studies on prostate cancer II. The effects of castration on advanced carcinoma of prostate gland. Arch of Sug 43:209-223.
21. Kokontis JM, Takakura K., Hay N, et al (1994) Increased androgen receptor activity and altered c-myc expression in prostate cancer cells after long-term androgen deprivation. Cancer Research 54:1566-1573.
22. Kokontis JM, Liao S (1999) Molecular action of androgen in the normal and neoplastic prostate. Vitam Horm 55:219-307.
23. Kokontis JM, Hay N, Liao S (1998) Progression of LNCaP prostate tumor cells during androgen deprivation: hormone-independent growth, repression of proliferation by androgen, and role for $p27^{kip1}$ in androgen-induced cell cycle arrest. Mol Endocrinol 12:941-953.
24. Umekita Y, Hiipakka RA, Kokontis JM, et al (1996) Human prostate tumor growth in athymic mice: inhibition by androgens and stimulation by finasteride. Proc Natl Acad Sci USA 93:11802-11807.
25. Liao S, Umekita Y, Guo J, et al (1995) Growth inhibition and regression of human prostate and breast tumors in athymic mice by tea epigallocatechin gallate. Cancer Letter 96:239-243.
26. Kao YH, Hiipakka RA, Liao S (2000) Modulation of endocrine systems and food intake by green tea epigallocatechin gallate. Endocrinology 141:980-987.
27. Kao YH, Hiipakka RA, Liao S (2000) Modulation of obesity by a green tea catechin. Am J Clin Nutr 72:1232-1233.
28. Gupta S, Hastak K, Ahmad N, et al (2001) Inhibition of prostate carcinogenesis in TRAMP mice by oral infusion of green tea polyphenols. Proc Natl Acad Sci USA 98:10350-10355.
29. Thompson IM (2003) The influence of finasteride on the development of prostate cancer. N Engl J Med 349:211-220.

17

Androgen Receptor and Interleukin-6 Signaling In PCA Progression

Zoran Culig, Hannes Steiner, Sonia Godoy-Tundidor, Barbara Comuzzi, Georg Bartsch, and Alfred Hobisch

Introduction

The key element in control of growth of benign and malignant prostate is the androgen receptor (AR). It belongs to the superfamily of nuclear receptors, which transmit extracellular hormonal signals to the nucleus. The AR regulates the expression of genes required for male sexual development and maintenance of function of accessory sexual organs. The structure of the AR is similar to that of other steroid receptors; it consists of highly conserved DNA- and ligand-binding domains, which are separated by a hinge region. The length of the N-terminal AR region varies because of differences in polymorphic polyglycine and polyglutamine repeats (1). African-American (AA), population with the highest prostate cancer (PCA) risk, has a reduced number of N-terminal polyglycine repeats. Increased PCA risk in these individuals may thus occur because of high AR activity.

Detection of PCA has considerably improved in the last decade due to introduction of screening programs. PCA in early stages could be cured by radical surgery or radiotherapy. However, it should be emphasized that many PCAs remain latent during patients' life time. They are frequently detected on autopsy in patients who died from other illnesses. At present, little is known about prognostic factors that will help to discriminate between potentially indolent and aggressive tumours.

Non-organ confined PCAs require endocrine therapy, which is palliative. Orchiectomy is still a golden standard in assessment of survival; however, there are also alternative approaches, such as chemical castration [use of luteinizing hormone-releasing hormone (LHRL) analogues] and anti-androgens. Nonsteroidal anti-androgens are widely used in PCA treatment to block initial increase in testosterone levels during treatment with LHRL agonists, or as a mono-therapy. Hydroxyflutamide and bicalutamide are nonsteroidal compounds that bind to the AR with a low affinity and prevent acquisition of a transcriptionally active conformation of the AR. Bicalutamide prevents interactions between the N- and C-terminal of the AR (2). In the last decade, it became clear that anti-androgens frequently exhibit agonistic properties in PCA because of changes in expression and activity of the AR.

This chapter will focus on the following issues essential to understand the role of the AR in PCA:

a) AR regulation in PCA models and tissue specimens.
b) Impact of AR mutations on PCA progression.
c) AR interaction with signaling pathways of cytokines and growth factors.
d) Inappropriate expression and function of AR cofactors

In addition, a summary of the current knowledge on IL-6 expression and signaling in human PCA, and a discussion of possible novel experimental therapies on the basis of these findings are described.

Expression of the Androgen Receptor in Prostate Tumor Models and Clinical Specimens

Among human PCA cell lines widely used, only LNCaP and MDA PCa 2a and 2b cells express AR. These cell lines are derived from PCA metastatic lesions (3, 4). The reasons for AR down-regulation in other cell lines are not completely clear. It is assumed that some peptide hormones present in serum are responsible for down-regulation of the AR. For example, basic fibroblast growth factor is a potent inhibitor of AR expression (5). PC-3 and DU-145 cells, which do not express the AR, could be used for transient and stable transfections with AR cDNA. PC-3 cells stably transfected with the AR proliferate less rapidly than the cells transfected with an empty vector (6). Also, these cells are less invasive than their counterparts. AR expression is low in rat Dunning tumour sublines that rapidly metastasize (7). For this reason, it was believed for a long time that AR expression decreases during tumor progression. AR expression was studied with respective antibodies in clinical specimens obtained during endocrine therapy. From those studies, it was concluded that the AR is expressed in most prostate tumour specimens as well as in distant metastases obtained from patients who failed endocrine treatment (8, 9). In LNCaP cells, AR mRNA is negatively regulated by androgen (10). However, androgenic hormones stabilize the protein, thus leading to increased expression of the AR following short-term treatment. Regulation of AR expression during long-term endocrine treatment is, however, different. Those cells adapt to an environment with low androgen supply and up-regulate AR mRNA and protein. Those findings were originally reported by Liao, *et al.* (11). They were confirmed by other researchers whose work led to the conclusion that PCA cells can overcome the effect of endocrine therapy by modulating AR expression. In one cell subline developed during prolonged steroid withdrawal, high AR activity was measured even in the absence of androgen (12). This increase in transcriptional activity of the AR could be in part explained by the influence of serum factors. To understand the cellular events in advanced PCA, it is important to emphasize that the AR in LNCaP derivatives established by long-term androgen ablation could be activated by much lower concentrations of androgens than the AR in parental LNCaP cells. The non-steroidal anti-androgen bicalutamide caused small, but reproducible increase in reporter gene activity in long-term ablated cells. This was, however, sufficient for stimulation of growth *in vitro* and *in vivo*. The sequence of the AR in the hypersensitive subline, LNCaP-abl, was the same as that in parental LNCaP cells. There was no evidence for AR amplification in LNCaP-abl cells. Although the prostate-specific antigen gene is a classic example for an AR-regulated gene, its expression decreased in sublines developed after long-term androgen depletion. It was demonstrated that inhibition of expression of the AR in androgen-independent LNCaP sublines retards proliferation of those cells (13). This finding can have several consequences on development of novel clinical treatments for metastatic PCA.

AR expression increases in a subgroup of patients who underwent endocrine therapy due to amplification of the AR gene (14). Interestingly, some of these patients are better candidates for complete androgen ablation (orchiectomy or administration of luteinizing hormone releasing hormone analogues in combination with an anti-androgen) (15). In most cases, complete androgen withdrawal is not recommended for endocrine treatment since large clinical trials failed to show an advantage over orchiectomy or medical castration alone.

AR expression in PCA may be also elevated due to stabilization of the AR protein. This phenomenon was observed in the xenograft CWR 22 (16). This prostate tumour responds to castration by regression, but it relapses after some time. Expression of AR-regulated genes also rebounds, thus indicating an important role for the AR in tumour progression.

Mutations of the Androgen Receptor in PCA

In contrast to AR mutations detected in patients with androgen insensitivity syndromes, several structural changes of the AR in PCA lead to gain of function. This was first observed in LNCaP cells, whose mutant AR is efficiently activated by adrenal androgens, anti-androgens, and estrogenic or progestagenic steroids (17). In LNCaP cells, threonine at position 877 is mutated to alanine. The mutation changes the stereochemistry of the binding pocket of the AR. The first two mutations discovered in PCA patients have very similar functional consequences; in each of these two ARs, amino acid valine (at positions 715 and 730, respectively) is substituted by methionine (18, 19). These ARs are increasingly activated by adrenal androgens, products of dihydrotestosterone (DHT) metabolism, and hydroxyflutamide. Higher AR activity was induced by these hormones, although there were no major changes in AR binding affinity. A more efficient AR activation by adrenal androgens and DHT metabolism may influence natural course of the disease. Hydroxyflutamide is a mixed antagonist/agonist whose agonistic properties may be enhanced in presence of mutated AR (20). There is evidence that anti-androgen treatment results with appearance of specific mutations in the AR. Some patients who failed endocrine treatment with hydroxyflutamide were subjected to a second-line treatment with bicalutamide and showed a time-limited response (21). Missense mutations of the AR were also discovered in patients who were treated with bicalutamide (22). Thus, knowledge on AR structure in PCA patients may be relevant to determine an appropriate endocrine treatment. More recent studies on AR structure revealed that, in some patients, AR point mutations that lead to loss of function or have no functional consequences also occur.

In general, the frequency of AR point mutations increases in late stages PCA (23). Most mutations detected in PCA are somatic, in contrast to inherited mutations associated with syndromes of androgen insensitivity. Mutant ARs were detected in bone specimens from patients who failed endocrine therapy (24). However, bone metastases are only infrequently used for AR studies because of ethical concerns.

Enhancement of Androgen Receptor Function by Nonsteroidal Activators

The recognition that the AR could be activated in a ligand-independent manner was somewhat surprising. Human AR is similar to progesterone and glucocorticoid receptors, which do not exhibit ligand-independent activation. In contrast, they are activated by ligands and nonsteroidal compounds in a synergistic manner. This cross-talk between peptide hormones and the AR is still not completely understood; however, in some cases intermediate protein kinase pathways involved in ligand-independent activation of the AR have been identified. AR activation by peptide growth factors, such as insulin-like growth factor or epidermal growth factor, is most probably relevant to regulation of cellular events in advanced PCA (25). These growth factors stimulate AR activity in the absence of ligand or synergistically in combination with low androgen doses. In patients who receive endocrine therapy for PCA, activation of the AR by low androgen doses is thus enhanced by peptide growth factors. Erb B2, which is related to the epidermal growth factor receptor, was proven to be a potent AR activator (26). It could stimulate tumour growth in vivo and the expression of prostate-specific antigen (PSA). Ligand-independent and synergistic AR activation by growth factors depend on phosphorylation by mitogen-activated protein kinases (27). In those experiments, the nonsteroidal anti-androgen bicalutamide was less efficient in inhibition of AR activity.

Importantly, ligand-inhibited activation of the AR is not necessarily associated with accelerated tumour growth. The AR is also involved in regulation of cellular differentiation. High doses of androgen thus retard proliferation of LNCaP cells in parallel with increased expression of the tumor suppressor p27 (28). In contrast, treatment of LNCaP cells with low doses of androgen stimulates cyclin-dependent kinases and phosphorylation of pRb (29). Activation of the AR by the pro-differentiation compound, phenylbutyrate, has been described (30). This process is an example how AR activation is associated with a beneficial effect on prostate tumour. The AR is also activated by stimulators of the protein kinase A pathway, such as forskolin or cAMP analogues (31). These are pleiotropic compounds that differentially affect tumor growth. Interestingly, an inhibitor of the protein kinase A pathway reversed not only the effect of forskolin on AR activity, but also partially reduced androgen-stimulated AR transcriptional function. This indicates that the protein kinase A pathway is implicated in both steroidal and non-steroidal activation of the AR. In cells transfected with AR cDNA, non-steroidal regulators do not change AR levels. In contrast, in LNCaP cells effects on AR expression by non-steroidal compounds were observed.

Interleukin-6 and Related Cytokines in Regulation of PCA Cell Growth and Androgen Receptor Activity

An important activator of the AR is interleukin-6 (IL-6), a pleiotropic cytokine that regulates growth of various tumors following binding to the IL-6 receptor. In several studies, it was revealed that IL-6 levels increase in sera from patients with metastatic PCA (32, 33). PCA cells PC-3 and DU-145 also produce high levels of IL-6 into their supernatants (34). IL-6 was not detectable in LNCaP cells. In patients with organ-confined PCA, both IL-6 and IL-6 receptor levels increase (35). It was demonstrated that primary prostate epithelial cell cultures, a cell line derived from prostate intraepithelial neoplasia, PC-3, and DU-145 cells are growth-stimulated by IL-6 (34, 35).

IL-6 binding to the IL-6 receptor that is composed of a ligand-binding and signal-transducing subunit leads to activation of the signaling pathways of JAK/STAT and/or mitogen-activated protein kinases. In some cell lines, oncogenic role of STAT3 has been documented. In case of IL-6, ligand-independent and synergistic effects on AR transcriptional activity have been described. Signaling pathways of JAK/STAT and mitogen-activated protein kinases are required for ligand-independent activation of the AR by IL-6. An inhibitory effect of activators of the signaling pathway of PI 3-kinase on IL-6-induced AR activity was reported (36). Therefore, an outcome of ligand-independent activation of the AR may depend on a balance between signaling pathways in a particular cell line. IL-6 effects on AR were observed in DU-145 cells, transfected with AR cDNA, and in LNCaP cells in which treatment with IL-6 stimulates the expression of the PSA gene mRNA and protein. Non-steroidal blockers of the AR hydroxyflutamide and bicalutamide down-regulated IL-6-induced AR activity (37). The AR is also activated by oncostatin M, which in some cell lines causes a more persistent signal than IL-6 (38). Paracrine stimulation of growth of PCA cells by oncostatin M was described (39). The most important difference between AR activations by IL-6 and oncostatin M is that anti-androgens do not inhibit AR activity measured in the presence of the latter cytokine. It is especially interesting that hydroxyflutamide acted as an agonist in the presence of oncostatin M (Figure 1) (38). Also the inhibitors of signaling pathways of JAK/STAT and mitogen-activated protein kinase were less effective in DU-145 cells in the presence of oncostatin M. The receptors for oncostatin M were detected by RT-PCR in AR-negative PC-3 and DU-145 cells. In the latter cell line, growth stimulation by oncostatin M was observed (39).

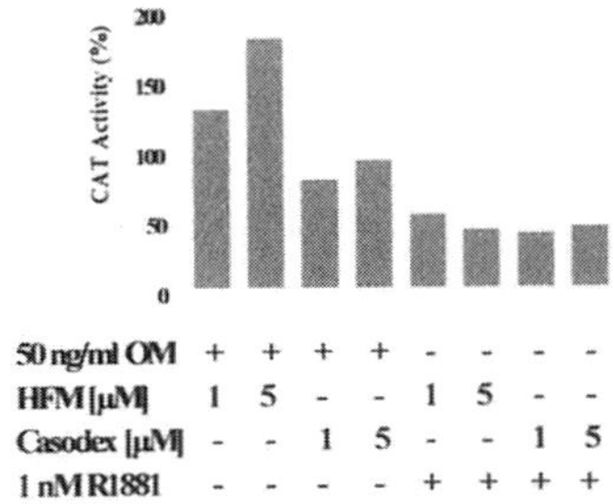

Figure 1. Effects of non-steroidal anti-androgens hydroxyflutamide (HFM) and bicalutamide (Casodex) in the presence of oncostatin M (OM) or the synthetic androgen methyltrienolone (R1881) in DU-145 cells transfected with AR cDNA and an androgen-inducible reporter gene. Reporter gene activity measured after treatment with OM or R188, respectively, was set as 100% and reporter gene activities measured in the presence of an anti-androgen were expressed in relation to those values.

For LNCaP cells, contrasting results as to IL-6 responsiveness were reported (35, 40). These data may be explained by the use of different LNCaP passages in various laboratories. An important issue in IL-6 signaling in prostate are changes in recruitment of signaling pathways during continuous IL-6 treatment. We have selected LNCaP cells for prolonged treatment with IL-6 (41). After continuous exposure to IL-6, a new cell subline, LNCaP-IL-6+, was established. These cells do not display a growth-inhibitory response to IL-6, are responsive to androgen, and up-regulate endogenous IL-6. They confer a growth advantage *in vitro* and *in vivo*. LNCaP-IL-6+ tumours were significantly larger than their control counterparts (Figure 2) (42). Their accelerated tumor growth could be explained by changes in the expression of several key cell cycle regulators; cyclin-dependent kinase (cdk) 2 was up-regulated, while the tumour suppressors p27 and pRb were expressed at a low level. Loss of p27 and pRb is a common feature of clinical PCAs. In parental LNCaP cells, in contrast to the IL-6-producing subline, IL-6 inhibits the expression and activity of cdks 2 and 4, and causes increased expression of p27. Our results as well as those published by some other researchers indicate importance of the phosphorylation of STAT3 for tumour differentiation (43). STAT3 was phosphorylated in control LNCaP-IL-6- cells, but not in the LNCaP-IL-6+ subline (42). The inhibitor of mitogen-activated protein kinase was partially effective in retardation of growth of LNCaP-IL-6+ cells. From those studies, we have concluded that significant changes in activation of signaling pathways during prolonged stimulation with IL-6 occur. Up-regulation of other growth factors in LNCaP-IL-6+ cells may be also relevant to growth advantage. In PCA, IL-6 is involved in the regulation of survival and neuroendocrine differentiation. For example, in PC-3 cells autocrine production of IL-6 leads to increased survival by activation of the PI 3-kinase pathway (44).

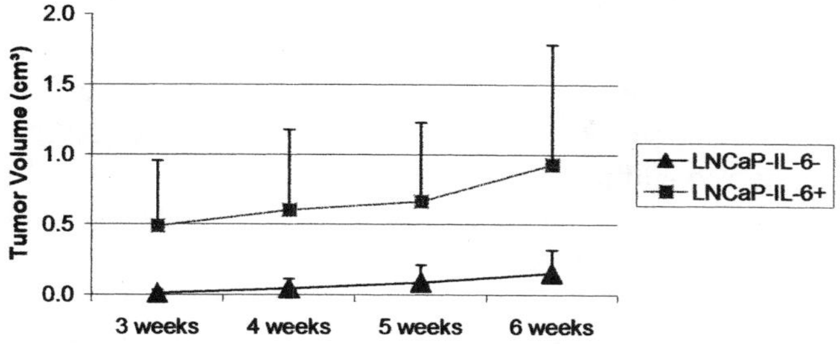

Figure 2. Growth of LNCaP xenografts in nude mice. LNCaP-IL-6 and LNCaP-IL-6+ cells were sc injected with Matrigel in mice treated with slow-release testosterone (T) pellets in two independent trials. Tumour size was measured weekly using a calliper. Tumor volume was calculated according to the formula length x width^2 of a tumor-area/2. The results of a representative experiment are shown. ($p < 0.01$, Mann-Whitney t-test; bars represent the mean ± SE) (Reprinted by permission from the American Society for Investigative Pathology.)

The Role for Androgen Receptor Coactivators in PCA

There is an increasing research interest for AR cofactors in PCA. AR cofactors have either histone acetylase or deacetylase activity, therefore, they are classified as co-activators or co-repressors. Some observations in this field are controversial, especially those concerning AR-specific enhancement of activation. More work is needed to identify specific cofactors involved in PCA progression. Studies on AR cofactors were hampered because of a lack of appropriate antibodies.

We focused on the role of the cofactor CBP in carcinoma of the prostate (45). It potentiates activation of the AR by androgen and non-steroidal anti-androgens, hydroxyflutamide and bicalutamide. The effects on AR-induced activity by hydroxyflutamide were more pronounced that those stimulated by bicalutamide. These effects were observed in DU-145 and LNCaP cells with the wild-type and two mutant receptors. CBP is expressed in PCA cell lines LNCaP, PC-3, and DU-145, and also in clinical specimens. Regulation of CBP in PCA by peptide and steroid hormones is being investigated. In other studies, it was shown that the expression of the cofactor RAC3 correlates with tumor grade and stage (46). Interestingly, nuclear localization of the cofactor Tip60 is enhanced during androgen withdrawal (47). Gregory, *et al.* investigated the expression of AR cofactors in relapsed PCA and found that SRC 1 and TIF 2 are up-regulated (48).

It was suggested that the AR cofactor ARA70 is involved in potentiation of action of androgens and anti-androgens in PCA (49, 50). In studies with ARA70, there was no considerable difference in the extent of activation of the AR by hydroxyflutamide or bicalutamide, respectively.

Conclusions and Directions for Future Research

Both the AR and IL-6 have an important role in PCA development and progression. The AR function could be better understood if more studies on AR cofactors will be carried out. The recognition that anti-androgens, currently used for endocrine therapy, switch to agonists led to establishment of novel experimental therapy strategies. Besides conventional androgen ablation, several agents that inhibit AR expression have been tested. Inhibition of AR expression is in most cases associated with retardation of PCA cell growth. Treatment of PCA cells with AR antisense oligonucleotides increased expression of the cell cycle inhibitor p21 and partially reversed androgen-independent phenotype (51). Concerning these novel therapies, there is an important unanswered question: whether these new treatments are efficient during a longer period of time.

IL-6 is clearly implicated in the development and progression of PCA, and therefore, it could be considered a target for novel therapies. PC-3 cells were treated with an anti-IL-6 antibody *in vitro* and *in vivo* (52). This treatment was efficient when combined with etoposide *in vitro*. Induction of apoptosis in PC-3 cells was observed after anti-IL-6 antibody treatment. Also, growth of PC-3 xenografts was inhibited *in vivo* by this treatment. It would be of importance to determine how anti-IL-6 based therapies may be combined with established treatments in PCA.

References

1. Irvine RA, Yu MC, Ross RK, et al (1995) The CAG and GGC microsatellites of the androgen receptor gene are in linkage disequilibrium in men with prostate cancer. Cancer Res 55:1937-1940.
2. Masiello D, Cheng S, Bubley GJ, et al (2002) Bicalutamide functions as an androgen receptor antagonist by assembly of a transcriptionally inactive receptor. J Biol Chem 277:26321-26326.
3. Horoszewicz JS, Leong SS, Kawinski E, et al (1983) LNCaP model of human prostatic carcinoma. Cancer Res 43:1809-1818.
4. Navone NM, Olive M, Ozen M, et al (1997) Establishment of two human prostate cancer cell lines derived from a single bone metastasis. Clin Cancer Res 3:2493-2500.
5. Cronauer MV, Nessler-Menardi C, Klocker H, et al (2000) Androgen receptor protein is down-regulated by basic fibroblast growth factor in prostate cancer cells. Br J Cancer 82:39-45.
6. Heisler LE, Evangelou A, Lew AM, et al (1997) Androgen-dependent cell cycle arrest and apoptotic death in PC-3 prostatic cell cultures expressing a full-length human androgen receptor. Mol Cell Endocrinol 126:59-73.

7. Quarmby VE, Beckman WCJ, Cooke DB, et al (1990) Expression and localization of androgen receptor in the R-3327 Dunning rat prostatic adenocarcinoma. Cancer Res 50:735-739.
8. Van der Kwast TH, Schalken J, Ruizeveld de Winter JA, et al (1991) Androgen receptors in endocrine-therapy-resistant human prostate cancer. Int J Cancer 48:189-193.
9. Hobisch A, Culig Z, Radmayr C, et al (1995) Distant metastases from prostatic carcinoma express androgen receptor protein. Cancer Res 55:3068-3072.
10. Quarmby VE, Yarbrough WG, Lubahn DB, et al (1990) Autologous down-regulation of androgen receptor messenger ribonucleic acid. Mol Endocrinol 4:22-28.
11. Kokontis J, Takakura K, Hay N, et al (1994) Increased androgen receptor activity and altered c-myc expression in prostate cancer cells after long-term androgen deprivation. Cancer Res 54:1566-1573.
12. Culig Z, Hoffmann J, Erdel M, et al (1999) Switch from antagonist to agonist of the androgen receptor blocker bicalutamide is associated with prostate tumour progression in a new model system. Br J Cancer 81:242-251.
13. Zegarra-Moro OL, Schmidt LJ, Huang H, et al (2002) Disruption of androgen receptor function inhibits proliferation of androgen-refractory prostate cancer cells. Cancer Res 62:1008-1013.
14. Visakorpi T, Hyytinen E, Koivisto P, et al (1995) In vivo amplification of the androgen receptor gene and progression of human prostate cancer. Nature Gen 9:401-406.
15. Palmberg C, Koivisto P, Kakkola L, et al (2000) Androgen receptor gene amplification at primary progression predicts response to combined androgen blockade as second line therapy for advanced prostate cancer. J Urol 164:1992-1995.
16. Gregory CW, Hamil KG, Kim D, et al (1998) Androgen receptor expression in androgen-independent prostate cancer is associated with increased expression of androgen-regulated genes. Cancer Res 58:5718-5724.
17. Veldscholte J, Ris-Stalpers C, Kuiper GGJM, et al (1990) A mutation in the ligand binding domain of the androgen receptor of human LNCaP cells affects steroid binding characteristics and response to anti-androgens. Biochem Biophys Res Commun 17:534-540.
18. Newmark JR, Hardy DO, Tonb DC, et al (1992) Androgen receptor gene mutations in human prostate cancer. Proc Natl Acad Sci USA 89:6319-6323.
19. Culig Z, Hobisch A, Cronauer MV, et al (1993) Mutant androgen receptor detected in an advanced stage of prostatic carcinoma is activated by adrenal androgens and progesterone. Mol Endocrinol 7:1541-1550.

20. Wong CI, Kelce WR, Sar M, et al (1995) Androgen receptor antagonist versus agonist activities of the fungicide vinclozolin relative to hydroxyflutamide. J Biol Chem 270:19998-20003.
21. Taplin ME, Bubley GJ, Ko YJ, et al (1999) Selection for androgen receptor mutations in prostate cancers treated with androgen antagonist. Cancer Res 59:2511-2515.
22. Haapala K, Hyytinen ER, Roiha M, et al (2001) Androgen receptor alterations in prostate cancer relapsed during a combined androgen blockade by orchiectomy and bicalutamide. Lab Invest 81:1647-1651.
23. Marcelli M, Ittmann M, Mariani S, et al (2000) Androgen receptor mutations in prostate cancer. Cancer Res 60:944-949.
24. Taplin ME, Bubley GJ, Shuster TD, et al (1995) Mutation of the androgen receptor gene in metastatic androgen-independent prostate cancer. N Engl J Med 332:1393-1398.
25. Culig Z, Hobisch A, Cronauer MV, et al (1994) Androgen receptor activation in prostatic tumor cell lines by insulin-like growth factor-I, keratinocyte growth factor, and epidermal growth factor. Cancer Res 54: 5474-5478.
26. Craft N, Shostak Y, Carey M, et al (1999) A mechanism for hormone-independent prostate cancer through modulation of androgen receptor signaling by the HER-2/neu tyrosine kinase. Nat Med 5:280-285.
27. Yeh S, Lin HK, Kang HY, et al (1999) From HER2/Neu signal cascade to androgen receptor and its coactivators: a novel pathway by induction of androgen target genes through MAP kinase in prostate cancer cells. Proc Natl Acad Sci USA 96:5458-5463.
28. Tsihlias J, Zhang W, Bhattacharya N, et al (2000) Involvement of p27(Kip1) in G1 arrest by high dose 5alpha-dihydrotestosterone in LNCaP human prostate cancer cells. Oncogene 19:670-679.
29. Hofman K, Swinnen JV, Verhoeven G, et al (2001) E2F activity is biphasically regulated by androgens in LNCaP cells. Biochem Biophys Res Commun 283:97-101.
30. Sadar MD, Gleave ME (2000) Ligand-independent activation of the androgen receptor by the differentiation agent butyrate in human prostate cancer cells. Cancer Res 60:5825-5831.
31. Nazareth LV, Weigel NL (1996) Activation of the human androgen receptor through a protein kinase A signaling pathway. J Biol Chem 271:19900-7.
32. Twillie DA, Eisenberger MA, Carducci MA, et al (1995) Interleukin-6: a candidate mediator of human prostate cancer morbidity. Urology 45:542-549.
33. Adler HI, McCurdy MA, Kattan MW, et al (1999) Elevated levels of circulating interleukin-6 and transforming growth factor-ß1 in patients with metastatic prostatic carcinoma. J Urol 161:182-187.

34. Chung TD, Yu JJ, Spiotto MT, et al (1999) Characterization of the role of IL-6 in the progression of prostate cancer. Prostate 38:199-207.
35. Giri D, Ozen M, Ittmann M (2001) Interleukin-6 is an autocrine growth factor in human prostate cancer. Am J Pathol 159:159-165.
36. Yang L, Wang L, Lin HK, et al (2003) Interleukin-6 differentially regulates androgen receptor transactivation via PI3K-Akt, STAT3, MAPK, three distinct signal pathways in prostate cancer cells. Biochem Biophys Res Commun 305:462-469.
37. Hobisch A, Eder IE, Putz T, et al (1998) Interleukin-6 regulates prostate-specific protein expression in prostate carcinoma cells by activation of the androgen receptor. Cancer Res 58:4640-4645.
38. Godoy-Tundidor S, Hobisch A, Pfeil K, et al (2002) Acquisition of agonistic properties of nonsteroidal antiandrogens after treatment with oncostatin M in prostate cancer cells. Clin Cancer Res 8:2356-2361.
39. Mori S, Murakami-Mori K, Bonavida B (1999) Oncostatin M (OM) promotes the growth of DU 145 human prostate cancer cells, but not PC-3 or LNCaP, through the signaling of the OM specific receptor. Anticancer Res 19:1011-1015.
40. Degeorges A, Tatoud R, Fauvel Lafeve F, et al (1996) Stromal cells from human benign prostate hyperplasia produce a growth-inhibitory factor for LNCaP prostate cancer cells, identified as interleukin-6. Int J Cancer 68:207-214.
41. Hobisch A, Ramoner R, Fuchs D, et al (2001) Prostate cancer cells (LNCaP) generated after long-term interleukin-6 treatment express interleukin-6 and acquire an interleukin-6-partially resistant phenotype. Clin Cancer Res 7:2941-2948.
42. Steiner H, Godoy-Tundidor S, Rogatsch H, et al (2003) Accelerated in vivo growth of prostate tumors that up-regulate interleukin-6 is associated with reduced retinoblastoma protein expression and activation of the mitogen-activated protein kinase pathway. Am J Pathol 162:655-663.
43. Spiotto MT, Chung TD 2000 STAT3 mediates IL-6-induced growth inhibition in the human prostate cancer cell line LNCaP. Prostate 42:88-98.
44. Chung TD, Yu JJ, Kong TA, Spiotto MT, et al (2000) Interleukin-6 activates phosphatidylinositol-3 kinase, which inhibits apoptosis in human prostate cancer cell lines. Prostate 42:1-7.
45. Comuzzi B, Lambrinidis L, Rogatsch H, et al (2003) The transciptional co-activator cAMP response element-binding protein-binding protein is expressed in prostate cancer and enhances androgen- and anti-androgen-induced androgen receptor function. Am J Pathol 162:233-241.
46. Gnanapragasam VJ, Leung HY, Pulimood AS, et al (2001) Expression of RAC3, a steroid hormone receptor co-activator in prostate cancer. Br J Cancer 85:1928-1936.

47. Halkidou K, Gnanapragasam VJ, Mehta PB, et al (2003) Expression of Tip60, an androgen receptor coactivator, and its role in prostate cancer development. Oncogene 22:2466-2477.
48. Gregory CW, He B, Johnson RT, et al (2001) A mechanism for androgen receptor-mediated prostate cancer recurrence after androgen deprivation therapy. Cancer Res 61:4315-4319.
49. Yeh S, Chang C (1996) Cloning and characterization of a specific coactivator, ARA70, for the androgen receptor in human prostate cells. Proc Natl Acad Sci USA 93:5517-5521.
50. Miyamoto H, Yeh S, Wilding G, et al (1998) Promotion of agonist activity of antiandrogens by the androgen receptor coactivator, ARA70, in human prostate cancer DU145 cells. Proc Natl Acad Sci USA 95:7379-7384.
51. Wang LG, Ossowski L, Ferrari AC (2001) Overexpressed androgen receptor linked to p21WAF1 silencing may be responsible for androgen independence and resistance to apoptosis of a prostate cancer cell line. Cancer Res 61:7544-7551.
52. Smith PC, Keller ET (2001) Anti-interleukin-6 monoclonal antibody induces regression of human prostate cancer xenografts in nude mice. Prostate 48:47-53.

18

Role of the Androgen Receptor and PI3K/Akt in the Survival of Androgen-Refractory Prostate Cancer Cells

Haojie Huang and Donald J. Tindall

Introduction

Surgical or pharmaceutical ablation of testicular androgens has been the most effective treatment of metastatic prostate cancer (PCA) since 1941 (1). Both normal and malignant epithelial cells of the prostate undergo programmed cell death (apoptosis) in the absence of androgens (2, 3). However, a majority of PCA patients usually relapse with tumors becoming refractory to androgen ablation therapy, suggesting that relapsed cells are able to resist androgen withdrawal-induced cell death (4). Overexpression of the anti-apoptotic protein Bcl-2 appears to promote the survival of androgen-refractory PCA cells (5-7). Bcl-2 protein is expressed intensely in 77% of androgen-refractory prostate tumors compared to 32% in androgen-sensitive tumors (5, 8, 9).

Bcl-2 belongs to a family of proteins that includes the anti-apoptotic subfamily (*i.e.,* Bcl-2, Bcl-xL, and Mcl-1), the pro-apoptotic subfamily (*i.e.,* Bax, Bak, and Bok), and the BH3-only subfamily (*i.e.,* Bid, Bim, and Noxa) (10). The pro- and anti-apoptosis counterparts of this family form heterodimers and titrate out each other's functions, suggesting that their relative concentrations play a pivotal role in execution of programmed cell death. Overexpressed Bcl-2 protein increases the level of Bcl-2/Bax heterodimer, stabilizes the integrity of mitochondria, and protects cells from apoptosis (11).

It is possible that the increased expression of Bcl-2 in androgen–refractory PCA may result from outgrowth of cells expressing high levels of Bcl-2 before androgen ablation treatment. It is also possible, but not necessarily mutually exclusive that androgen manipulation may reset the programmed regulation of Bcl-2 expression in prostate epithelial cells. In fact, levels of Bcl-2 transcripts increase in the rat prostate following castration. Moreover, this increase is abrogated in castrated rats that receive testosterone (5). Expression of Bcl-2 mRNA is elevated in LNCaP xenografts in castrated mice. In contrast, androgen treatment of LNCaP cells results in down regulation of Bcl-2 mRNA (12). These data suggest that androgen regulation of Bcl-2 expression is a common event in different types of cells, although the molecular basis of the regulation is unknown.

In addition to androgen action, growth factors are also believed to play an essential role in the androgen-refractory progression of PCA (13). Stimulation of cells with growth factors such as IGF-I, BDNF, and NGF up regulates Bcl-2 expression in a variety of cell types (14-16). Growth factor-elicited survival signalings are negatively regulated by the PTEN tumor suppressor gene. Frequent loss of PTEN through mutation, deletion, or methylation has been reported in late-stage PCA (17). PI3K/Akt signaling has been shown to be constitutively active in the PTEN-mutated PCa cell lines LNCaP and PC-3 (18-21). Therefore, loss of PTEN and/or activation of PI3K/Akt may contribute to the overexpression of Bcl-2 in PCA cells.

Androgen-refractory Outgrowth of LNCaP Cells under Androgen-deprived Conditions

In order to understand the molecular mechanism for the survival of PCA cells under androgen ablation conditions, LNCaP cells were cultured in charcoal-stripped serum (CSS)-medium. Very limited death of LNCaP cells was detected under these culture conditions. Indeed, cells began to undergo neuroendocrine cell differentiation (NED) three days after androgen withdrawal. Cells at this stage are called LNCaP-tNED since NED is a transient process. After being cultured in CSS-medium for 10 weeks, LNCaP cells started to re-grow, and an androgen-refractory subline LNCaP-Rf was established eventually from parental LNCaP cells (Figure 1).

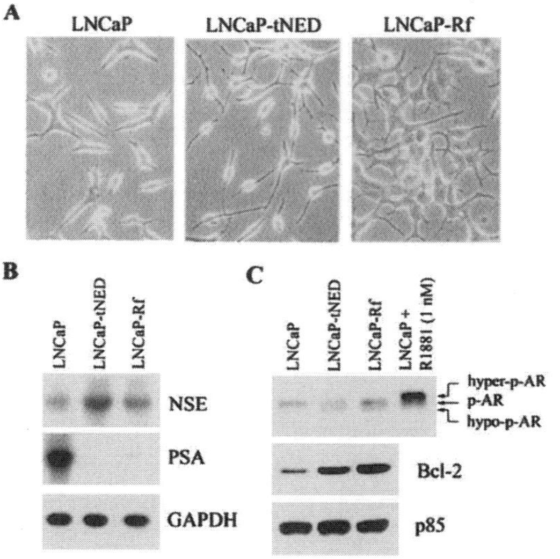

Figure 1. Morphological and biochemical analyses of LNCaP cells grown under acute and chronic androgen-deprived culture conditions.
A. Morphology of LNCaP, LNCaP-tNED, and LNCaP-Rf cells.
B. Northern analysis of NSE and PSA in LNCaP, LNCaP-tNED, and LNCaP-Rf cells.
C. Western analysis of AR and Bcl-2 in LNCaP, LNCaP-tNED, and LNCaP-Rf cells.

The parental LNCaP, LNCaP-tNED and LNCaP-Rf cells showed distinct morphological features (Fig. 1A). LNCaP-tNED cells exhibited neuroendocrine-like features, including long slender processes and compact cell body, while the untreated LNCaP cells show their typical fusiform morphology. ,LNCaP-Rf cells maintained an intermediate morphological appearance between parental LNCaP and LNCaP-tNED cells. Consistent with morphological observations, LNCaP-Rf cells expressed more mRNA of neuron-specific enolase (NSE), a specific neuroendocrine marker, than parental LNCaP cells, but less than LNCaP-tNED cells (Figure 1B). In Figure 1C, levels of the androgen receptor (AR) protein were transiently decreased in LNCaP-tNED cells. However, LNCaP-Rf showed comparable AR protein levels to untreated LNCaP cells. It is worth noting that more AR protein became hypophosphorylated following androgen ablation (lane 2 and 3, Figure 1C). In contrast, the AR protein was present mainly in a phosphorylated state under normal culture conditions (lane 1, Fig. 1C). The AR protein became hyperphosphorylated after androgen treatment (lane 4, Fig. 1C). The expression levels of prostate-specific antigen (PSA) were decreased largely in LNCaP-Rf cells in comparison to those in LNCaP cells (Fig. 1B). These experiments indicate that following androgen ablation, LNCaP cells transiently differentiate into neuroendocrine-like cells, and eventually evolve into androgen-refractory cells. Establishment of LNCaP-tNED and LNCaP-Rf from LNCaP cells provides excellent cellular models to explore the mechanism for survival and re-growth of LNCaP cells under androgen deprivation conditions.

Activation of Survival Signalings under Androgen Ablation Conditions in LNCaP Cells

In order to define the survival signaling pathways that allow LNCaP cells to avoid cell death induced by androgen withdrawal, the levels of Bcl-2 in LNCaP and its derivatives were examined. Shown in Figure 1C, expression of Bcl-2 was increased under acute and chronic androgen ablation conditions. Additionally, PI3-Kinase activity in LNCaP and its derived cells was determined. Shown in Figure 2, PI3-kinase was active in parental LNCaP cells when cultured in regular culture medium. However, higher activity of PI3-Kinase was seen in the LNCaP-tNED cells. Approximately 4.0-fold higher level of PI3-kinase activity was found in the LNCaP-Rf cells as that found in the parental LNCaP cell line. PI3K-dependent activation of Akt has been shown to require phosphorylation at residues T308 and S473. To assess Akt activation, a specific antibody for phosphorylated Akt was used. In Figure 2, Akt phosphorylation at serine 473 was dramatically increased in LNCaP-tNED and LNCaP-Rf cells but no changes were detected in total Akt protein levels. Therefore, consistent with previous reports (18, 20), we found that PI3K and Akt signaling pathway is constitutively activated in LNCaP cells when cultured in medium containing whole serum. Thus, our data demonstrate that PI3-Kinase activity is increased by androgen withdrawal.

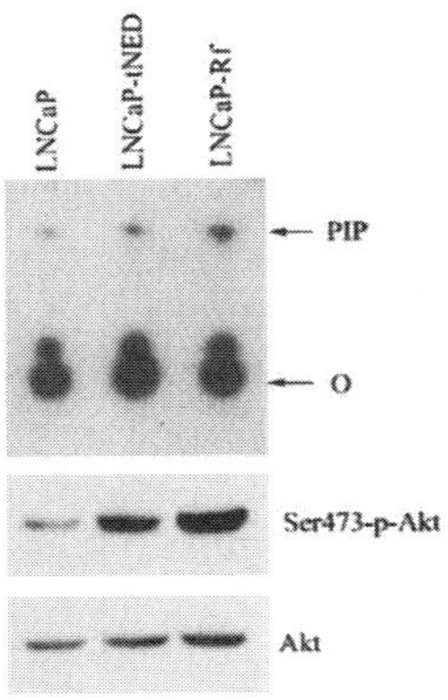

Figure 2. Activities of PI3K and Akt in LNCaP and its derivatives. PIP, phophatidylinositol monophophate; O, origin of sample loading.

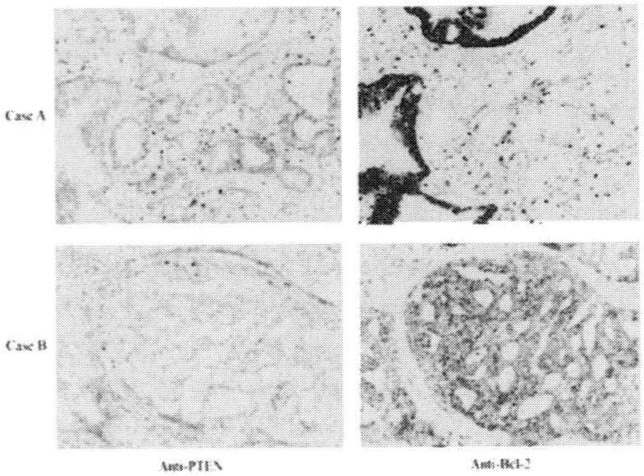

Figure 3. Immunohistochemistry of PTEN and Bcl-2 proteins in primary prostate tumors.

Regulation of Bcl-2 Expression by the PI3K/PTEN/Akt Pathway PCA Cells

Next, it was determined whether there is a causal relationship between activation of the PI3K/Akt pathway and upregulation of Bcl-2. A number of studies have reported either loss of PTEN or overexpression of Bcl-2 in advanced prostate tumors. However, no study has examined expression of both proteins in the same tumors. Therefore, we utilized immunohistochemistry to measure expression of PTEN and Bcl-2 in 17 cases of high-grade (Gleason score 7 or higher) prostate tumors. The findings indicate an inverse correlation of PTEN and Bcl-2 staining in 82.4% of the tumors examined (41.2% exhibited negative staining for PTEN but positive staining for Bcl-2, whereas 41.2% exhibited positive staining for PTEN but negative staining for Bcl-2). Examples of each type of staining pattern can be seen in Figure 3, where case A exhibits PTEN (+) and Bcl-2 (-), and case B exhibits PTEN (-) and Bcl-2 (+). Taken together, these studies show that Bcl-2 expression is inversely correlated with PTEN loss in many malignant prostate tissues and cell lines.

Ectopic expression of PTEN in PTEN-null cells suppresses Bcl-2 expression. Although LNCaP and PC-3 cells lack PTEN protein, both cell lines display an intact PTEN signaling pathway (18, 47-49). To determine a causal relationship between the PTEN loss and Bcl-2 overexpression, a PTEN expression vector was transiently transfected into LNCaP and PC-3 cells. As shown in Figure

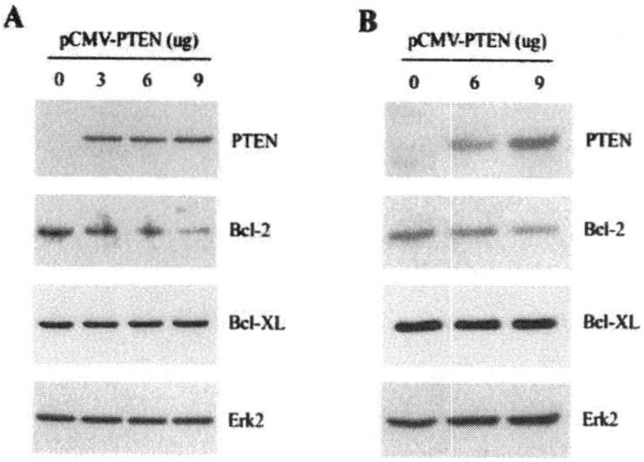

Figure 4. Ectopic expression of PTEN suppresses Bcl-2 expression in:
(**A**) PTEN-mutated PCA cell lines LNCaP, and (**B**) PC-3.

4, ectopic expression of PTEN in both LNCaP and PC-3 cells resulted in decreased levels of Bcl-2 protein in a dose-dependent manner. In contrast, PTEN did not affect the expression of Bcl-XL, another anti-apoptotic member of the Bcl-2 family, in either LNCaP or PC-3 cells (Figure 4).

Further analyses indicate that PTEN down regulates Bcl-2 expression by decreasing the Akt-depedent phosphorylation of CREB, thereby modulating its transcriptional activity (21).

Androgen Regulation of Bcl-2 Expression in PCA Cells

As it was demonstrated above, androgen withdrawal resulted in an increase in Bcl-2 expression in both transient and refractory stages (Figure 1C), suggesting that under physiological conditions, androgens suppress Bcl-2 expression in LNCaP cells. Therefore, to elucidate the molecular basis of androgen regulation of Bcl-2 expression, LNCaP cells were incubated in culture medium, +/- physiological concentrations of androgens. Treatment of LNCaP cells with 1 and 5 nM of the synthetic androgen R1881 resulted in a decreased expression of Bcl-2 protein (Figure 5A). To determine whether the effect of androgens on Bcl-2 is mediated by the AR, AR-negative PC-3 PCA cells were transiently transfected with a human AR expression vector. As shown in Figure 5B, androgens not only stabilized the AR protein, but also caused a decrease in Bcl-2 expression in the presence of the AR. However, there was no effect of AR alone on Bcl-2 expression even at a higher level of the protein (Figure 6B). These findings suggest that the androgenic effect on Bcl-2 expression is mediated by the AR.

Next, it was sought to determine whether androgens affect transcription of the *Bcl-2* gene. As shown in Figure 6A, treatment of LNCaP cells with 1 nM of R1881 resulted in a decrease in Bcl-2 messenger RNA. To determine whether androgens repress Bcl-2 expression via a direct mechanism, cells were treated with R1881 and cycloheximide, an inhibitor of *de novo* protein synthesis. As shown in Figure 6A, the effect of androgens on Bcl-2 expression was abolished by pre-treatment of the cells with cycloheximide, suggesting that the androgenic effect on Bcl-2 mRNA is mediated through an indirect mechanism. This finding is consistent with previous findings indicating that there is no perfect consensus sequence for an androgen response element (ARE) in the promoter of the human *Bcl-2* gene (22-24). This conclusion was confirmed indirectly by the blockage of androgenic induction of expression of fatty acid synthase (data not shown), a gene that is coordinately regulated by androgens (25).

By using promoter reporter assays, we sought to determine whether androgen-induced downregulation of Bcl-2 mRNA is caused by a decreased transcription of the *Bcl-2* gene. Previous studies have shown that there are two promoters (P1 and P2) that regulate transcription of the *Bcl-2* gene, and that fewer than 5% of all Bcl-2 transcripts are initiated from P2 in many different cell systems (23, 26-28). This suggests that P1 is the predominant promoter of the *Bcl-2* gene. Indeed, results from us and others have shown previously that *PTEN* and *WT1* inhibit Bcl-2 expression through the P1 promoter in LNCaP cells (21, 29). Therefore, in this study we focused on the regulation of P1 promoter activity by androgens. LNCaP cells were transiently transfected with a panel of luciferase reporter constructs containing various regions of the *Bcl-2* P1 promoter.

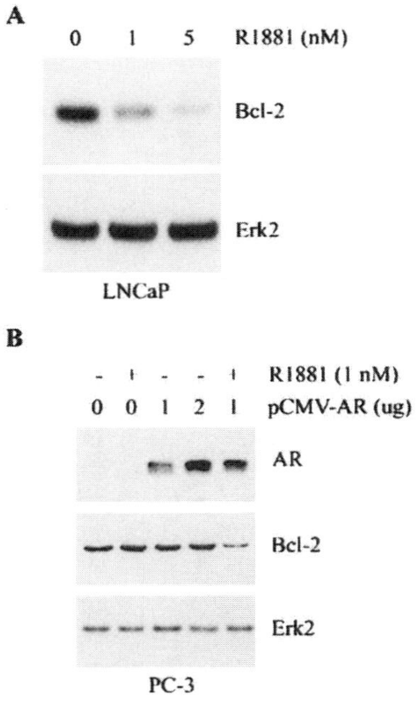

Figure 5. The effects of androgens on expression of Bcl-2 protein in: **(A)** LNCaP and **(B)** PC-3 cells.

As shown in Figure 6B, luciferase activities of reporter constructs LB124 and LB334 were inhibited approximately 50% in the cells treated with 1 nM of R1881, whereas, the activity of the reporter plasmids pBcl2m-Luc and LB360 were unaffected. These data suggest that transcriptional activity of the *Bcl-2* gene is downregulated by androgens, and that an androgen regulatory region exists between nucleotides −1532 and −1343 of the *Bcl-2* P1 promoter. When this region was examined for potential regulatory elements, we found an element (-1452 to −1445) with a perfect match to a functional variant of the E2F binding sequence defined previously in the promoter of the *p21WAF1* gene (30). However, no consensus sequence for an androgen response element was found in this region of the *Bcl-2* gene (GenBank accession number NT_033907). Since expression of the E2F1 protein is known to be regulated by androgens in the CWR22 human PCA xenograft model, we sought to determine whether the E2F binding site is involved in androgen regulation of Bcl-2 expression. We found that luciferase activity of a reporter plasmid containing the E2F site [pBCL2(E2F-v)] was reduced by 50% in cells treated with 1 nM of R1881 (Figure 6C). This inhibition was similar to that observed with constructs LB124 and LB334 (Figure 6B). As shown in Figure 6C, mutations in the E2F site [pBcl2(mE2F-v)] abrogated the androgenic inhibition of this reporter construct. These data suggest that regulation of *Bcl-2* expression by androgens is mediated by E2F proteins.

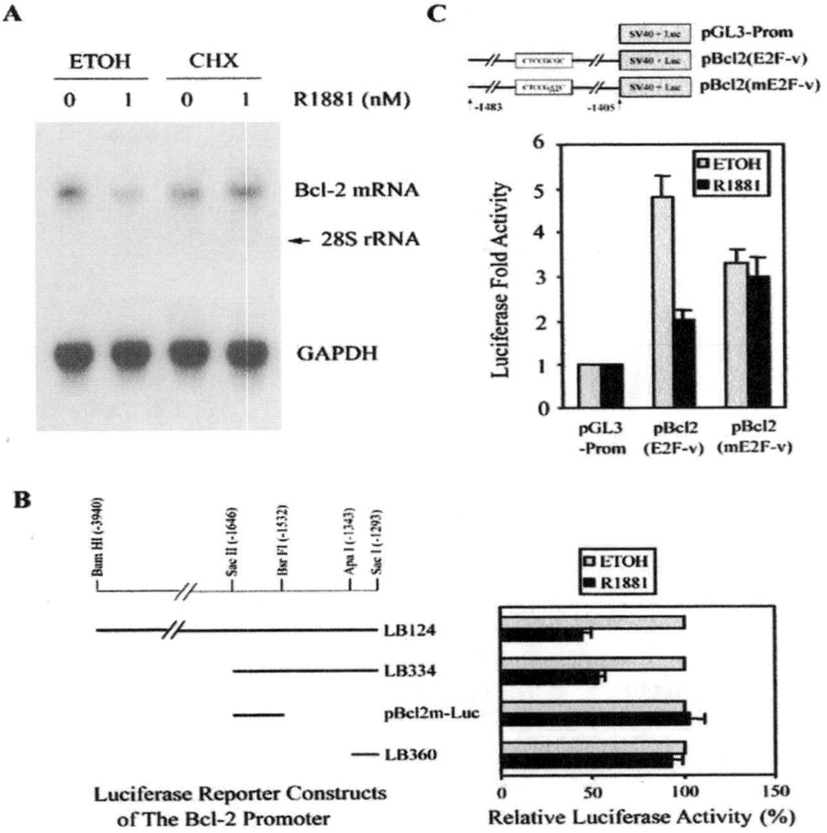

Figure 6. The effect of androgens on transcription of the Bcl-2 gene in LNCaP cells. **A.** Androgens inhibit Bcl-2 mRNA expression in LNCaP cells. **B**. Androgen treatment of LNCaP inhibits the activity of the Bcl-2 promoter reporter genes. **C**. Mutations in a putative E2F binding site within the androgen-regulated region in the Bcl-2 promoter abolish inhibitory effect of androgens on Bcl-2 expression.

To verify that the E2F site in the *Bcl-2* promoter is involved in the androgen regulation of Bcl-2 expression, we sought to determine whether E2F1 regulates expression of the *Bcl-2* gene in PCA cells. As shown in Figure 7A and 7B, ectopic expression of E2F1 in LNCaP cells resulted in increases in Bcl-2 protein and mRNA. Moreover, transcriptional activity of the *Bcl-2* promoter was increased approximately five-fold by ectopic expression of E2F1, and this activity was markedly decreased when mutations were introduced into the E2F binding site

(Figure 7C). Chromatin immunoprecipitation (ChIP) assays demonstrated that E2F1 binds to the promoter of the *Bcl-2* gene *in vivo* (Figure 7D). Thus, these data provide direct evidence that E2F1 binds to and transactivates the promoter of the *Bcl-2* gene.

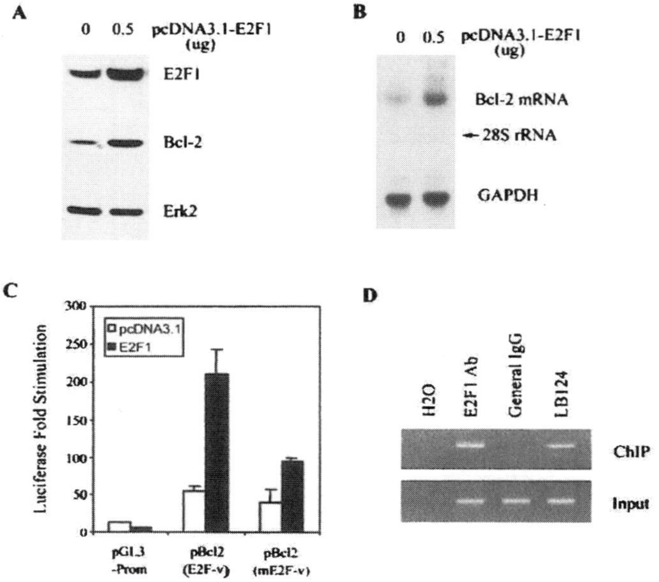

Figure 7. E2F1 regulation of Bcl-2 expression in LNCaP cells. Ectopic expression of E2F1 upregulates Bcl-2 protein (**A**) and mRNA (**B**). (**C**). Mutations in the putative E2F binding site in the Bcl-2 promoter abolish E2F1-mediated stimulation on the activity of the Bcl-2 exprespromoter. (**D**). E2F1 binds to the Bcl-2 promoter in vivo as demonstrated by Chromatin immunoprecipitation (ChIP) assays.

Next we sought to determine whether the E2F1 protein is a mediator of the androgen-induced suppression of Bcl-2 expression. Androgen treatment of LNCaP cells resulted in a decrease in levels of E2F1 protein (Figure 8A). Similarly, ectopic expression of the androgen receptor in PC-3 cells diminished the expression of the E2F1 protein (Figure 8B). *E2F1* mRNA was also downregulated by treatment of LNCaP cells with 1 nM of R1881 as demonstrated by Northern blot analyses (Figure 8C). The effect was abrogated completely by pre-treatment of cells with cycloheximide, an inhibitor of protein synthesis, or actinomycin D, an inhibitor of RNA synthesis (Figure 8C). These data suggest that androgen-induced downregulation of *E2F1* is not mediated directly by the androgen receptor. These findings also rule out the possibility that the androgenic effect is mediated by a mechanism of mRNA stability.

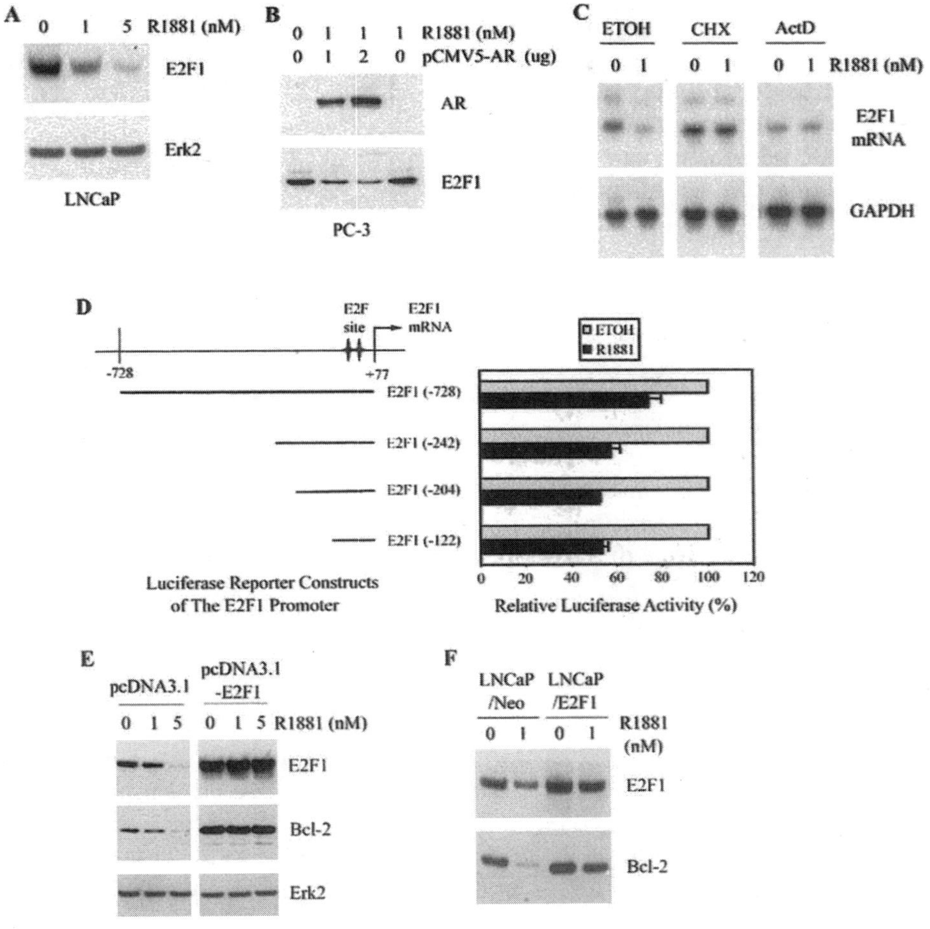

Figure 8. The effect of E2F1 expression on androgen-induced suppression of Bc-2. **(A)**. Effect of androgens on E2F1 expression in LNCaP cells. **(B)**. Effect of androgens on E2F1 expression in PC-3 cells. **(C)**. Androgenic effect on expression of E2F1 mRNA in LNCaP cells. **(D)**. Effect of androgens on transcription of E2F1 in LNCaP cells. **(E)**. Transient transfection of LNCaP cells with E2F1 abrogated androgenic inhibition of Bcl-2 expression. **(F)**. Stable expression of E2F1 in LNCaP cells diminished androgenic suppression of Bcl-2.

We further examined the effects of androgens on the transcriptional activity of *E2F1* promoter reporter constructs. We found that the activities of a number of reporter constructs containing up to 728 bp of the *E2F1* promoter, were inhibited by androgen treatment (Figure 8D). All of these constructs include a region that contains two E2F consensus-binding sites (see Figure 8D) that is known to mediate autoregulation of *E2F1* expression. Thus, androgens may downregulate E2F1, which in turn downregulates Bcl-2 expression. To test this possibility, we transiently transfected an E2F1 expression vector into LNCaP cells. As shown in Figure 8E, transfection of an E2F1 expression plasmid increased the levels of E2F1 protein. Importantly, elevated levels of E2F1 blocked the androgen-induced decrease in Bcl-2 expression (Figure 8E), suggesting that E2F1 is mediating the androgenic effects on Bcl-2 expression. To further test this under more physiological conditions, we established a subline of LNCaP, which stably expresses exogenous E2F1. We found that the androgen-induced decrease in Bcl-2 expression was largely abrogated by high levels of E2F1 protein (Figure 8F), supporting the results obtained from the transient transfection experiments. Thus, these data confirm that E2F1 is a mediator of androgenic suppression of Bcl-2 expression.

Conclusions

In summary, our data demonstrate that following androgen withdrawal, a majority of LNCaP androgen-sensitive PCA cells did not undergo apoptosis, but rather transient neuroendocrine differentiation, and eventually relapsed to an androgen-refractory phenotype. The anti-apoptotic protein Bcl-2 was overexpressed during this complex process. Additionally, the PI3K/Akt survival pathway was further activated due to androgen withdrawal. Moreover, we found that loss of PTEN and overexpression of Bcl-2 are inversely correlated in primary prostate tumors. Ectopic expression of PTEN resulted in decreased expression of Bcl-2 in PTEN-null PCA cells by negatively regulating the Akt and CREB signaling pathway. Furthermore, expression of Bcl-2 in PCA cells is androgen-regulated. Androgen treatment of PCA cells downregulates Bcl-2 expression in an E2F1-dependent manner. Therefore, our data suggest that deregulation of PI3K/Akt/CREB and E2F1 pathways caused by loss of PTEN function and/or androgen deprivation contributes to the overexpression of Bcl-2 in androgen-refractory prostate tumors.

References

1. Huggins C, Hodges CV (1941) Studies on prostatic cancer. I. The effect of castration, and of androgen injection on serum phosphatases in metastatic carcinoma of the prostate. Cancer Res 1:293-297.
2. Isaacs JT (1984) Antagonistic effect of androgen on prostatic cell death. Prostate 5:545-557.
3. Kyprianou N, English HF, Isaacs JT (1990) Programmed cell death during regression of PC-82 human PCA following androgen ablation. Cancer Res 50:3748-3753.
4. Denmeade SR, Lin XS, Isaacs JT (1996) Role of programmed (apoptotic) cell death during the progression and therapy for PCA. Prostate 28:251-265.

5. McDonnell TJ, Troncoso P, Brisbay SM, et al (1992) Expression of the protooncogene bcl-2 in the prostate and its association with emergence of androgen-independent PCA. Cancer Res 52:6940-6944.
6. Colombel M, Symmans F, Gil S, et al (1993) Detection of the apoptosis-suppressing oncoprotein bc1-2 in hormone-refractory human PCAs. Am J Pathol 143:390-400.
7. Krajewska M, Krajewski S, Epstein JI, et al (1996) Immunohistochemical analysis of bcl-2, bax, bcl-X, and mcl-1 expression in PCAs. Am J Pathol 148:1567-1576.
8. Westin P, Stattin P, Damber JE, Bergh A (1995) Castration therapy rapidly induces apoptosis in a minority and decreases cell proliferation in a majority of human prostatic tumors. Am J Pathol 146:1368-1375.
9. McDonnell TJ, Navone NM, Troncoso P, et al (1997) Expression of bcl-2 oncoprotein and p53 protein accumulation in bone marrow metastases of androgen independent PCA. J Urol 157:569-574.
10. Green DR (2000) Apoptotic pathways: paper wraps stone blunts scissors. Cell 102:1-4.
11. Haldar S, Basu A, Croce CM (1997) Bcl2 is the guardian of microtubule integrity. Cancer Res 57:229-233.
12. Gleave M, Tolcher A, Miyake H, et al (1999) Progression to androgen independence is delayed by adjuvant treatment with antisense Bcl-2 oligodeoxynucleotides after castration in the LNCaP prostate tumor model. Clin Cancer Res 5:2891-2898.
13. Grossmann ME, Huang H, Tindall DJ (2001) Androgen receptor signaling in androgen-refractory PCA. J Natl Cancer Inst 93:1687-1697.
14. Pugazhenthi S, Miller E, Sable C, et al (1999) Insulin-like growth factor-I induces bcl-2 promoter through the transcription factor cAMP-response element-binding protein. J Biol Chem 274:27529-27535.
15. Bonni A, Brunet A, West AE, et al (1999) Cell survival promoted by the Ras-MAPK signaling pathway by transcription-dependent and -independent mechanisms. Science 286:1358-1362.
16. Riccio A, Ahn S, Davenport CM, et al (1999) Mediation by a CREB family transcription factor of NGF-dependent survival of sympathetic neurons. Science 286:2358-2361.
17. Ali IU, Schriml LM, Dean M (1999) Mutational spectra of PTEN/MMAC1 gene: a tumor suppressor with lipid phosphatase activity. J Natl Cancer Inst 91:1922-1932.
18. Carson JP, Kulik G, Weber MJ (1999) Antiapoptotic signaling in LNCaP PCA cells: a survival signaling pathway independent of phosphatidylinositol 3'-kinase and Akt/protein kinase B. Cancer Res 59:1449-1453.

19. Wu X, Senechal K, Neshat MS, et al (1998) The PTEN/MMAC1 tumor suppressor phosphatase functions as a negative regulator of the phosphoinositide 3-kinase/Akt pathway. Proc Natl Acad Sci USA 95:15587-15591.

20. Lin J, Adam RM, Santiestevan E, et al (1999) The phosphatidylinositol 3'-kinase pathway is a dominant growth factor-activated cell survival pathway in LNCaP human prostate carcinoma cells. Cancer Res 59:2891-2897.

21. Huang H, Cheville JC, Pan Y, et al (2001) PTEN induces chemosensitivity in PTEN-mutated PCA cells by suppression of Bcl-2 expression. J Biol Chem 276:38830-38836.

22. Lapointe J, Fournier A, Richard V, et al (1999) Androgens down-regulate bcl-2 protooncogene expression in ZR-75-1 human breast cancer cells. Endocrinology 140:416-421.

23. Seto M, Jaeger U, Hockett RD, et al (1988) Alternative promoters and exons, somatic mutation and deregulation of the Bcl-2-Ig fusion gene in lymphoma. Embo J 7:123-131.

24. Bruckheimer EM, Spurgers K, Weigel NL, et al (2003) Regulation of Bcl-2 expression by dihydrotestosterone in hormone sensitive LNCaP-FGC PCA cells. J Urol 169:1553-1557.

25. Swinnen JV, Ulrix W, Heyns W, et al (1997) Coordinate regulation of lipogenic gene expression by androgens: evidence for a cascade mechanism involving sterol regulatory element binding proteins. Proc Natl Acad Sci USA 94:12975-12980.

26. Tsujimoto Y, Croce CM (1986) Analysis of the structure, transcripts, and protein products of bcl-2, the gene involved in human follicular lymphoma. Proc Natl Acad Sci USA 83:5214-5218.

27. Miyashita T, Harigai M, Hanada M, et al (1994) Identification of a p53-dependent negative response element in the bcl-2 gene. Cancer Res 54:3131-3135.

28. Wilson BE, Mochon E, Boxer LM (1996) Induction of bcl-2 expression by phosphorylated CREB proteins during B-cell activation and rescue from apoptosis. Mol Cell Biol 16:5546-5556.

29. Cheema SK, Mishra SK, Rangnekar VM, et al (2003) PAR-4 transcriptionally regulates Bcl-2 through a WT1 binding site on the bcl-2 promoter. J Biol Chem 278:19995-20005.

30. Hiyama H, Iavarone A, Reeves SA (1998) Regulation of the cdk inhibitor p21 gene during cell cycle progression is under the control of the transcription factor E2F. Oncogene 16:1513-1523.

PART 7.
ENDOMETRIUM/OVARIAN/COLON

19

The BDII Inbred Rat: A Model for Genetic Analysis of Endometrial Carcinoma

Karin Klinga-Levan and Göran Levan

Introduction

Analysis based on family studies has been extremely successful in identifying the gene defects causing simple monogenetic diseases. But much of human discomfort and lethality is due to complex diseases, examples are cardiovascular disease, diabetes, rheumatoid arthritis, and cancer. From twin studies, it is known that there must be significant hereditary components acting within each of these disease groups, but inheritance is non-mendelian and there is also a considerable influence on the disease course from the environment and life style. The conclusion is that in the development of these conditions many genes cooperate, but the effect of each individual gene defect is neither necessary nor sufficient to generate the observed phenotype. It follows that complex diseases are always genetically heterogeneous, *i.e*, two patients presenting with the same symptoms are likely to suffer for completely different reasons with respect to both hereditary and environmental factors. This heterogeneity constitutes a serious obstacle to efficient genetic analysis of complex disease in humans, where factors such as inherent genotype variability and differences in life style are completely uncontrollable.

To some extent, this obstacle can be eliminated by the use of inbred animal models, in which both the genetic constitution and environmental influences can be carefully controlled (1). Model studies complement and facilitate the analysis of human disease, and, provided that the necessary molecular tools are available, they can be very appropriate. Some 200 different inbred rat strains have been carefully characterized with respect to physiological and other characteristics, and many of them constitute excellent models of human complex diseases. In the last ten-year period, important scientific tools have been developed for the rat, such as large insert genomic libraries, expression libraries, physical and genetic maps, as well as radiation hybrid and comparative maps. Information about these resources is readily available to the interested researcher via several databases (Table 1). The efforts to make the rat a useful model organism for genomics and biomedical research have culminated in the fall of 2002, when the first draft DNA sequence of the complete rat genome was released.

Table 1. List of Some Important Rat Genomics-related Internet Sites.

Internet URL	Designation	Rat Resources	Database Location
http://www.ratmap.org	RatMap	World consortium chromosomal map, nomenclature, rat-mouse-human homologies	Göteborg University, Sweden
http://rgd.mcw.edu	RGD	Rat genome database	Medical College of Wisconsin, USA
http://ratmap.ims.u-tokyo.ac.jp/	OLETF Project	Radiation hybrid Server, rat-mouse-human comparative map	Otsuka GEN Research Institute, Japan
http://www.well.ox.ac.uk	Wellcome Trust	Oxford consortium genetic map, rat mapping resources	Wellcome Trust Centre for Human Genetics, Oxford, UK
http://www.tigr.org/tdb/ tgi/rgi/index.html	TIGR	TIGR Rat Gene Index	The Institute for Genomic Research, USA
http://resgen.com	Research Genetics	Commercial supplier of genetic markers, YAC and BAC clones	Research Genetics Inc., Huntsville, AL, USA
http://www.ensembl.org/ Rattus_norvegicus/	Ensembl	Rat genome server	EMBL/EBI - Sanger Institute, UK
http://www.ncbi.nlm.nih. gov/genome/guide/rat/index.html	NCBI	Rat genome information, rat unigenes	National Center for Biotechnology Information, NIH, USA
http://www.hgsc.bcm.tmc .edu/projects/rat	Sequencing Center	Rat Genome Project, BAC sequences	Baylor College of Medicine, Houston, TX, USA

Endometrial Cancer Susceptibility in Inbred BDII Rats

Females of the BDII inbred rat strain have been shown to spontaneously develop endometrial adenocarcinomas (EAC) at high incidence. These tumors are particularly frequent in virgin females, where more than 90% of the animals develop EAC before the age of 18 months (2, 3). Because they are typically estrogen-dependent, no EACs will develop in ovariectomized BDII rats. In most other inbred rat strains EACs are very rare tumors, and it is therefore clear that BDII rats must be genetically predisposed for this tumor type. In humans, EAC is now the most common malignant tumor in the female reproductive organs, and it is known that hereditary predisposition to EAC exists also among humans. Thus, the BDII rat may be used to model human EAC.

In order to study genetic aspects of EAC development in rats, we made crosses, in which BDII rats were used as one parent and rats from two different inbred strains, BN and SPRD, as the other. The latter two strains are known to have very low or non-existing incidences of EAC, although they may present with other types of tumors [reviewed in (4)]. The primary crosses generated F1 hybrids, which were subsequently intercrossed to produce F2 progeny or back-crossed to BDII females to produce N1 progeny. All animals were palpated weekly to look for tumors and when a tumor was suspected the animal was sacrificed and subjected to necropsy. Samples of all tumors were submitted to pathological analysis and tumor tissue was secured for molecular studies. Whenever possible, cell cultures were established from the tumors. Normal tissues (liver, spleen) were secured from all animals for DNA extraction and molecular analysis.

The tumor incidence is shown in Table 2. As seen, 110 (31%) animals developed malignant tumors and 89 (82%) of them were diagnosed as EAC tumors. Since the animals were followed during their entire life span (up to 750 days) the EAC tumors were subdivided into two groups, those with a latency typical of the BDII strain (<700 days) were designated "early onset tumors", and constituted 54% of the EAC tumors, and those developing with longer latency. Normal tissue DNA (liver and/or spleen) from all animals was subjected to a genome-wide screen with polymorphic microsatellite markers (5, 6). A total of 328 markers were used, 225 were informative in the BN cross and 214 were informative in the SPRD cross yielding an average distance between adjacent makers of 10 and 11 cM, respectively. By means of TDT (transmission disequilibrium tests) on F2 and N1 data, we detected statistically significant association of the early onset EAC phenotype to the BDII alleles in certain chromosomal regions. In the SPRD crosses the major associations were to segments on rat chromosome (RNO) 1q35-q36, on RNO11q23 and on RNO17p11-q12.1. In the BN crosses, we detected association to the BDII allele in chromosomal segments on RNO12q12-q15 and on RNO20p12. Thus, we found that the tumor phenotype showed the strongest association to different chromosome regions in the two crosses. Our interpretation is that the development of the cancer phenotype is determined by overall genotype interactions in the different genetic settings provided by the contribution of the BN and SPRD strain animals, respectively.

Table 2. Animal Material and Tumor Incidence in Crosses Between BDII Animals (predisposed to EAC) and Animals From Two Non-predisposed Inbred Strains (BN and SPRD).

	BN Cross				**SPRD Cross**				**Grand Total**
Animal group	**F1**	**F2**	**BC**	**Total**	**F1**	**F2**	**BC**	**Total**	
Total animals	18	59	105	182	17	54	103	174	356
Animals without tumor	6	36	65	107	10	23	59	92	199
Animals with benign tumor	1	5	9	15	5	20	7	32	47
Animals with malignant tumor	11	18	31	60	2	11	37	50	110
-- out of them EAC	10	11	26	47	2	9	32	43	90
-- out of them early onset EAC	5	9	12	26	1	6	16	23	49

By combining the results from both F2 and N1 animals, each of the associated segments could be estimated to be 2-5 centimorgan in size. On average this corresponds to 3-10 Mb of DNA sequence, which means that the number of genes in each segment can be estimated at 30-100, figures that are corroborated when the draft sequence of the actual segments are scrutinized in the Ensembl database. Among the genes presented in the rat sequence database most are predicted based on structure, and have not yet been assigned any distinct function. But by combining information from the rat database with comparative data from mouse and humans it was possible to identify 5-10 genes with known cancer-relation in each of the segments. Further experimentation will be necessary to determine which of these genes actually influence EAC susceptibility in BDII rats.

Patterns of Genetic Change in Rat Endometrial Adenocarcinoma

In order to obtain an overview of the genetic changes occurring in the EAC tumor cells themselves, we employed cytogenetics and comparative genome hybridization (CGH) (7, 8). The cytogenetic analyses of tumor cell cultures revealed that rat

EACs usually have near-diploid or near-triploid chromosome sets. In most cases the karyotypes exhibited a range of chromosome aberrations, both numerical and structural. Signs of gene amplification, so-called HSR (homogeneously staining regions) were detected in several of the tumors. CGH analysis was applied to both primary tumor DNA and to DNA from tumor cultures and it was revealed that certain chromosome regions were repeatedly involved in copy number changes in rat EACs. Most commonly, the following regions were affected: recurrent copy number gains were seen in RNO4 (proximally), RNO6 (proximally), RNO10 (distally), RNO12 (q arm) and RNO15 (q arm), whereas recurrent reduction in copy number was seen in RNO5 (middle), RNO10 (proximally) and RNO15 (p arm). In certain other regions, e g parts of RNO1, 2, 16 and 18, copy number gains were also rather common but less specific. It could be concluded that every tumor exhibited one or more of the common cytogenetic changes detected, suggesting that the development of EAC tumors in the model at least to some extent followed a common predictable pattern, probably involving similar genetic changes.

Specific Genetic Change During EAC Development

The next step, then, will be to characterize specific genetic changes occurring in the chromosomal regions identified. Since each region is quite large, this work is a substantial undertaking that is open to several different approaches. Here, we will discuss some of our results from two of the chromosome regions, situated on RNO4 and RNO10, respectively. Aberrations in the proximal part of RNO4 were the most common among those detectable with CGH, and gains in RNO4 occurred in 34 out of 56 tumors (61%). In 12 cases these increases were designated as gene amplification, since the number of copies gained was in excess of five. For the analysis of the proximal part of RNO4 we started by constructing a detailed physical map of the region (9). Subsequently, a subset of 12 tumors was subjected to analysis with FISH and Southern blotting (10). Fifteen gene probes covering the proximal 75 Mb of RNO4 (Table 3; the total length of RNO4 is presently estimated at 193 Mb) were tested in Southern blot analysis and it was shown that in 6 of the 12 tumors there was amplification of subsets of the genes (Figure 1). Notably, both the *Met* oncogene and the cell cycle regulating *Cdk6* gene were amplified in 5 tumors. These two genes are located only about 20 Mb apart (Table 3). Still, it was found that the amplified copies of the two genes were situated in different HSR chromosomes in some of the tumors, making it obvious that, at least in these cases, each gene had been subjected to an independent amplification event. Furthermore, two genes located between *Cdk6* and *Met* in native rat DNA (*Tac1* and *Asns*) were never amplified, in fact, in some cases there was actually a reduction in copy number of these genes. Our conclusion from these findings was that increased levels of *Cdk6* and *Met* gene products somehow interacted to stimulate tumor development. Clearly, stimulation by such interaction is just one of several possibilities open in BDII EAC development, since coamplification of *Cdk6* and *Met* was seen only in a subset of the tumors (5 out of 12 tumors analyzed). In addition, these results strongly suggest that an important tumor suppressor gene (TSG) must be situated somewhere in the interval between the two genes, since this region was never included in the amplicons.

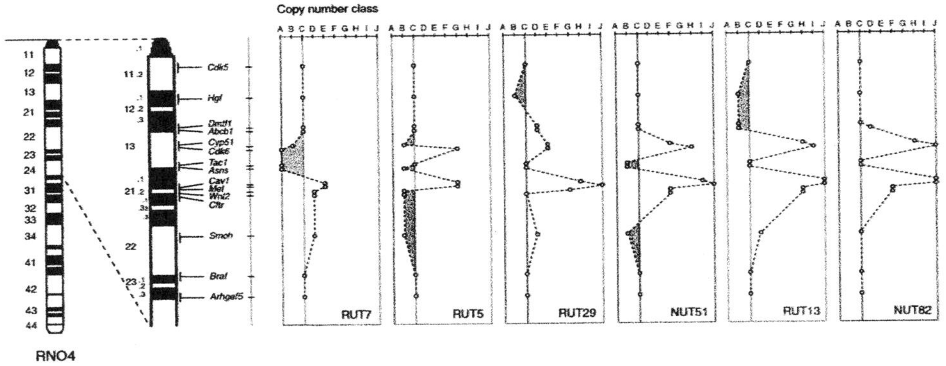

Figure 1. On the left, the ideogram of RNO4 is shown, including higher magnification of the proximal part. The cytogenetic localization of the 15 genes analyzed is indicated. The notches on the thin line immediately to the right of the ideogram shows the approximate physical positions of the genes (not completely to scale, for details see Table 4). The diagrams in the right part of the figure show the copy number level for each studied locus in the six tumors (circles connected by dashed lines). Copy number classes refer to the number of copies per diploid chromosome set equivalent (DCSE), thus, when there is no copy number change this value will be 2, when there is amplification the value should be ≥5. A value ≤1.5 is considered to be indicative of copy number reduction (deletion). Division into classes: A = <1; B = 1-1.5; C = 1.6-4; D = 5-10; E = 11-25; F = 26-40; G = 41-55; H = 56-70; I = 71-90; K = >90 copies/DCSE, of which class C represents the normal range. The diagrams generated resemble the patterns seen in the CGH analysis, but give much greater detail. Note that there are some regions, which actually show reduction in copy numbers (shaded) suggestive of chromosomal deletion, perhaps occurring during the actual amplification process. The data strongly suggest the involvement of two distinct targets for gene amplification on RNO4 in these tumors.. [Reprinted by permission of Cancer Res (10)].

Even though all available indications were that *Cdk6* and *Met* were the targets of amplification, obviously we cannot be sure of this, since in the entire 75 Mb region 461 genes have already been predicted in Ensembl based on structural criteria, and if one extrapolates available data to an estimated total of about 33000 genes in the rat, this region should harbor about 715 genes. Still, when this entire

DNA becomes fully annotated, it seems that the number of possible candidate genes in the region will be quite limited, and that it will be a reasonable task to determine which ones among them are involved in EAC development, using available standard molecular methodology.

Table 3. Cytogenetic and Draft DNA Sequence Position of 15 Cancer-related Genes in the Proximal Part of RNO4 (see Figure 1). The present estimate is that this region contains about 715 genes (extrapolated from the Ensembl database).

Gene Symbol	Gene Description	Cytogenetic Position	DNA Sequence Position (Mb)
Cdk5	Cyclin-dependent kinase 5	*4q11.2*	5.0
Hgf	Hepatocyte growth factor (scatter factor)	*4q12.1*	14.6
Dtmf1	Cyclin D binding myb-like transcription factor 1	*4q12-13*	21.6
Abcb1	ATP-binding cassette, sub-family B (MDR/PGY), member 1	*4q12-13*	22.0
Cyp51	Cytochrome P450, 51 (lanosterol 14-alpha-demethylase)	*4q13*	26.7
Cdk6	Cyclin-dependent kinase 6	*4q13*	27.5
Tac1	Tachykinin, precursor 1 (substance P)	*4q13-21*	37.3
Asns	Asparagine synthetase	*4q13-21*	37.4
Cav1	Caveolin 1	*4q21.1-21.2*	47.7
Met	MET proto-oncogene (hepatocyte growth factor receptor)	*4q21.2*	47.8
Wnt2	Wingless-type MMTV integration site family member 2	*4q21.3-21.31*	48.6
Cftr	Cystic fibrosis transmembrane conductance regulator	*4q21.3-21.31*	48.9
Smoh	Smoothened (Drosophila) homolog	*4q22*	60.9
Braf	V-raf murine sarcoma viral oncogene homolog B1	*4q22-23*	71.4
Arhgef5	Rho guanine nucleotide exchange factor (GEF) 5, TIM proto-oncogene	*4q23.3*	75.6

In one of the tumors (RUT7) there was amplification only of the *Met* oncogene. A possible explanation was obtained from expression studies in the 12 tumors. The gene product of the *Met* oncogene is the hepatocyte growth factor receptor, and in RUT7 there was expression of the corresponding ligand (*Hgf*, hepatocyte growth factor). The *Hgf* gene is normally turned off in endometrial tissue – the fact that it was activated in RUT7 suggests that in this tumor a different pathway involving an autocrine loop stimulating cell growth was active. Taken together, our findings suggest that there are several genes in the proximal part of RNO4 that play important roles in BDII EAC development. These genes probably include *Met*, *Cdk6* and *Hgf* and at least one still undiscovered tumor suppressor gene, which is located not far from the *Tac1/Asns* genes.

In analyzing genetic changes in RNO10 we took a slightly different approach. Because indications from the CGH analysis were that there was loss of chromosomal material in the proximal part of RNO10, we started out with an analysis of allelic imbalance looking for signs of LOH (loss of heterozygosity) (11, 12). We found indications that allelic imbalance was particularly common in at least four rather well-defined regions. One region, situated in RNO10q24, was shown to be affected by LOH in virtually all informative tumors, and in this region the *Tp53* gene is an obvious target candidate gene. Two other regions exhibiting recurrent allelic imbalance were located in 10q11-q12 and in 10q22, but here there were no obvious specific candidate suppressor genes. It is known that RNO10, which is mostly homologous to human chromosome (HSA) 17, contains many cancer-related genes. We are currently in the process of scrutinizing the RNO10 draft sequence in search of potential candidate genes in these regions. A fourth region showing allelic imbalance was pinpointed in 10q32.1, but the CGH study had shown that in BDII EAC tumors there was copy number increase rather than decrease in this region. Thus, we will be looking for activated oncogenes rather than suppressor genes in this particular region. One thing was very clear: if there is a relative copy number decrease in the proximal part of RNO10, and a copy number increase in the distal part there must, by necessity, occur chromosomal breaks within a rather specific segment between the two regions. To examine this notion we have performed FISH studies using an RNO10 chromosome paint in conjunction with gene-specific probes (13). Thirty EACs were subjected to cytogenetic analysis, which was combined with RNO10-specific painting. It was found that the chromosome numbers of EAC tumors were in the diploid or triploid regions, modal numbers ranging between 39 and 71. The RNO10 chromosome paint revealed that the number of chromosomes entirely made up of RNO10 material or, in a few instances, translocation chromosomes with distinct RNO10 segments, in 16 of the tumors was equal to that expected from ploidy (*i.e.,* two in diploid and three in triploid cells). Fourteen EACs had a total of 20 RNO10-derived chromosome segments in excess of those expected from ploidy. It should be mentioned that in most tumors only 1-2 normal-looking RNO10 chromosomes were present, additional RNO10-derived segments were mostly smaller than ordinary RNO10 chromosomes.

To further analyze the RNO10 involvement, two gene probes from different parts of the chromosome were used in dual-color FISH in order to determine which chromosome segments the smaller RNO10-derived elements represented. Thus, one PAC (P1 artificial chromosome) probe corresponding to *Tp53* (at 10q24) and another PAC probe for the *Thra1* gene (at 10q32.1) were labeled with red and green fluorescence, respectively, and the two probes were simultaneously hybridized to metaphase preparations from the EAC cultures. The hybridization pattern was compared to the previously determined pattern of RNO10-derived chromosomes. Based on the combined analysis it was found that EAC tumors neatly fall into two separate groups, those with no cytogenetic involvement of RNO10 (11/30 tumors; 37%) and those exhibiting deleted and/or translocated RNO10-derived chromosomes (19/30 tumors; 63%). In the latter subgroup, the analysis showed that 35 out of 37 deletion or translocation chromosomes had arisen after a break in the region between bands 10q24 and 10q32.1, leading to a situation in which they contained either *Tp53* or *Thra1,* but never both (the remaining two rearranged chromosomes were small and contained neither *Tp53* nor *Thra1*). Furthermore, among the rearranged chromosomes those derived from the distal part and containing *Thra1* dominated (30/35 chromosomes; 86%). Thus, this analysis showed that compared to the number of copies expected from the ploidy, there was a relative deficit of the proximal part of RNO10 and a relative overrepresentation of the distal part, just as would be expected based on the CGH-results. In what way this imbalance in copy number affects EAC development remains to be determined. Based on the findings from the allelic imbalance analysis the possibility that one or more TSG loci need to be inactivated in the proximal part seems valid. Similarly, increases in the distal part may lead to overrepresentation of an oncogene. However, an interesting additional possibility is that a cancer-related gene located at the chromosomal break point is of importance in EAC development in this EAC subgroup. Only further experimentation can lead to a distinction between the different possibilities.

Concluding Remarks

It can be concluded that the inbred BDII rat represents an interesting cancer model, which is available to detailed molecular analysis. The strain is genetically predisposed to spontaneous EAC development and a genetic approach such as the one taken by us should ultimately lead to the identification of the susceptibility genes. As a step in this direction, we have already identified five small chromosome regions likely to harbor genes important in EAC susceptibility. Somewhat unexpectedly, these susceptibility genes appear to be different depending on what other strain is brought in for outcrossing purposes. This finding suggests that susceptibility depends on a delicate interplay among several different genes. Still, one might perhaps have expected that one or a few major susceptibility genes would play the main role in BDII EAC susceptibility.

We have also studied the pattern of genetic changes during EAC development. Our data show that certain recurrent genetic and cytogenetic changes occur during EAC development. However, there appears to be considerable heterogeneity among individual tumors, in spite of the fact that the underlying variation in genotypes and environmental influences is so much more limited that among human subjects. Our results suggest that several different pathways can be selected, each pathway leading to the same end, each with reasonably great probability of occurring. This is as if during tumor development several choices occur (e g amplification of *Met* or not, activation of *Hgf* or not, deletion of *Tp53* or not), the outcome of each choice essentially depending on stochastic factors, but the cancerogenic process is flexible and presses on along each alternate path. It may be, however, that the variation among tumors is less than what appears, since quite different changes on the gene structural level may have the same effect on the functional level (*i.e.,* deletion may have the same effect as methylation, overexpression the same as gene amplification etc). The detailed answers to these questions can only come from further experimentation.

Acknowledgments

This project originated with support from the European Commission (contract ERBBio4CT960562) and the Swedish Cancer Society (1090-B02-26XCC). Financial support was also received from the IngaBritt and Arne Lundberg Research Foundation, the Erik Philip-Sörensen Foundation, and the Adlerbertska Research Foundation.

References

1. Gill III TJ, Smith GJ, Wissler RW, et al (1989) The rat as an experimental animal. Science 245:269-276.
2. Deerberg F, Kaspareit J (1987) Endometrial carcinoma in BDII/Han rats: model of a spontaneous hormone-dependent tumor. J Natl Cancer Inst 78:1245-1251.
3. Kaspareit-Rittinghausen J, Deerberg F, Rapp K (1987) Mortality and incidence of spontaneous neoplasms in BDII/Han rats. Zeitschr Vesuchstierkunde 30:209-216.
4. Hedrich HJ (ed) (1990) Genetic monitoring of inbred strains of rats. Gustav Fischer Verlag, Stuttgart-New York, 540.
5. Roshani L, Wedekind D, Szpirer J, et al (2001) Genetic identification of multiple susceptibility genes involved in the development of endometrial carcinoma in a rat model. Int J Cancer 94:795-799.
6. Roshani L, Wedekind D, Szpirer J, et al (2004) Genetic analysis of susceptibility to endometrial adenocarcinoma in a rat model, submitted.
7. Helou K, Walentinsson A, Beckmann B, et al (2001) Analysis of genetic changes in rat endometrial carcinomas by means of comparative genome hybridization. Cancer Genet Cytogenet 127:118-127.

8. Adamovic T, Hamta A, Helou K, et al (2004) Recurrent patterns of cytogenetic aberrations in experimenetal endometrial carcinomas revealed by comparative genome hybridization CGH, submitted.
9. Walentinsson A, Helou K, Levan G (2001) A dual-color FISH gene map of the proximal region of rat chromosome 4 and comparative analysis in human and mouse. Mamm Genome 12:900-908.
10. Walentinsson A, Helou K, Wallenius V, et al (2001) Independent amplification of two gene clusters on chromosome 4 in rat endometrial cancer: identification and molecular characterization. Cancer Res 61:8263-8273.
11. Behboudi A, Levan G, Hedrich HJ, et al (2001) High-density marker loss of heterozygosity analysis of rat chromosome 10 in endometrial adenocarcinoma. Genes Chromosomes Cancer 32:330-341.
12. Nordlander C, Behboudi A, Levan G, et al (2004) Four segments of chromosome 10-specific allelic imbalance in rat endometrial adenocarcinoma, submitted.
13. Behboudi A, Nordlander C, Levan G, et al (2004) Recurrent chromosome 10 aberrations and *Tp53* mutations in rat endometrial cancer, submitted.

20

Potential Role of Gonadotropin-Releasing Hormone and Estrogen in Ovarian Cancer

Peter C. K. Leung, Kyung-Chul Choi, and Nelly Auersperg

Introduction

Epithelial ovarian cancer is a prime cause of death from gynecological malignancies in the Western world. It appears to arise from the ovarian surface epithelium (OSE) composed of a simple squamous-to-cuboidal mesothelium covering the ovary (1). Women with a family history of ovarian and breast cancer have a higher risk of developing ovarian cancer (2). Although the exact mechanism of ovarian carcinogenesis remains to be uncovered, repeated ovulation appears to be a contributing factor in the (pre)neoplastic transformation of OSE. According to the incessant ovulation theory, the healing process of ruptured OSE cells following ovulation may contribute to the changes of normal OSE to its neoplastic counterpart (3). Recent epidemiological data from an Australian case-control study of 791 ovarian cancer cases and 853 controls indicate that repeated ovulations during the 20s in women are associated with high risk of this disease (4). There is increasing evidence that hormonal influences may play a pivotal role in the occurrence of ovarian tumors in women (5-7). We have recently investigated the effects of several key reproductive hormones on the transformation and progression of ovarian cancers using normal OSE, immortalized OSE transfected by the SV-40 t and T antigens (pre-neoplastic), and representative ovarian cancer cell lines. The results of these studies indicate that hormonal influences in normal OSE and its neoplastic counterpart are much more complex than previously predicted. The findings support the hypothesis that reproductive hormones are important regulators of cell proliferation and apoptosis in these cells. Specifically, the influences of gonadotropin-releasing hormone (GnRH) and estrogen (E), in the regulation of OSE and ovarian cancer cell proliferation/apoptosis, are summarized in this review.

Gonadotropin-Releasing Hormone (GnRH) and its Receptors

In addition to its well-documented role in the regulation of gonadotropin synthesis and secretion in the anterior pituitary gland, GnRH has been suggested to act as an autocrine/paracrine regulator in several extra-pituitary tissues, especially in carcinomas. Many human malignancies including ovarian, endometrial, breast, and prostate carcinomas have been shown to express GnRH and its receptor (8). In the human ovary, the majority of epithelial ovarian tumors and ovarian cancer cell lines have been shown to express GnRH receptor (GnRH-R), implicating that the use of GnRH analogs as a potential therapeutic approach for epithelial ovarian cancer (9-14). In a recent study, we demonstrated that GnRH and a full sequence of GnRH-R are expressed in normal OSE cells and the treatment with GnRH analogs resulted in a direct growth inhibitory effect in these cells (15). This anti-proliferative effect of the GnRH agonist was receptor-mediated, as co-treatment with antide abolished the growth inhibitory effects of the GnRH agonist, strongly suggesting that GnRH can act as an autocrine/paracrine regulator in normal OSE cells (15). It is well known that GnRH and its analogs inhibit the growth of a number of GnRH-R-bearing ovarian cancer cell lines *in vitro*. Treatment with GnRH agonist, [D-Trp^6]GnRH, reduced the growth of two ovarian cancer cell lines, EFO-21 and EFO-27 in a dose- and time-dependent manner (9), and in OVCAR-3 (16). It is of interest that an antagonistic analog of GnRH, SB75 also induced an inhibition of cell proliferation in OV-1063 ovarian cancer cells in a dose-dependent manner (17). In an animal model, the treatment with GnRH agonist suppressed the endogenous FSH and LH secretion in the pituitary gland and resulted in a growth inhibition of hetero-transplanted ovarian cancers, indicating that GnRH agonist may have an indirect effect to inhibit gonadotropin-stimulated ovarian cancer growth in *in vivo* (16). A combined treatment with [D-Trp^6]-GnRH and cisplatin was performed and enhanced therapeutic results of the combined treatment were obtained as compared to chemotherapy alone with cisplatin in a clinical trial (18). On the other hand, a multi-center, randomized trial failed to confirm the benefit of GnRH agonist (Triptorelin) for patients with late-stage ovarian cancer previously subjected to cytoreduction and adjuvant chemotherapy (19). More recently, Goserelin, a GnRH agonist, was tested in pre-clinical trials and shown to be effective in one-third of the patients with advanced and refractory ovarian cancer, indicating that Goserelin is one of best strategies to treat the patients with late stages of the disease (20). Also, combined therapy with GnRH analogs plus chemotherapy can produce favorable results in late-stage ovarian cancer to enhance therapeutic strategy (21).

Recently, a second form of GnRH, now referred to as GnRH-II (pGln-His-Trp-Ser-His-Gly-Trp-Tyr-Pro-Gly) that was first discovered in chicken (22), has been cloned in the brain extracts from rhesus monkeys (23) and in humans (24). It has been demonstrated that GnRH-II encodes a different gene and is expressed at higher levels outside the brain, including kidney, bone marrow, and prostate, suggesting that GnRH-II may have multiple functions in these organs (24). In extra-pituitary tissues, it appears that GnRH-II binds to GnRH-R more effectively than the classical GnRH (GnRH-I) (25). However, the possible functions of GnRH-II in different tissues remain unknown. Recently, we have demonstrated that GnRH-II mRNA is expressed in the human ovary including normal OSE, immortalized OSE (IOSE) cells, primary cultures of ovarian tumors and ovarian cancer cell lines. In addition, treatment with increasing doses (10-9 – 10-7 M) of GnRH–II resulted in growth inhibition in both non-tumorigenic and tumorigenic OSE cell lines. These results demonstrate for the first time the expression and potential anti-proliferative effect of GnRH-II, suggesting that GnRH-II, similar to GnRH-I, may have a growth-inhibitory effect in normal and neoplastic OSE cells (26). The presence of a second type of GnRH-R (type II GnRH-R) with low affinity and high capacity has been proposed in a study in ovarian cancer cell lines, EFO-21, OVCAR-3, and SKOV-3 (27). These results were subsequently confirmed by RT-PCR using specific primers derived from human pituitary type II GnRH-R mRNA (28) and Southern blot analysis (29). Treatment with both native GnRH-II or the GnRH-I agonist Triptorelin resulted in an inhibition of cell number in a time- and dose-dependent manner. Native GnRH-II showed a greater anti-proliferative effect in EGO-21 and OVCAR-3 cells which have both types of GnRH-R, whereas only native GnRH-II but not the GnRH-I agonist inhibited cell growth in SKOV-3 cell line which is type II GnRH-R positive but type I GnRH-R negative ovarian cancer cell line (29).

The mechanism underlying the GnRH-induced growth inhibitory effect in ovarian cancer cells has been investigated. Treatment with GnRH agonists resulted in a significant inhibition of ovarian cancer growth. This effect of GnRH was completely blocked by co-treatment with a specific GnRH antagonist, indicating that ovarian GnRH receptors may be involved in a direct anti-proliferative effect of GnRH analogs (15). Interestingly, both agonistic and antagonistic analogs have been reported to exert a similar anti-proliferative effect on ovarian cancer cells (17, 30). For instance, treatment with the GnRH-I antagonist, Cetrorelix, resulted in an inhibition of ovarian cancer cell growth. The mechanism of the Cetrorelix effect may involve one or more steps in cell cycle progression, including G1 phase cell cycle arrest coupled with down-regulation of cyclin A-Cdk2 complex levels, suggesting an up-regulation of p53 and p21 protein levels and apoptosis (30). Phospholipase C (PLC) may mediate the coupling of GnRH-GnRH-R signaling pathway to its anti-proliferative action in ovarian cancer cells (9). In addition, an activation of phosphotyrosine phosphatase (PTP) is involved in the GnRH induced signaling pathway in GnRH-R positive tumors, suggesting a possible function of GnRH to dephosphorylate the activation of receptor tyrosine kinase by growth factors (31, 32). GnRH-R has been shown to couple to multiple G proteins, but the antiproliferative signal transduction is mediated through the PTX-sensitive G protein alpha(i) in ovarian and endometrial cancers. Further, the tumor GnRH-R activates a PTP counteracting EGF-induced tyrosine auto-phosphorylation of EGF receptor, resulting in down-regulation of mitogenic signal transduction and cell proliferation in these cells (33). GnRH analogs have been demonstrated to counteract the EGF-induced growth stimulatory effect in ovarian cancer cells, suggesting that GnRH may down-regulate its receptor numbers and/or mRNA levels (11). In addition, EGF-induced c-*fos* mRNA expression and protein synthesis was dose-dependently down-regulated with GnRH agonists and antagonists in the GnRH-R expressing cell lines (34). Other mechanisms of GnRH/GnRH-R action in the inhibition of cell proliferation may be involved, such as an increase in the cell portion of the resting phase, G_0-G_1 (35) and induction of cell death or apoptosis by the Fas ligand-Fas system in ovarian cancer cells (36). Also, GnRH may modulate the growth of ovarian cancer cells by inhibiting telomerase activity (37) and by modulating JunD activation (38).

The role of the MAPK family in the anti-proliferative effect of GnRH in CaOV-3 ovarian cancer cell line has been demonstrated (39). Treatment of CaOV-3 cells with GnRH resulted in an activation of ERK at 5 min, reached the highest activation at 3 h and sustained until 24 h, whereas GnRH had no effect on the activation of the JNK. In addition, the ERK kinase was also activated and an increase in phosphorylation of son of sevenless (Sos), and Shc was observed following GnRH treatment. Treatment with an inhibitor of mitogen-activated protein/ERK kinase, PD98059 reversed the anti-proliferative effect of GnRHa and the GnRH-induced de-phosphorylation of the retinoblastoma protein. These results indicate that an activation of ERK may play an important role in the anti-proliferative effect of GnRH (39). Furthermore, we have shown that an agonist of GnRH, (D-Ala^6)-GnRH, induced a biphasic pattern of ERK-1/-2 activation. A low concentration of GnRH (10^{-10} M) resulted in a significant decrease of MAPK activity, whereas high concentrations (10^{-7} and 10^{-6} M) induced an activation of MAPK pathway in ovarian and placental cells (40).

Estrogen and its Receptors

Estrogens and estrogen receptor (ER) have been implicated in the pathogenesis and progression of ovarian cancer based on epidemiological and experimental observations (41-43). Even though estrogens plus progestins taken as an oral contraceptive during the premenopausal period in women are protective, when used in postmenopausal years as hormone replacement therapy, estrogens may increase the risk of ovarian cancer (7). Although no difference in risk was observed in women who had used estrogens continuously supplemented by progestins, an elevated risk of epithelial ovarian cancer in women who had used estrogen, either unopposed or combined with sequential progestins, was observed in an epidemiological study (41).

Both a classical ER (ERα) and second form of ER (ERβ) have been demonstrated to be expressed in normal OSE and ovarian tumors including ovarian cancer cell lines (44-46). Although the precise role of these two ERs in ovarian carcinogenesis is unknown, more than 50 % ovarian tumors express ER. The mRNA level of ERβ was decreased, whereas the level of ERα mRNA was similar or slightly higher in ovarian tumors when compared to normal ovaries (46). The role of these ERs in normal OSE and malignant cells remains elusive. There is some suggestion that the loss of ERα mRNA expression in normal OSE cells may be related to neoplastic transformation, whereas ERβ level is not affected in ovarian tumors compared to their normal counterpart (45). It is of interest that the ERα mRNA mutation with a 32-bp deletion in exon 1 was found in the SKOV-3 cell line, which is insensitive to E with respect to cell proliferation and regulation of gene expression (45). However, the differential splicing or major deletions in a coding region of ERβ were not found in either normal ovaries or ovarian tumors (46).

It has been reported that E is produced in primary cell cultures derived from 25 patients suffering from epithelial ovarian tumors, suggesting that E from primary ovarian tumors may play an active role in the proliferation and/or survival of these cells and alter the hormonal environment for promoting tumor progression (47). Although E appears to be ineffective in affecting the growth of normal OSE (15, 48), E induces growth stimulation of ER-positive ovarian carcinoma cell lines including OVCAR-3 and BG-1 cells, which have twice as many estrogen receptors as MCF-7 breast cancer cell line (49, 50). Treatment with E for 6 days resulted in growth stimulation in pre-neoplastic OSE cells immortalized by transfecting SV-40 t and T antigens into human normal OSE cells. Co-treatment with tamoxifen, an estrogen antagonist, abolished the E-induced stimulatory effect, suggesting that E action is mediated by ERs (44).

The mechanism of E in the regulation of cell proliferation and survival in neoplastic OSE cells has been examined. An increase of c-Myc protein is involved in the E-induced ovarian cancer cell growth (49). Several growth factors including epidermal growth factor (EGF), transforming growth factor (TGF)-α, and insulin-like growth factor (IGF) have been implicated to E-induced interaction in the normal ovary and ovarian cancer cells. For instance, the growth stimulatory effect of estrogen in the ovarian cancer cell line, PE01, was blocked by the use of an EGF receptor-targeted antibody, suggesting that an E-induced signaling pathway appears to be linked with an activation of EGF pathway (51). In addition, treatment with E resulted in a significant increase in TGF-α in the media and regulated phosphorylation of the EGF receptor, indicating that E may play a role in the production of TGF-α and activation of EGF receptor in ovarian cancer cells. The effect of EGF and IGF-1 on cell proliferation was potentiated by E via an increased binding affinity to EGF receptor and IGF-I receptor number (52). In our laboratory, we have demonstrated that E resulted in a prevention of tamoxifen induced-apoptosis through ERs in pre-neoplastic OSE cells, and a mechanism of E action may be involved to up-regulation of bcl-2 (anti-apoptotic gene) at the mRNA (Figure 1) and protein levels (Figure 2). In contrast, no significant difference was observed in the mRNA and protein levels of bax (pro-apoptotic gene) in Tag-immortalized OSE cells (44). These data indicate that E may play a role in the prevention of apoptosis, and bcl-2 appears to be a dominant regulator in the E-induced anti-apoptotic signal in pre-neoplastic OSE cells. Results in a recent study have indicated that the BRCA1 gene can be a ligand-reversible barrier to transcriptional activation by promoter-bound ER, and functional inactivation of this gene may promote tumorigenesis through inappropriate hormonal regulation of mammary and ovarian epithelial cell proliferation (53).

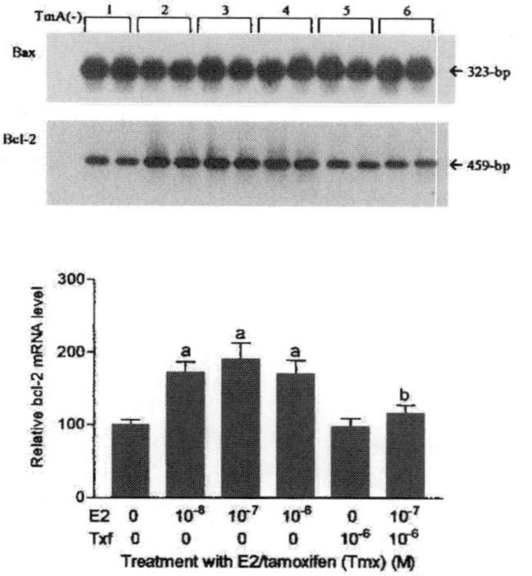

Figure 1. Effect of E on bax and bcl-2 mRNA levels. Data are shown as the means of three individual experiments, and are presented as the mean ± SD. **a**, P<0.05 *vs.* untreated control; **b**, P < 0.05 *vs.* E (10^{-7} M). 1, untreated control; 2, E (10^{-8} M; 3, E (10^{-7} M); 4, E (10^{-6} M); 5, tamoxifen (10^{-6} M; 6, E (10^{-7} M) plus tamoxifen (10^{-6} M). [Reproduced with permission from the Endocrine Society: (44) Choi K-C, Kang SK, Tai C-J, Auersperg N, Leung PCK (2001) Estradiol up-regulates anti-apoptotic bcl-2 mRNA and protein in tumorigenic ovarian surface epithelium (OSE). Endocrinology 142:2351-2360.]

Crosstalk Between the E/ER and GnRH/GnRH-R Systems

There is increasing evidence that E may play a role in the regulation of GnRH and GnRH-R in in vivo studies. For instance, E may be an important regulator in the events leading to ovulation by modulating GnRH and its receptor levels at the hypothalamus-pituitary level (54, 55). Previous studies in our laboratory have demonstrated the presence of GnRH and GnRH-R system in the ovary, suggesting a potential role of E in the regulation of this system (56, 57). More recently, we have shown that the ovary possesses an intrinsic GnRH axis which is regulated during luteinization in vitro, and that E is capable of regulating GnRH and its receptor in human granulosa-luteal cells (58). In a previous study, GnRH-I mRNA levels were significantly decreased by E in a time-dependent manner, whereas GnRH-II mRNA levels were significantly increased in a time-dependent fashion in human granulosa-luteal cells, indicating that expression of GnRH-I and GnRH-II at the transcriptional level is differently regulated by E and P4 in these cells (59).

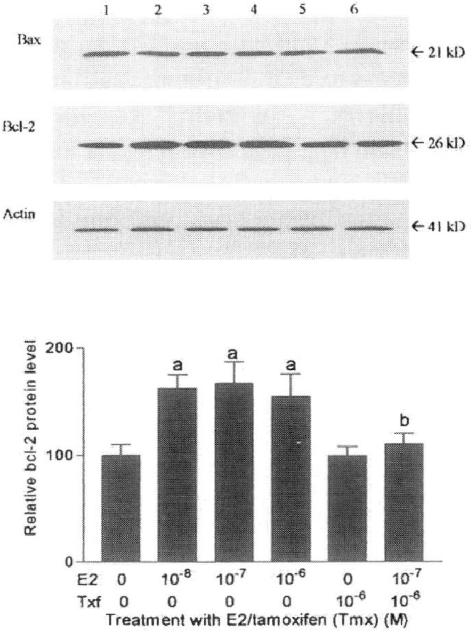

Figure 2. Effect of E on bax and bcl-2 proteins. Data are shown as the means of three individual experiments, and are presented as the mean ± SD. **a,** $P < 0.05$ *vs.* untreated control; **b,** $P < 0.05$ *vs.* E (10^{-7} M). 1, untreated control; 2, E (10^{-8} M); 3, E (10^{-7} M); 4, E (10^{-6} M); 5, tamoxifen (10^{-6} M); 6, E (10^{-7} M) + tamoxifen (10^{-6} M). [Reproduced with permission from the Endocrine Society: (46) Choi K-C, Kang SK, Tai C-J, Auersperg N, Leung PCK (2001) Estradiol up-regulates anti-apoptotic bcl-2 mRNA and protein in tumorigenic ovarian surface surface epithelium (OSE).Endocrinology 142:2351-2360.]

Based on these findings, the next question was whether estrogens influence the GnRH and GnRH-R expression in normal OSE and its neoplastic counterpart. Considering that GnRH is a potent autocrine regulator in normal OSE and ovarian cancer cells (15, 60), we investigated the relationship between E and GnRH/GnRH-R system in these cells (57). Treatment with E resulted in a significant decrease in the GnRH mRNA in OVCAR-3 cells, whereas no significant down-regulation of GnRH mRNA was observed in normal OSE cells. In addition, treatment with E resulted in a significant down-regulation of GnRH-R mRNA in both normal OSE and OVCAR-3 cells. To determine whether E action is mediated through the estrogen receptor, the cells were treated with E together with tamoxifen, an estrogen antagonist. Cotreatment with tamoxifen abolished the down-regulation of GnRH in OVCAR-3 cells and GnRH/GnRH-R mRNAs in both cell types. The molecular mechanism underlying the negative E regulation on GnRH/GnRH-R is not understood. Our preliminary data indicate that E-activated ER□ represses human GnRH-R gene transcription via an indirect mechanism including competition for a limiting amount of CREB-binding protein with AP-1 transcription factor (Cheng CK, Chow BK and Leung PCK, unpublished observation).

As estradiol down-regulated GnRH and GnRH-R mRNA levels, we investigated whether estrogen treatment antagonizes the growth inhibitory effect of GnRH in OVCAR-3 and normal OSE cells (57). Treatment with GnRH induced a significant growth inhibition of OVCAR-3 cells as early as the second day of pre-treatment with E for 24 h partially blocked the growth inhibitory effect of GnRH on day 2, but failed to block the effect of GnRH on days 4 and 6 of treatment. Pre-treatment for 24 h and co-treatment with E on a daily basis induced a significant attenuation of growth inhibitory effect of GnRH. In contrast, neither pre- nor cotreatment with E blocked the growth inhibitory effect of GnRH in normal OSE cells. These results indicate that a potential cross-talk between the E/ER and GnRH/GnRH-R systems exist in the human ovary, which may be important in the growth regulation of neoplastic OSE cells (57).

Concluding Remarks

In addition to its well-documented role in the reproductive hormone cascade, GnRH may play a role as an autocrine/paracrine regulator in the inhibition of cell proliferation and/or induction of apoptosis in normal and neoplastic OSE cells. The use of GnRH and its analogs is being considered for pre-clinical trials as a possible therapeutic agent in gynecological tumors such as ovarian and endometrial cancers. The exact mechanisms of GnRH action in the inhibition of cell growth need to be delineated using in-vitro and in-vivo models for ovarian cancer. The recent discovery of a second form of GnRH (GnRH-II) is posing a new challenge in understanding the role of the GnRH/GnRH-R system in normal OSE and its neoplastic counterpart. It appears that GnRH-II may have a more potent effect in the inhibition of cell proliferation when compared to that of GnRH-I in ovarian cancer cells.

On the other hand, E and ER may play a role in ovarian carcinogenesis as indicated by epidemiological and experimental observations. Further studies are needed to elucidate the influence of E and other hormones such as P4, gonadotropins and growth factors in normal and neoplastic OSE. Interestingly, it appears that E may be involved in the regulation of GnRH and GnRH-R genes in normal OSE and ovarian cancer cells. This potential cross-talk between the E and GnRH systems in the ovary may be important in the multi-faceted regulation of cell proliferation, apoptosis and/or differentiation of normal OSE and its neoplastic counterparts.

Acknowledgments

Research was supported by the Canadian Institutes of Health Research. PCKL is the recipient of a Distinguished Scholar Award from the Michael Smith Foundation for Health Research.

References

1. Auersperg N, Maines-Bandiera SL, Kruk PA (1995) Human ovarian surface epithelium: growth patterns and differentiation. In: Sharp F, Mason P, Blacket T, Berek J (eds), Ovarian Cancer 3. London:Champman & Hall, 157-169.

2. Negri E, Pelucchi C, Franceschi S, et al (2003) Family history of cancer and risk of ovarian cancer. Eur J Cancer 39:505-510.
3. Fathalla MF (1971) Incessant ovulation – a factor in ovarian neoplasia? Lancet 2:163.
4. Purdie DM, Bain CJ, Siskind V, et al (2003) Ovulation and risk of epithelial ovarian cancer. Int J Cancer 104:228-232.
5. Godwin AK, Testa JR, Hamilton TC (1993) The biology of ovarian cancer development. Cancer 71:530-536.
6. Rao BR, Slotman BJ (1991) Endocrine factors in common epithelial ovarian cancer. Endocr Rev 12:14-26.
7. Risch HA (1998) Hormonal etiology of epithelial ovarian cancer, with a hypothesis concerning the role of androgens and progesterone. J Natl Cancer Inst 90:1774-1786.
8. Schally AV, Nagy A (1999) Cancer chemotherapy based on targeting of cytotoxic peptide conjugates to their receptors to tumors. Eur J Endocrinol 141:1-14.
9. Emons G, Müller V, Ortmann O, et al (1998) Effects of LHRH-analogues on mitogenic signal transduction in cancer cells. J Steroid Biochem Mol Biol 65:199-206.
10. Emons G, Ortmann O, Schulz KD, et al (1997) Growth inhibitory actions of analogues of luteunizing hormone-releasing hormone on tumor cells. Trends Endocrinol Metab 8:355-362.
11. Emons G, Müller V, Ortmann O, et al (1996) Luteinizing hormone-releasing hormone agonist triptorelin antagonizes signal transduction and mitogenic activity of epidermal growth factor in human ovarian and endometrial cancer cell lines. Int J Oncol 9:1129-1137.
12. Emons G, Ortmann O, Becker M, et al (1993) High affinity binding and direct antiproliferative effects of LHRH analogues in human ovarian cancer cell lines. Cancer Res 53:5439-5446.
13. Miyazaki M, Nagy A, Schally AV, et al (1997) Growth inhibition of human ovarian cancers by cytoxic analogues of luteinizing hormone-releasing hormone. J Natl Cancer Inst 89:1803-1809.
14. Volker P, Grundker C, Schmidt O, et al (2002) Expression of receptors for luteinizing hormone-releasing hormone in human ovarian and endometrial cancers: frequency, autoregulation, and correlation with direct antiproliferative activity of luteinizing hormone-releasing hormone analogues. Am J Obstet Gynecol 186:171-179.
15. Kang SK, Choi KC, Cheng KW, et al (2000) Role of gonadotropin-releasing hormone as an autocrine growth factor in human ovarian surface epithelium. Endocrinology 141:72-80.

16. Peterson CM, Jolles CJ, Carrell DT, et al (1994) GnRH agonist therapy in human ovarian epithelial carcinoma (OVCAR-3) heterotransplanted in the nude mouse is characterized by latency and transience. Gynecol Oncol 52:26-30.
17. Yano T, Pinski J, Radulovic S, et al (1994) Inhibition of human epithelial ovarian cancer cell growth in vitro by agonistic and antagonistic analogues of luteinizing hormone-releasing hormone. Proc Natl Acad Sci USA 91:1701-1705.
18. Medl M, Peters-Engel C, Fuchs G, et al (1993) Triptorelin (D-Trp-6-LHRH) in combination with carboplatin-containing polychemotherapy for advanced ovarian cancer: a pilot study. Anticancer Res 13:2373-2376.
19. Emons G, Ortmann O, Teichert HM, et al (1996b) Luteinizing hormone-releasing hormone agonist Triptorelin in combination with cytotoxic chemotherapy in patients with advanced ovarian carcinoma. Cancer 78:1452-1460.
20. Zidan J, Zohar S, Mijiritzky I, et al (2002) Treating relapsed epithelial ovarian cancer with luteinizing hormone-releasing agonist (goserelin) after failure of chemotherapy. Isr Med Assoc J 4:597-599.
21. Rzepka-Gorska I, Chudecka-Glaz A, Kosmider M, et al (2003) GnRH analogues as an adjuvant therapy for ovarian cancer patients. Int J Gynaecol Obstet 81:199-205.
22. Miyamoto K, Hasegawa Y, Nomura M, et al (1984) Identification of the second gonadotropin-releasing hormone in chicken hypothalamus: evidence that gonadotropin secretion is probably controlled by two distinct gonadotropin-releasing hormones in avian species. Proc Natl Acad Sci USA 81:3874-3878.
23. Lescheid DW, Terasawa E, Abler LA, et al (1997) A second form of gonadotropin-releasing hormone (GnRH) with characteristics of chicken GnRH-II is present in the primate brain. Endocrinology 138:5618-5629.
24. White RB, Eisen JA, Kasten TL, et al (1998) Second gene for gonadotropin-releasing hormone in humans. Proc Natl Acad Sci USA 95:305-309.
25. King JA, Millar RP (1991) Vertebrate Endocrinology. In: Fundamentals and Biomedical Implications. Pang P, Schreibman M (eds), San Diego: Academic, 1-32.
26. Choi KC, Auersperg N, Leung PCK (2001) Expression and antiproliferative effect of a second form of gonadotropin-releasing hormone in normal and neoplastic ovarian surface epithelial cells. J Clin Endocrinol Metab 86:5075-5078.
27. Grundker C, Grunthert AR, Westphalen S, et al (2002) Biology of the gonadotropin-releasing hormone (GnRH) system in gynecological cancers. Eur J Endocrinol 146:1-14.

28. Faurholm B, Millar RP, Katz AA (2001) The genes encoding the type II gonadotropin-releasing hormone receptor and the ribonucleoprotein RBM8A in humans overlap in two genomic loci. Genomics 78:15-18.
29. Grundker C, Gunthert AR, Millar RP, et al (2002) Expression of gonadotrotropin-releasing hormone II (GnRH-II) receptor in human endometrial and ovarian cancer cells and effects of GnRH-II on tumor cell proliferation. J Clin Endocrinol Metab 87:1427-1430.
30. Tang X, Yano T, Osuga Y, et al (2002) Cellular mechanisms of growth inhibition of human epithelial ovarian cancer cell line by LH-releasing hormone antagonist Cetrorelix. J Clin Endocrinol Metab 87:3721-3727.
31. Imai A, Takagi H, Furui T, et al (1996) Evidence for coupling for phosphotyrosine phosphatase to gonadotropin-releasing hormone receptor in ovarian carcinoma membrane. Cancer 77:132-137.
32. Lee MT, Liebow C, Kramer AR, et al (1991) Effects of epidermal growth factor and analogues of luteinizing hormone-releasing hormone and somatostatin on phosphorylation and dephosphorylation of tyrosine residues of specific protein substrates in various tumors. Proc Natl Acad Sci USA 88:1656-1660.
33. Grundker C, Volker P, Emons G (2001) Antiproliferative signaling of luteinizing hormone-releasing hormone in human endometrial and ovarian cancer cells through G protein alpha(I)-mediated activation of phosphotyrosine phosphatase. Endocrinology 142:2369-2380.
34. Grundker C, Volker P, Schulz KD, et al (2000) Luteinizing hormone-releasing hormone agonist triptorelin and antagonist cetrorelix inhibit EGF-induced c-fos expression in human gynecological cancers. Gynecol Oncol 78:194-202.
35. Thomson MA, Adelson MP, Kaufman LM (1991) Lupron retards proliferation of ovarian epithelial tumor ells cultured in serum free medium. J Clin Endocrinol Metab 72:1036-1041.
36. Nagata S, Golstein P (1995) The Fas death factor. Science 267:1449-1456.
37. Ohta H, Sakamoto H, Satoh K (1998) In vitro effects of gonadotropin-releasing hormone (GnRH) analogue on cancer cell sensitivity to cis-platinum. Cancer Lett 134:111-118.
38. Gunthert AR, Grundker C, Hollmann K, et al (2002) Luteinizing hormone-releasing hormone induces JunD-DNA binding and extends cell cycle in human ovarian cancer cells. Biochem Biophys Res Commun 294:11-15.
39. Kimura A, Ohmichi M, Kurachi H, et al (1999) Role of mitogen-activated protein kinase/extracellular signal-regulated kinase cascade in gonadotropin-releasing hormone-induced growth inhibition of a human ovarian cancer cell line. Cancer Res 59:5133-5142.

40. Kang SK, Tai CJ, Cheng KW, et al (2000) Gonadotropin-releasing hormone activates mitogen-activated protein kinase in human ovarian and placental cells. Mol Cell Endocrinol 170:143-151.
41. Riman T, Dickman PW, Nilsson S, et al (2002) Hormone replacement therapy and the risk of invasive epithelial ovarian cancer in Swedish women. J Natl Cancer Inst 94:497-504.
42. Moll F, Katsaros D, Lazennec G, et al (2002) Estrogen induction and overexpression of fibulin-1C mRNA in ovarian cancer cells. Oncogene 21:1097-1107.
43. Stopper H, Schmitt E, Gregor C, et al (2003) Increased cell proliferation is associated with genomic instability: elevated micronuclei frequencies in estradiol-treated human ovarian cancer cells. Mutagenesis 18: 243-247.
44. Choi K-C, Kang SK, Tai C-J, et al (2001) Estradiol up-regulates anti-apoptotic bcl-2 mRNA and protein in tumorigenic ovarian surface epithelium (OSE). Endocrinology 142:2351-2360.
45. Lau K-M, Mok SC, Ho S-M (1999) Expression of human estrogen receptor-α and -β, progesterone receptor, and androgen receptor mRNA in normal and malignant ovarian epithelial cells. Proc Natl Acad Sci USA 96:5722-5727.
46. Brandenberger AN, Tee MK, Jaffe RB (1998) Estrogen receptor alpha (ER-alpha) and beta (ER-beta) mRNAs in normal ovary, ovarian serous cystadenocarcinoma and ovarian cancer cell lines: down-nepletion of ER-beta in neoplastic tissues. J Clin Endocrinol Metab 83:1025-1028.
47. Taube M, Hockenstrom T, Isaksson M, et al (2002) Low sex steroid environment affects survival and steroid secretion of ovarian tumour cells in primary cultures. Int J Oncol 20:589-594.
48. Karlan BY, Jones J, Greenwald M, et al (1995) Steroid hormone effects on the proliferation of human ovarian surface epithelium in vitro. Am J Obstet Gynecol 173:97-104.
49. Chien C, Wang F, Hamilton TC (1994) Transcriptional activation of c-myc proto-oncogene by estrogen in human ovarian cancer cells. Mol Cell Endocrinol 99:11-19.
50. Baldwin WS, Curtis SW, Cauthen CA, et al (1998) BG-1 ovarian cell line: an alternative model for examining estrogen-dependent growth in vitro. In Vitro Cell Dev Biol Anim 34:649-654.
51. Simpson BJB, Langdon SP, Rabiasz GJ, et al (1998) Estrogen regulation of transforming growth factor-α in ovarian cancer. J Steroid Biochem Molec Biol 64:137-145.
52. Wimalasena J, Meehan D, Dostal R, et al (1993) Growth factors interact with estradiol and gonadotropins in the regulation of ovarian cancer cell growth and growth factor receptors. Oncol Res 5:325-337.

53. Zheng L, Annab LA, Afshari CA, et al (2001) BRCA1 mediates ligand-independent transcriptional repression of the estrogen receptor. Proc Natl Acad Sci USA 98:9587-9592.
54. Conn PM (1994) The molecular mechanism of gonadotropin-releasing hormone action in the pituitary. In: Knobil E, Neill JD (eds), The Physiology of Reproduction. New York:Raven Press, 1815–1826.
55. Braden TD, Conn PM (1993) GnRH and its mechanism of action. In: Leung PCK, Hsueh AJW, Friesen HG (eds), Molecular Basis of Reproductive Endocrinology. New York:Springer-Verlag, 12–38.
56. Peng C, Fan NC, Ligier M, et al (1994) Expression and regulation of gonadotropin-releasing hormone (GnRH) and GnRH receptor messenger ribonucleic acids in human granulosa-luteal cells. Endocrinology 135:1740–1746.
57. Kang SK, Choi KC, Tai CJ, et al (2001) Estradiol regulates gonadotropin-releasing hormone (GnRH) and its receptor gene expression and modulates the growth inhibitory effects of GnRH in human ovarian surface epithelial and ovarian cancer cells. Endocrinology 142:580-588.
58. Nathwani PS, Kang SK, Cheng KW, et al (2000) Regulation of gonadotropin-releasing hormone (GnRH) and its receptor gene expression by 17b-estradiol in cultured human granulosa-luteal cells. Endocrinology 141:1754-1763.
59. Khosravi S, Leung PC (2003) Differential regulation of gonadotropin-releasing hormone (GnRH)I and GnRHII messenger ribonucleic acid by gonadal steroids in human granulosa luteal cells. J Clin Endocrinol Metab 88:663-672.
60. Kang SK, Cheng KW, Nathwani PS, et al (2000) Autocrine role of gonadotropin-releasing hormone (GnRH) and its receptor in ovarian cancer cell growth. Endocrine 13:297-304.

COMMUNICATIONS

Session I. Epidemiology/Human Derived Studies

Dose of Progestogen in Postmenopausal Combined Hormone Therapy and Risk of Endometrial Cancer

Susan D. Reed, Lynda F. Voigt, Shirley A. Beresford, Deirdre A. Hill, Jennifer A. Doherty, and Noel S. Weiss

Summary

Progestogens are given to diminish the risk of endometrial cancer (ECA) observed with postmenopausal estrogen therapy. The pattern of progestogen administration and number of days that the progestogen is given per month appear to affect ECA risk. We studied the impact of dose of progestogen, specifically medroxyprogesterone acetate (MPA), on ECA risk in a population-based case control study of 647 cases and 1,209 controls, ages 45-74 years. Among women who took a combined hormone regimen with MPA for less than10 days/mo, the overall adjusted risk of ECA (relative to that of hormone non-users) was 4.2 [95%, Confidence Interval (CI) 2.0, 8.9] in those with an MPA dose less than70 mg/mo; and the adjusted risk of ECA was 2.8 (95%, CI 0.9, 8.6) in those with a higher monthly MPA dose. Among women who took MPA cyclically for 10-24 days/mo or as continuous combined therapy, the risk of ECA was similar to non-users, irrespective of the monthly dose. Our findings showed that among the combined hormone regimens most commonly used by postmenopausal women in the USA today, monthly dose of MPA bears little or no relation to risk of ECA.

Introduction

The use of unopposed estrogen in postmenopausal women is associated with an increased risk of developing endometrial adenocarcinoma, the size of which increases with increasing duration of use (1). Epidemiologic studies have verified that progestogen administration assists in diminishing the stimulatory effect of estrogen on the endometrium, thereby decreasing the risk of endometrial cancer (ECA) associated with unopposed estrogen. The association of ECA with postmenopausal combined hormone therapy has been examined in several large case control studies performed by groups in Seattle, Los Angeles, and Stockholm,

Sweden (2-5). These studies have provided information about different patterns of progestogen delivery (cyclical for less than10 days, cyclical for 10-24 days, and continuous); and duration of use. In addition, the Swedish study[5] evaluated the risk of ECA by type of progestogen, comparing risks among women using C-21 derived progestogens, [principally medroxyprogesterone acetate (MPA)] to non-users and the nor-testosterone derived progestogens to non-users. There are no large studies that have reported on the dose of progestogen as it relates to the risk of ECA.

Progestogen dose becomes relevant because the balance between decreasing ECA risk with an adequate progestogen dose must be weighed against potential progestogen side effects seen with higher dosing regimens. Progestogen side effects lead to non-compliance with postmenopausal combined hormone therapy in at least one third of patients (6, 7). In addition, and potentially most importantly, the addition of a progestogen to postmenopausal estrogen therapy appears to decrease the favorable lipid benefits of estrogen (8, 9) and increase the risk of breast cancer (9-12). It is unknown whether progestogen has a major contribution to the increased risk of stroke and cardiovascular events observed with combined continuous hormone therapy (9). Identification of the dose of progestogen for use in combined postmenopausal hormone therapy that maximizes efficacy but also has a low rate of adverse consequences is therefore desirable.

Material and Methods

We conducted a population-based case-control study in Washington State, interviewing women 45-74 years of age with ECA diagnosed during 1985-1999. Controls identified by random digit dialing (93%) and the Health Care Financing Administration files (7%), were frequency matched for age and county and interviewed. After excluding women who took unopposed estrogen, used more than one combined hormone regimen, a regimen not including at least 6 months of MPA or had an unknown dose, type or duration of progestogen, 647 cases and 1,209 controls remained. Women were categorized as follows: 1) Hormone non-users. 2) Users of cyclical MPA for less than 10 days/mo. 3) Users of cyclical MPA for 10-24 days/mo. 4) Users of MPA daily for at least 25 days/mo. Combined hormone therapy was analyzed by low and high monthly MPA dose in the three different categories of combined hormone regimens. Each group was compared to the group of women who either did not use hormones or who used for less than six months.

Logistic regression analysis was performed and odds ratios with 95% confidence intervals (CI) calculated, adjusting for age, county, reference year, prior oral contraceptive exposure and body mass index. Potential confounding by duration of progestogen use and dose of estrogen used were also evaluated. A sub-analysis on the interaction of body mass index (BMI) and the association of progestogen dose and risk for ECA was performed.

Results

The distribution of established risk factors for ECA among cases and controls, *i.e.*, nulliparity, obesity, diabetes, and non-use of oral contraceptives, corresponded to what would be expected from the epidemiologic literature.

The dose of MPA had little bearing on risk of ECA with the exception of women who took the progestogen for less than10 days/mo. Among women who took a combined hormone regimen with MPA for less than10 days/mo, the overall adjusted risk of ECA, relative to that of hormone non-users, was 4.2 (95%, CI 2.0, 8.9) in those with an MPA dose less than70 mg/mo and was 2.8 (95% CI, 0.9, 8.6) in those with a higher monthly MPA dose. Among women who took MPA cyclically for 10-24 days/mo or as combined continuous therapy, the risk of ECA was similar to non-users, irrespective of the monthly dose.

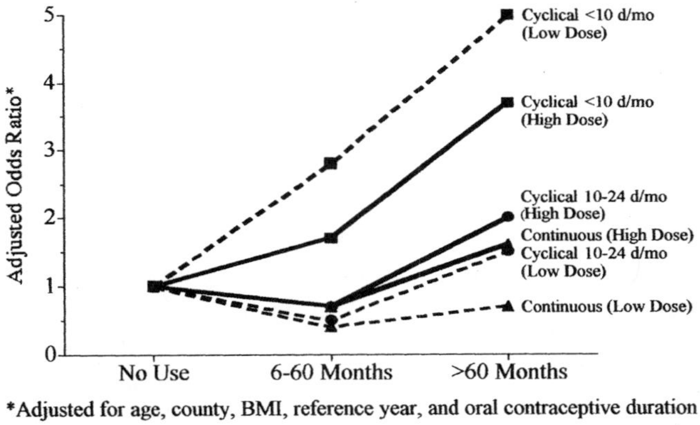

Figure 1. Risk of endometrial cancer associated with monthly dose and duration of medroxyprogesterone acetate in combined postmenopausal hormone therapy.

The effect of dose and duration by pattern of use is shown in Figure 1. The greatest risk of ECA, among women who took combined hormone therapy relative to non-users, was observed among women who took lower monthly doses of MPA (less than70 mg/mo for less than 10 days/mo), but for longer durations (over 5 years), adjusted OR 5.0 (95%, CI 2.1, 12.2). The effect of estrogen dose or BMI appeared to have little or no impact on the risk of ECA associated with dose of progestogen, although small numbers of overweight and obese women who took combined hormone therapy limit the interpretation of our results.

Conclusions

Our findings showed that among the combined hormone regimens most commonly used by postmenopausal women in the USA today, monthly dose of MPA bears little or no relation to the risk of ECA, irrespective of estrogen dose or body mass index. Given the possible adverse effects of progestogen use, there appears to be little rationale for the prescription of more than 75 mg of continuous MPA/mo, or more than 100 mg of cyclical MPA/mo, or their equivalents, for postmenopausal women taking estrogen therapy. Future studies on the impact of progestogen dose on the risk of ECA among women taking different formulations of progestogens are warranted so that optimal dosing regimens for progestogens other than MPA can be defined.

References

1. Grady D, Gebretsadik T, Kerlikowske K, et al (1995) Hormone replacement therapy and endometrial cancer risk: a meta-analysis. Obstet Gynecol 85:304-313.
2. Beresford SA, Weiss NS, Voigt LF, et al (1997) Risk of endometrial cancer in relation to use of oestrogen combined with cyclic progestagen therapy in postmenopausal women. Lancet 349:458-461.
3. Hill DA, Weiss, NS, Beresford SA, et al (2000) Continuous combined hormone replacement therapy and risk of endometrial cancer. Am J Obstet Gynecol 183:1456-1461.
4. Pike MC, Peters RK, Cozen W, et al (1997) Estrogen-progestin replacement therapy and endometrial cancer. J Natl Cancer Inst 89:1110-1115.
5. Weiderpass E, Adami H-O, Baron JA, et al (1999) Risk of endometrial cancer following estrogen replacement with and without progestins. J Natl Cancer Inst 91:1131-1137.
6. Greendale GA, Reboussin BA, Hogan P, et al (1998) Symptom relief and side effects of postmenopausal hormones: results from the Postmenopausal Estrogen/Progestin Interventions Trial. Obstet Gynecol 92:982-988.
7. Hill DA, Weiss NS, LaCroix AZ (2000) Adherence to postmenopausal hormone therapy during the year after the initial prescription: A population-based study. Am J Obstet Gynecol 182:270-276.
8. The Writing Group for the PEPI trial (1995) Effects of estrogen or estrogen/progestin regimens on heart disease risk factors in postmenopausal women. The Postmenopausal Estrogen/Progestin Interventions (PEPI) Trial. JAMA 273:199-208.
9. Writing Group for the Women's Health Initiative Investigators (2002) Risks and benefits of estrogen plus progestin in healthy postmenopausal women: principal results from the Women's Health Initiative randomized controlled trial. JAMA 288:321-333.
10. Ross RK, Paganini-Hill A, Wan PC, et al (2000) Effect of hormone replacement therapy on breast cancer risk: estrogen versus estrogen plus progestin. J Natl Cancer Inst 92:328-332.
11. Schairer C, Lubin J, Troisi R, et al (2000) Estrogen-progestin replacement and risk of breast cancer. JAMA 284:691-694.
12. Li CI, Weiss NS, Stanford JL, et al (2000) Hormone replacement therapy in relation to risk of lobular and ductal breast carcinoma in middle-aged women. Cancer 88:2570-2577.

Mammographic Densities and Urinary Hormones in Healthy Women with Different Ethnic Backgrounds

Gertraud Maskarinec, Andrew E. Williams, Sabina Rinaldi, Rudolph Kaaks

Summary

An association between sex steroids and mammographic density (MD), a predictor of breast cancer (BC) risk, is supported by findings of increased densities after hormone therapy and reduced densities after tamoxifen treatment. Herein, ethnic differences in urinary hormone levels and their relation to MDs were investigated. Because 2-hydroxyestrone (2-OH-E_1) is considered less carcinogenic than 16α-hydroxyestrone (16α-OH-E_1), we hypothesized an inverse relation between the 2-OH-E_1/16α-OH-E_1 (2/16) ratio and breast densities. Women recruited completed a questionnaire and donated urine during the luteal phase. Urinary estrone (E_1), estradiol (E_2), testosterone (T), and 5α-androstane-3α, 17β-diol (ADIOL) were measured by indirect radioimmunoassay (RIA) and 16α- and 2-OH-E_1 by competitive immunoassays. MDs were assessed with a computer-assisted method and applied multiple linear regressions. The total number of subjects was 305 (35-75 years, $\bar{x} = 47.2$ years). Their ethnic distribution was Caucasian (110), Japanese (86), Hawaiian (35), Chinese (28), and mixed/other ethnicity (46). The data indicate that the body mass index (BMI), 2-OH-E_1, androgens (A), total hormones, and the 2/16 ratio were significantly lower in Asians than in Caucasians, but the % MDs were 22% higher in Asians. None of the individual hormones was associated with MDs. However, contrary to the initial hypothesis, the 2/16 ratio was directly related to MDs. The ratio was 25% lower in the lowest MD category as compared to the highest category. The data suggest that the effects of endogenous hormones on BC risk may not be mediated through MDs in adult women.

Introduction

MD patterns refer to the distribution of fat, connective, and epithelial tissue in the breast, and are strong predictors of BC risk. A high percentage of dense parenchyma on mammographic images confers a 4.0- to 6.0-fold risk to develop BC (1). Endogenous estrogens (Es) and perhaps As are important in the etiology of BC (2-4). An association of MDs and hormone levels is supported by several observations. Hormone replacement therapy (HRT) increases MD (5). In the other hand, tamoxifen treatment improves MDs (6) by suppressing ovarian function through a gonadotropin-releasing hormone agonist (7). Moreover, MDs returned to baseline after tamoxifen was discontinued (8). However, two cross-sectional investigations detected no strong relation between MDs and serum E and progesterone (P) levels (9, 10).

Because some metabolites of endogenous Es may have more estrogenic effects than others, we hypothesized that differences in E metabolic pathways may be related to MDs. The metabolism of E_2 follows two major competing pathways, C2- and C16α-OH-lation, and a minor C4-OH-lation (11-13). It has been proposed that women who metabolize a larger proportion of their endogenous Es through 16α-OH-lation are at greater BC risk because 16αOH-E_1 has genotoxic effects, damages DNA, and enhances breast cell growth, whereas 2-OH-E_1 inhibits breast cell proliferation. However, before transformation into methoxy compounds by the enzyme catechol methyl transferase (COMT), 2-OH compounds have some estrogenic and growth promoting effects (13). The evidence on the association of the 2/16 ratio with BC is inconsistent (14-18), and a previous study on MDs reported results contrary to the original hypothesis (19). Women in the highest tertile of the 2/16 ratio were 6.0 times more likely to have a high risk mammographic pattern. In disagreement with the hypothesis that women from ethnic groups with lower BC risk have a higher 2/16 ratio, a comparison between Finnish and Asian women (20) reported a higher ratio in Finnish women. Herein, ethnic differences in urinary E, A, and E metabolite levels and their relation with MD were investigated.

Materials and Methods

Study Design and Population. This study included 305 women, participants of three studies involving MD, 97 were from a cross-sectional study (21), 7 from an isoflavone intervention study (22), and 201 from a soy intervention study (23). For the latter two studies, baseline mammograms and urine samples were used for the analysis. After the approval of the Committee on Human Studies at the University of Hawaii, all subjects provided informed consent. Women for all three studies were recruited at mammography clinics in Honolulu. Eligible women had no previous history of BC and required a normal mammogram at baseline. The subjects for the intervention studies were free of serious medical conditions, had regular menstrual periods, intact uterus and ovaries, were not on oral contraceptives or other hormones, and had no intention of becoming pregnant within a year. Due to the nature of the nutritional intervention, only women who reported a dietary consumption of ≤ 7 servings of soy food/week were eligible. All subjects completed a validated food frequency questionnaire (24), which included questions on reproductive, medical, and anthropometric factors. In the cross-sectional study, self-reported information was used to classify women's menopausal status and hormone use (HRT). The majority of subjects (n = 208) donated an overnight urine specimen, while those of the cross-sectional study (n = 97) only a spot urine sample.

Mammogram Density Assessment. Cranio-caudal views of the mammogram were obtained from the clinics after complete radiologic evaluation and ruled out any malignancy. The films were scanned into a PC using a Kodak LS-85 X-ray digitizer with a pixel size of 260 μm (resolution = 98 pixels/inch). One of the authors (GM) performed computer-assisted MD assessment using a Canadian method (25). The reader chooses a threshold value that defines the outline of the breast, and then selects the best threshold to identify the regions that represent MDs. The pixel count corresponding to the dense area is determined by the computer, as is the total area within the outline of the breast. Percent MD was calculated as the ratio of the dense area to the total area of the breast multiplied by 100. A random sample of 58 mammograms was read in duplicate. The intra-class correlation coefficients (26) were 0.95 (95% CI, 0.92-0.97), for the size of the dense areas, and 0.98 (95% CI, 0.97-0.99) for the total breast area, and of 0.97% density (95% CI, 0.95-0.98).

Urinary Hormone Analysis. Urine concentration of E_1, E_2, T, and ADIOL was performed with slight modifications previously published (27). Briefly, 1.0-ml urine samples were hydrolyzed, purified by solid phase extraction, and HPLC. Hormone concentrations were measured by RIA on the dried extracts. All measurements were done in duplicate, including hydrolysis, solid phase extraction, HPLC, and RIA. For quality control, two control samples containing known amounts of steroids were included for all the analytical steps in each sample batch. The detection limits were: 0.02 ng/ml for E_1 and E_2, 0.08 ng/ml for ADIOL, and 0.02 ng/ml for T. Intra- and inter-batch coefficients of variations were 4.7 & 16%, respectively, for E_1 (at 3.3 ng/ml), 1.7% & 14% for E_2 (at 0.32 ng/ml), 4.7% & 14% for T (at 3.3 ng/ml), and 7.0% & 11% for ADIOL (at 15.5 ng/ml). 2-OH- and 16α-OH-E_1 were measured by solid-phase enzyme immunoassays after enzymatic hydrolysis with Helix Pomatia (Estramet, Immunacare Co., Bethlehem, USA). Mean intra- and inter-batch coefficients of variations were 10 & 15%, respectively, for both analytes.

Statistical Analysis. The SAS statistical software package version 8.2 (SAS Institute Inc., Cary, NC) was used for data management and statistical analyses. In the questionnaire, subjects marked all ethnic backgrounds that applied to themselves and to their parents. Summary categories were assigned according to the following rules: A woman was classified as Caucasian if both parents had some Caucasian ancestry and shared no other ethnic background. Subjects with no more than three ethnic backgrounds were classified as Chinese, Japanese, or Filipino, if both parents were of the same ethnicity or if the mother was of the respective ethnic background and the parents shared no other ethnic background. Because of the similarity in percent MDs, the 86 Japanese, 28 Chinese, and 9 Filipino women were combined into one Asian category. In agreement with rules applied in the State of Hawaii (28), women with any Hawaiian background were classified as Native Hawaiian. Because of their mixed ancestries, the Native Hawaiian women (n = 35) were included into the other category containing Pacific Islanders, African-Americans, Latinas (n = 24), and women with mixed ethnic backgrounds that did not fit any of the above categories (n = 13).

Body Mass Index. BMI was calculated as the ratio of weight in kg divided by the square of the height in m. Non-normally distributed variables were transformed using their natural logarithm. Percent MD was classified into five categories: <10%, 10 to 24.9%, 25 to 49.9%, 50 to 74.9%, and ≥75%. To explore associations between MD and urinary hormone measurements, we computed Spearman correlation coefficients (r_s) and included potential confounders (29). Then, we applied analysis of variance to test for associations between ethnicity, hormone levels, and mammographic characteristics with adjustment for confounding variables (30). In addition, we computed least-squares means for the urinary hormone levels by category of percent MD using the proc glm procedure in the SAS software package (30). Finally, we performed trend tests to investigate a possible relation between the different hormones and percent MD. We regressed the mean level of the hormones onto the mean MD of each of the five MD categories.

Results

Characteristics of study population. Most of the 305 women were premenopausal; only 25% were postmenopausal (Table 1). The mean age was 47.2 years. Caucasians were slightly younger than Asian women. The BMI was lowest among Asian, intermediate in Caucasians, and highest in the mixed/other category (p = 0.0004). Percent MD differed significantly by ethnicity (p = 0.003), even after adjustment for age, menopausal status, and BMI (p = 0.03). Percent MD was highest among Asians and lowest among Caucasians, 45.2% vs. 34.9%. Of the 305 women, 35, 48, 108, 87, and 27 belonged to the five MD categories. Close to 40% of Caucasians were in the two lowest MD categories, but only 20% of Asians were classified that way. MD density was significantly higher for pre than for postmenopausal women (44.2 ± 23.0% vs. 28.1 ± 19.2%). The MD difference between Caucasians and Asians was greater after menopause than before (16% vs. 9%).

Excretion of all combined hormones was lowest among Asians and similar in the two other groups, but this difference was not statistically significant (p = 0.09). The difference in hormone levels was primarily due to 2-OH-E_1, ADIOL, and T, which differed significantly by ethnicity, even after stratification for menopausal status or HRT use. While A levels did not vary by HRT use, E levels were approximately 2.0-fold higher among postmenopausal women on HRT than among non-users. After stratification by menopause and HRT use, A levels remained lower among Asian than Caucasian women in each subgroup. Levels of 16α-OH-E_1, E_1, and E_2 were similar in the three ethnic groups. The 2/16 ratio was approximately 25% lower in Asian women than in the other two groups (p = 0.04). The ethnic difference in the 2/16 ratio was greater before (1.56 vs. 2.11) than after menopause (1.55 vs. 1.72) for Asian and Caucasian women, respectively. The 2/16 ratio did not differ by HRT use (p = 0.57).

Table 1. Characteristics of the Study Population.

Variable	Asian	Mixed/ Others	Caucasian	All Mean	All Std	p-value[1]
Number	123	72	110	305		--
Menopausal (%)	28	18	24	25		0.32
Age (years)	48.50	45.40	46.80	47.20	8.00	0.03
BMI (kg/m^2)	23.40	26.40	25.70	24.90	5.10	0.0004
Percent MD	45.16	39.90	34.92	40.23	23.13	0.003
E_1 (ng/ml)	11.82	14.27	11.90	12.43	19.78	0.68
E_2 (ng/ml)	3.90	4.85	3.84	4.11	5.03	0.43
2-OH-E_1 (ng/ml)	16.39	19.54	20.99	18.80	14.07	0.04
16α-OH-E_1 (ng/ml)	11.60	12.52	11.08	11.63	8.45	0.52
2/16 ratio	1.56	1.81	2.02	1.79	1.00	0.0006
T (ng/ml)	2.83	4.06	4.16	3.60	2.94	0.0007
ADIOL (ng/ml)	19.94	25.78	21.82	22.00	17.26	0.07
All hormones (ng/ml)	67.06	82.27	75.28	73.62	48.77	0.09
E/A ratio	2.71	2.23	2.73	2.61	3.41	0.50

[1] χ^2-test for categorical and ANOVA for continuous variables.

All hormone levels declined with age (Table 2). The correlation coefficients for all combined hormones was r_s = -0.28, while the individual correlations varied between r_s = -0.14 and -0.39. These relations were strongest for ADIOL and the two E_1 metabolites, followed by T, and weaker for E_2 and E_1. The 2/16 ratio was not significantly related to age. Only the A levels were associated with BMI; women with a higher BMI excreted more T and ADIOL. As a result, BMI showed a significant inverse relation with the E/A ratio. These associations did not change after excluding women on HRT, but they were stronger before menopause. Due to the small sample size, none of the associations between hormones and BMI were statistically significant after menopause. The correlation coefficients with T and ADIOL were slightly lower than in premenopausal women and, among those not using HRT, E_2 was weakly correlated with BMI (r_s = 0.31, p = 0.08). Percent MD was negatively related to age and BMI, but not for E or A levels. A weak positive association with 2-OH-E_1 disappeared after confounders adjustment, as did the correlation between the 2/16 ratio and MD was 0.07 (p =

0.22). Restricting the correlation analysis to premenopausal women did not change the results. After stratifying postmenopausal women by HRT use, MD was positively related to E_1, both OH metabolites, and T among women not taking HRT. The correlation coefficients were between 0.32 & 0.38, but were not statistically significant. However, postmenopausal women on HRT, E_1 and both OH metabolites were negatively associated with percent MD (r_s = -0.39).

Table 2. Relation of Hormones with Percent Densities, Age, and BMI.

Variable	Spearman Correlation Coefficients (*p*) with Age	BMI	Percent Density Unadjusted	Percent Density Adjusted[1]
Age	--	-0.09 *0.11*	-0.21 *0.0003*	--
BMI	-0.09 *0.11*	--	-0.42 *<.0001*	--
E_1 (ng/ml)	-0.14 *0.02*	0.05 *0.42*	0.10 *0.10*	0.07 *0.23*
E_2 (ng/ml)	-0.17 *0.004*	0.11 *0.07*	0.05 *0.41*	0.02 *0.72*
2-OH-E_1 (ng/ml)	-0.32 *<0.0001*	-0.04 *0.51*	0.13 *0.02*	0.02 *0.70*
16α-OH-E_1 (ng/ml)	-0.34 *<0.0001*	0.07 *0.23*	0.06 *0.30*	-0.04 *0.48*
2/16 ratio	-0.09 *0.12*	-0.07 *0.25*	0.09 *0.13*	0.07 *0.22*
T (ng/ml)	-0.20 *0.0005*	0.27 *<0.0001*	-0.03 *0.57*	0.02 *0.77*
ADIOL (ng/ml)	-0.39 *<0.0001*	0.21 *0.0002*	0.04 *0.49*	0.005 *0.92*
All hormones (ng/ml)	-0.28 *<0.0001*	0.09 *0.11*	0.08 *0.14*	0.03 *0.62*
E/A ratio	0.04 *0.48*	-0.19 *0.0007*	0.10 *0.08*	0.05 *0.44*

[1] Adjusted for age, menopausal status, HRT, BMI, ethnicity, age at menarche, age at first live birth, and number of children.

Mean levels of hormones by categories of percent MD (Figure 1) illustrate the lack of an association of MD with E_1, E_2, 2-OH-E_1, T, and ADIOL. The p-values for the respective trend tests were 0.94, 0.85, 0.54, 0.78, and 0.69. For 16α-OH-E_1, we observed a weak negative relation (p for trend = 0.19) that translated into a 2-5% higher 2/16 ratio for women in the highest MD category. The respective values for the five categories of percent MD were: 1.59, 1.68, 1.86, 1.85, and 1.99 with a p-value of 0.01 for the linear trend test.

This trend changed minimally when excluding the 37 women on HRT. However, stratification by menopausal status showed the association was stronger in pre- than postmenopausal women (p=0.08 vs. 0.97 for postmenopausal). HRT use did not affect the relation in postmenopausal women. Separate analysis for women with a BMI of ≤ 25 (p = 0.07) vs. women with a BMI ≥ 25 (p = 0.40) indicated that the relation was restricted to women with normal weight. Stratification by ethnicity showed similar trends in Caucasian and in Asian women (p = 0.07 and p = 0.04), but the mean 2/16 ratio was lower for Asians than for Caucasians in all MD categories.

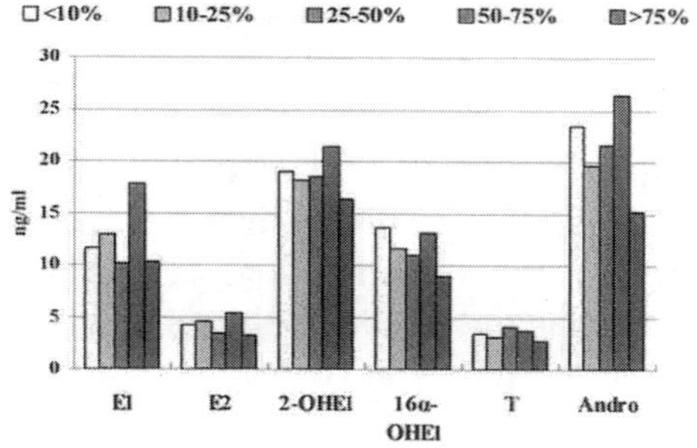

Figure 1. Levels of urinary hormones by categories of percent

Conclusions

In this cross-sectional investigation among women of different ethnicity, we observed higher percent MDs, lower urinary A levels, and a lower 2/16 ratio among Asian than Caucasian women. However, we did not observe any significant associations between urinary hormone levels and MD. The 2/16 ratio showed a relation with percent MD in a direction opposite to our initial hypothesis.

Women with percent MDs of 75% or greater had an approximately 25% higher 2/16 ratio than women with percent MDs below 10%. This relation was similar in Asian and Caucasian women, but it was not present among postmenopausal and overweight women. The ethnic differences in MD agree with previous studies (21, 31). Due to the smaller breast size of Asian women, percent MD is higher than in Caucasian women. However, the absolute MDs appears to be lower among Asian women (32).

Our findings agree with a previous report of MD and the 2/16 ratio (19). The study among postmenopausal women used a qualitative assessment method for MD assessment. The mean 2/16 ratio was 1.12 in the high risk group and 0.83 in the low risk group, a 35% difference. The results also agree with the higher 2/16 ratio in Finnish as compared to Asian women who have the lower BC risk (20). Although a number of studies have investigated the 2/16 ratios and BC, the evidence does not offer a definite answer to the question whether a higher 2/16 ratio reduces or increases BC risk. An association with postmenopausal BC was detected in one small case-control study (14), but a larger study did not support the hypothesis (33). Two cohort studies (17, 18) found a higher 2/16 ratio associated with a non-significantly reduced BC risk among premenopausal women. A recent study (16), conducted among Chinese women, reported a reduced BC risk with a higher urinary 2/16 ratio, but only when urine was collected prior to BC treatment.

The present study has a number of limitations. The strict eligibility criteria for the nutritional interventions may have introduced selection bias. The fact that our population was highly motivated, had a high proportion of subjects with a family history of BC (14.4%), and relatively few children suggests that the women in the study may have a higher than average risk to develop BC. Asian ethnicity at the age group of the women studied, at least second generation migrants, probably does not protect against BC. As shown in a large prospective study, BC incidence among Japanese women in Hawaii and California is at least as high as among Caucasians (34). The two urine collection protocols, overnight samples in premenopausal women vs. spot urine specimens in postmenopausal women, may have introduced additional bias. However, the possible effect on the results is difficult to assess. Although no information on urinary volume was collected, we are confident that the hormone concentrations reflect actual excretion patterns. For a subset of 196 premenopausal women who had creatinine levels available, we repeated the correlation analysis and adjusted for creatinine as a surrogate for urinary volume. None of the Es or As showed any association with percent MD.

Our results suggest an absence of a relationship between urinary sex steroids and MD. Given the negative results in studies of circulating endogenous hormones and MD (9, 10), it appears likely the effects of hormones are not mediated through MDs in premenopausal women. Unfortunately, the small number of postmenopausal women not taking HRT made it impossible to make any valid conclusions for that population. A larger study among women who have never taken HRT would be needed to elucidate this question in postmenopausal women. To understand the importance of different E metabolites in the etiology of BC may also require assessment of additional metabolites, such as 4-OH metabolites, as well as studies regarding the combined carcinogenic effects when different metabolites are present.

Acknowledgments

The project was funded by the U.S. Army Medical Research Materiel Command under DAMD17-00-1-0281, the National Cancer Center as project R01CA80843, and by a contract from the Pharmavite Corporation, San Fernando, CA.

References

1. Boyd NF, Lockwood GA, Byng JW, et al (1998) Mammographic densities and breast cancer risk. Cancer Epidemiol Biomarkers Prev 7:1133-1144.
2. Toniolo PG, Levitz M, Zeleniuch-Jacquotte A, et al (1995) A prospective study of endogenous estrogens and breast cancer in postmenopausal women. J Natl Cancer Inst 87:190-197.
3. Hankinson SE, Willett WC, Manson JE, et al (1998) Plasma sex steroid hormone levels and risk of breast cancer in postmenopausal women. J Natl Cancer Inst 90:1292-1299.

4. Cauley JA, Lucas FL, Kuller LH, et al (1999) Elevated serum estradiol and testosterone concentrations are associated with a high risk for breast cancer. Study of Osteoporotic Fractures Res Group. Ann Intern Med 130:270-277.
5. Greendale GA, Reboussin BA, Slone S, et al (2003) Postmenopausal hormone therapy and change in mammographic density. J Natl Cancer Inst 95:30-37.
6. Brisson J, Brisson B, Cote G, et al (2000) Tamoxifen and mammographic breast densities. Cancer Epidemiol Biomarkers Prev 9:911-915.
7. Spicer DV, Ursin G, Parisky YR, et al (1994) Changes in mammographic densities induced by a hormonal contraceptive designed to reduce breast cancer risk. J Natl Cancer Inst 86:431-436.
8. Gram IT, Ursin G, Spicer DV, et al (2001) Reversal of gonadotropin-releasing hormone agonist induced reductions in mammographic densities on stopping treatment. Cancer Epidemiol Biomarkers Prev 10:1117-1120.
9. Meyer F, Brisson J, Morrison AS, et al (1986) Endogenous sex hormones, prolactin, and mammographic features of breast tissue in premenopausal women. J Natl Cancer Inst 77:617-620.
10. Boyd NF, Stone J, Martin LJ, et al (2002) The association of breast mitogens with mammographic densities. Br J Cancer 87:876-882.
11. Nebert DW (1993) Elevated estrogen 16α-hydroxylase activity: is this a genotoxic or nongenotoxic biomarker in human breast cancer risk? J Natl Cancer Inst 85:1888-1891.
12. Bradlow HL, Davis DL, Lin G, et al (1995) Effects of pesticides on the ratio of 16alpha/2-hydroxyestrone: a biologic marker of breast cancer risk. Environ Health Perspect 103(Suppl 7):147-150.
13. Yager JD, Liehr JG (1996) Molecular mechanisms of estrogen carcinogenesis. Annu Rev Pharmacol Toxicol 36:203-232.
14. Kabat GC, Chang CJ, Sparano JA, et al (1997) Urinary estrogen metabolites and breast cancer: a case-control study. Cancer Epidemiol Biomarkers Prev 6:505-509.
15. Ursin G, London S, Stanczyk FZ, et al (1999) Urinary 2-hydroxyestrone/16alpha-hydroxyestrone ratio and risk of breast cancer in postmenopausal women. J Natl Cancer Inst 91:1067-1072.
16. Fowke JH, Qi D, Bradlow HL, et al (2003) Urinary estrogen metabolites and breast cancer: differential pattern of risk found with pre- versus post-treatment collection. Steroids 68:65-72.
17. Meilahn EN, De SB, Allen DS, et al (1998) Do urinary oestrogen metabolites predict breast cancer? Guernsey III cohort follow-up. Br J Cancer 78:1250-1255.
18. Muti P, Bradlow HL, Micheli A, et al (2000) Estrogen metabolism and risk of breast cancer: a prospective study of the 2:16alpha-hydroxyestrone ratio in premenopausal and postmenopausal women. Epidemiology 11:635-640.

19. Riza E, dos SS, I, De Stavola B, et al (2001) Urinary estrogen metabolites and mammographic parenchymal patterns in postmenopausal women. Cancer Epidemiol Biomarkers Prev 10:627-634.
20. Adlercreutz H, Gorbach SL, Goldin BR, et al (1994) Estrogen metabolism and excretion in Oriental and Caucasian women. J Natl Cancer Inst 86:1076-1082.
21. Maskarinec G, Meng L, Ursin G (2001) Ethnic differences in mammographic densities. Int J Epidemiol 30:959-965.
22. Maskarinec G, Williams AE, Stanczyk FZ, et al (2002) A randomized isoflavone intervention among premenopausal women. Cancer Epidemiol Biomarkers Prev 11:195-201.
23. Maskarinec G, Robbins C, Riola B, et al (2003) Three measures show high compliance in soy intervention among premenopausal women. J Am Diet Assoc 103:861-866.
24. Stram DO, Hankin JH, Wilkens LR, et al (2000) Calibration of the dietary questionnaire for a multiethnic cohort in Hawaii and Los Angeles. Am J Epidemiol 151:358-370.
25. Byng JW, Boyd NF, Fishell E, et al (1994) The quantitative analysis of mammographic densities. Phys Med Biol 39:1629-1638.
26. Shrout P, Fleiss J (1979) Intraclass correlations: uses in assessing rater reliability. Psychol Bull 86:420-428.
27. Rinaldi S, Onland-Moret N, Kaaks R, et al (2003) Reproducibility over time of measurements of androgens, estrogens and hydroxy estrogens in urine samples from post-menopausal women. Eur J Epidemiol 18:417-424.
28. Hawaii Department of Business and Economic Development (2002) 2001 State of Hawaii data book: a statistical abstract. http://www.hawaii.gov/dbedt/db01/index.html. Hawaii Department of Business and Economic Development. 9-19-2002.
29. Armitage P (1971) Statistical methods in medical research. Blackwell Scientific Publications, Oxford.
30. SAS OnlineDoc, Version 8 (2002) Cary, NC: SAS Institute Inc.
31. Ursin G, Ma H, Wu AH et al. (2003) Mammographic density and breast cancerin three ethnic groups. Cancer Epidemiol Biomarkers Prev 12:332-338.
32. Maskarinec G, Nagata C, Shimizu H, et al (2002) Comparison of mammographic densities and their determinants in women from Japan and Hawaii. Int J Cancer 102:29-33.
33. Ursin G, London S, Stanczyk FZ, et al (1997) A pilot study of urinary estrogen metabolites (16alpha-OHE1 and 2-OHE1) in postmenopausal women with and without breast cancer. Environ Health Perspect 105 Suppl 3:601-605.
34. Pike MC, Kolonel LN, Henderson BE, et al (2002) Breast cancer in a multiethnic cohort in Hawaii and Los Angeles: risk factor-adjusted incidence in Japanese equals and in Hawaiians exceeds that in whites. Cancer Epidemiol Biomarkers Prev 11:795-800.

Polymorphism of the Estrogen Receptor-β Gene in Breast Cancer

Cristina Caraion, Nadine Vincent, Claude Lambert, Constantin Caraion, Guorong Li, Pierre Seffert, Jean-Marc Dumollard, and Christian Genin

Summary

Microsatellite polymorphism may influence gene expression or receptor signaling function. In cancer, microsatellite sequences, even when located in non-coding regions of a gene, are capable of modifying the susceptibility to carcinogenic factors. The estrogen-receptors (ER) α and β are involved in breast oncogenesis, with different prognosis significance. They mediate different and complex pathways. Recently, the expression of the ERβ in breast tumors has been shown to be a marker for good prognosis for breast cancer (BC). Although this piece of information needs further testing, the involvement of the ERβ in BC signaling processes has been clearly demonstrated. The goal of the present study was to gain insight into the genetic polymorphism of ERβ, and its possible relationship to BC pathogenesis regarding estrogen dependence. We conducted a case-control study comprising 27 BC samples and 30 control samples, matched by age and hormonal status. Our results indicate that the length of the ERβ genotype variants oscillates between 149-169 bp. Clearly, the overall ERβ distribution had a bimodal distribution with a majority of alleles ranging between 161 and 165 pb. A group of ERβ alleles was significantly shorter (below 159 bp) with a predominant peak at 155 pb. A comparison of the length of the ERβ genotypes revealed that short alleles [below 21 cytosine (C) and adenosine (A) repeats - 159 bp] were significantly more frequent in BC patients (42.6%) compared to controls (18.3%; $t = -2.50$; $p = 0.013$). In conclusion, (CA)n polymorphism at the ERβ locus may represent a modulator factor of BC risk.

Introduction

In BC, high susceptibility gene mutations of BRCA1 and BRCA2 are associated with a 10.0-20.0-fold increased risk of the disease in females (1). However, the rarity of such carriers explains less than 10% of the BC cases. Therefore, it has been established that low-penetrance polymorphic genes may play an important role in BC development. Candidates for such potential high-prevalence genes are low risk genes as those involved in mediating steroid hormone metabolism, activation or detoxification of environmental carcinogens, DNA repair, signal transduction, and cell cycle control, leading to a multigenic model for BC susceptibility (2).

Particular polymorphic variants may affect gene expression and/or function, depending on their localization on the gene structure. If a coding region is involved, the polymorphism could result in a sense-mutation, and eventually to an amino acid substitution in the expressed protein. Also, polymorphism may modify the DNA transcription, the stability of primary transcripts, or the translation process of messenger RNA (3).

Exposure to estrogens (Es) is a modifier of BC risk. Es stimulate breast cell proliferation via their receptors, resulting in the likelihood of the occurrence or propagation of errors in DNA replication. The role of endogenous Es in BC etiology is also supported by prospective epidemiological studies. Hankinson (4) has shown that plasma E levels are higher in women who have developed BC compared to those who remained disease free.

The ERs are nuclear receptor proteins with an E- and a DNA-binding domain. They act as transcription factors and stimulate E-responsive genes. The presence of ER in BC is a marker of a favorable prognosis and responsiveness to adjuvant hormone therapy (5). About 66% of BCs are ER^+ (6); however, it is not yet known why 40% of those BCs are primarily resistant to hormonal treatment; neither whether ER^+ and ER^- BCs represent distinct forms of the disease with different risk factors patterns. There are two isoforms of the ER: α and β, encoded by different genes. The two ER isoforms have different histological distribution and mechanisms of action (7). This suggests that they may play different roles in gene regulation, and that their relative levels could modify the response to Es or anti-estrogens. It has been shown that the ERβ may have a role in tamoxifen resistance (8), therefore, ERβ may be considered of clinical use. The ERβ has a polymorphic region composed of variable numbers of CA repeats in the 3' end of the gene. In order to gain insight into the regulation of ERβ expression in BC, we analyzed a previously described genetic polymorphism in ERβ gene (9) in a BC case-control pilot study.

Materials and Methods

Study Subjects. We analyzed 27 peripheral blood samples from patients who had surgical treatment of primary BC. All cases were invasive ductal carcinomas. None had any familial history of BC. Histopathological characteristics (tumor stage and size, histological agressivity) were ascertained from medical records. The age of the patients ranged from 37-82 years old, with a mean age of 57.11 ± 12.98, and a median age of 54 years old. Control subjects (n = 30) were randomly selected from healthy female (free of diagnosed cancer) blood donors, matched by age and menopausal status. The age ranged from 19 to 81 years old, with a mean age of 51-96 (35 ± 11) and a median age of 52 years old.

Genotyping the Polymorphism. Genomic DNA was extracted from buffy coat obtained from EDTA-blood, using Qiagen QIAamp blood kit. The (CA)n microsatellite was amplified using a fluoresce labeled primer [5'-GGT AAA CCA TGG TCT GTA CC-3' (forward)], and an un-labeled primer [5'-AAC AAA ATG TTG AAT GAG TGG G-3' (reverse)]. Aliquots (~100 ng of DNA) were amplified in a 25 μl reaction volume, containing 1.5 mM $MgCl_2$, 5 pM of each primer, 200 nM dNTP, and 1.25 units of Taq-polymerase. PCR amplification was performed by an initial denaturation at 94°C/2 min, followed by a 35 cycle amplification (94°C/45 sec, 53°C/45 sec, and 72°C/45 sec) followed by a final extension at 72°C/2 min. The PCR products were separated on a polyacrylamide gel by automated fluorescence detection in a sequencer . Allele sizes were interpolated using a size standard, 20 DNA fragments from 60 to 400 bp, and with a CEQ™ 8000 DNA Analysis System. The results were confirmed by re-running quality control samples.

Statistical Analysis. Mantel-Haenszel χ^2 and t-test statistics across matched case-control sets were used to compare the distribution of the different alleles. All p-values are from two-tailed tests. The influence of the genotype length on the risk of BC was tested with an unconditional linear logistic regression. Because the results of significance tests may depend on the choice of cut-points for the continuous variable, we tested the differences among the allele size distributions by use of the Mann Whitney U test. All analyses were performed with the use of STATISTICA 6.0 software (StatSoft Inc, USA). Statistical significance level was set at a p value = 0.05.

Results and Discussion

Eleven alleles of different sizes, ranging from 16 to 26 CA repeats (149-169 bp, respectively), were found. A bimodal distribution of ERβ allele sizes was evident in both BC and control groups (Figure 1). The chosen cutoff value was 159 bp (21 CA repetitions). Alleles longer than 21 CA repeats (median 163) were termed "long alleles" (L), while those below 21 (median 155) were termed "short alleles" (S). The S allele was more frequently found in BC patients compared to controls (Figure 2). The mean CA repeat length was lower in BC patients [159.55 bp (median 161 bp; SD 4.98)] than in healthy controls [161.73 bp (median 163 bp, SD 4.30); t= -2.50; df = 112; p = 0.013].

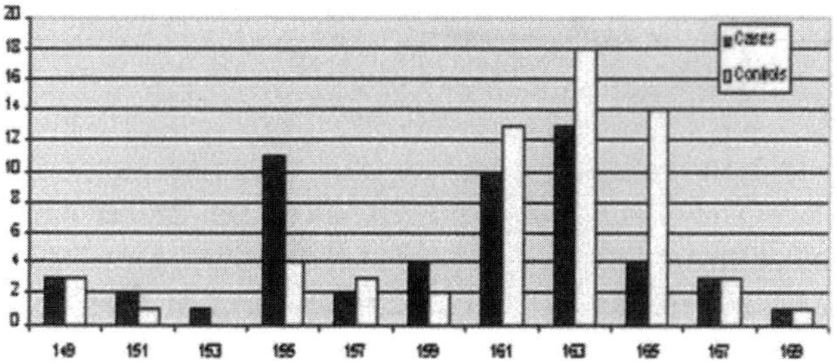

Figure 1. Size distribution of (CA)n repeat in the ERβ gene. A bimodal distribution was observed with two groups, shorter or longer than 159 pb (21CA). The distribution was different between BC patients (n = 54) and healthy age-matched controls (n = 60).

Figure 2. Cumulative distribution of (CA)n allele sizes in BC cases and controls. The (CA)n allele lengths were frequently shorter in BC cases than in controls.

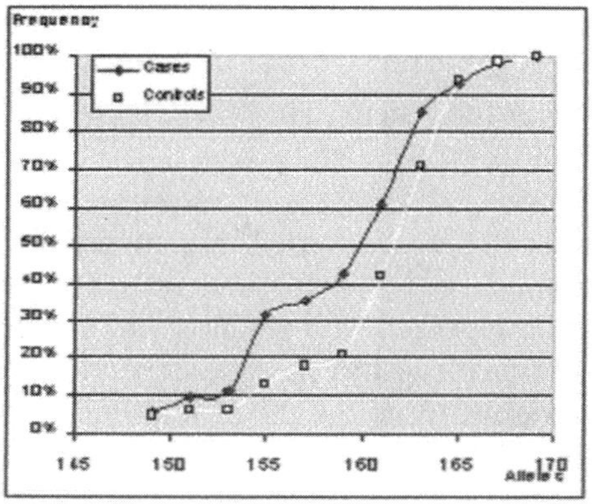

The data in Table 1 show that the S allele (≤ 21 repeats; 159 bp) was significantly more frequent in BC patients (42.6 %) compared to controls (18.3%; chi square = 6.45, df =1, p = 0. 011). These results were confirmed using the Mann Whitney U test that does not depend on cut-points (p = 0.019). The subjects with S allele sizes (≤ 159 bp) had a 1.75-fold increased risk of BC compared with those with L allele sizes (> 159 bp). When the influence of the genotype (SS, SL or LL) on BC risk was assessed, a trend for higher risk in SS and SL carriers (chi square = 5.05, df =1, p = 0.02) compared to the LL carriers, was found (Figure 3).

Table 1. Distribution of Number of CA Repeats in ERβ Alleles [1].

(CA)n repeats	Control	Patients
> 21	49 (81.6%)	31 (57.4%)
≤ 21	11 (18.3%)	23 (42.6%)
Total:	60 (100%)	54 (100%)

1 Two groups were defined by bimodal distribution (S and L than 21 repeats). Results were compared between 30 controls and 27 BC patients (χ^2 p = 0.011).

Moreover, to integrate the effect of both allele lengths of an individual genotype, we compared the means of the two individual alleles between BC cases and controls (sum of allele sizes/2). We found a more frequent "S" mean allele in the BC group compared to the controls group [chi square = 7.81; df=1 (p = 0.0052) and t-value = -2.29; df = 55 (p = 0.025)] suggesting that the effect of both alleles in a genotype is more significant to determine BC risk (Table 2). Because of the small size of the population tested, we could not calculate whether the allelic frequencies values with respect to the Hardy-Weinberg equilibrium.

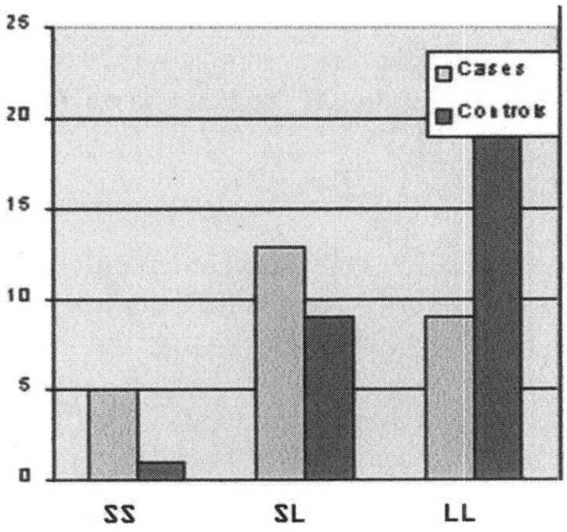

Figure 3. Distribution of genotypes SS, SL, and LL in BC cases vs controls The LL genotype is more frequent in the control group, while the SS and SL genotype are more frequent in the BC cases (Yates corrected χ^2, p = 0.02).

Table 2. Distribution of genotypes regarding CA repeats alleles in ERβ[1].

Genotypes	Subjects	Patients
LL	20 (66.6%)	9 (33.3%)
SL	9 (30.0%)	13 (48.1%)
SS	1 (3.3%)	5 (18.5%)
Total:	30 (100%)	27 (100%)

[1] Three groups were defined according to homozygous long (LL) or short (SS) alleles (shorter and longer than 21 repeats), and heterozygous (SL) genotype in 30 controls and 27 BC patients (χ^2 p = 0.02 comparing LL vs SL+SS).

Our results show that the (CA)n polymorphism in the ERβ gene may be associated with a higher risk for BC. We hypothesize that the different levels of steady-state expression of the ERβ variants, and/or their functionalities may influence BC risk possibly through different hormonal susceptibilities. Whether this polymorphism is functional or not, remains to be determined. Since it is located in the 3' region of the gene, it is reasonable to consider that differences in allele lengths may influence gene expression regulation. Studies of other genes have shown that the (CA)n region near the promoter or in the first intron of a gene affect its promoter activity (10). Indeed, a 1.75 relative risk is not high enough, but reveals a significant role of CA repeats among other BC risk factors, *i.e.,* other genes, gene-gene interactions, and gene-environment interactions. ERβ genetic variant interactions with known BC risk factors may provide new insights into BC etiopathology. Finally, a possible relationship of ERβ polymorphism with other E-dependant diseases (endometriosis, fibroma, and atherosclerosis) should be considered.

Conclusion

ERβ gene polymorphism appears to confer a BC risk; however, a larger study with integration of multiple BC risk factors, such as hormonal exposure, is needed to confirm these results.

References

1. Hopper JL, Southey MC, Dite GS, et al (1999) Population based estimate of the average age-specific cumulative risk of breast cancer for a defined set of protein-truncating mutations in BRCA1 and BRCA2. Cancer Epidemiol Biomarkers Prev 8:741-747.
2. Coughlin SS, Piper M (1999) Genetic polymorphisms and risk of breast cancer. Cancer Epidemiol Biomarkers Prev 8:1023-1032.
3. Comings DE (1998) Polygenic inheritance and micro/minisatellites. Mol Psychiatry 3:21-31.
4. Hankinson SE, Willett WC, Manson JE, et al (1998) Plasma sex steroid hormone levels and risk of breast cancer in postmenopausal women. J Natl Cancer Inst 90:1292-1299.
5. Osborne CK (1998) Steroid hormone receptors in breast cancer management. Breast Cancer Res Treat 5:227-238.
6. Roodi N, Bailey LR, Kao WY, et al (1995) Estrogen receptor gene analysis in estrogen receptor-positive and receptor-negative primary breast cancer. J Natl Cancer Inst 87:446-451.
7. Paech K, Webb P, Kuiper GCJM, et al (1997) Differential ligand activation of estrogen receptors ER alpha and ER beta at AP1 sites. Science 277:1508-1510.
8. Speirs V, Parkes AT, Kerin MJ, et al (1999) Co-expression of estrogen receptor -a and b: poor prognosis factors in human breast cancer? Cancer Res 59:1175- 1179.
9. Tsukamoto K, Inoue S, Hosoi T, et al (1998) Isolation and radiation hybrid mapping of dinucleotide repeat polymorphism at the human estrogen receptor beta locus. J Hum Genet 43:73-74.
10. Gebhardt F, Zanker KS, Brandt B (1999) Modulation of epidermal growth factor receptor gene transcription by a polymorphic CA repeats in intron 1. J Biol Chem 274:13176-13180.

Characterization of Genetic Polymorphism of Glycine N-Methyltransferase Gene in Hepatocellular Carcinoma

Yu-Chuen Huang, Yi-Ping Shih, Chun-Chih Li, Li-Ying Liao, and Yi-Ming A. Chen

Summary

Previously, we reported that glycine N-methyltransferase (GNMT) expression levels were diminished in human liver cancer. Herein, the polymorphisms of GNMT in 5 hepatocellular carcinoma (HCC) and 2 hepatoblastoma (HBL) cell lines were analyzed. In the HCC cell lines, 4/5 had homozygous genotypes in all the GNMT polymorphism markers expect PLC/PRF/5 which had a G/A genotype of SNP3. Among 6 GNMT polymorphism markers, HA22T/VGH had 3 homo and 3 heterozygous genotypes. In contrast, both Huh 6 and Hep G2 HBL cell lines had heterozygous genotypes in most of their GNMT polymorphism markers. The INS/DEL genotypes of GNMT in peripheral blood mononuclear cells of 308 HCC patients were analyzed. The data indicate that the rates of homozygous insertion (INS/INS), heterozygous (INS/DEL), and homozygous deletion (DEL/DEL) genotypes were 11.59%, 48.19% and 40.11%, respectively. When the SNP1 genotype was analyzed, we found that the genotype rates of T/T, T/C, and C/C in HCC patients were 1.30%, 25.65% and 73.05%, respectively. The frequency distribution of both INS/DEL and SNP1 genotypes in HCC patients was not statistically different from those of the 304 normal controls. When the interaction between GNMT genotypes and HBV infection were analyzed, the data showed that among patients without HBV infection, those with C/C genotype had a 1.55-fold higher risk of developing HCC. Further studies in the interaction between GNMT and risk factors for HCC besides HBV are needed.

Introduction

GNMT is a protein with multiple functions (1-3). In 1998, we reported that expression level was diminished both in human liver cancer cell lines and tissues (4). GNMT is expressed only in breast, prostate, liver, kidney, and pancreas. It regulates the ratio of SAM/SAH by catalyzing the synthesis of sarcosine from glycine and SAM (1). This GMNT mechanism may be involved in carcinogenesis since GNMT may affect the cellular levels of DNA methylation. SAM is the sole methyl donor for all the mammalian methyl-transferases, except for those involved in methionine biosynthesis. There is evidence that epigenetic events, *i.e.,* abnormal DNA methylation, may be critical in human diseases (5). DNA hypomethylation is an early event, both in human and experimental hepatocarcinogenesis, and its role in the activation of various genes, has been postulated (6, 7). In addition, previous observations in rodent models indicated that the administration of exogenous SAM greatly inhibits the development of pre-neoplastic lesions in liver (8). The human *GNMT* gene is localized at chromosome 6p12 and its genomic DNA has been isolated and characterized (9). In 2003, we identified 6 novel polymorphisms including two short tandem repeat (STRPs), 3 single nucleotide (SNPs), and 1 insertion/deletion (INS/DEL) around flanking region of exon 1 of the human *GNMT* gene (10) (Figure 1). Herein, we determine expression of *GNMT* SNP1 and INS/DEL in relation to HCC risk.

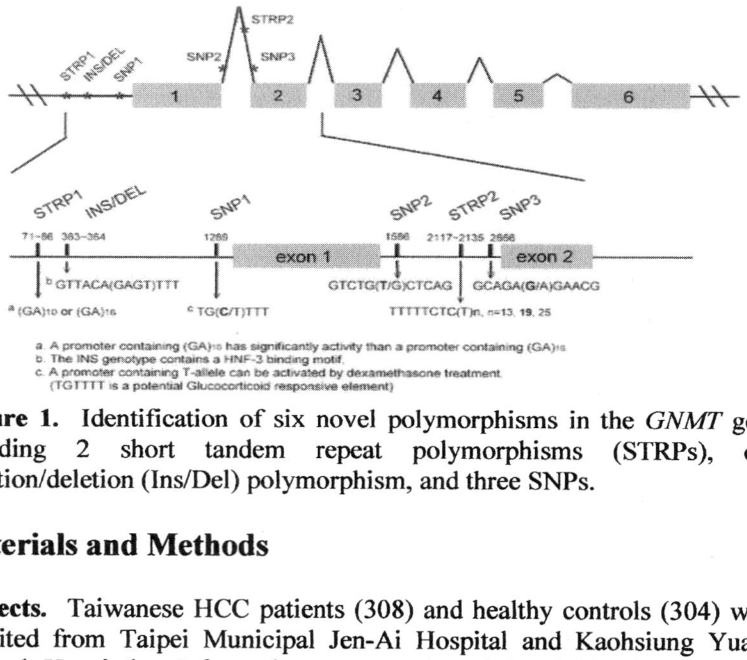

Figure 1. Identification of six novel polymorphisms in the *GNMT* gene including 2 short tandem repeat polymorphisms (STRPs), one insertion/deletion (Ins/Del) polymorphism, and three SNPs.

Materials and Methods

Subjects. Taiwanese HCC patients (308) and healthy controls (304) were recruited from Taipei Municipal Jen-Ai Hospital and Kaohsiung Yuan's General Hospital. Informed consent and peripheral blood (PBL) was obtained from all participants. A questionnaire was designed to collect relevant information regarding HCC risk factors, *i.e.,* age at diagnosis, family history of HCC, hepatitis B virus (HBsAg) or C virus infection (anti-HCV antibody), diabetes mellitus, as well as α-fetoprotein, AST and ALT, and cigarette smoking. The protocols were approved by the Institutional Reviewed Boards, Taipei Veterans General Hospital.

DNA Preparation. PBL was collected from each study participant. Buffy coat samples were prepared and stored at -80°C until use. Genomic DNA was prepared by conventional phenol/chloroform extraction (11), followed by ethanol precipitation, and stored at -20°C until used for genotype analysis.

***GNMT* INS/DEL GeneScan Analysis.** To identify the genotypes of INS/DEL, the primers were described in Tseng, *et al.* (10). PCR amplification was carried out by using 1.0 ng/μl PBL DNA, 1.0 unit of AmpliTaq Gold, 0.5μM of each primer, 0.5 mM deoxynucleotides, and 2.5 mM $MgCl_2$ in a total reaction volume of 20μl. The PCR conditions used 93°C for 10 min to activate the AmpliTaq Gold, follow by 35 cycles of denature of 94°C for 30 s, annealing at 54°C for 30 s, and extension at 72°C for 30 s. The final elongation is at 72°C for 60 min. PCRs are running in a GeneAmp PCR 9700 thermocycler. PCR products were electrophoresed on a 377 ABI PRISM sequencer, and the fluorescent signals from the different size alleles were recorded using GENOTYPER version 2.1 software.

TaqMan Assay. The *GNMT* SNP1 genotype was determined using a TaqMan assay. The procedures were described in Tseng, *et al.* (10).

Data Analysis. *GNMT* INS/DEL and SNP1 allele and genotype frequencies were calculated. Odds ratios (OR) were calculated as a measure of association between *GNMT* genotypes and HCC. OR and 95% confidence interval (CI) were estimated from unconditional logistic regression models. OR for *GNMT* INS/DEL and SNP1 genotypes were calculated after subdividing cases based upon HBV infection and HCC.

Results

Characteristics of Liver Cancer Cell Lines and their GNMT Polymorphisms. The properties of 5 HCC and 2 HBL cell lines and their genotypes of different polymorphisms of *GNMT* gene are summarized in Table 1. All the 7 cell lines were hyperdiploid except Huh 7. Two cell lines (Hep 3B and PLC/PRF/5) had HBV genome integration, while 4/5 HCC cell lines had homozygous genotypes in all the GNMT polymorphism markers, expect that PLC/PRF/5 had a G/A genotype of SNP3. Among 6 GNMT polymorphism markers, HA22T/VGH had 3 homozygous and 3 heterozygous genotypes. In contrast, both Huh 6 and Hep G2 HBL cell lines had heterozygous genotypes in most of their GNMT polymorphism markers.

Frequencies of INS/DEL and SNP1 Genotypes of *GNMT* in HCC Patients. The INS/DEL genotypes of *GNMT* in PBL mononuclear cells of 308 HCC patients were analyzed using the GeneScan assay. The rates of homozygous insertion (INS/INS), heterozygous (INS/DEL) and homozygous deletion (DEL/DEL) genotypes were 11.59%, 48.19% and 40.11%, respectively. The distribution of genotype INS/DEL frequencies in HCC patients was not statistically different from that of the 304 normal controls (Table 2). When comparing patients with the INS/INS GNMT genotype, the OR of developing HCC in patients with INS/DEL or DEL/DEL were 0.88 (95% confidence interval [CI] = 0.51-1.53) and 0.78 (95% CI = 0.45-2.24), respectively. The SNP1 genotype was analyzed in the same cohort mentioned above using TaqMan assay.

Table 1. Summary of the Characteristics of Different Human Liver Cancer Cell Lines and Their Genotypes of GNMT.

Cell Lines	Cancer Type	Chromosome Ploidy	HBV DNA	Origins	Tumori-genecity	STRP1[1]	INS/DEL[2]	SNP1	SNP2	STRP2[3]	SNP3
HA22T/VGH	HCC	Heperdiploid	-	Taiwan	+	10GA/10GA	INS/INS	C/T	G/G	13T/25T	G/A
Huh 7	HCC	Hypotetraploid	-	Japan	+	16GA/16GA	DEL/DEL	C/C	T/T	19T/19T	G/G
Hep 3B	HCC	Heperdiploid	+	US	+	10GA/10GA	INS/INS	T/T	G/G	13T/13T	A/A
Sk-Hep-1	HCC	Heperdiploid	-	US	+	10GA/10GA	INS/INS	T/T	G/G	13T/13T	A/A
PLC/PRF/5	HCC	Heperdiploid	+	South Africa	+	10GA/10GA	INS/INS	T/T	G/G	13T/13T	G/A
Huh 6	Hepto-blastoma	Heperdiploid	-	Japan	No data	10GA/16GA	INS/DEL	C/C	T/G	13T/19T	G/A
Hep G2	Hepto-blastoma	Heperdiploid	-	Argentina	+	10GA/17GA	INS/DEL	C/C	T/G	19T/25T	G/G

1 Short tandem repeat polymorphism 1 is located in the up-stream gene-regulatory region of *GNMT* (Figure 2). According to a previous study, the more GA repeats, the lower promoter activity of GNMT (10).

2 The INS-allele contained 4-nucleotide (GAGT) insertion in the up-stream region of *GNMT* gene (Figure 2). The promoter containing INS-allele has higher activity than the promoter with the DEL-allele (10).

3 Short tandem repeat polymorphism 2 is located in intron 1 (Figure 2). The numbers of T are 13, 19, and 25.

Table 2. Distribution of *GNMT* Gene Polymorphisms and Estimated OR for HCC risk.

Genotype	HCC PBMC (%)	Normal PBMC (%)	OR	95%CI	
INS/DEL[1]					
INS/INS	32 (11.59)	30 (9.87)	1	Ref	1.00 (Ref)
INS/DEL	133 (48.19)	141 (46.38)	0.88	0.51-1.53	
DEL/DEL	111 (40.22)	133 (43.75)	0.78	0.45-2.24	0.86 (0.62-1.20)
SNP1					
T/T	4 (1.30)	5 (1.64)	1	Ref	1.00 (Ref)
C/T	79 (25.65)	83 (27.30)	1.19	0.31-4.59	
C/C	225 (73.05)	216 (71.05)	1.30	0.34-4.91	1.10 (0.78-1.57)

[1] The INS-allele contained 4-nucleotide (GAGT) insertion.

Table 3. Estimated OR of HCC Risk Associated Between GNMT Polymorphism and HBV Infection Status.

	HCC PBMC (%)	Normal PBMC (%)	OR	95%CI
With HBV infection				
INS/DEL				
INS/DEL+INS/INS	67 (59.82)	20 (60.61)	1	Ref
DEL/DEL	45 (40.18)	13 (39.39)	1.03	0.43-2.33
SNP1				
C/T+T/T	38 (30.89)	10 (30.30)	1	Ref
C/C	85 (69.11)	23 (69.70)	0.97	0.46-2.25
Without HBV infection				
INS/DEL				
INS/DEL+INS/INS	22 (53.66)	151 (55.72)	1	Ref
DEL/DEL	19 (46.34)	120 (44.28)	1.09	0.56-2.11
SNP1				
C/T+T/T	9 (17.65)	64 (24.90)	1	Ref
C/C	42 (82.35)	193 (75.10)	1.55	0.70-3.43

The results showed that the rates of T/T, T/C, and C/C genotypes in HCC individuals were 1.3% (4/308), 25.65% (79/308) and 73.05% (225/308), respectively.

The distribution of the SNP1 genotype frequencies between HCC and normal groups was not statistically different (Table 2). The OR of developing HCC in individuals with C/T or C/C genotype vs. persons with T/T genotype were 1.19 (95% CI= 0.31-3.59) and 1.30 (95% CI= 0.34-4.91), respectively (Table 2). If we combined the data of persons with T/T and C/T genotypes and used as reference, then the OR of developing HCC for persons with C/C was 1.10 (95% CI= 0.78-1.57).

Interaction Between HBV Infection and GNMT Polymorphism in HCC Patients. We next analyzed the combined effect of the *GNMT* genotype polymorphism and HBV infection on the risk of HCC development. As shown in Table 3, the OR of HCC in persons with HBV infection and

DEL/DEL genotype was 1.03 (95% CI= 0.45-2.33) when we used both DEL/INS and INS/INS genotypes as a reference. For persons with DEL/DEL genotype but without HBV infection, the OR was 1.09 (95% CI= 0.56-2.11). In terms of the interaction between SNP-1 and HBV infection, the OR for persons with C/C genotype and HBV infection was 0.97 (95% CI= 0.46-2.25) when we used both C/T and T/T genotypes as a reference. Among HCC patients without HBV infection, the results showed that for persons with C/C genotype, the OR was 1.55 (95% CI=0.70-3.43) when we combined the data of the C/T and T/T groups (Table 3, lower panel).

Discussion

All in all, we analyzed the polymorphisms of GNMT in different human liver cancer cell lines including 5 HCC and 2 hepatoblatoma cell lines. The high percentage of homozygous *GNMT* genotypes in the 5 HCC cell lines implies that there may be high rate of somatic mutation of *GNMT* gene during liver tumorigenesis. In terms of the HA22T/VGH cell line, among its 6 GNMT polymorphism markers, 3 were heterozygous and 3 homozygous. Haplotypic analysis showed that somatic mutation had happened. In contrast, the 2 HBL cell lines had heterozygous genotypes in most of their GNMT polymorphism markers.

We used 2 genotypic markers of GNMT-INS/DEL and SNP1 to determine the association between GNMT polymorphisms and HCC development. No significant differences of the GNMT INS/DEL genotype frequencies between HCC and normal control groups were observed. However, in individuals with 1 or 2 DEL-alleles, the OR of developing HCC decreased. The risk for HCC among heterozygous individuals was intermediate to those who were homozygous for the DEL-allele (OR of 0.88 and 0.78, respectively; Table 2). In terms of the SNP1, there was no significant association between the SNP-1 genotype and HCC risk, while increased risk for HCC was observed in individuals with at least one copy of the C-allele. The risk for HCC among heterozygous individuals was intermediate to those who were homozygous for the C-allele (OR of 1.19 and 1.30, respectively; Table 2). Studies with larger sample size are needed to elucidate the usefulness of these GNMT polymorphism markers for the genetic counseling of high-risk populations.

Finally, we investigated the interaction between GNMT genotypes and HBV infection in HCC patients. After data stratification by status of HBV infection, we found that among patients without HBV infection, those with C/C genotype had a 1.55-fold higher risk of developing HCC (Table 3, lower panel). This suggests that GNMT and HBV may be independent risk factors for HCC development. The interaction between GNMT and other risk factors for HCC, *i.e.*, environmental carcinogens or HCV infection should be further investigated in the future.

Acknowledgements

We thank the Genotyping Lab of the National Yang-Ming University Genome Research Center (YMGC) for their services, and the medical staff of the Taipei Municipal Jen-Ai Hospital and Yuan's General Hospital for their help in collecting specimens and questionnaires. This study was supported by a grant (NSC 92-3112-B-010-001) from the National Research Program for Genomic Medicine of Taiwan and VTY Joint Research Program (VTY92-P5-32), Tsou's Foundation.

References

1. Kerr SJ (1972) Competing methyltransferase system. J Biol Chem 247:4248-4252.
2. Yeo EJ, Wagner C (1994) Tissue distribution of glycine N-methyltransferase, a major folate-binding protein of liver. Proc Nat Acad Sci USA 9:210-214.
3. Bhat R, Wagner C, Bresnick E (1997) The homodimeric form of glycine N-methyltransferase acts as a polycyclic aromatic hydrocarbon-binding receptor. Biochem J 36:9906-9910.
4. Chen YM, Shiu JY, Tzeng SJ, et al (1998) Characterization of glycine-N-methyltransferase-gene expression in human hepatocellular carcinoma. Int J Cancer 75:787-793.
5. Robertson KD, Wolffe AP (2000) DNA methylation in health and disease. Nature Rev 1:11-19.
6. Pascale R, Simile MM, Ruggiu ME, et al (1991) Reversal by 5-azacytidine of the S-adenosyl-L-methionine-induced inhibition of the development of putative preneoplastic foci in rat liver carcinogenesis. Cancer Lett 56:259-265.
7. Feo F, Pascale RM, Simile MM, et al (2000) Genetic alterations in liver carcinogenesis: implications for new preventive and therapeutic strategies. Crit Rev Oncogenesis 11:19-62.
8. Garcea R, Daino L, Pascale R, et al (1989) Inhibition of promotion and persistent nodule growth by S-adenosyl-L-methionine in rat liver carcinogenesis: role of remodeling and apoptosis. Cancer Res 49:1850-1856.
9. Chen YM, Chen LY, Lin JR, et al (2000) Genomic Structure, Chromosomal Localization and Functional Study of the Human Glycine N-methyltransferase Gene. Abstract from the Annual Meeting of American Society of Human Genetics.
10. Tseng TL, Shih YP, Huang YC, et al (2003) Genotypic and phenotypic characterization of a putative tumor susceptibility gene, GNMT, in liver cancer. Cancer Res 63:647-654.

Role of Soy Phytoestrogens Genistein and Daidzein in Focal Adhesion Assembly and Focal Adhesion Kinase Activity in Breast Cancer Cells

Nicolas G. Azios and Suranganie F. Dharmawardhane

Summary

Recently, our laboratory and others have shown that estrogen receptor (ER)α. negative (-) breast cancer (BC) cells respond to estrogenic compounds via rapid signaling mechanisms. To understand the role of soy phytoestrogens in BC cell invasion, we analyzed focal complex formation and focal adhesion kinase (FAK) activity, as detected by phospho-specific anti-FAK antibodies to tyrosine residues 397 or 925. Stimulation of $ER\alpha^+$ T47D and $ER\alpha^-$ MDA-MB-231 BC cell lines for 10 min with 17β-estradiol (E_2), daidzein, or genistein (GEN) increased the number of focal adhesions and FAK activity. This data implicate E_2 and soy phytoestrogens in the regulation of cellular interactions with the ECM via novel signaling mechanisms.

Introduction

The steroid hormone E_2 mediates establishment and maintenance of female reproductive physiology, as well as the initiation and progression of breast and gynecological cancers via transcriptional activation of both α and β isoforms of the ER. Loss of ERα expression is common in malignant progression of BC leading to the metastatic state (1). In addition to this well-characterized genomic mode of action, recent data have shown that E_2 can exert rapid effects on cell function, possibly via plasma membrane ERs. Such rapid signaling events have been implicated in mechanisms of cross talk with other signaling pathways such asG protein coupled receptors and tyrosine kinase type growth factor receptors (2). These novel mechanisms may help explain ERα-independent effects of estrogenic compounds on cell functions relevant to BC progression, such as invasion and survival.

Recent data, including a report from our laboratory using both $ER\alpha^+$ and $ER\alpha^-$ BC cell lines, have demonstrated that E_2 and related plant-derived estrogenic compounds (phytoestrogens) modulate the activity of signaling proteins that regulate cell survival and invasion (3, 4). The soybean phytoestrogens GEN and daidzein, used in the present study, bind to and transactivate both ERα and β to exert estrogenic and anti-estrogenic effects. The soy phytoestrogens bind ERβ with high affinity (5), which may explain activation of signaling in malignant BCs that have lost ERα expression (3).

During metastatic progression, BC cells invade the ECM both by degradation of matrix components and by modulation of focal adhesions with the ECM. Focal adhesions are multi-molecular complexes of structural and signaling proteins that are formed as a result of clustering integrin receptors by ECM proteins (6). Intracellular signaling by activated integrins and actin polymerization regulates cell motility and cancer cell invasion, including BC metastasis (7). Actively migrating cells contain smaller, more dynamic focal complexes at their leading edge that are thought to generate forces necessary for forward movement. Thus, cell migration is an integrated process that requires coordinated assembly and disassembly of contacts with the ECM (8).

FAK is a key signaling intermediate recruited to the membrane in response to integrin and tyrosine-kinase type receptor activation. Recent data have shown that effects on cell migration via phosphorylation of residue 397 of FAK can be uncoupled from activation of cell invasion via phosphorylation of residue 925 to modulate Rho GTPase activity. FAK also activates matrix metalloproteinase secretion, which facilitates ECM degradation. Thus, FAK-mediated increases in cell proliferation, motility, and invasion have been correlated with tumor malignancy (9).

E_2 has been shown to effect tyrosine phosphorylation of FAK in BC cells (10). GEN has been shown to promote cell spreading and adhesion concomitant with accumulation of FAK in focal adhesions, and transient FAK activation in prostrate and breast cancer cell lines (11,12). These effects of GEN on FAK activity and modulation of cell adhesion to the ECM are thought to be correlated with the preventive effects of GEN in prostrate cancer metastasis (12). To understand the relevance of estrogenic compounds to BC cell invasion, we investigated the role of E_2, GEN, and daidzein in FAK activity and focal complex formation.

Materials and Methods

Cell Stimulation. T47D and MDA-MB-231 human BC cells were cultured in DMEM with 10% FBS and starved for 48 h prior to experiments in phenol red minus media. Quiescent cells were stimulated for 10 min with vehicle (DMSO), 50 ng/ml epidermal growth factor (EGF) (Upstate Biotechnology Inc., NY), 0.01 or 0.1 μM E_2 (Sigma, MO); 10, 25, 50, or 100 μM GEN or daidzein (LKT Labs, MN).

Western Blot Analysis. Cells were disrupted in lysis buffer (20 mM Tris HCl (pH 7.5), 150 mM NaCl, 1 mM EGTA, 1 mM EDTA, 2.5 mM sodium pyrophosphate, 50 mM sodium fluoride, 10% glycerol, 1% NP-40, 1 mM DTT, 0.5% deoxycholate and protease inhibitors) at 4°C. Lysates were centrifuged at 14,000 rpm, the proteins in the supernatant eluted with Laemmli sample buffer, equally loaded, and separated on 10% SDS-PAGE gels. Proteins were transferred to PVDF membranes, blocked with 5% bovine serum albumin (BSA), and probed with specific primary antibodies. Positive bands were detected using an alkaline phosphatase conjugated secondary antibody and developed with NBT/BCIP reagent (Pierce, IL). Anti-FAK (against the N-terminus), anti-phospho-FAK (Tyr-397), and anti-phospho-FAK (Tyr-925) antibodies were purchased from Upstate Biotechnology Inc., NY. The density of positive bands was quantified using NIH Image software. The relative FAK activity was calculated as the density of phospho-FAK in stimulated cell lysate/density of FAK in stimulated cell lysate divided by density of phospho-FAK in un-stimulated cell lysate/density of FAK in un-stimulated cell lysates. The results are representative of three separate experiments.

Immunocytochemistry. Cells were seeded on coverslips and serum-starved in phenol red free media for 48 h. Cells were treated for 10 min with DMSO (control), E_2 (0.1 μM), EGF (50 ng/ml), genistein, or daidzein (10, 50, 100 μM). Cells were fixed in 3.7% formaldehyde, permeablized with 0.2% TritonX100, and blocked with 5% BSA. Fixed cells were stained with rhodamine phalloidin to visualize the actin cytoskeleton and a mouse monoclonal anti-phosphotyrosine antibody (Santa Cruz Biotechnol., CA), followed by fluorescein-conjugated goat anti-mouse secondary antibody (Cappel, PA), to visualize focal complexes. Ten microscopic fields per treatment were photographed using a SPOTII digital camera and quantified for two separate experiments.

Results

To determine the effect of estrogenic compounds on focal complex assembly and the actin cytoskeleton, we monitored the changes in focal adhesions and polymerized actin (F-actin) in response to E_2 or soy phytoestrogens. The concentrations of E_2 and soy phytoestrogens used were in agreement with those of these compounds that induce gene transcription via both ERα and β (13). GEN is commonly used as a tyrosine kinase inhibitor at concentrations above 50 μM (14). Therefore, low (10 μM), medium (50 μM), and high (100 μM) concentrations of phytoestrogens were used to differentiate their effects as estrogenic, anti-estrogenic, or tyrosine inhibitory compounds.

As has been shown previously, EGF stimulation resulted in lamellipodia extension with the assembly of a large number of smaller focal contacts along the leading edge of polarized cells (15). For $ER\alpha^+$ T47D and $ER\alpha^-$ MB-231 cell lines, E_2 and soy phytoestrogens increased the total number of focal adhesions compared to controls (Figure 1A, B). Daidzein treatment at 10, 50 and 100 μM concentrations demonstrated a statistically significant increase in focal adhesions in MDA-MB-231 cells (Figure 1B). GEN also increased focal adhesions at 50 μM. These results are similar to previous reports that documented GEN-induced cell spreading, adhesion, and FAK activity in MCF-7, an $ER\alpha^+$ BC cell line (11).

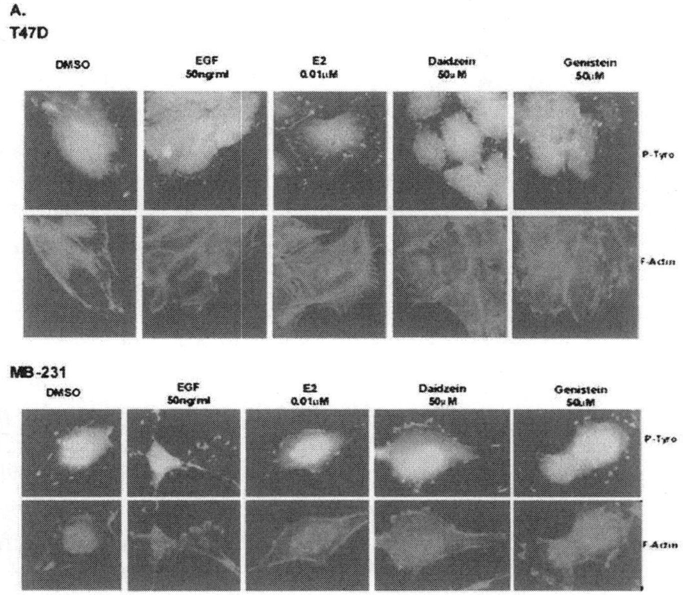

Figure 1A. Immunofluorescence microscopy of BC cell lines in response to E_2 and soy phytoestrogens. $ER\alpha^+$ T47D non-metastatic human BC cells (**top panel**) or $ER\alpha^-$ MDA-MB-231 metastatic human BC cells (**bottom panel**) were serum-starved in phenol red free culture media for 48 h and stimulated for 10 min with DMSO (vehicle), 50 ng/ml EGF, 0.1 μM E_2, or 50μM daidzein or GEN. Cells were stained for F-actin with rhodamine phalloidin and for focal adhesions with anti-phospho-tyrosine followed by a FITC-tagged secondary antibody.

FAK is a key signaling intermediate that is recruited to focal adhesions and activated by auto-phosphorylation in response to integrin clustering (9). Therefore, FAK activation in response to E_2 and phytoestrogens was determined by Western blotting of stimulated cell lysates with phospho-specific antibodies. Data presented demonstrate that phospho-FAK is altered in response to GEN and daidzein in both $ER\alpha^+$ and $ER\alpha^-$ BC cell lines. In $ER\alpha^+$ T47D cells, phospho-FAK levels increased in a concentration dependent manner up to 50 μM in response to daidzein and genistein (Figure 2). Our previously published results were quantified using a C-terminal anti-FAK antibody, which detected only the 125-kD form of full length FAK (3). Herein, this assay has been improved by the use of an anti-FAK antibody which specifically detects the N-terminus of FAK. This improved method allowed the quantification of an often detected N-terminal fragment (90-kD) of FAK, which contains the phosphorylated tyrosine residue 397 (16).

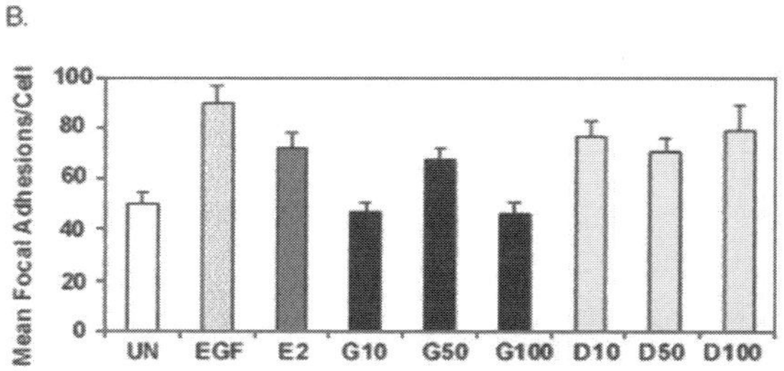

Figure 1B. Mean focal adhesions/cell assembled in response to E_2 and phytoestrogens. MB-231 cells were stimulated with DMSO (UN), 50 ng/ml EGF, 0.1 μM E_2, or 10, 50, or 100 μM GEN (G) or daidzein (D). Following immunofluorescence microscopy, 10 random microscopic fields per coverslip were photographed, and the number of focal adhesions/cell quantified. Results shown are from two separate experiments for 25-40 cells/treatment, and represent the mean ± SEM.

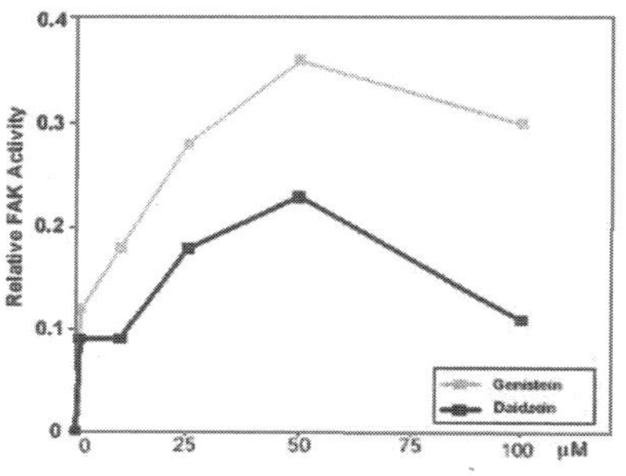

Figure 2. FAK activity in response to genistein and daidzein in T47D cells. Serum-starved cells were stimulated with 0, 1, 10, 25, 50, or 100 μM GEN or daidzein for 10 min and immediately lysed. Cell lysates were Western blotted for total FAK (using an anti-FAK N-terminus antibody) or anti-phospho-FAK (Tyr-397) antibody. The relative FAK activity was calculated and graphed relative to un-stimulated controls, (*i.e.*, the difference between "P-FAK/FAK ratio" with stimulation and the "P-FAK/FAK ratio" without stimulation, divided by the "P-FAK/FAK ratio" without stimulation), for each compound tested. Results are representative of three separate experiments.

FAK phosphorylation as a response to E_2 and phytoestrogens was also determined in ERα$^-$ MDA-MB-231 BC cells. Similar to ERα$^+$ cells, in MDA-MB-231 ERα$^-$ cells, daidzein increased FAK activity in a concentration dependent fashion, with peak activities at 10 μM and decreased at 50 and 100 μM (Figure 3).

The ERα$^-$ MDA-MB-231 cells responded to all phytoestrogenss tested by enhanced FAK phosphorylation at low and medium concentrations (Figures 3 and 4). The effect of GEN on the phosphorylation of both FAK tyrosine residues 397 and 925 was tested. As indicated in Figure 4, phosphorylation of each tyrosine residue demonstrated differential concentration dependence. Phosphorylation of tyrosine residue 397 was maximal at 10 μM, and decreased slightly at 50 μM in response to GEN, suggesting that this effect is not dependent on the tyrosine kinase inhibitory action of GEN. GEN enhanced FAK phosphorylation in residue 925 with increasing concentration, again acting more as a stimulator of FAK activity rather than as a tyrosine kinase inhibitor. Thus, our results demonstrate novel effects of soy phytoestrogens on cell signaling.

Figure 3. FAK activity in response to daidzein in MDA-MB-231 cells. Serum-starved cells were stimulated with DMSO (control), EGF (50 ng/ml) or 10, 50, or 100 μM daidzein for 10 min and immediately lysed. **A)** Cell lysates were Western blotted for total FAK (using an anti-FAK N-terminus antibody), or phospho-FAK (using an anti-phospho-FAK tyr-397 antibody). **B)** Relative FAK activity is calculated as the density of phospho-FAK in stimulated cell lysate/density of FAK in stimulated cell lysate divided by density of phospho-FAK in un-stimulated cell lysate/density of FAK in un-stimulated cell lysates. Results are representative of three separate experiments.

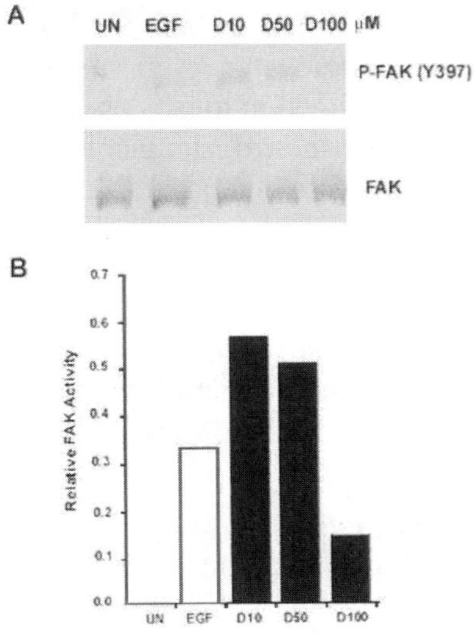

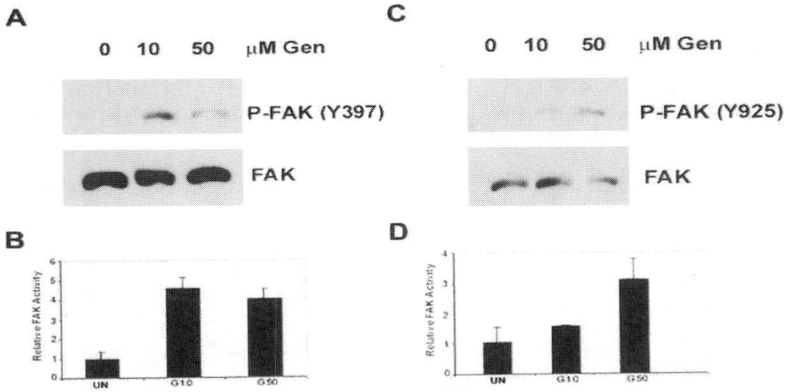

Figure 4. FAK activity in response to genistein in MDA-MB-231 cells. Serum-starved cells were stimulated with DMSO (control), 10, or 50 μM genistein for 10min and immediately lysed. Cell lysates were Western blotted for total FAK (using an anti-FAK N-terminus antibody) or phospho-FAK with an anti-phospho-FAK (Tyr-397) antibody **(A, B)** or anti-phospho-FAK (Tyr-925) antibody **(C, D)**. Relative FAK activity **(B, D)** is calculated as the density of phospho-FAK in stimulated cell lysate/density of FAK in stimulated cell lysate divided by density of phospho-FAK in un-stimulated cell lysate/density of FAK in un-stimulated cell lysates. Error bars represent the mean ± S.D. n = 4.

Conclusions

Soy phytoestrogens affect apoptosis, growth, differentiation, and have anti-oxidant and anti-angiogenic properties (17). The present data clearly demonstrate an additional function for soy phytoestrogens in modulating cell-ECM interactions. E_2 and soy phytoestrogens increased focal adhesion assembly and FAK activity in $ER\alpha^+$ and $ER\alpha^-$ human BC cell lines. Our novel data on the role of GEN in enhancing phosphorylation of both 397 and 925 tyrosine residues of FAK reflect a role for soy phytoestrogens in activation of multiple FAK-regulated signaling pathways that promote both cancer cell proliferation and invasion. Since the effects of soy phytoestrogens were similar to those of E_2, we predict that GEN and daidzein act as estrogenic compounds via the same signaling pathways as E_2 to activate FAK and regulate focal adhesion assembly.

References

1. Fuqua SA (2001) The role of estrogen receptors in breast cancer metastasis. J Mammary Gland Biol Neoplasia 6:407-417.
2. Levin ER (2003) Bidirectional Signaling between the estrogen receptor and the epidermal growth factor receptor. Mol Endocrinol 17:309-317.
3. Brownson DM, Azios NG, Fuqua BK, et al (2002) Flavonoid effects relevant to cancer. J Nutr 132(11 Suppl):3482S-3489S.
4. Stoica GE, Franke TF, Wellstein A, et al (2003) Estradiol Rapidly Activates Akt via the ErbB2 Signaling Pathway. Mol Endocrinol 17:818-830.
5. Kuiper GG, Lemmen JG, Carlsson B, et al (1998) Interaction of estrogenic chemicals and phytoestrogens with estrogen receptor beta. Endocrinology 139:4252-4263.
6. Sastry SK, Burridge K (2000) Focal adhesions: a nexus for intracellular signaling and cytoskeletal dynamics. Exp Cell Res 261:25-36.
7. Felding-Habermann B, O'Toole TE, Smith JW, et al (2001) Integrin activation controls metastasis in human breast cancer. Proc Natl Acad Sci USA 98:1853-1858.
8. Webb DJ, Parsons JT, Horwitz AF (2002) Adhesion assembly, disassembly and turnover in migrating cells -- over and over and over again. Nat Cell Biol 4:E97-100.
9. Hsia DA, Mitra SK, Hauck CR, et al (2003) Differential regulation of cell motility and invasion by FAK. J Cell Biol 160:753-767.
10. Bartholomew PJ, Vinci JM, DePasquale JA (1998) Decreased tyrosine phosphorylation of focal adhesion kinase after estradiol treatment of MCF-7 human breast carcinoma cells. J Steroid Biochem Mol Biol 67:241-249.
11. Bergan R, Kyle E, Nguyen P, et al (1996) Genistein-stimulated adherence of prostate cancer cells is associated with the binding of focal adhesion kinase to beta-1-integrin. Clin Exp Metastasis 14:389-398.

12. Liu Y, Kyle E, Lieberman R, et al (2000) Focal adhesion kinase (FAK) phosphorylation is not required for genistein-induced FAK-beta-1-integrin complex formation. Clin Exp Metastasis 18:203-212.
13. An J, Tzagarakis-Foster C, Scharschmidt TC, et al (2001) Estrogen receptor beta-selective transcriptional activity and recruitment of coregulators by phytoestrogens. J Biol Chem 276:17808-17814.
14. Kim YR, Cha HY, Lim K, et al (2003) Activation of epidermal growth factor receptor is responsible for pervanadate-induced phospholipase D activation. Exp Mol Med 35:118-124.
15. Bailly M, Condeelis JS, Segall JE (1998) Chemoattractant-induced lamellipod extension. Microsc Res Tech 43:433-443.
16. Carragher NO, Levkau B, Ross R, et al (1999) Degraded collagen fragments promote rapid disassembly of smooth muscle focal adhesions that correlates with cleavage of p125(FAK), paxillin, and talin. J Cell Biol 147:619-630.
17. Adlercreutz H (2002) Phytoestrogens and breast cancer. J Steroid Biochem Mol Biol 83:113-118.

Identification of Genes Regulated by the Antiprogestin, Onapristone, in Breast Cancer Cells using Microarray Analysis

Simon J. Crook, J. David Brook, Anette Sommer, Joern Kraetzschmar, and John F.R. Robertson

Summary

Progesterone receptor (PR) antagonists are a potential new therapeutic strategy for the treatment of patients with invasive breast cancer (BC). One such progesterone antagonist is Onapristone. Onapristone has been reported to have strong anti-progestational and anti-tumor activity. A tumor remission rate of approximately 66% was found in a phase II clinical trial of Onapristone as a first-line endocrine therapy in post-menopausal patients with primary BC. However, during this trial, a minority of the patients developed abnormal liver function test results. As a consequence, the further therapeutic use off this drug has stop. Although the mechanism of action of Onapristone has been suggested by animal model experiments, the genes regulated by the drug in the breast are not known. In the present study, the BC cell line T47D was treated with Onapristone and examined with Affymetrix microarrays and real-time RT-PCR to identify genes which were differentially expressed. Cells were treated with Onapristone either alone or in the presence of progesterone, and microarray analysis carried out upon the RNA extracted from the cells. The expression of a selection of up- or down-regulated genes observed was further investigated using real-time RT-PCR. In addition, the expression of these genes was investigated in the PR^- cell line, MDA-MB-231 treated with Onapristone, to ensure that any changes in gene expression observed, were due to Onapristone effects via the PR.

Introduction

PR antagonists are a potential new therapeutic strategy for the treatment of patients with invasive BC. One such progesterone antagonist is Onapristone (ZK98.299) (1, 2). Onapristone was reported to have strong anti-progestational and anti-tumor activity. In the MXT mammary tumor model in mice, and the DMBA- and NMU-induced rat mammary tumors, the anti-tumor activity of Onapristone was equal to or even greater than that of tamoxifen and oophorectomy (3, 4). Furthermore, in the Onapristone treated tumors differentiation of mitotically active polygonal tumor cells towards dysplastic glandular structures was noted. This was accompanied by a massive sequestering of secretory products (5). Morphometric data of tumor tissues indicated that apoptotic cell death is enhanced by Onapristone treatment (5). Additionally, tumor cells were found to accumulate in G_0G_1 phase of the cell cycle following Onapristone treatment, with an accompanying significant and biologically relevant reduction in the number of cells in G_2M and S phases, using flow cytometry (3). Thus, the accumulation of cells in G_0G_1 may result from the induction of differentiation. Although, mechanisms by which Onapristone exerts its effects have been suggested, it is not known which genes are affected by the drug treatment in BC cells. The aim of this study was to identify genes which are affected by Onapristone *in-vitro* using microarray technology.

Materials and Methods

Cell Culture. The routine culture of the T47D and MDA-MB-231 BC cell lines was carried out as follows: Cell lines were grown in Dubelecco's Modified Eagle Media (DMEM) containing 10% fetal bovine serum (FBS), and 10 μg/ml bovine insulin for the T47D cell line, or RPMI 1640 media containing 10% FBS for the MDA-MB-231. The media was supplemented for three days with Onapristone (a gift from Schering AG) and/or progesterone (Sigma Aldrich) as follows: 10^{-6}M Onapristone (O), 10^{-6}M progesterone (P6), 10^{-6}M progesterone + 10^{-6}M Onapristone (OP6), 10^{-9}M progesterone (P9), 10^{-9}M progesterone + 10^{-6}M Onapristone (OP9), and ethanol, the hormone vehicle as control (C).

RNA Extraction and Probe Generation. Total RNA was isolated from treated BC cells using the PUREscript RNA Isolation Kit (Flowgen). Double stranded cDNA was generated using 10 μg of total RNA with the SuperScript Choice System. A T7 tagged oligo-dT primer was used with this system to introduce a T7 RNA polymerase promoter site in the cDNA. This site was subsequently used to produce biotin-labelled cRNA by *in-vitro* transcription using the BioArray™HighYield™ RNA Transcript Labelling Kit (Ambion). Biotin-labelled CTP and UTP, in addition to unlabelled nucleotides, were used in this reaction. After the reaction, excess nucleotides were removed by RNeasy columns (Qiagen).

Array Hybridization and Scanning. To facilitate microarray analysis, a 50 μg aliquot of cRNA was fragmented for hybridization to a GeneChip® HuGeneFL Array (Affymetrix). Fragmentation was carried out for 35 min at 94°C in fragmentation buffer (40 mM Tris-acetate pH 8.1, 100 mM potassium acetate, and 30 mM magnesium acetate). Hybridization of 15 μg of fragmented cRNA was carried out in a 2X MES hybridization buffer (200 mM MES, 2M [Na^+], 40 mM EDTA, 0.02% Tween 20) which was heated to 94°C for 30 min. Subsequently, the cRNA was loaded onto the Affymetrix probe array cartridge and incubated at 45°C for 16 h with constant rotation (60 rpm). After the hybridization, the arrays were washed on a GeneChip® Fluidics Station. The arrays were washed ten times in 6X SSPE-T at 25°C, and then for 4 cycles in MES-T at 50°C. The biotinylated cRNA was labeled using streptavidin-phycoerythrin conjugate (final concentration 2 μg/ul) in Stain Buffer for 10 min at 25°C, and then washed ten times in 6X SSPE-T at 25°C. Antibody amplification of the reaction was carried out by incubation with biotinylated goat anti-streptavidin antibody (3 μg/ml), and goat IgG antibody (0.1 μg/ml) as a blocking reagent, for 10 min. The array was then re-stained with the streptavidin-phycoreythrin conjugate, and washed fifteen times in 6X SSPE-T. The array was scanned at 560 nm using a Hewlett Packard GeneArray™ Scanner. Analysis of the quantitative scanning data was carried out using the GeneChip® Analysis Suite, and an independently constructed analysis program. The programs were used to carry out the following pairwise comparisons; control vs 10^{-6}M Onapristone; 10^{-6}M progesterone vs 10^{-6}M progesterone + 10^{-6}M Onapristone; and 10^{-9}M progesterone vs 10^{-9}M progesterone + 10^{-6}M Onapristone. Bioinformatic analysis of gene expression profiles was carried out with the Expressionist TM software (GeneData, Basel)

Analysis of Differential Expression by TaqMan RT-PCR. To confirm the differential expression seen in the microarray analysis, TaqMan RT-PCR was performed. These reactions were carried out upon cDNA prepared from RNA isolated from the BC cell lines, T47D and MDA-MB-231, after the hormone treatments detailed above by the PUREscript RNA Isolation Kit. The RNA was reverse-transcribed using the Reverse-iT™ 1st strand synthesis kit (ABGene). TaqMan real-time PCR was carried out using qPCR™ Mastermix (Eurogentec). Amplicons of 100 bp were amplified and the reactions were mediated by the cleavage of TaqMan probe oligonucleotides containing FAM and TAMRA fluorophores. The reactions were carried out upon an ABI Prism 7700 Sequence Detector.

Results

Microarray Analysis of Onapristone Treated Breast Cancer Line, T47D. The effect of Onapristone upon gene expression in the BC cell line, T47D was monitored by microarray analysis. Correlation analysis of the data generated identified a set of 18 genes that were consistently up-regulated by Onapristone in each of the comparisons tested, while another set of 8 genes was consistently down-regulated. The results are summarized in Table 1.

TaqMan RT-PCR Analysis. The expression of 6/26 genes that were consistently differentially expressed in the three comparison experiments were examined by

TaqMan real-time PCR. Three of the up-regulated genes were investigated, namely intestinal trefoil factor, testosterone-repressed prostate message-2 (TRPM-2), and interferon-inducible protein 9-27. In addition, three of the down-regulated genes were investigated; prolactin-inducible protein, zinc-α2-glycoprotein, and CD9. The putative down regulated genes investigated had expression profiles similar to that found with the microarray data (Figure 1) when RNA isolated from the T47D cell line was investigated using TaqMan RT-PCR. This reinforces that Onapristone is capable of inducing differential expression of genes.

Table 1. Fold change in mRNA expression after treatment with Onapristone.

Gene	Fold Change in Expression[1]	Gene	Fold Change in Expression[1]
Histone H2B.1	1.9, 3.3, >3.6	MAR/SAR DNA Binding Protein	1.3, >1.6, >1.5
Histone H2A.1	1.7, 1.6, 2.1	Methionine Aminopeptidase	1.3, 1.3, 1.5
Histone H2B/g	1.6, 1.8, 1.6	OCRL1	1.3, 1.3, 1.5
Neurofibromin	1.6, 1.7, >2.6	TGF-β Superfamily Protein	1.3, 1.3, 1.4
TRPM-2	1.6, 3.5, 1.8	RNA Helicase A	>1.2, >1.4, 1.3
Butyrophilin	>1.5, >1.3, 1.4	Zinc-α2-Glycoprotein	<-1.7, <-5.7, -1.7
CL100	1.5, 1.3, 1.4	Prolactin-Inducible Protein	-1.6, -3.9, -3.9
GL105	1.5, 1.7, 2.0	Desmoplakin	-1.5, -2.0, -3.0
Histone H1.2	1.5, 2.4, 2.3	FACL1	-1.5, -1.5, -2.0
Intestinal Trefoil Factor	1.4, 3.7, 1.6	MAD-3	-1.4, -2.0, -1.5
Neuromedin U	1.4, 1.3, 1.4	CD9	-1.3, -1.8, -1.7
Interferon-inducible Protein 9-27	1.3, 1.4, 1.4	RAB4	-1.3, -1.4, -2.0
MAGE5b Antigen	1.3, >1.4, 1.3	Squalene Epoxidase	-1.3, -1.6, -1.5

[1] The figures given are in the order of fold change difference in expression of the genes in T47D cells treated with 10^{-6}M Onapristone alone against those treated with control vehicle, 10^{-6}M progesterone + 10^{-6}M Onapristone vs those treated with 10^{-6}M and 10^{-9}M progesterone + 10^{-6}M Onapristone vs those treated with progesterone, 10^{-9}M alone, respectively.

Effect of Progesterone Receptor Upon Differential Gene Expression. To examine whether the effects of Onapristone identified in the microarray experiments are regulated by the progesterone receptor, real-time PCR was carried out upon cDNA made from RNA extracted from the progesterone receptor negative breast cancer cell line MDA-MB-231. The same putative down regulated genes were examined as for the T47D cell line. Onapristone had no significant effect upon the expression of any of the genes within the MDA-MB-231 cell line (Figure 2). These data support the theory that Onapristone has its effects through the PR.

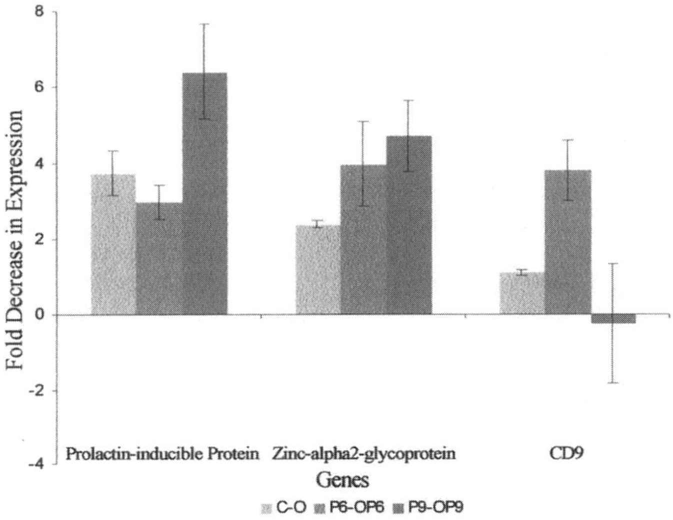

Figure 1. Average fold increases of putative down regulated genes in the BC cell line, T47D. The data presented are based on four independent reactions. The bars represent the mean ± 1 SD.

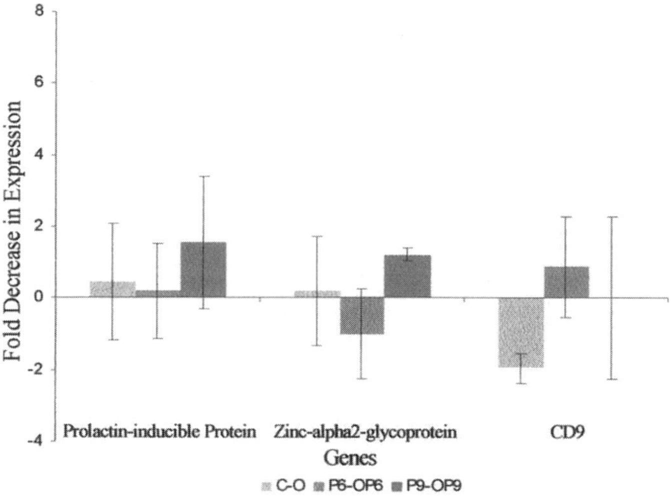

Figure 2. Average fold increases of putative down regulated genes in the BC cell line, MDA-MB-231. The data presented are based on four independent reactions. The bars represent the mean ± 1 SD.

Conclusions

The anti-progestin, Onapristone is capable of influencing the expression of genes in the BC cell line T47D, both in the presence and absence of progesterone. Knowledge of the genes regulated by Onapristone, should provide a better understanding of the molecular mechanisms by which anti-progestins inhibit tumor cell proliferation, mediate cell differentiation, and induce apoptosis.

References

1. Neef G, Beier S, Elger W, et al (1984) New steroids with antiprogestational and antiglucocorticoid activities. Steroids 44:349-372.
2. Henderson D (1987) Antiprogestational and antiglucocorticoid activities of some novel 11-aryl substitute steroids. In: Furr B, Wakeling A (eds), Pharmacological and Clinical Uses of Hormone Secretion and Action. London:Bailliere Tindal, 184-211.
3. Michna H, Schneider M, Nishino Y, et al (1989) Antitumor activity of the antiprogestins ZK 98.299 and Ru 38.486 in hormone dependent rat and mouse mammary tumors: mechanistic studies. Breast Cancer Res Treat 14:275-288.
4. Schneider M, Michna H, Nishino Y, et al (1989) Antitumour activity of the progesterone antagonist ZK 98.299 and RU 38.486 in the hormone-dependent MTX mammary tumour of the mouse and the DMBA- and the MNU-induced mammary tumours of the rat. Euro J Cancer Clin Oncol 25:691-701.
5. Gehring S, Michna H, Kuhnel W, et al (1991) Morphometrical and histochemical studies on the differentiation potential of progesterone antagonists in experimental mammary carcinomas. Acta Endo 124:177.
6. Robertson J, Willsher P, Winterbottom L, et al (1999) Onapristone, a progesterone receptor antagonist, as first-line therapy in primary breast cancer. Euro J Cancer 35:214-221.

Expression of Estrogen α and β Variants, Androgen, and Progesterone Receptors in Human Normal and Neoplastic Endometrium

Maciej Skrzypczak, Sylwia Zagulska-Szymczak, Rosette Lidereau, Ivan Bieche, Sebastian Lewandowski, Katarzyna Radwanska, Cezary Szczylik, Michael Vidaud, Jerzy A. Jakowicki, and Leszek Kaczmarek

Summary

The recent cloning of the estrogen receptor β (ERβ) has been followed by the discovery of a variety of its isoforms. Since endometrial cancer (ECA) is an estrogen-dependent cancer, an analysis of its expression of ERα and ERβ isoforms may contribute to our understanding of the mechanism of its development. Using the real-time RT-PCR assay, the quantitative expression of ERα, ERβ1, ERβ2 (βcx), ERβ3, ERβ4, and ERβ5, as well as androgen (AR) and progesterone receptors (PR) was determined and compared in normal and neoplastic endometrium. Our data demonstrates the co-expression of all the ERβ isoforms in the normal endometrium, and an up-regulation of the ERβ5 transcript in malignant endometrium. A decrease in the levels of AR mRNA between normal and neoplastic endometrium was also noted. With respect to clinical parameters, the decreased expression of PR mRNA correlates inversely with endometrium tumor grade. The expression profiles of ERα and AR were different between pre and postmenopausal endometrium, with a significant increase during menopause.

Introduction

Estrogens influence growth, differentiation, and the function of many target tissues in female and male reproductive organs. Their best-described mechanism of action is through ER-mediated gene transcription. This ligand-activated transcription factor is known to exist in two forms, α and β, and multiple splice variants of both are present in a variety of tissues (1-3). The presence of these ERs and their isoforms adds a high degree of complexity to their action, which may be cell/tissue specific, or modulating a specific estrogen response.

Estrogen has long been known to stimulate uterine cellular proliferation. In fact, hyperestrogenism is strictly connected with a high risk for ECA (4). ERα and ERβ have been identified in normal and neoplastic endometrium. Yet no conclusive data regarding alterations in their expression has been reported, as in breast and ovarian tumors, where the levels of ERβ decline in invasive breast and ovarian cancers (5-7). As our understanding of the mechanism of steroid receptors has increased, the analysis of isoforms, products of alternative splicing of ERα and β, has steadily gained acceptance as a marker for tumor prognosis and a guide for best choice of treatment. For example, ERα lacking exon 5 has been associated with breast cancers (7), as the ERβ isoforms, ERβ2 (βcx) and ERβ2 splice variant, devoided of exon 5 sequences. Leygue, *et al.*, Moore, *et al.* (3, 8) have co-localized another isoform, ERβ5, with other ERβ isoforms in human breast tumors. Moore, *et al.* (3, 8) have shown that ERβ correlates inversely with breast tumor grade, and that the tumor expression level of βcx and ERβ5 exceeds the expression of ERβ1. These data suggest that changes in the transcript levels of ER isoforms take place during tumorigenesis, and that they may be important both in the etiology and the assessment of cancer invasion.

Materials and Methods

Patients and Samples. Samples of ECA tissue were collected from women (ages 39 to 69) who underwent total abdominal hysterectomy. Normal endometrium was obtained from patients undergoing surgery for reasons other than the endometrium pathology. We examined 21 normal endometrium (6 post and 15 premenopausal) and 19 malignant endometrium samples [10 in stage greater than IA and 8 greater than IC, according to International Federation of and Gynecology and Obstetrics (FIGO) 1988 criteria].

Semiquantitative RT-PCR. Semiquantitative RT-PCR was performed on all the samples, prior to their Real-time PCR analysis. Equal amounts of RNA (1 μg) were reverse transcribed with Expand Reverse Transcriptase (Roche Diagnostics) and subjected to PCR, in accordance with the protocol provided with the enzyme. The following primers were used: for ERα (GenBank M12674): 5' AATTCTGACAATCGACGCCAG and 3' GCTTCAACATTCTCCCTCCTC; forward primer common to ERβ (GenBank AB006590), ERβ2 (GenBank AF051428), and ERβ2 splice variant lacking exon 5 (GenBank AF124790): ATGATGATGTCCCTGACCAAG; and reverse primers specific for: ERβ1 3'GCCCTCTTTGCTTTTACTGTC, and the ERβ2 and ERβ2 splice variant: 3'CTTTAGGCCACCGAGTTGGATT; primers for housekeeping gene G3PDH -5' CAATGACCCCTTCATTGAC and 3'GCAGTGATGGCATGGACTGAGGTC. (Figure 1). The cycling conditions were as follows: 94°C for 3 min, followed by 30 cycles of 94°C for 30s, 60°C for 50s, 72°C for 1 min. Digitized images of ERβ1, ERβ2, ERβ2 splice variant, ERα, and GAPDH products were imported into Scion Image program (NIH) and the optical density of each band was measured. The background density was then subtracted from each measurement, and all PCR products were normalized against G3PDH to obtain a ratio of target product to the G3PDH housekeeping gene for every sample.

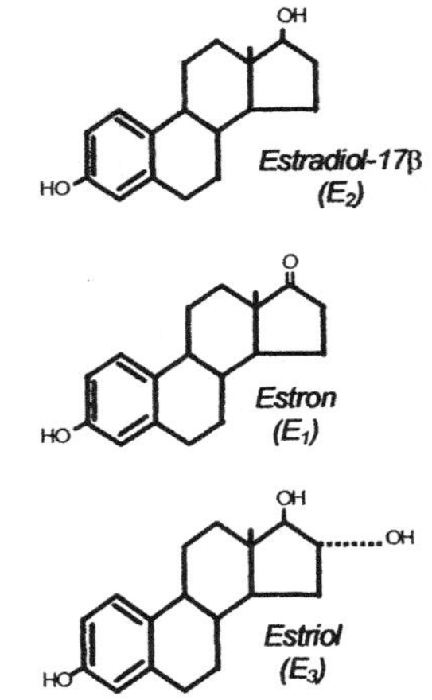

Figure 1. Schematic representation of estrogens.

Real-time PCR. Total RNA was isolated following the protocol supplied with TRIZOL reagent (GIBCO Life Technologies) and 1 μg was reverse transcribed in final volume of 20 μl RT-PCR Buffer (500 mM each dNTP, 3 mM MgCl, 75 mM KCl, 50 mM Tris-HCl pH 8.3), 10 units RNAsin RNAse inhibitor (Promega, Madison WI), 10 mM DTT, 50 units Superscript II RNAse- H Reverse Transcriptase (Life Technologies, Inc, Gaithesburg, MD), and 1.5 mM random hexamers. Incubations: 20°C 10 min, 42°C 30 min, 99°C 5 min, 5°C 5 min. All PCR amplifications were performed using ABI Prism 7700 Sequence Detection System (Perkin-Elmer Applied Biosystems), with SYBR® Green PCR Core Reagents kit (Perkin-Elmer Applied Biosystems). Thermal cycling conditions: 95°C 10 min and 50 cycles at 95°C for 15 s and 65°C 1 min. Quantitative values were obtained from the cycle number (Ct value), at which the increase in fluorescent signal associated with an exponential growth of PCR products starts to be detected by the laser detector of the ABI Prism 7700 Sequence Detection System, using the PE Biosystems analysis software according to the manufacturer's manuals. The higher the starting copy number of the target mRNA, the earlier a significant increase in fluorescence was observed.

Statistical Analysis. To determine the statistical significance of any differences observed between the normal and malignant samples P values were calculated using the χ^2 analysis, Kruskal-Wallis and Mann-Whitney U test.

Results

ERα and total ERβ expression was observed in all examined samples of normal and ECA. ERβ expression was further subdivided into its constituent isoforms. The expression of transcripts for ERβ1, ERβ2 (βcx), ERβ4, and ERβ5 was observed. From these isoforms, only ERβ5 differed significantly ($P = 0.010$) between normal and ECA samples showing an increase in the latter (Table 1). The levels of expression of ERβ1 mRNA and the "wild-type" ERβ, also increased, but were not significant. Although the expression of ERβ4 in the healthy as well as in ECA was observed, the level of expression was at the detection margins and its quantification unreliable. ERβ3 expression was at such low levels of target gene mRNA, that was neither detected nor reliably quantified.

Table 1. Expression Levels of ERα, ERβ Variants, PR, and AR in Human Normal and Neoplastic Endometrium.

	Normal Endometrium (n = 21)		ECA (n = 19)		P values Normal vs ECA
Genes	M ± SD[1]	Range Values	M ± SD	Range Values	$p^{2,3}$
ERα	1.0 ± 0.69	0.28 – 2.65	0.79 ± 0.47	0.02 – 1.89	NS[2] (0.59)[3]
ERβ	1.0 ± 0.61	0.11 – 2.49	0.90 ± 0.59	0.14 – 2.46	NS[2] (0.62)[3]
ERβ1	1.0 ± 0.74	0.08 – 3.16	1.56 ± 1.16	0.20 – 4.76	NS[2] (0.12)[3]
ERβ2	1.0 ± 0.58	0.14 – 2.08	0.85 ± 0.48	0.15 – 1.73	NS[2] (0.39)[3]
ERβ3	NE[4]	-	NE[4]	-	-
ERβ4	5 WE[5]	-	5 WE[5]	-	NS[2] (0.85)[6]
ERβ5	1.0 ± 1.07	0.11 – 4.21	1.95 ± 1.58	0.23 – 5.96	P = 0.010[3]
PR	1.0 ± 0.86	0.16 – 3.43	0.82 ± 0.81	0 – 3.34	NS[2] (0.46)[3]
AR	1.0 ± 0.57	0.29 – 2.10	0.56 ± 0.36	0.04 – 1.10	P = 0.036[3]

[1] Mean ± SD [2] NS, Not Significant [3] P value, Kruskal-Wallis test
[4] NE, Not expressed [5] WE, Without expression [6] P value, χ^2 analysis

The expression levels of all ER transcripts were additionally analysed in pre and postmenopausal normal endometria, and in two stages of increasing tumor malignancy indicated by the degree of infiltration of myometrium, where < ½ infiltration, was indicated as IC; and > ½ infiltration, as IB (Figure 2). The ERα mRNA expression was increased in normal postmenopausal endometrium as compared to healthy premenopausal samples (P = 0.01) (Figure 3). Besides estrogen, progestagens and androgens affect the biological function of the female reproductive tract. Thus, we also assessed the profile of PR and AR mRNA in normal and ECA tissue samples. We observed that the AR mRNA underwent significant (P, 0.036) down-regulation in the ECA samples. With respect to its profile in normal endometrium, the AR was significantly (P, 0.043) up-regulated in the postmenopausal patients. The PR mRNA, on the other hand, showed a decline (P, 0.041) in expression in more malignant ECA (1C), compared to ECA with less or no infiltration of the myometrium (1A and 1B). (Figure 2).

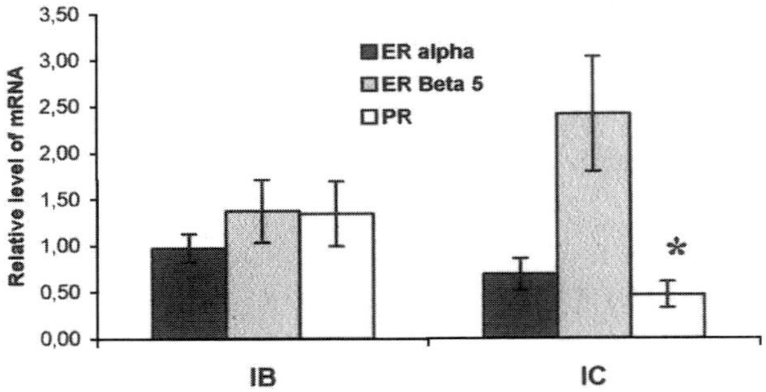

Figure 2. Expression of ERα, ERβ5, and PR in relation to FIGO. Myometrium infiltration > ½, was IC; and < ½ IB. PR down regulation was observed in IC grade (P, 0.0041).

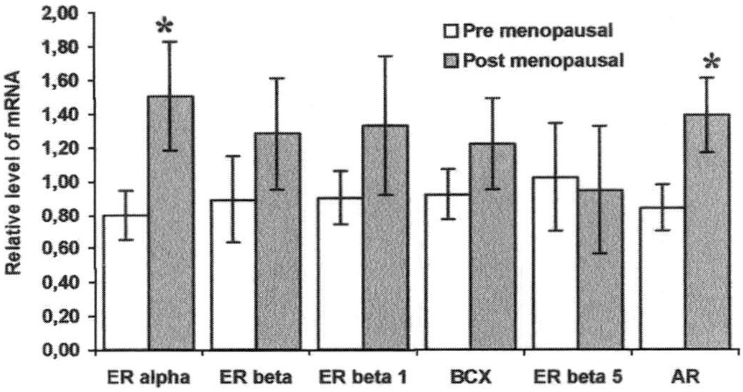

Figure 3. Expression of ERα, ERβ, ERβ1, Bβcx, ERβ5, and AR in pre and postmenopausal endometrium. Significantly increasing of ERα (P, 0.001) and AR (P, 0.0043) in postmenopausal endometrium.

Discussion

The endometrium is the least-studied tissue in terms of changes in the cellular status of ERs, especially the β subtype. Matsuzaki, *et al.* (9) who analyzed ER α and β mRNA expression in the uterus throughout the human menstrual cycle, concluded that both ERs show cycle-related changes in their mRNA expression. Although ERα is the prevalent ER in the endometrium, a disruption of ERβ function results in physical abnormalities of the uterus and poor reproductive capacity, as demonstrated by Weihua, *et al.* (10) in ERβ knockout mice. Comparing the profile of the ERs and the AR in normal pre and postmenopausal endometrium, we observed significant changes (P, 0.016 and P, 0.043, respectively) in their transcript levels, most probably in response to the changing steroid environment.

We demonstrate a co-expression of not only the principal subtypes of ERs, but also of the isoforms of the less studied ERβ varients. Furthermore, we demostrated that these isoforms are differentially expressed in ECA. Changes in levels of the various ERβ mRNA isoforms, as well as their protein, have been documented in breast (3) and ovarian cancer (5, 6). The authors related such changes to the progression of the neoplastic condition and furthermore, suggest that ERβ status is a significant predictor of survival in patients with breast cancer. *In-vitro* studies suggest that the protein products of these alternatively spliced mRNAs, often lacking key functional domains, may possess dominant negative activities, and thus may alter tissue response to estrogen or antiestrogens. This seems to be the case of ERβ2 (βcx), a ligand-independent supressor of ERα and ERβ1-mediated transciption.

Both 17β-estradiol and tamoxifen, a leading anti-estrogen in breast cancer treatment, increase uterine cell proliferation and incidence of endometrial tumors. Although the exact mechanism has not been established, the profile (the types and cellular levels) of ERs might play a significant role. ERα is a partial agonist while ERβ1 responds with a pure antagonistic effect to the administration of tamoxifen. Weihua, *et al.* (10) have shown that when ERβ is disabled, as in ERβ knockout mice, there is an unusual uterine cell proliferation. Therefore, the natural function of ERβ appears to be anti-proliferative, when this effect is block by tamoxifen, it may contribute to excessive endometrial cellular proliferation. Thus, the down-regulation of any ERβ isoform, that is a ligand-independent antagonist of the proliferative actions of ERα, may contribute to uterine cell proliferation.

We observe a decline in ERβcx and ERβ2 splice variant through both the traditional and real-time RT-PCR methods. Both ER subtypes form DNA-binding homo-and heterodimers in tissues like the endometrium, which co-express all ER subtypes, the estrogen signaling may occur through heterodimers formed by ERα and β, and perhaps also through their co-present isoforms. The α and β ERs have different gene transcription properties, as do their respective dimers and heterodimers, under conditions of varying levels of estrogens and anti-estrogens. Hence, it appears that not only the co-expression of the ERs and their isoforms, but also variations in their cellular levels, may influence a tissue response to both agonists and antagonists. ERβ2 splice variant, shown by our traditional RT-PCR to be present in the endometrium and down-regulated in its neoplasms, is devoid of exon 5 sequences. (ligand binding domain) yet retains other essential receptor regions like a dimerization function and a ligand-independent transactivation function (AF-1). Endometrial neoplasms are estrogen-dependent, and changes in the mRNA levels of their various ERs reflect the tissue's changing response to this hormone. Since several reports indicated that ECA is influenced by androgens mediated through their receptors, the observed down-regulation of the AR in ECA may be instrumental in greater cell proliferation, favoring a malignant profile.

Our hypothesis is that a similar pattern of differential ER or PR subtype expression may distinguish ECA from normal endometrium. Furthermore, such privileged expression may explain why adenocarcinomas tend to be susceptible to hormonal treatment, and assume significant anti-proliferative effect on endometrial tissue, and changes in expression of PR can lead to discoveries of new drugs and explain the use of gestagens in therapy of advanced ECA.

Cancers of ovary, colon, and endometrium are all characterized by a relative loss of normally predominant ER subtype. In the ovary and colon, ERβ is the predominant ER form, while in the endometrium, ERα is the main subtype. The functional significance of these changes remains to be elucidated. However, such a role has yet to be described for either ER isoform.

References

1. Kuiper GGJM, Enmark E, Pelto-Huikko M, et al (1996) Cloning of a novel estrogen receptor expressed in rat prostate and ovary. Proc Natl Acad Sci USA 93:2920-2930.
2. Mosselman S, Polman J, Dijkema R (1996) ERβ: Identification and characterization of a novel human estrogen receptor. FEBS Lett 392:49-53.
3. Leygue E, Dotzlaw H, Watson PH et al (1999) Expression of estrogen receptor β1, β2, and β5 messenger RNAs in human breast tissue. Cancer Res 59:1175-1179.
4. McDonald TW, Malkasian GD, Gaffey TA (1977) Endometrial cancer associated with feminizing ovarian tumor and polycystic ovarian disease. Obstet Gynecol 6:654-658.
5. Brandenberger AW, Tee MK, Jaffe RB (1998) Estrogen receptor alpha (ER-α) and beta (ER-β) mRNAs in normal ovary, ovarian serous cystadenocarcinoma and ovarian cancer cell lines: Down-regulation of ER-β in neoplastic tissues. J Clin Endocrinol Metab 83:1025-1028.
6. Rutherford T, Brown WD, Sapi E, et al (2000) Absence of estrogen receptor-β expression in metastatic ovarian cancer. Obstet Gynecol 96:417-421.
7. Vladusic EA, Hornby AE, Guerra-Vladusic FK, et al (1998) Expression of estrogen receptor β messenger RNA variant in breast cancer. Cancer Res 58:210-214.
8. Moore JT, McKee DD, Slentz-Kesler K, et al (1998) Cloning and characterization of human estrogen receptor β isoforms. Biochem Biophys Res Commun 247:75-78.
9. Matsuzaki S, Fukaya T, Suzuki T, et al (1999) Oestrogen receptor alpha and beta mRNA expression in human endometrium throughout the menstrual cycle. Mol Hum Reprod 5:559-564.
10. Weihua Z, Saji S, Makinen S, et al (2000) Estrogen receptor (ER) β, a modulator of ERα in the uterus. Proc Natl Acad Sci USA 97:5936-5941.

Toremifene (ACAPODENE^{TM}) Reduces High Grade Prostatic Intraepithelial Neoplasis in a Phase IIa Clinical Trial

Mitchell S. Steiner

Summary

Men with high grade prostatic intraepithelial neoplasia (HG PIN) on prostate biopsy are at high risk for prostate cancer (PCA). The ability to reverse HG PIN, a precursor of PCA, may reduce the incidence or delay the development of PCA. ACAPODENE ^{TM} (toremifene) is a selective estrogen receptor modulator (SERM) that has been shown in preclinical models to both eliminate HG PIN and reduce the incidence of PCA. Toremifene was evaluated for safety and efficacy in men diagnosed with HG PIN. An open labeled, Phase IIa clinical trial enrolled 21 men (mean age 64.7 years) who had HG PIN only on biopsy within 6 mo of entry into the study. From those men, 18 (86%) completed toremifene treatment (60 mg/day PO for 4 mo) followed by a prostate biopsy to determine HG PIN status. Serum prostate specific antigen (PSA), % free PSA, testosterone (T), estradiol (E_2), and quality of life were measured. Following toremifene treatment, 72% of the HG PIN men compared to 17.9% of historical controls had no HG PIN on subsequent prostate biopsies. Mean PSA was trended higher and % free PSA was elevated. Quality of life was not significantly affected, and there were no serious adverse events. Toremifene appears to reduce HG PIN, and was well tolerated in this small, exploratory trial. A double blind, dose finding, randomized, and placebo controlled Phase IIb/III study is currently in progress in 516 men with HG PIN to further study toremifene's activity against PCA incidence.

Introduction

HG PIN is a well established pre-malignant lesion of the prostate because of its potential to progress to PCA (1, 2). PIN is the abnormal proliferation within the prostatic ducts of pre-malignant foci of cellular dysplasia and carcinoma *in-situ* without stromal invasion (1). In the USA, approximately 1,300,000 prostate biopsies are performed annually to detect 189,000 new cases of PCA (3). The incidence of HG PIN averages 9% (range 4-16%) of prostate biopsies representing 115,000 new cases diagnosed each year (4). There is estimated to be 1,000,000 USA men with histologic proven HG PIN. The support for HG PIN as a PCA pre-malignant lesion is based on several lines of evidence derived from PCA animal models, epidemiological, morphologic, genetic, and molecular studies (4).

HG PIN is in the direct causal pathway to PCA, and its presence increases PCA risk (5). When HG PIN is found on needle biopsy, there is a 39.5%, 57.1%, and 80% risk of finding PCA on subsequent biopsies after 1 (6), 2 (7-9), and 10 years (10), respectively. The diagnosis of HG PIN changes the patient's quality of life, as there is great patient and physician anxiety about the concern that PCA may be imminent. Although there is no medical consensus for the standard of care of HG PIN, urologists recognize that HG PIN is a dangerous lesion and it should be aggressively managed. The patient must be subjected to more frequent biopsies and physician visits, as currently, the only way to diagnose whether HG PIN has progressed to PCA is by prostate biopsy. Although some urologists have advocated saturation biopsies (over 12 biopsy cores obtained) of the prostate following the diagnosis of HG PIN, the more common recommendation is repeated prostate biopsies every 3-6 mo for 2 years, then annually (8, 11).

Evidence for estrogen's role in PCA has emerged not only from animal studies, but also from epidemiological studies on diet in man (12). Phytoestrogens like soy are non-steroidal substances that have weak estrogenic activity. Essentially, they act like ER partial antagonists. Soy is consumed daily in large amounts in China and Japan, regions where the PCA incidence is low (13). A direct inverse correlation has been observed between serum levels of isoflavonoids from soy and PCA incidence (14). In addition to its anti-estrogenic effects, isoflavonoids have many other mechanisms that may account for this decrease in PCA (14). Generally, SERMs are considered, like phytoestrogens, to be "weak estrogens" because they possess both agonist and antagonist activities depending on the specific tissue type. SERMs possess the ability to suppress prostate carcinogenesis. Animal toxicity studies of toremifene and raloxifene (LY156758) have revealed a dose dependent regression of the androgen dependent ventral prostate and seminal vesicles in SD rats (15). Wang, *et al.* (16) treated wild type mice with T propionate and E_2 for 4 mo. These mice developed prostatic hyperplasia, HG PIN, and invasive PCA. Under the same treatment, α-ERKO mice (mice with ERα genetically knocked out) develop prostatic hyperplasia, but not HG PIN or invasive PCA (16). Similarly, a prospective, placebo controlled study of transgenic adenocarcinoma mouse prostate (TRAMP), a PCA mouse model, has been shown to mimic human PCA. In toremifene-treated TRAMP mice a reduction in HG PIN, significant decrease in PCA incidence, and an increase in animal survival were observed (17). Thus, estrogenic signally through ERα may play a key role in the development of HG PIN and PCA.

The purpose of this study was to investigate whether agents that block ERα may potentially prevent HG PIN and decrease PCA risk. Toremifene is a SERM that antagonizes ERα in the prostate, and is safer than tamoxifen because it does not form DNA adducts (18). An exploratory Phase IIa trial was conducted in men who have HG PIN to determine whether toremifene can reduce HG PIN.

Materials and Methods

Study Design. This study was an open-label, historically controlled, single center phase IIa study in which men diagnosed with HG PIN were treated with 60 mg/day of toremifene for 120 days. The study primary objectives were to determine whether toremifene was able to reduce HG PIN in men with HG PIN, to evaluate its effect on other intermediate endpoints including serum total PSA and % free PSA, as well as its safety and impact on male hormonal status. The effects of toremifene on quality of life issues, *i.e.,* changes in libido, erectile function, and hot flashes were also assessed. Once informed consent was obtained, subjects were referred to the study if any prostate biopsy within the last 6 mo had HG PIN (at least 6 prostate cores were needed to be included in the study). The prostate pathology was re-evaluated to confirm HG PIN. Subjects who had HG PIN and fulfilled all of the eligibility requirements were enrolled in the study. At study day 120, subjects underwent transrectal ultrasound guided biopsies (8 prostate cores).

Exclusion/Inclusion Criteria. Subjects were eligible for the study if they met the following criteria: Willing to participate in the study and sign an informed consent, be a ≥ 30 years old male, have a histologically confirmed Grade II and/or III PIN on biopsy; serum PSA ≤ 20 ng/ml, Zubrod performance status ≤ 1, adequate bone marrow, liver, and renal function. Subjects were excluded from the study if they had any of the following: Underwent prior experimental therapy for chemoprevention, evidence of PCA on initial evaluation (local, regional, and/or distant metastasis), active systemic viral, bacterial, or fungal infections requiring treatment, a serious concurrent illness or psychological, familial, sociological, geographical, or other concomitant conditions which did not permit adequate follow-up and compliance with the study protocol, concurrent treatment with other investigational agents, were taking finasteride or T, herbal medicine or dietary supplements for prostate health, and had a history of thromboembolic disease.

Efficacy and Safety Measures. Prostate biopsy and a quality of life questionnaire were evaluated at baseline and at day 120. Serum PSA was evaluated at baseline, days 60 and 120. Final prostate biopsies were obtained under transrectal ultrasound guidance and standardized with 8 total biopsy cores taken from right and left apex, right and left midbase, right and left base, and right and left lateral base. All samples were reviewed by a single pathologist. Investigators questioned subjects on any signs/symptoms experienced since the previous visit to determine whether the subject experienced any adverse event (AE). AEs were assessed using the revised NCI Common Toxicity Criteria.

Statistics. The primary efficacy variable was the determination of the response of HG PIN by prostate biopsy. T-tests were performed for serum PSA, hormones, hematology, biochemistry, and quality of life to determine whether changes from baseline were different from zero.

Results

Patient Characteristics. Of the 21 subjects enrolled, 57% were white and 43% black. Their age ranged from 49-76 years (average, 64.7). All received toremifene, 60 mg/day. Three subjects discontinued the study early.

HG PIN Status. 86% (18/21) of the subjects completed 120 days of treatment, while 90% (19/21) completed at least 60 days. The majority of subjects (72%) had a complete response (complete disappearance) at day 120, while 22% had a partial response (≥ 25% reduction). HG PIN remained stable at day 120 in only one subject (6%). In the patients that had a limited response or stable disease, there was not a correlation among number of prostate cores or location of HG PIN, and the subsequent amount or location of HG PIN in follow-up biopsies.

Table 1. Summary of PSA Mean Change from Baseline to Days 60 and 120.

Variable	**Toremifene (n = 21, all enrolled)**		
PSA[1] (ng/mL)	**Baseline**	**Change at Day 60**	**Change at Day 120**
n	19	19	18
Mean (SD[2])	8.8 (5.4)	+ 2.9 (10.0)	+ 3.7 (8.1)
% Free PSA[1]			
n	17	17	15
Mean (SD[2])	18.9 (9.7)	+ 2.3 (5.8)	+ 5.5 (9.6)[3]

[1] PSA, prostate-specific antigen [2] SD, standard deviation
[3] Statistically significant from baseline ($p < 0.050$).

PSA Response. A summary of mean change from baseline to days 60 and 120 for PSA and % free PSA values is presented in Table 1. There was an increase in mean PSA at days 60 and 120 compared to baseline, but was not statistically significant. Five subjects had baseline PSA values >10. Mean % free PSA increased at days 60 and 120 compared to baseline; at day 120 it was statistical significant ($p = 0.045$).

Quality of Life. A decrease in score indicates improvement. At baseline, the mean quality of life score was 54.7 and increased by + 1.7 at day 120, but was not statistically significant. There were no changes for libido, erectile function, and hot flashes.

Safety. The incidence of AEs was low (only 2 subjects experienced 3 AEs). One subject experienced 2 AEs (anemia and skin rash), both of which were mild in severity. The anemia was considered possibly related to toremifene and the rash was unrelated. Both were ongoing when the subject completed the trial. The second subject experienced hematospermia that resolved approximately 5 mo after onset. None of the 3 AEs reported in the study required any alteration in toremifene administration. There were no sudden adverse effects or deaths during the study.

A summary of mean change from baseline to days 60 and 120 for hormonal values is presented in Table 2. Mean total T was increased at day 60, and significantly increased at day 120 ($p = 0.002$) compared to baseline. Free T and E_2 were significantly increased ($p < 0.050$) at days 60 and 120 compared to baseline.

Table 2. Summary of Serum Hormone Levels Mean Change from Baseline to Days 60 and 120

Variable	**Toremifene (n = 21, all enrolled)**		
Serum Hormones	**Baseline**	**Change at Day 60**	**Change at Day 120**
Total T (ng/mL), n	18	15	16
Mean (SD[1])	3.9 (1.7)	+ 2.6 (4.8)	+ 1.7 (1.8)[2]
Free T (ng/mL), n	17	14	16
Mean (SD[1])	10.1 (3.2)	+ 3.8 (3.2)[2]	+ 5.2 (4.9)[2]
E_2 (pg/mL), n	14	13	13
Mean (SD[1])	25.4 (19.9)	+ 14.3 (16.6)[2]	+ 18.2 (16.4)[2]

[1] SD, standard deviation [2] Statistically significant from baseline ($p < 0.050$).

Discussion

Treatment of pre-cancerous lesions would appear to be of clinical benefit notwithstanding the potential for cancer prevention. These clinical benefits would reduce morbidity, enhance quality of life, delay surgery or radiation, and increase the interval for surveillance requiring invasive procedures (5). Examples of treated pre-malignant lesions include: cervix- cervical intraepithelial neoplasia, breast ductal carcinoma *in-situ,* colon-adenomatous polyps, bladder-carcinoma *in- situ,* skin- actinic keratosis, esophagus-Barrett's esophagus, and oral mucosa-dysplastic oral leukoplakia (5). Like HG PIN, these all represent types of intraepithelial neoplastic lesions (5).

HG PIN is also a pre-malignant lesion that merits treatment. The clinical implications of HG PIN are very different than that for an elevated serum PSA. Serum PSA is prostate, not PCA specific. Other conditions besides PCA more commonly raise serum PSA levels including BPH, ejaculation, prostatitis, and prostatic infarct. In fact, over 70% of men with an elevated PSA over 4 ng/ml will ultimately be found to have no evidence of PCA. Interestingly, HG PIN does not contribute to serum PSA or serum free PSA, which is not surprising since HG PIN, unlike PCA, has not yet invaded the vasculature of the prostate to leak secreted PSA into the blood stream (19). Thus, HG PIN appears to precede even PCA related serum PSA elevations.

The ideal chemopreventive agent must have minimal or no side affects or toxicity to be accepted by otherwise healthy men who are at risk for PCA. One class of compounds called SERMs has this profile. In this exploratory Phase IIa open label single center study conducted in men with HG PIN, toremifene appeared to be well tolerated. The primary purpose of this single center, pilot study was to assess the efficacy of toremifene for the treatment of HG PIN. Twenty-one men with HG PIN were enrolled in the trial and assigned to receive toremifene at a dose of 60 mg/day for 120 days. In general, subjects had comparable demographic characteristics and no significant medical histories. Toremifene (60 mg/day) eliminated HG PIN in the final prostate biopsies in 13 of 18 men (72%). Four of 18 men (22%) had partial response (≥ 25% reduction in HG PIN). A single subject (6%) did not have a reduction in HG PIN in response to toremifene therapy.

The real question is what does this change in HG PIN mean? First, what is the limitation of prostate biopsy sampling and the prevalence of HG PIN compared to PCA? HG PIN and PCA are morphometrically and phenotypically similar. HG PIN occurs primarily in the peripheral zone and is seen in areas that transition to PCA (20). HG PIN and PCA are multifocal and heterogeneous (10). In fact, HGPIN is more prevalent as the precursor lesion than PCA. As a "field defect," random biopsies are performed to sample the "field." This is no different than prostate cancer where a focal lesion in a biopsy core does not reliably predict the amount of the disease in the prostate. The detection rate of 8 random biopsies, therefore, really becomes the critical issue. Prostate biopsy sampling error has been shown to be average of 17.9 % for men who have HG PIN based on 6 studies (Table 3). Up to 80% of men who have HG PIN had HG PIN or PCA when subjected to 6 core sample repeat biopsy which is similar to prostate cancer detection rates using 6 core sample biopsies (8, 11). In PCA series, if 8 core samples are obtained the detection rate for PCA increases to 95% (21).

In our study, the finding that 72% of patients treated with toremifene had no HG PIN in follow-up prostate biopsies compared to 17.9% expected based on historical data suggests that toremifene may be reducing HG PIN and that the reduction of HG PIN cannot be accounted by detection error alone (6-9, 19,23).

The results of this pilot study support the preliminary efficacy, safety, and tolerability of toremifene for treatment of HG PIN. These results provide rationale for future studies to assess the effectiveness of a range of toremifene doses for treatment of HG PIN in a larger number of subjects, and to examine closely the effects of toremifene. The clinical significance of the reduction of HG PIN should be correlated with a reduction of PCA incidence. This current trial provided the proof-of-concept support behind a currently fully enrolled 515 patient placebo controlled, randomized dose finding Phase IIb/III clinical trial that is currently being conducted in 65 sites in the USA. This trial is investigating the efficacy of toremifene to reduce PCA incidence in men who have HG PIN. Men with HG PIN are treated for 12 mo with placebo or various doses of toremifene and undergo a repeat prostate biopsy at 6 and 12 mo (additional information may be found at GTxinc.com). The placebo arm will prospectively assess the true natural history of HG PIN.

Acknowledgments

The following physicians participated in this clinical trial: Charles Pound, Lynn A. Patterson, Robert Wake, Jeffrey Gingrich, Lynn Conrad, Michael McSwain, Lawrence Seymour, and Thomas Shelton.

References

1. Bostwick DG (1989) Prostatic intraepithelial neoplasia (PIN). Urology 34:16-22.
2. Stanford JL (1999) Epidemiology of Prostate Cancer. Prostate Cancer, WHO 2nd International Consultation on Prostate Cancer 23-55.
3. Jemal A, Thomas A, Murray T, et al (2002) Cancer statistics. CA Cancer J Clin 52:23-47.
4. Steiner MS (2001) High grade prostatic intraepithelial neoplasia is a disease. Curr Urol Rep 2:195-198.
5. O'Shaughnessey JA, Kelloff GJ, Gordon GB, et al (2002) Treatment and prevention of intraepithelial nepolasia: An important target for accelerated new agent development. Clin Cancer Res 8:314-346.
6. Davidson D, Bostwick DG, Qian J, et al (1995) Prostatic intraepithelial neoplasia is a risk factor for adenocarcinoma: predictive accuracy in needle biopsies. J Urol 154:1295-1299.
7. Weinstein MH, Epstein JI (1993) Significance of high-grade prostatic intraepithelial neoplasia on needle biopsy. Hum Path 24:624.
8. Pasqualotto F, Zippe C, Sekhon V, et al (2000) Repeat sextant TRUS-guided needle biopsy for prostatic intraepithelial neoplasia (PIN) does not require transition zone sampling. J Urol 163:279.
9. Absoseif A, Shinohara K, Weidner N, et al (1995) The significance of prostatic intra-epithelial neoplasia. Br J Urol 76:355-359.
10. Bostwick DG (2000) Prostatic intraepithelial neoplasia. Current Urology Reports 1:65-70.
11. Leibovich B, Kuntzman R, Wilson T, et al (2000) Sextant biopsy of the prostate is inadequate for cancer detection in men with high grade prostatic intraepithelial neoplasia. J Urol 163:279.
12. Castle EP, Thrasher JB (2002) The role of soy phytoestrogens in prostate cancer. Urol Clin North Am 29:71-81.
13. Price KR, Fenwick GR (1985) Naturally occurring oestrogens in foods--a review. Food Addit Contam 2:73-106.

14. Denis L, Morton MS, Griffiths K (1999) Diet and its preventive role in prostatic disease. Eur Urol 35:377-387.
15. Neubauer BL, Best KL, Clemens JA, et al (1993) Endocrine and antiprostatic effects of Raloxifene (LY 156758) in the male rat. Prostate 23:246-262.
16. Wang YZ, Hayward SW, Cao M, et al (2001) Role of estrogen signaling in prostatic hormonal carcinogenesis. J Urol 165:132-133.
17. Raghow S, Hooshdaran MZ, Katiyar S, et al (2002) Toremifene prevents prostate cancer in the transgenic adenocarcinoma of mouse prostate model. Cancer Res 62:1370-1376.
18. Montandon F, Williams GM (1994) Comparison of DNA reactivity of polyphenylethylene hormonal agents diethylstilbestrol, tamoxifen and toremifene in rat and hamster liver. Arch Toxicol 68:272-275.
19. Morote J, Encabo G, Encabo G, et al (1999) Influence of high-grade prostatic intra-epithelial neoplasia on total and percentage free serum prostatic specific antigen. BJU Int 84:657-660.
20. Sakr WA, Grignon D, Haas G, et al (1995) Epidemology of high grade prostatic intraepithelial neoplasia. Path Res Pract 191:838-841.
21. Presti JC, Jr., Chang JJ, Bhargava V, et al (2000) The optimal systematics prostate biopsy scheme should include 8 rather than 6 biopsies: results of a prospective clinical trial. J Urol 163:163-167.
22. Galan MC (2000) High grade intraepithelial neoplasia and prostate cancer. Arch Esp Urol 53:227-229.

Stereological Study of Mean Nuclear Volume Weighted by Volume in Normal Prostate, Prostatic Intraepithelial Neoplasia, and Adenocarcinoma

Fernando Teba, Rocio Martín, Vicente Gómez, and Luis Santamaría

Summary

Prostatic interepithelial neoplasia (PIN) is considered a precursor of prostate cancer (PCA). In the majority of the PINs, the atypia is more marked in basal than luminal nuclei. The aim of this study was to quantitate the differences in the stereological estimation of mean nuclear volume weighted by volume (ν_V nuc) between basal and luminal cells of PIN, in relation to normal prostate epithelium, and PCA. Ten PIN cases were compared with 50 PCA, and 5 normal prostates. All the specimens were routinely processed and stained with haematoxylin-eosin. The mean ± SD of ν_V nuc was evaluated for all the groups. The epithelium of both normal and PIN specimens was segmented in basal and luminal compartments, and the ν_V nuc measured in both strata. Comparisons among groups were performed by ANOVA ($p < 0.05$). The value of ν_V nuc was significantly lower in normal epithelium compared to both PIN and PCA. Basal layer PIN ν_V nuc was significantly higher in comparison to the luminal stratum. Luminal PIN ν_V nuc was similar to ν_V nuc of PCA. The similarities in nuclear size between PIN and PCA are according to the pre-malignant character commonly assigned to PIN. The increase of basal layer PIN ν_V nuc may indicate that the changes heralding the progression from PIN to PCA are produced in this layer, whereas the nuclear features of luminal layer are the same to those of the carcinoma.

Introduction

The relationship between PIN and PCA of the peripheral prostatic zone in men has been well documented (1, 2). A number of studies concluded that high-grade PIN (HG PIN) represents a pre-malignant lesion that could evolve into invasive PCA

(2). Molecular and immunohistochemical markers have been used in the characterization of human PIN progression. These markers include the increase expression of anti-apoptotic proteins (bcl-2, bcl-X, and bax) (3-5), changes in several genes that encode cell proliferation regulators (p53, bcl-2, p16, p22, and c-erbB2) (5, 6), an increase in cell proliferation, and genetic instability (7, 8). Nevertheless, few studies have evaluated the nuclear size and its variability in the PIN pre-neoplastic epithelium. Some authors have suggested that the cytologic atypia and nuclear pleomorphism are more remarkable in the basal layer in comparison to the apical or columnar stratum of PIN lesions (9). The aim of this study was to quantitate the differences in the stereological estimation of mean nuclear volume weighted by volume (ν_V nuc) between basal and luminal (columnar) PIN cells, in comparison to normal prostate epithelium, and PCA.

Materials and Methods

Samples. Ten cases of HG PIN, and 50 specimens of PCA were selected either from prostate biopsies or surgical pieces (namely radical prostatectomies). Five prostate specimens obtained from autopsies of young men without prostatic disease were used as controls. The tissues were fixed immediately after surgery in 10% paraformaldehide for 24 h. Afterwards, the samples were paraffin embedded, and serially sectioned at 5 μm. The sections were stained with haematoxylin-eosin (H&E).

Evaluation of ν_V nuc. All the pieces were used for stereologic evaluation of the ν_V nuc. The estimation was carried out on three systematically random sampled H&E stained sections per specimen, using the stereologic software GRID (Interactivision, Silkeborg, Denmark) (10). This software allows the selection of fields to study per section by random systematic sampling after the input of an appropriate sampling fraction. Since the number of nuclei per specimen needed to obtain reliable results was within 70-100 nuclei (11), in the present study, an average of 100 nuclei were point sampled per case. The sampling protocol for the normal prostate and PIN was designed to estimate separately the ν_V nuc of the basal and columnar layers. The epithelial lining of both normal and PIN glands was segmented in two strata: a basal compartment, 7 μm wide from basal membrane; and a columnar layer from the limit of the basal compartment to the glandular lumen. In both strata, ν_V nuc was independently estimated. All the measurements were carried out using an Olympus microscope equipped with a 100 X oil immersion lens (numerical aperture of 1.4) at a final magnification of 1200 X. The software used to evaluate the ν_V nuc enables the generation of random test-line directions that were superimposed onto the microscope images (12). The nuclear intercepts were measured along these test-lines. The length of nuclear intercepts (l_0) was processed to obtain $\pi\ l_0^3/3$, an unbiased estimate of ν_V nuc independent of nuclear shape, which, because of point sampling, emphasizes larger nuclei rather than smaller ones. In addition, estimates of v_V nuc combine information regarding the three-dimensional nuclear size with knowledge of variability of nuclear size (13). The measurements obtained in PCA and PIN cases were compared to those obtained in controls. In control and PIN epithelium, the mean nuclear size measured in basal layer was also compared with the measurements performed in columnar stratum.

Statistical Analysis. All comparisons made were carried out by ANOVA using Fisher and Behren`s test. The contribution of each of the three sampling levels (*i.e.,* nuclear intercepts and their measurements, fields of vision, and individual cases) to the totally observed variance associated with estimates was investigated by nested analysis of variance (14).

Results

The prostatic acini affected by PIN, showed a remarkable enlargement of the epithelial layer with pseudo-estratification and crowding of the nuclei in comparison to normal acini from control samples (Figures 1, 2). The nuclei from PIN lesions were larger on size than those observed in normal epithelium (Figure 1) or in PCA (Figure 3), showing cytological atypias and frequent nucleoli (Figure 2). The nuclear size in PIN lesions decreases towards the glandular lumen (Figure 2). No mitotic figures were visualized in any of the groups studied.

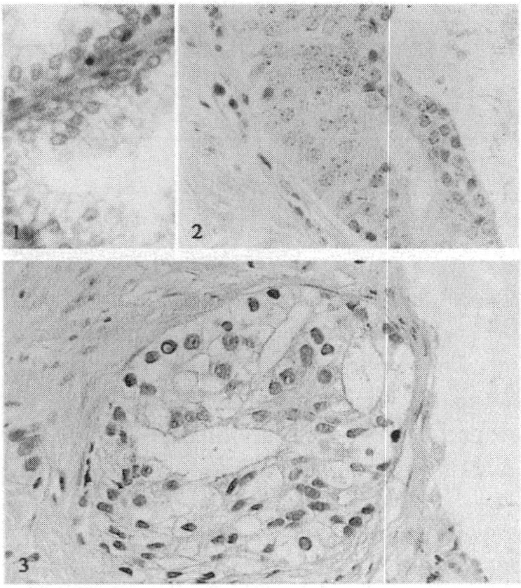

Figures 1-3. Morphological characteristics. **1. Normal acini** from a control prostate: The epithelium is columnar, with a well defined basal layer, the apical cytoplasm is abundant and clear. **2. PIN lesion**: The epithelium shows pseudo-estratification with nuclear crowding. Some isolated basal cells are seen. The nuclei are increased in size, showing atypias and frequent nucleoli. A negative gradient of nuclear size was observed towards lumen. **3. PCA**: Tumoral acinus with cribriform pattern. The nuclei were irregular, and heterogeneous in size. Some enlarged nucleoli are seen. H&E. X 400.

The PIN v_V nuc of nuclei from basal compartment was significantly higher than that of all the other groups examined including PCA. However, PCA v_V nuc was significantly higher than in the controls, and without differences when comparing to the columnar compartment of PIN. Moreover, both estimates from PIN columnar and PCA were significantly higher than those observed in controls basal and columnar. There were not significant differences between v_V nuc of controls (basal and columnar). A summary of the results is depicted on Table 1.

Table 1. Nested Analysis of Variance for v_V nuc.

Cases[1]	Sampling Level[2]	Observed Variance	Variance of the Mean	Estimated Variance of the Mean	Contribution to Total Variance (%)
Control Basal ($112^3 \pm 44$)	Measurements (11)	12144	1101	1101	56
	Fields (8)	8692	1086	-15	0
	Cases (5)	1958	1958	872	44
Control Columnar ($167^3 \pm 40$)	Measurements (36)	13118	364	364	23
	Fields (16)	11989	749	385	24
	Cases (5)	1604	1604	855	53
PIN Basal ($450^3 \pm 82$)	Measurements (27)	49516	1834	1834	29
	Fields (14)	25114	1794	-40	0
	Cases (10)	6395	6395	4601	71
PIN Columnar ($249^3 \pm 51$)	Measurements (27)	16562	613,4	613,4	28
	Fields (14)	11032	788	174,6	8
	Cases (10)	2161	2161	1373	64
PCA ($281^3 \pm 51$)	Measurements (100)	33159	331,6	331,6	15
	Fields (45)	23296	518	186,4	8
	Cases (50)	2212	2212	1694	77

[1] Mean ± SD. The measurements of v_V nuc are expressed in μm^3
[2] Average number of nuclei, fields or cases
[3] Values are statistically significantly different ($p < 0.05$)

In columnar stratum of both control and PIN, basal compartment of PIN, and PCA, the greatest contribution (53-77%) to the totally observed variance was provided by the highest level of sampling, *i.e.,* by the biologic variation among the cases. However, when the basal stratum of the controls was considered, the contribution to the total variance provided by the nuclear intercepts and their measurements was higher (56%) than that provided by the individual cases (44%) (Table 1).

The relative contribution to overall variance is estimated by regarding the observed variance at each level of sampling as the sum of the true variance at that level plus the variance of the mean (SEM^2), at the lower level of sampling. Thus, large variances at the lower levels are diminished in their contribution to totally observed variance by the number of observations at that particular level. The individual values of each measurement are indicated in Figure 4. In comparison with controls, a greater dispersion around the mean was observed in both PIN and PCA measurements.

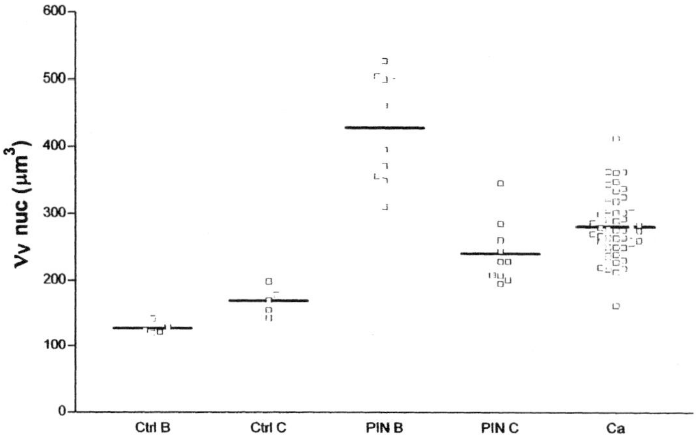

Figure 4. Scatter plot of individual values for v_V nuc, in basal stratum of controls (Ctrl B), columnar stratum of controls (Ctrl C), basal stratum of PIN (PIN B), columnar stratum of PIN (PIN C), and carcinoma (Ca). The horizontal lines in the graph indicate the correspondent mean.

Discussion

The increase of v_V nuc in a number of neoplasias is related to several factors: histological grade (15, 16), prognosis (15, 17, 18), or the response to the therapy (19). In the present study, a significant increase of v_V nuc was detected in PIN when compared to normal prostatic epithelium, this fact can be put in relation to the presumptive pre-malignant character of this lesion. Similar data, a progressive increase of nuclear volume from normal prostate acini to low grade PIN, high grade PIN, and PCA have been reported by others (20).

It is interesting to note that the basal layer of PIN showed variability and the highest values for both v_V nuc. This might indicate that basal cells of PIN correspond to cells in a stage of transient proliferation (with features of pluripotential cells). These cells, when they reach the luminal compartment, decrease in nuclear size, maintaining their potential for proliferation (21, 22) and malignancy. In fact, they are similar to those observed in invasive PCA. Thus, in contrast to the description in other neoplasias (23, 24) the invasive features in PCA are not immediately related to the increase of nuclear volume.

Conclusion

The similarities in nuclear size between PIN and PCA are according to the pre-malignant character commonly assigned to PIN. The increase of basal ν_V nuc in PIN might indicate that the changes heralding the progression from PIN to PCA are produced in this layer, whereas the nuclear features of luminal layer are the same to those of the carcinoma.

References

1. Bostwick DG (1992) Prostatic intraepithelial neoplasia (PIN): current concepts. J Cell Biochem (Suppl) 16:10-19.
2. Haggman MJ, Macoska JA, Wojno KJ, et al (1997) The relationship between prostatic intraepithelial neoplasia and prostate cancer: critical issues. J Urol 158:12-22.
3. Stattin P, Damber JE, Karlberg L et al (1996) Bcl-2 immunoreactivity in prostate tumorigenesis in relation to prostatic intraepithelial neoplasia, grade, hormonal status, metastatic growth and survival. Urol Res 26:257-264.
4. Krajewska M, Krajewski S, Epstein JI, et al (1996) Immunohistochemical analysis of bcl-2, bax, bcl-X and mcl-1 expression in prostate cancers. Am J Pathol 148:1567-1576.
5. Bostwick DG (1996) Progression of prostatic intraepithelial neoplasia to early invasive adenocarcinoma. Eur Urol 30:142-145.
6. Myers RB, Grizzle WE (1996) Biomarker expression in prostatic intraepithelial neoplasia Eur Urol 30:153-166.
7. Montironi R, Magi-Galluzzi C, Marina S, et al (1994) Quantitative characterization of the frequency and location of cell proliferation and death in prostate pathology. J Cell Biochem (Suppl) 19:238-245.
8. Takahashi S, Qian J, Brown JA, et al (1994) Potential markers of prostate cancer aggressiveness detected by fluorescence in situ hybridization. Cancer Res 54:3574-3579.
9. Epstein J I (1994) Prostatic intraepithelial neoplasia. Adv Anat Pathol 1:123-134.
10. Santamaría L, Martín R, Martín JJ (1998) Volume-weighted mean nuclear volume in dysplastic glands from the ventral prostate of rats treated with cadmium chloride, and resembling human prostatic intraepithelial neoplasia. Acta Stereol 17:377-382.
11. Martín R, Nieto S, Santamaría L (1997) Stereologic estimates of volume-weighted mean nuclear volume in colorectal adenocarcinoma: correlation with histologic grading, Dukes′ staging, cell proliferation activity and p53 protein expression. Gen Diag Pathol 143:29-38.
12. Sørensen FB (1991) Stereological estimation of mean and variance of nuclear volume from vertical sections. J Microsc 162:203-222.

13. Gundersen HJG, Bendtsen TE, Korbo L, et al (1988) Some new, simple and efficient stereological methods and their use in pathological research and diagnosis APMIS 96:379-394.
14. Gundersen HJG, Osterby R (1981) Optimizing sampling efficiency of stereological studies in biology or "Do more less well" J Microsc 121:65-73.
15. Artacho-Pérula E, Roldan-Villalobos R, Blanco-Rodriguez A (1995) Application of recent stereological tools for unbiased three-dimensional estimation of number and size of nuclei in renal cell carcinoma samples. Anal Cell Pathol 9:295-309.
16. Fujikawa K, Sasaki M, Aoyama T, et al (1995) Prognostic criteria in patients with prostate cancer: correlation with volume weighted mean nuclear volume. J Urol 154:2123-2127.
17. Roldan-Villalobos R, Artacho-Pérula E, Ruiz-Moruno JR (1996) Grading and prognosis of infiltrating ductal breast carcinoma by mean nuclear volume estimates. Anal Quant Cytol Histol 18:158-166.
18. Martín R, Sneige N, Vásquez M, et al (1999) Stereologic estimates of volume-weighted mean nuclear nolume in aspiration smears of ductal breast carcinoma. Correlation with cytologic grade, tumor size and lymph node status. Anal Quant Cytol Histol 21:185-193.
19. Fukuzawa S, Hashimura T, Sasaki M, et al (1995) Nuclear morphometry for improved prediction of the prognosis of human bladder carcinoma. Cancer 76:1790-1796.
20. Lopez-Beltran, A, Artacho-Perula E, Roldan-Villalobos R, et al (2000) Nuclear volume estimates in prostatic intraepithelial neoplasia. Anal Quant Cytol Histol 22:37-44.
21. Montironi R, Galluzi CM, Diamanti L, et al (1993) Prostatic intraepithelial neoplasia: expression and location of proliferating cell nuclear antigen (PCNA) in epithelial, endothelial, and stromal nuclei. Virchows Arch (A) 422:185-192.
22. Tsuji M, Kanda K, Murakami Y, et al (1999) Biologic markers in prostatic intraepithelial neoplasia: immunohistochemical and cytogenetic analyses. J Med Invest 46:35-41.
23. Sørensen FB (1989) Stereological estimation of nuclear volume in benign melanocytic lesions and cutaneous malignant melanomas. Am J Dermatopathol 11:517-523.
24. Ladekarl M, Boek-Hansen T, Henrik-Nielsen R, et al (1995) Objective malignancy grading of squamous cell carcinoma of the lung. Stereologic estimates of mean nuclear size are of prognostic value, independent of clinical stage of disease. Cancer 786:797-802.

Constitutively Active Androgen Receptor Variant Detected in a Human Prostate Cancer

Jocelyn Céraline, Marion D. Cruchant, Eva Erdmann, Philippe Erbs, Jean-Emmanuel Kurtz, Brigitte Duclos, Didier Jacqmin, Dominique Chopin, Patrick Dufour, and Jean-Pierre Bergerat

Summary

Human androgen receptor (AR) mutation is a possible alternative used by prostate adenocarcinoma (PCA) cells to escape androgen dependence. These mutations may broaden the specificity and/or the sensitivity of the AR to other hormones, resulting in the inappropriate receptor activation, and thus, affecting the PCA response to hormonal therapies. We have developed a yeast-based functional assay to detect mutant ARs in PCA by analyzing their transactivation capacities in response to different ligands. We report herein the detection of two different mutant ARs within the same metastatic tumor sample. Concomitantly to the T877A mutant AR, we identified an additional double mutant AR harboring the nonsense mutation Q640Stop, just downstream of the DNA binding domain together with the T877A point mutation. This mutation leads to a C-terminal truncated AR. This study is the first description of this type of mutation in PCA. We demonstrated that this truncated AR exhibited constitutive transactivation properties. In conclusion, our data suggest that mutation-induced constitutive activation of the AR may be an alternative mechanism used by PCA cells to escape androgen deprivation.

Introduction

Androgen ablation is the standard first line therapy for advanced PCA (1, 2). The benefit of this therapy is only transitory, lasting some months or years, and all patients will eventually relapse with tumor cells whose growth escapes the androgen deprivation (3). AR mutations, selected during hormonal treatments, are one of the possible mechanisms to explain the development of these androgen-independent PCAs (4). Indeed, some AR mutations can expand the specificity and/or the affinity of the AR to other hormones and allow it to respond to different steroids as well as anti-androgens (5, 6). The presence of tumor clones with such AR mutations should be considered as hormonal therapies that may ultimately exert a positive selection of these tumor cells. We have previously described a yeast-based functional assay to simultaneously detect and analyze AR mutations in PCA (7). We used a *Saccharomyces cerevisiae* yeast strain which is deficient in aminoimidasole ribonucleotide carboxylase (ADE2), an enzyme that intervenes in purine biosynthesis. In these yeast cells, we reintroduced the wild type *ADE2* open reading frame placed under the control of an androgen–dependent promoter. After transformation with a plasmid encoding the human AR, the expression of the *ADE2* gene, and consequently, the growth of the yeast cells in an adenine-depleted selective medium depends on the specificity and the affinity of the AR for the ligand added in the medium. Therefore, mutant ARs can be distinguished from wild-type AR (AR_{WT}) by comparing their specificity and affinity for a range of steroid hormones and anti-androgens.

In the present study, we described the results obtained with a hormone refractory PCA from a bone marrow aspirate of a consenting patient. Two different mutant ARs, T877A and Q640Stop/T8777A, were detected within this metastatic PCA sample. The analysis of functional properties of the new Q640Stop/T8777A double AR mutant revealed a constitutive activity in the absence of androgens. This report is the first identification of such a constitutively active mutant AR in PCA, and our data suggest a previously unknown mechanism for androgen-independent growth for PCA after androgen ablation.

Materials and Methods

Tissue Sample, RT-PCR, and AR Yeast-based Functional Assay. The patient was diagnosed with T3NxMo PCA at age 76, and was first treated with leuprorelin and flutamide. The patient responded to this complete androgen blockade for 17 months, and relapsed thereafter. Using bone scanning, diffuse bone marrow metastases including bilateral posterior iliac metastases were revealed at that time. He was later treated by alternative hormonal (cyproterone acetate, cortisol, medroxyprogesterone) and chemotherapy. The bone marrow sample was aspirated at the site of a hyperfixation area on the right posterior iliac crest, five years after the initial diagnosis, and cytologic examination showed the presence of numerous metastatic PCA cells. Total RNAs were isolated from the bone marrow sample, and an AR fragment (1786 bp, 2311 - 4097) was amplified by RT-PCR, and cloned into a yeast expression plasmid. The AR functional assay has been described (7).

Results

A Yeast-based Functional Assay to Detect Mutant ARs in PCA. We used a previously described AR yeast-based functional assay (7) (Figure 1) to detect and characterize mutant AR in PCA from patients who failed androgen ablation therapy. Total RNAs were extracted from bone marrow samples from patients with metastatic PCA, and an AR fragment was amplified by RT-PCR. The PCR product was cloned by homologous recombination into a yeast expression plasmid containing the N- and the C-terminal parts of the AR. The expression of the full-length AR leads to yeast growth in the presence of specific agonists. In this assay, mutant ARs were distinguished from the AR_{WT} by comparing their responsiveness to a panel of steroid and non steroid ligands tested.

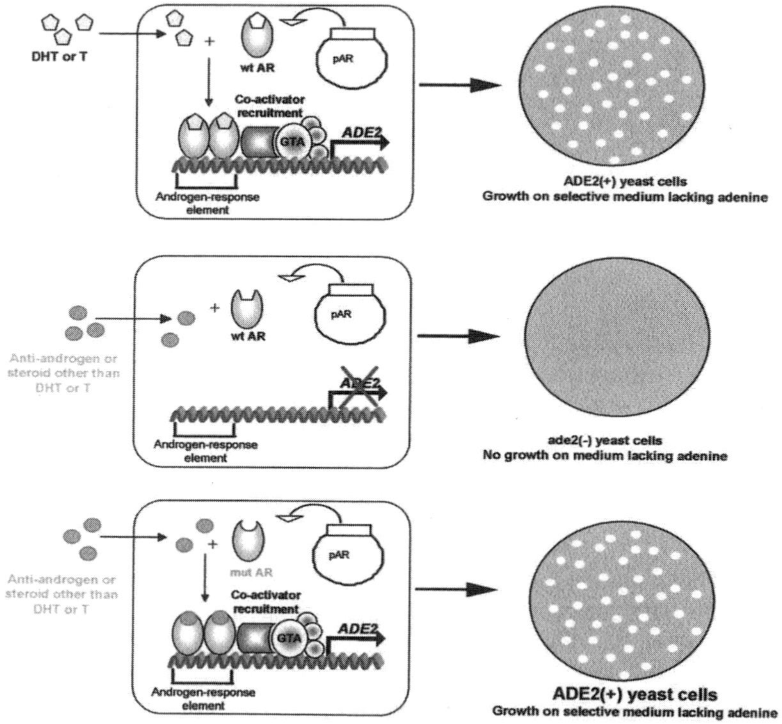

Figure 1. Functional AR assay in yeast. The expression of the *ADE2* gene is androgen-dependent. **A.** Activation of the AR_{WT} by dihydrotestosterone (DHT) or testosterone (T) leads to the expression of the *ADE2* gene resulting in yeast growth on selective adenine-deprived medium. **B.** There is no expression of *ADE2* gene and consequently no yeast growth in the presence of steroid molecules that AR_{WT} binds to with low affinity. **C.** Mutant AR with a broaden specificity can be activated by different ligands resulting in yeast growth on selective adenine-deprived medium.

Detection of AR Mutations in a Metastatic PCA. Figure 2 depicts data obtained from a patient with advanced PCA who had escaped androgen-ablation therapy. When compared to the AR_{WT} (Figure 1A), an abnormal yeast growth response was observed in the presence of 100 nM androstenedione, 100 μM flutamide, 100 nM progesterone (Prog) and 100 nM medroxyprogesterone (MProg) (Figure 2B). This profile suggests the presence of tumor clones expressing ARs that exhibit a higher affinity for androstenedione, Prog, and MProg than the AR_{WT}, and a paradoxical response to flutamide. Moreover, a significant number of yeast colonies were also observed in the absence of any ligand (lane *control*, Figure 2B), suggesting the presence of mutant AR molecules with a constitutive activity. To further characterize these AR molecules with modified transactivation capacities, yeast colonies were picked from both the control and the 100 nM Prog containing plates, and re-assayed.

The AR expressed in yeast cells isolated from the Prog-containing plate demonstrated a strong response to low concentrations of androstenedione, Prog, and MProg, and a paradoxical response to flutamide, confirming the data obtained from the first screening. The sequencing of plasmid DNA rescued from these yeast cells revealed the presence of the mutation ACT (threonine) to GCT (alanine; T877A) in codon 877 within the ligand binding domain of the AR.

As expected, yeast cells from the control plate exhibited a ligand independent growth. Yeast proliferation was observed independently of the addition of a ligand in the medium, suggesting a constitutive activation of the AR expressed in these yeast cells. The sequencing of plasmid DNA rescued from these yeast cells revealed the presence of two mutations within the same AR molecule, the CAG(Gln) → TAG(Stop) mutation at codon 640 (Q640Stop) in addition to the T877A mutation. As these data represent the first description of the Q640Stop/T877A mutation, we further investigated the functional analysis of this double mutant AR.

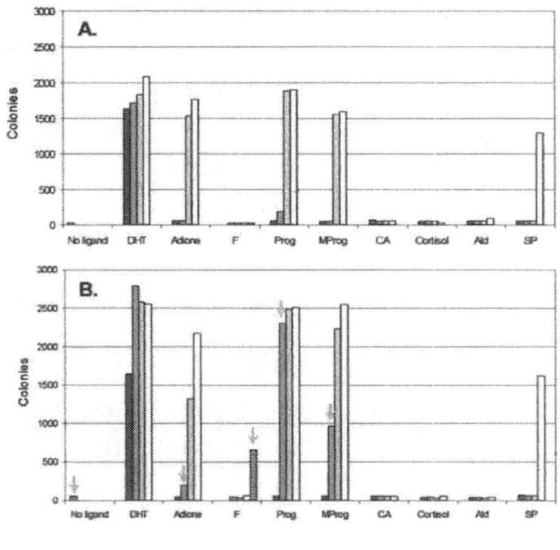

Figure 2. AR mutations detection in a hormone refractory PCA. Histograms represent the number of colonies obtained with AR_{WT} (**A**) or AR from the metastatic PCA (**B**), in response to increasing doses (10 nM to 10 μM) of the indicated ligand, except flutamide (100 nM to 100 μM). Arrows indicate changes in specificity and/or sensitivity of the ARs expressed in the yeast cells for the indicated ligand. Androstenedione (Adione); dehydroandrosterone (DHEA); flutamide (F); cortisone acetate (CA); aldosterone (Ald); spironolactone (SP).

Western Blot Analysis of the Q640Stop/T877A Mutant AR. The amino acid Q640 is located in the hinge region of the AR, between the DNA binding domain and the ligand binding domain. To confirm that the Q640Stop mutation leads to a truncated AR protein, western blot analysis was performed on protein extracts from yeast cells expressing this mutant AR. As shown in Figure 3, the Q640Stop mutant AR was highlighted by an immunological band with an approximate size of 75 kDa. In the controls, AR_{WT} was expressed in our yeast strain at the expected size of 110 kDa, No AR expression was detectable in the non-transformed parental yeast strain EJ250.

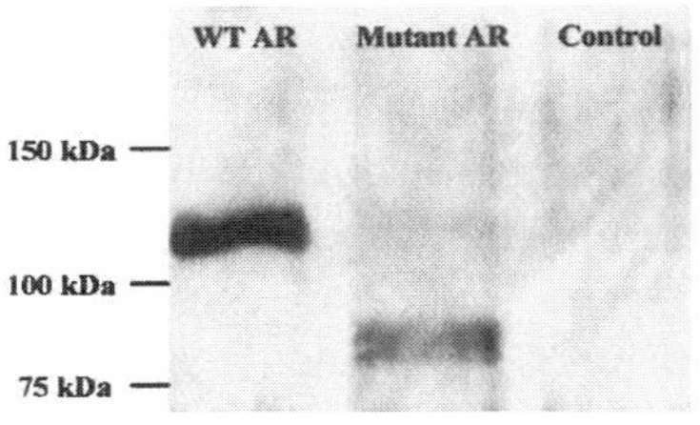

Figure 3. The Q640Stop mutation leads to a truncated AR molecule. Western-blot analysis of the AR_{WT} and the Q640Stop AR expressed in the transformed yeast strain. The parental yeast strain is shown as control. Protein extracts were prepared from yeast cells as previously described (7), loaded on an 7.5 % SDS-PAGE and transferred onto nitrocellulose membranes. Blots were probed with mouse IgG_{2a} mAb G122-434 against human AR and peroxidase-conjugated secondary antibodies, and specific AR signal was detected by chemiluminescence.

The Q640Stop Mutant AR is Exclusively Localized in the Nucleus. Because the Q640Stop mutation was located just after the nuclear localization signal (NLS1) within the second Zn finger of the DBD and the hinge region, we examined the cellular compartmentalization of this truncated mutant AR by using the EGFP technique in CV1 cells. As presented in Figure 4, the nuclear import of the EGFP-AR_{WT} fusion protein from the cytoplasm was androgen-mediated. In steroid hormone-free medium, EGFP-AR_{WT} was evenly distributed between the cytoplasm and the nucleus (Figure 4a). After DHT treatment, the fluorescence was concentrated within the nucleus of the transfected cells (Figure 4b). In contrast, the nuclear import of the EGFP-$AR_{Q640Stop}$ was ligand-independent. All the transfected cells showed exclusively nuclear fluorescence whether DHT was added in the medium or not (Figure 4c, d). The DHT treatment did not modify the localization of EGFP alone (Figure 4e, f).

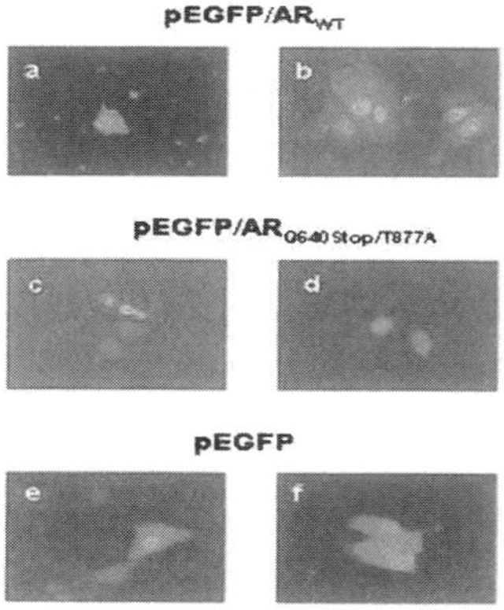

Figure 4. Mutant AR Compartmentalization in CV-1 Cells. CV-1 cells were transiently transfected with 1μg pEGFP-AR_{WT} or pEGFP-$AR_{Q640Stop}$. Twenty four hours after transfection, cells received fresh medium containing 10% charcoal-stripped FCS and 100 nM DHT or vehicle. Slides of transfected cells were rinsed with PBS 24 h later, fixed in freshly made PBS/4% paraformaldehyde and mounted for visualization by fluorescence microscopy.

The Q640Stop Mutant AR Exhibits Constitutive Transactivation Properties. CV-1 cells were transiently co-transfected with wild type (pARo) or mutant ($pAR_{Q640Stop/T877A}$) AR expression vectors and the reporter plasmid pMMTV-Luc. The expression of the luciferase reporter gene was regulated by the androgen-responsive elements present in the MMTV-LTR. The AR_{WT} showed minimal luciferase activity in the absence of androgen, but were highly stimulated (~120.0-fold induction) in the presence of 100 nM DHT (Figure 5). On the other hand, the truncated mutant AR exhibited a strong luciferase activity, which was androgen-independent. In the absence of androgen, the Q640Stop mutant AR showed maximal activity (~275.0-fold induction). DHT treatment did not improve the transactivation capacities of this double mutant AR. These results support the conclusion that this double mutant AR exhibits constitutive transactivation properties.

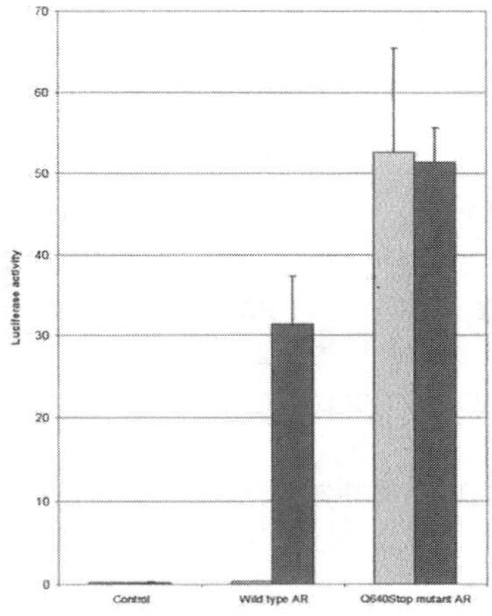

Figure 5. Constitutive transactivation properties of the Q640Stop mutant AR. CV-1 cells were transfected with 200 ng pMMTV-Luc, 20 ng of pARo or $pAR_{Q640Stop/T877A}$, and 50 ng of pEGFP-C3 as the internal control. Cells received fresh medium containing 10% charcoal-stripped FCS and 100 nM DHT or vehicle, 24 h after transfection. Thereafter, the luciferase activity was assayed using the Luciferase Assay System (Promega). The induction of luciferase activity is indicated in arbitrary units. All points were done in quadruplicate, and the results represent the mean ± 1SD.

Conclusions

Herein, we describe the presence of two tumor cell clones harboring different AR mutations within the same hormone refractory metastatic PCA. These findings provide direct evidence for the clonal outgrowth of androgen-independent PCA cells with AR mutations in response to selective pressure from hormonal therapies. The coexistence of tumor clones expressing mutant ARs harboring either the T877A mutation or the Q640Stop/T877A double mutation argues that the event Q640Stop has occurred later during the disease history probably in response to a new selective pressure from one of the treatments that followed flutamide.

This study is the first report of a constitutively activated mutant AR in PCA and describes a new way by which PCA cells may escape androgen deprivation. In addition, this study demonstrates the great efficacy of our yeast functional assay for the detection of mutant ARs in PCA, and the analysis of their transactivation properties. By performing the assay on total RNA isolated from primary tumor samples, from hormone-sensitive metastases, and ultimately from hormone-resistant metastatic foci, we should be able to detect the clonal outgrowth of PCA cells with mutant AR during the disease evolution. Data obtained from these studies will provide a better understanding of the AR functions in the metastatic progression of the disease, and in the escape to androgen-dependent growth after androgen ablation therapies.

Acknowledgements

This work was supported by the Programme Hospitalier de Recherche Clinique National 2002, the Hôpitaux Universitaires de Strasbourg, the Association pour la Recherche en Onco-Hematologie (ARECOH), the Association pour la Thérapie Génique des Cancers (ATGC), the Faculté de Médecine of the Université Louis Pasteur, and the IRCAD. This study was approved by the Ethic Committee, CCPPRB (Comité Consultatif de Protection des Personnes se Prêtant à la Recherche Biomédicale) – Alsace1.

References

1. Huggins C, Hodges CV (1941) Studies on prostatic cancer; effect of castration, of estrogen and of androgen injection on serum phosphatases in metastatic carcinoma of the prostate. Cancer Res 1:293-297.
2. Denis LJ, Griffiths K (2000) Endocrine treatment in prostate cancer. Semin Surg Oncol 18:52-74.
3. Feldman BJ, Feldman D (2001) The development of androgen-independent prostate cancer. Nat Rev Cancer 1:34-45.
4. Gottlieb B, Lehvaslaiho H, Beitel LK, et al (1998) The Androgen Receptor Gene Mutations Database. Nucleic Acids Res 26:234-238. http://ww2.mcgill.ca/androgendb/
5. Taplin ME, Bubley GJ, Ko YJ, et al (1999) Selection for androgen receptor mutations in prostate cancers treated with androgen antagonist. Cancer Res 59:2511-2515.
6. Zhao XY, Boyle B, Krishnan AV, et al (1999) Two mutations identified in the androgen receptor of the new human prostate cancer cell line MDA PCa 2a. J Urol 162:2192-2199.
7. Céraline J, Erdmann E, Erbs P, et al (2003) A yeast-based functional assay for the detection of mutant androgen receptor in prostate cancer. Eur J Endocrinol 148:99-109.

Cancer Prevention by Green Tea via EGCG-Mediated Inhibition of Fatty Acid Synthase

Koen Brusselmans, Ellen De Schrijver, Walter Heyns, Guido Verhoeven, and Johannes V. Swinnen

Summary

Green tea is widely accepted to lower the risk of developing cancer, including hormone-responsive cancers, but the precise mechanism of its cancer-preventive effect is not fully understood. Recently, the green tea component epigallocatechin-3-gallate (EGCG) was demonstrated to inhibit *in-vitro* enzymatic activity of chicken liver fatty acid synthase (FAS), an enzyme that is frequently overexpressed in many epithelial tumors. Since chemical FAS inhibitors such as cerulenin and C75 are known to inhibit growth and to induce apoptosis of several cancer cell lines *in vitro* and tumor xenografts *in vivo*, it was investigated whether EGCG also inhibited FAS activity in cultured prostate cancer (PCA) cells *in vivo* and how this inhibition affected lipogenesis, cell proliferation, and cell viability. EGCG significantly inhibited FAS activity in PCA cells (with high FAS expression levels). This FAS inhibition was paralleled by decreased lipogenesis, growth inhibition, and apoptosis. In contrast, epicatechin, another closely related catechin that does not influence FAS activity, had no effect on PCA cell proliferation or survival. EGCG also inhibited FAS activity and proliferation of normal fibroblasts (with low FAS expression), but did not induce fibroblast apoptosis. Taken together, it can be concluded that EGCG efficiently inhibits FAS in cultured cells, and specifically induces apoptosis in PCA cells but not in normal fibroblasts, thereby providing interesting perspectives for using EGCG in anti-neoplastic therapies.

Green Tea and Cancer Chemoprevention. Epidemiological studies have indicated that multiple health benefits may be accredited to green tea, such as preventive effects against atherosclerosis, coronary heart disease, stroke, obesity, but also cancer, the second most common cause of death in the western world (1, 2). Frequent consumption of green tea has been suggested to lower the risk for PCA (1, 2). In addition, Chinese men, who drink green tea on a very frequent basis, show the lowest incidence of PCA worldwide. Frequent consumption of green tea has also been demonstrated to lower the risk of developing cancer of the breast, stomach, colon, esophagus, and pancreas (1, 2). Although the cancer-preventive effect of green tea is generally accepted, the precise mechanism by which green tea exerts its anti-carcinogenic activity is not fully understood.

Cancer-Preventive Effects of Green Tea are Primarily Mediated by Epigallocatechin-3-Gallate (EGCG). Most of the cancer-preventive effects of green tea have been ascribed to EGCG (Figure 1) (1, 2). EGCG has been shown to inhibit proliferation and induce apoptosis of several human tumor cell lines *in vitro*, including PCA and breast cancer (BC) cells (3, 4). In addition, experiments with prostate tumor and breast tumor xenografts in mice have demonstrated that administration of EGCG decreases the tumor size (1, 5, 6). Several molecules have been proposed as a target for EGCG. Interestingly, EGCG inhibits urokinase and matrix metalloproteinases, enzymes that are essential for matrix degradation and tumor invasion (1, 7). Furthermore, EGCG decreases expression of vascular endothelial growth factor, thereby reducing tumor angiogenesis, which may limit the vascular supply of oxygen and nutrients (7). EGCG has also been shown to block protein tyrosine kinase activity of EGF receptor, PDGF receptor, and FGF receptor, all important signal transduction mediators that often show an abnormal regulation in tumor cells (1, 7). In addition, EGCG may also influence tumor cell proliferation and viability directly, as it causes G0/G1 cell cycle arrest in epidermoid carcinoma cells (8) and inhibits Bcl-XL phosphorylation in PCA cells, a process associated with apoptosis (9). Finally, with regard to hormone-sensitive tumors, it is worth mentioning that EGCG administration reduced blood levels of testosterone and estradiol in rats (10). Nevertheless, the precise contribution of these and other potential targets to the anti-carcinogenic effects of EGCG remains elusive.

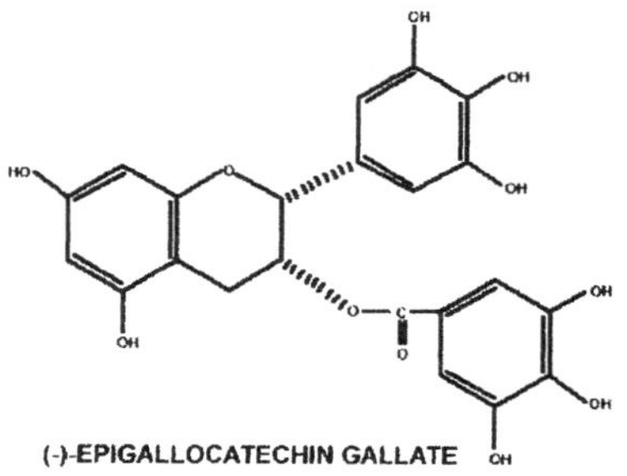

Figure 1. Chemical structure of EGCG.

Fatty Acid Synthase, a New Target for Anti-neoplastic Therapy. FAS is a key metabolic enzyme that catalyzes the synthesis of long-chain fatty acids (12). Expression of FAS is low in most human tissues, as the majority of the required lipids is obtained from the diet, but is markedly elevated in a variety of human cancers, such as cancer of the prostate, breast, ovary, oesophagus, stomach, colon, kidney, bladder and lung (13, 14). Moreover, FAS up-regulation occurs early during tumorigenesis and is more pronounced in more advanced tumors, whereas high tumor FAS levels often correlate with a poor prognosis.

Studies in our laboratory have revealed that androgens stimulate FAS expression in PCA cells via a mechanism involving activation of the sterol regulatory element binding protein (SREBP) pathway (15, 16). SREBPs are transcription factors that control the expression of enzymes belonging to the two major lipogenic pathways: fatty acid synthesis and cholesterol synthesis. Accordingly, androgen-mediated SREBP activation did not only stimulate FAS expression in prostate tumor cells *in vitro*, but also expression of other lipogenic enzymes including ATP-citrate lyase, acetyl-CoA carboxylase and malic enzyme (15). FAS upregulation in tumors is most likely part of a general stimulation of the fatty acid synthesis pathway, since prostate and breast cancers with high FAS levels also showed increased acetyl-CoA carboxylase expression (17, 18). Others observed that SREBPs also regulate FAS overexpression in BC cells (19).

Remarkably, several studies have shown that the chemical FAS inhibitors, cerulenin and C75, are growth inhibitory and cytotoxic for various tumor cell lines *in vitro*, including BC and PCA cells (20, 21). In addition, administration of cerulenin or C75 delay tumor progression in xenograft models of BC and ovarian cancer in mice (22, 23). Since EGCG was recently shown to inhibit *in-vitro* enzymatic activity of FAS (11), our group investigated whether EGCG also blocked fatty acid synthesis and lipogenesis in cultured human PCA cells, and whether this EGCG-mediated FAS inhibition also influenced growth and viability of tumor cells.

EGCG is a Potent Inhibitor of FAS in Cultured PCA Cells. We analyzed the impact of EGCG on the FAS activity of LNCaP cells, a PCA cell line with high FAS expression levels. Exposure of LNCaP cells to different concentrations of EGCG significantly reduced the enzymatic activity of FAS (24). The EGCG-mediated reduction of FAS activity was caused by chemical inhibition, not by a decrease in FAS expression. The decreased FAS activity in EGCG-treated LNCaP cells resulted in a reduced lipid biosynthesis as revealed by measuring the incorporation of [^{14}C]-labeled acetate. TLC analysis showed that EGCG severely reduced synthesis of triglycerides and phospholipids (including phosphatidylethanolamine, phosphatidylcholine, phosphatidylserine, phosphatidyl-inositol and sphingomyelin), lipid fractions that are rich in fatty acids (Table 2) (24). EGCG also inhibited the synthesis of cholesterol, most likely not through FAS inhibition but via inhibition of squalene epoxidase, a rate-limiting enzyme in the cholesterol synthesis (25). Taken together, it can be concluded that EGCG is an efficient FAS inhibitor in intact cells.

EGCG Inhibits Proliferation and Induces Apoptosis in PCA Cells. The chemical FAS inhibitors cerulenin and C75 inhibit proliferation and are cytotoxic to various cancer cell lines *in vitro* (20, 21). Since EGCG inhibits FAS as efficiently as cerulenin and C75, we analyzed whether FAS inhibitory EGCG concentrations also affected growth and viability of PCA cells. EGCG concentrations, that significantly inhibited FAS activity in LNCaP cells, also inhibited cell growth and induced cell death of LNCaP cells (Table 1), as revealed by trypan blue dye exclusion assays. Stainings with Hoechst 33342 and Annexin V-EGFP/propidium iodide demonstrated that cell death of LNCaP cells was primarily due to apoptosis (24). The parallelism between the EGCG dose-response curves reflecting FAS inhibition on the one hand and those reflecting growth inhibition and cell death on the other hand strongly suggests that EGCG-mediated tumor cell death is a result of FAS inhibition.

Table 1. Effect of EGCG on Viability of LNCaP Cells[1].

EGCG concentration Exposure time	0 μM	40 μM	100 μM	150 μM
24 h	6 ± 2 %	9 ± 4 %	17 ± 5 %[2]	50 ± 1 %[2]
48 h	4 ± 1 %	10 ± 2 %	19 ± 3 %[2]	43 ± 4 %[2]
72 h	5 ± 1 %	13 ± 2 %[2]	25 ± 5 %[2]	41 ± 3 %[2]

[1] % dead (trypan-blue positive) LNCaP cells (mean ± SD; n = 4).
[2] Significantly different ($P < 0.05$) from control (0 μM EGCG).

Epicatechin (EC), Which Does Not Inhibit FAS, Has No Effect on Growth or Viability of PCA Cells. The correlation between EGCG-induced FAS inhibition and cytotoxicity was further illustrated by treatment of LNCaP cells with another green tea polyphenol closely related to EGCG, epicatechin (EC). Contrary to EGCG, EC did not inhibit FAS activity neither lipogenesis in LNCaP cells. Interestingly, EC had no effect on cell growth or viability, as LNCaP cells continued proliferating and did not show any signs of cell death (24), thereby illustrating that the cytotoxicity caused by EGCG was most likely due to the FAS inhibitory effect of EGCG.

The FAS Inhibitor EGCG Selectively Induces Cell Death in PCA Cells But Not in Normal Human Fibroblasts. To explore the potential of EGCG as a chemotherapeutic compound that selectively targets cancer cells but not normal cells, the effect of EGCG was also analyzed on non-malignant human fibroblasts, which show low FAS expression levels similar to that *in vivo* conditions. Baseline FAS activity levels were about 20 times lower in fibroblasts than in LNCaP cells (24). EGCG also inhibited FAS activity and proliferation of normal fibroblasts in a dose-dependent way. However, in contrast with LNCaP cells, EGCG did not induce cell death of fibroblasts (24). Therefore, it can be concluded that the FAS inhibitor EGCG is selectively cytotoxic for PCA cells with high FAS expression levels, but not for normal human cells. Taken together, these findings may provide interesting perspectives for the possible use of EGCG as a therapeutic agent in future anti-cancer treatment.

By Which Mechanism Does EGCG-mediated FAS Inhibition Result in Apoptosis of PCA Cells? The precise mechanism by which inhibition of FAS induces growth retardation and cell death of cancer cells is not fully understood. Possibly, FAS inhibition inhibits tumorigenesis by limiting the amount of lipids that are necessary for the membrane synthesis of rapidly dividing tumor cells. It has also been proposed that the cytotoxicity mediated by FAS inhibition in cancer cells with high FAS expression levels is caused by toxic accumulation of malonyl-CoA (22). As the up-regulation of FAS in tumor cells is part of a general stimulation of fatty acid synthesis (15, 17, 18), also enzymes catalyzing early steps in fatty acid synthesis are up-regulated in tumor cells, such as acetyl-CoA carboxylase which converts acetyl-CoA to malonyl-CoA. When FAS is inhibited in cancer cells, the increased activity of acetyl-CoA carboxylase may cause intracellular accumulation of the fatty acid precursor malonyl-CoA, which may become cytotoxic. However, it can not be excluded that the cytotoxic effects of FAS inhibition towards cancer cells are caused by alterations in the membrane composition. Indeed, FAS plays a major role in the synthesis of phospholipids partitioning into detergent-resistant membrane microdomains, also known as 'rafts' (26). These membrane micro-domains have been implicated in various cellular processes such as signal transduction, intracellular trafficking, cell polarization and cell migration (27), all of which play key roles during tumor progression and metastasis. FAS inhibition may significantly change the membrane lipid composition of cancer cells, thereby possibly causing dysregulation of particular membrane functions.

Conclusion

We have demonstrated that EGCG is a potent natural inhibitor of FAS in intact cells and that the EGCG-induced inhibition of the high FAS activity in PCA cells is accompanied by inhibition of cell proliferation and induction of apoptosis (24). Remarkably, growth inhibition and apoptosis are most pronounced at EGCG concentrations that also efficiently inhibit FAS activity. This striking parallelism between the dose-response curves reflecting FAS inhibition and those reflecting growth inhibition and apoptosis indicates that EGCG-mediated tumor cell death is most likely a result of FAS inhibition. However, it can not be excluded that EGCG, in addition to the induction of cell death via FAS inhibition, may also influence tumor development via other effects. Indeed, EGCG has also been demonstrated to inhibit various tumor-promoting factors including matrix metalloproteinases, vascular endothelial growth factor, receptor tyrosine kinases and sex steroids (1, 7, 10). In contrast with PCA cells, EGCG does not induce apoptosis of normal fibroblasts (24), which display low FAS activity, a feature shared by most human cells *in vivo*. Therefore, it can be concluded that the FAS inhibitor EGCG is selectively cytotoxic for PCA cells but not for non-malignant fibroblasts. Taken together, these data strongly suggest that FAS inhibition may substantially contribute to the tumor preventive and anti-neoplastic effects of EGCG and green tea, thereby providing interesting perspectives for future anti-cancer therapies.

Acknowledgments

This work was supported by Cancer Research Grants from FB Insurance (Fortis Bank) and VIVA, by a grant 'Geconcerteerde Onderzoeksactie van de Vlaamse Gemeenschap', by research grants from the Fund for Scientific Research-Flanders (Belgium) (FWO), and by a grant Interuniversity Poles of Attraction Programme-Belgian State, Prime Minister's Office, Federal Office for Scientific Affairs.

References

1. Mukhtar H, Ahmad N (2000) Tea polyphenols: prevention of cancer and optimizing health. Am J Clin Nutr 71:1698S-1702S; discussion 1703S-4S.
2. Bushman JL (1998) Green tea and cancer in humans: a review of the literature. Nutr Cancer 31:151-159.
3. Yang GY, Liao J, Kim K, et al (1998) Inhibition of growth and induction of apoptosis in human cancer cell lines by tea polyphenols. Carcinogenesis 19:611-616.
4. Paschka AG, Butler R, Young CY (1998) Induction of apoptosis in prostate cancer cell lines by the green tea component, (-)-epigallocatechin-3-gallate. Cancer Lett 130:1-7.
5. Liao S, Umekita Y, Guo J, et al (1995) Growth inhibition and regression of human prostate and breast tumors in athymic mice by tea epigallocatechin gallate. Cancer Lett 96:239-243.
6. Sartippour MR, Heber D, Ma J, et al (2001) Green tea and its catechins inhibit breast cancer xenografts. Nutr Cancer 40:149-156.
7. Jung YD, Ellis LM (2001) Inhibition of tumour invasion and angiogenesis by epigallo-catechin gallate (EGCG), a major component of green tea. Int J Exp Pathol 82:309-316.
8. Ahmad N, Cheng P, Mukhtar H (2000) Cell cycle dysregulation by green tea polyphenol epigallocatechin-3-gallate. Biochem Biophys Res Commun 275:328-334.
9. Kazi A, Smith DM, Zhong Q, et al (2002) Inhibition of bcl-x(l) phosphorylation by tea polyphenols or epigallocatechin-3-gallate is associated with prostate cancer cell apoptosis. Mol Pharmacol 62:765-771.
10. Kao YH, Hiipakka RA, Liao S (2000) Modulation of endocrine systems and food intake by green tea epigallocatechin gallate. Endocrinology 141:980-987.
11. Wang X, Tian W (2001) Green tea epigallocatechin gallate: a natural inhibitor of fatty-acid synthase. Biochem Biophys Res Commun 288:1200-1206.
12. Wakil SJ (1989) Fatty acid synthase, a proficient multifunctional enzyme. Biochemistry 28:4523-4530.
13. Swinnen JV, Roskams T, Joniau S, et al (2002) Overexpression of fatty acid synthase is an early and common event in the development of prostate cancer. Int J Cancer 98:19-22.

14. Kuhajda FP (2000) Fatty-acid synthase and human cancer: new perspectives on its role in tumor biology. Nutrition 16:202-208.
15. Swinnen JV, Ulrix W, Heyns W, et al (1997) Coordinate regulation of lipogenic gene expression by androgens: evidence for a cascade mechanism involving sterol regulatory element binding proteins. Proc Natl Acad Sci USA 94:12975-12980.
16. Swinnen JV, Esquenet M, Goossens K, et al (1997) Androgens stimulate fatty acid synthase in the human prostate cancer cell line LNCaP. Cancer Res 57:1086-1090.
17. Milgraum LZ, Witters LA, Pasternack GR, et al (1997) Enzymes of the fatty acid synthesis pathway are highly expressed in in situ breast carcinoma. Clin Cancer Res 3:2115-2120.
18. Swinnen JV, Vanderhoydonc F, Elgamal AA et al (2000) Selective activation of the fatty acid synthesis pathway in human prostate cancer. Int J Cancer 88:176-179.
19. Yang YA, Morin PJ, Han WF, et al (2003) Regulation of fatty acid synthase expression in breast cancer by sterol regulatory element binding protein-1c. Exp Cell Res 282:132-137.
20. Pizer ES, Jackisch C, Wood FD, et al (1996) Inhibition of fatty acid synthesis induces programmed cell death in human breast cancer cells. Cancer Res 56:2745-2747.
21. Kuhajda FP, Pizer ES, Li JN, et al (2000) Synthesis and antitumor activity of an inhibitor of fatty acid synthase. Proc Natl Acad Sci USA 97:3450-3454.
22. Pizer ES, Thupari J, Han WF, et al (2000) Malonyl-coenzyme-A is a potential mediator of cytotoxicity induced by fatty-acid synthase inhibition in human breast cancer cells and xenografts. Cancer Res 60:213-218.
23. Pizer ES, Wood FD, Heine HS, et al (1996) Inhibition of fatty acid synthesis delays disease progression in a xenograft model of ovarian cancer. Cancer Res 56:1189-1193.
24. Brusselmans K, De Schrijver E, Heyns W, et al (2003) Epigallocatechin-3-gallate is a potent natural inhibitor of fatty acid synthase in intact cells and selectively induces apoptosis in prostate cancer cells. Int J Cancer 106:856-862.
25. Abe I, Seki T, Umehara K, et al (2000) Green tea polyphenols: novel and potent inhibitors of squalene epoxidase. Biochem Biophys Res Commun 268:767-771.
26. Swinnen JV, Van Veldhoven PP, Timmermans L, et al (2003) Fatty acid synthase drives the synthesis of phospholipids partitioning into detergent-resistant membrane microdomains. Biochem Biophys Res Comm 302:898-903.
27. Simons K, Toomre D (2000) Lipid rafts and signal transduction. Nat Rev Mol Cell Biol 1:31-39.

Silencing of the Fatty Acid Synthase Gene by RNA Interference Inhibits Growth and Induces Apoptosis of LNCaP Prostate Cancer Cells

Ellen De Schrijver, Koen Brusselmans, Walter Heyns, Guido Verhoeven, and Johannes V. Swinnen

Summary

A screening for androgen-regulated genes in prostate cancer (PCA) cells revealed that androgenes, apart from their well known effects on cell survival, proliferation, and differentiation, stimulate the expression of several lipogenic genes, including the gene encoding fatty acid synthase (FAS). FAS, a key enzyme in the biosynthesis of fatty acids, is markedly overexpressed in many epithelial cancers including cancer of the prostate, breast, ovary and endometrium. To gain more insight into the role of FAS in cancer cells and to explore its potential as a novel target for antineoplastic therapy, we have used the potent and highly sequence-specific technique of RNA interference (RNAi) to silence FAS in LNCaP prostate cancer cells. RNAi-mediated down-regulation of FAS expression resulted in a major decrease in the synthesis of triglycerides and of phospholipids partitioning into detergent-resistant membrane microdomains. These effects were accompanied by marked morphological changes, including a reduction in cell volume, a loss of cell-cell contacts, and the formation of spider-like extrusions. Furthermore, silencing of the FAS gene by RNAi significantly inhibited LNCaP cell growth and ultimately resulted in induction of apoptosis. In striking contrast with LNCaP cells, RNAi-mediated inhibition of FAS did not influence viability of nonmalignant fibroblasts. The data presented herein suggest that overexpression of FAS induced by hormones, growth factors, or other mechanisms may play an important role in cancer cell biology, and that RNA interference, particularly targeting lipogenic genes, constitutes a promising tool for the development of new cancer treatments.

Introduction

Steroid hormones have a strong influence on the biology of several common human cancers, including cancer of the prostate, breast, endometrium, and ovarium. To gain more insight into this process, a screening for androgen-regulated genes was set up in PCA cells (1). In addition to their well known effects on cell survival, proliferation, and differentiated function, androgens markedly stimulate the expression of several lipogenic enzymes, including the one encoding fatty acid synthase (FAS) (2).

FAS is a key enzyme in the biosynthesis of fatty acids from small carbon substrates such as acetyl-CoA and malonyl-CoA (3). In most human tissues, FAS expression is low, as the bulk of the required lipids are obtained from the diet, with the exception of lactating breast and cycling endometrium (4). In contrast, elevated expression of FAS and FAS activity are frequent phenotypic alterations in many human epithelial cancers including cancer of the prostate, breast, endometrium, ovary, lung, colon, stomach, tongue, oral cavity, and esophagus (5-8). Expression of FAS increases very early in cancer development, and is further elevated in more advanced tumors, particularly those with a poor prognosis. Although the underlying mechanism of FAS overexpression in tumor cells is not completely understood, recently, it was shown that FAS overexpression is part of a more general coordinate up-regulation of multiple lipogenic genes (9, 10). Factors that have been shown to contribute to FAS overexpression include steroid hormones and alterations in growth factor production or signaling. Effects of both factors are mediated by activation of sterol regulatory binding proteins (SREBPs), lipogenic transcription factors that play a key role in cellular lipid homeostases (11-15).

To gain more insight into the role of FAS in cancer cells, and to explore its potential as a novel target for antineoplastic therapy, chemical inhibitors of FAS, such as cerulenin and c75, have been used (5, 16-20). However, these inhibitors have several shortcomings. Cerulenin, a natural mycotoxin, harbors a very reactive epoxide group that may interact with other proteins, and, therefore, affect processes other than fatty acid synthesis. In this respect, cerulenin suppresses protein palmitoylation (21), cholesterol synthesis (22), and proteolysis (23, 24). In addition, the use of cerulenin as FAS inhibitor is limited because of its chemical instability (5). Recently, more stable FAS inhibitors, such as c75, have become available (17), but the specificity of these compounds requires further investigation.

To avoid potential non-specific effects related to chemical inhibitors, we have explored the use of the potent and highly sequence specific technique of RNA interference (RNAi) to silence FAS in LNCaP PCA cells. RNAi is a cellular process resulting in enzymatic cleavage and breakdown of mRNA, guided by sequence-specific double stranded RNA oligonucleotides (siRNAs) (Figure 1) (25-27).

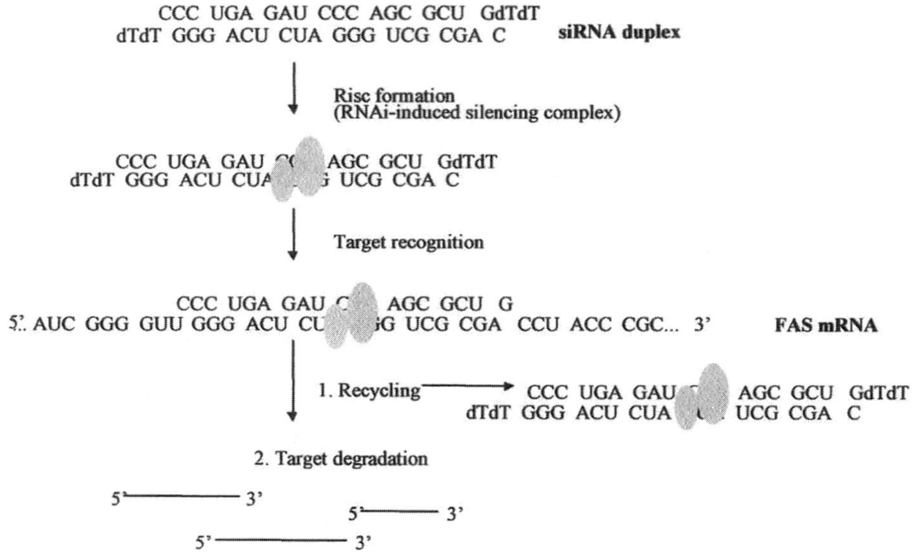

Figure 1. Principle of siRNA-mediated RNA interference. Small interfering RNA (siRNA) interacts with helicase (circle) and nuclease (oval) to form a complex termed RNA-induced silencing complex (RISC). Helicase in RISC uses ATP to unwind siRNA, enabling the antisense strand to bind to its target in messenger RNA. Nuclease in RISC cleaves the mRNA, which is then quickly degraded by other RNAses.

Results and Discussion

To specifically silence the FAS gene, LNCaP cells were transfected with FAS siRNA as previously described (27). As a control for specificity of RNAi, LNCaP cells were transfected with siRNA-targeting luciferase, which is not expressed by LNCaP cells. Western blot analysis demonstrated that expression of FAS was severely decreased in LNCaP cells transfected with FAS siRNA compared to control cells transfected with luciferase siRNA (Figure 2).

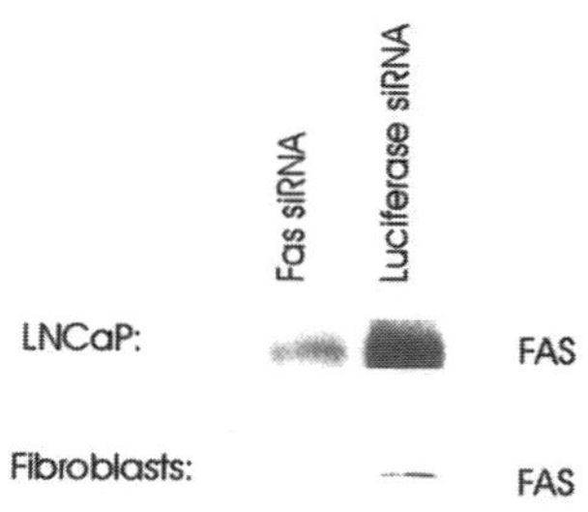

Figure 2. Effects of FAS RNAi on FAS expression. LNCaP cells were transfected with luciferase siRNA or FAS siRNA as previously described (27). FAS protein levels (and CK18 as a control for equal loading) were determined by western blot analysis 72 h after transfection.

Commensurate with the lower FAS protein expression, FAS RNAi resulted in a 4.0-fold decrease of FAS activity in LNCaP extracts, and reduced total lipogenesis as quantified by measuring the incorporation of 2-[^{14}C]-labeled acetate into cellular lipids of growing LNCaP cells (27). To investigate the impact of the decreased lipid synthesis by silencing the different lipid species, lipid extracts were analyzed by TLC for phospholipids and neutral lipids, respectively. The majority of [^{14}C]-label in control cells was incorporated into phospholipids (80%), specifically in phospholipids partitioning into detergent-resistant membrane microdomains (28). FAS RNAi caused 2.0- to 3.0-fold decrease in the synthesis of these phospholipids, suggesting that FAS plays an important role in the control of membrane microdomain biology. Smaller amounts of label were found in triglycerides and free cholesterol (Figure 3). FAS RNAi caused a 7.0-fold decrease in the synthesis of triglycerides in LNCaP cells. In contrast, no effect was observed on the total cellular cholesterol content, illustrating the specificity of the effects of RNAi.

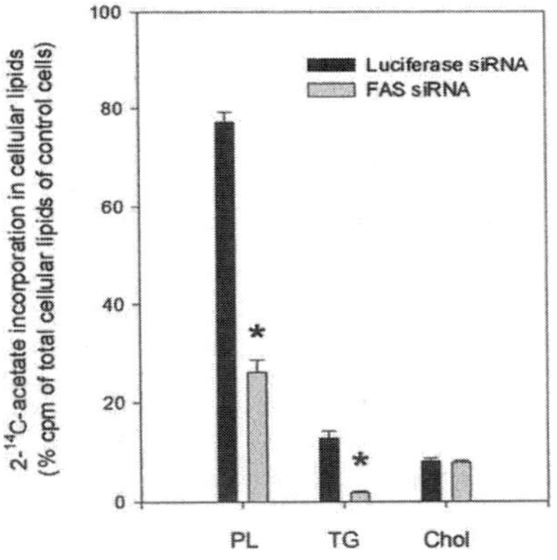

Figure 3. Effects of FAS RNAi on lipid biosynthesis. LNCaP cells transfected with siRNA-targeting luciferase or FAS. After 72 h, cells were treated with 2-[^{14}C]-labeled acetate for 4 h. Lipids were extracted as previously described (1). Radioactivity was measured by scintillation counting. Acetate incorporation was analyzed after TLC lipid separation, and [^{14}C]-incorporation was quantitated by PhosphorImager screens. PC, phosphatidyl-choline; PS, -serine; PI, -inositol; PE, -ethanolamine; TG, Triglycerides; C, cholesterol; CE cholesterol esters. *Significantly different from the luciferase control condition.

As a result of FAS RNAi, LNCaP cells underwent striking morphological changes reminiscent of those seen after androgen-deprivation. The cells became smaller, made poor cell-cell contacts, and displayed multiple spider-like extensions (27). In addition, inhibition of FAS by RNAi inhibited growth of LNCaP cells, and ultimately, resulted in apoptosis and cell death (Table 1) (27). To explore whether FAS RNAi may cause a growth disadvantage selectively to cancer cells expressing high levels of FAS, we evaluated the effects of RNAi in a control line with low levels of FAS expression. Since normal epithelial cells (which show low FAS expression *in vivo*) expressed high levels of FAS when cultured *in-vitro*, we turned to nonmalignant human fibroblasts, as a model of low FAS-expressing cells. Interestingly, in these fibroblasts, siRNA-targeting FAS still reduced FAS expression and activity, but this decrease did not result in either growth inhibition or induction of apoptosis (27) (Table1).

Table 1. Impact of FAS RNAi on Proliferation of LNCaP Cells and Normal Human Fibroblasts[1].

Time (h) After Transfection	LNCaP cells Luciferase siRNA	LNCaP cells FAS siRNA	Normal Human Fibroblast Luciferase siRNA	Normal Human Fibroblast FAS siRNA
0	4.0 ± 0.3	4.0 ± 0.3	2.04 ± 0.48	2.0 ± 0.38
72	15.0 ± 0.6	6.8 ± 1.3[2]	8.70 ± 0.93	7.9 ± 0.74
120	28.5 ± 4.0	6.7 ± 0.8[2]	13.46 ± 1.45	11.93 ± 1.17

1 LNCaP cells or normal human fibroblasts were transfected with siRNA-targeting luciferase or FAS. At the indicated time points, the number of viable cells were counted using the trypan blue dye exclusion assay. Data represent the mean ± SD (n = 6) of the number of viable cells (x10^5).

2 Significantly different from control (luciferase siRNA-transfected cells).

Taken together, these data suggest that overexpression of FAS induced by hormones, growth factors, or other mechanisms may play an important role in cancer cell biology, and that RNA interference targeting lipogenic genes constitutes a promising tool for the development of new cancer treatments.

Acknowledgments

This work was supported by Cancer Research Grants FB Insurance (Fortis Bank) and VIVA, a grant 'Geconcerteerde Onderzoeksactie van de Vlaamse Gemmenschap', research grants from the Fund for Scientific Research-Flanders (Belgium), FWO, and a grant Interuniversity Poles of Attraction Programme-belgian State, Prime Minister's Office, Federal Office for Scientific, Technical and Cultural Affairs.

References

1. Swinnen JV, Van Veldhoven PP, Esquenet M, et al (1996) Androgens markedly stimulate the accumulation of neutral lipids in the human prostatic adenocarcinoma cell line LNCaP. Endocrinology 137:4468-4474.
2. Swinnen JV, Esquenet M, Goossens, et al (1997) Androgens stimulate fatty acid synthase in the human prostate cancer cell line LNCaP. Cancer Res. 57: 1086-1090.
3. Wakil SJ (1989) Fatty acid synthase, a proficient multifunctional enzyme, Biochemistry 28:4523-4530.
4. Weiss L, Hoffmann GE, Schreiber R, et al (1986) Fatty-acid biosynthesis in man, a pathway of minor importance. Purification, optimal assay conditions, and organ distribution of fatty-acid synthase. Biol Chem Hoppes-Seyer 367:905-912.

5. Kuhajda FP (2000) Fatty-acid synthase and human cancer: new perspectives on its role in tumor biology. Nutrition 16:202-208.
6. Piyathilake CJ, Frost AR, Manne, et al (2000) The expression of fatty acid synthase (FASE) is an early event in the development and progression of squamous cell carcinoma of the lung. Hum Pathol 31:1068-1073.
7. Swinnen JV, Roskams T, Joniau S, et al (2002) Overexpression of fatty acid synthase is an early and common event in the development of prostate cancer. Int J Cancer 98:19-22.
8. Kusakabe T, Nashimoto A, Honma, et al (2002) Fatty acid synthase is highly expressed in carcinoma, adenoma and in regenerative epithelium and intestinal metaplasia of the stomach. Histopathology 40:71-79.
9. Swinnen JV, Vanderhoydonc F, Elgamal AA, et al (2000) Selective activation of the fatty acid synthesis pathway in human prostate cancer. Int J Cancer 88:176-179.
10. Milgraum LZ, Witters LA, Pasternack GR, et al (1997) Enzymes of the fatty acid synthase are highly expressed in in situ breast carcinoma. Clin Cancer Res 3:2115-2120.
11. Swinnen JV, Heemers H, Deboel, et al (2000) Stimulation of tumor-associated fatty acid synthase expression by growth factor activation of the sterol regulatory element-binding protein pathway. Oncogene 19:5173-5181.
12. Li JN, Mahmoud MA, Han WF, et al (2000) Sterol regulatory element-binding protein-1 participates in the regulation of fatty acid synthase expression in colorectal neoplasia. Exp Cell Res 261:159-165.
13. Swinnen JV, Ulrix W, Heyns W, et al (1997) Coordinate regulation of lipogenic gene expression by androgens: evidence for a cascade mechanism involving sterol regulatory element binding proteins. Proc Natl Acad Sci USA 94:12975-12980.
14. Heemers H, Maes B, Foufelle, et al (2001) Androgens stimulate lipogenic gene expression in prostate cancer cells by activation of the sterol regulatory element-binding protein cleavage activating protein/sterol regulatory element-binding protein pathway. Mol Endocrinol 15:1817-1828.
15. Van de Sande T, De Schrijver E, Heyns W, et al (2002) Role of the phosphatidylinositol 3'-kinase/PTEN/Akt kinase pathway in the overexpression of fatty acid synthase in LNCaP prostate cancer cells. Cancer Res 62:642-646.
16. Pizer ES, Chrest FJ, DiGiuseppe JA, et al (1998) Pharmacological inhibitors of mammalian fatty acid synthase suppress DNA replication and induce apoptosis in tumor cell lines. Cancer Res 58:4611-4615.
17. Kuhajda FP, Pizer ES, Li JN, et al (2000) Synthesis and antitumor activity of an inhibitor of fatty acid synthase. Proc Natl Acad Sci USA 97:3450-3454.
18. Pizer ES, Wood FD, Heine HS, et al (1996) Inhibition of fatty acid synthesis delays disease progression in a xenograft model of ovarian cancer. Cancer Res 56:1189-1193.

19. Pizer ES, Jackisch C, Wood FD, et al (1996) Inhibition of fatty acid synthesis induces programmed cell death in human breast cancer cells. Cancer Res 56:2745-2747.
20. Li JN, Gorospe M, Chrest FJ, et al (2001) Pharmacological inhibition of fatty acid synthase activity produces both cytostatic and cytotoxic effects modulated by p53. Cancer Res 61:1493-1499.
21. Lawrence DS, Zilfou JT, Smith CD (1999) Structure-activity studies of cerulenin analogues as protein palmitoylation inhibitors. J Med Chem 42:4932-4941.
22. Malvoisin E, Wild F (1990) Effect of drugs which inhibit cholesterol synthesis on syncytia formation in vero cells infected with measles virus. Biochim Biophys Acta 1042:359-364.
23. Moelling K, Schulze T, Knoop MT, et al (1990) In vitro inhibition of HIV-1 proteinase by cerulenin. FEBS Lett 261:373-377.
24. Ikuta K, Luftig RB (1986) Inhibition of cleavage of Moloney murine leukemia virus gag and env coded precursor polyproteins by cerulenin. Virology 154:195-206.
25. Zamore PD, Tuschl T, Sharp PA, et al (2000) RNAi: double-stranded RNA directs the ATP-dependent cleavage of mRNA at 21 to 23 nucleotide intervals. Cell 101:25-33.
26. Elbashir SM, Harborth J, Lendeckel W, et al (2001) Duplexes of 21-nucleotide RNAs mediate RNA interference in cultured mammalian cells. Nature 411:494-498.
27. De Schrijver E, Brusselmans K, Heyns W, et al (2003) RNA interference-mediated silencing of the fatty acid synthase gene attenuates growth and induces morphological changes and apoptosis of LNCaP prostate cancer cells. Cancer Res 63:3799-3804.
28. Swinnen J, Van Veldhoven P, Timmermans L, et al (2003). Fatty acid synthase drives the synthesis of phospholipids partioning into detergent-resistant membrane microdomains. Biochem Biophys Res Com 302:898-903.

Androgens Stimulate the SREBP Pathway in Prostate Cancer Cells by Inducing a Shift in the SCAP-Retention Protein(s) Balance

Hannelore Heemers, Walter Heyns, Guido Verhoeven, and Johannes V. Swinnen

Summary

In human prostate cancer (PCA) cells, androgens (As) coordinately stimulate the expression of genes involved in the synthesis of fatty acids and cholesterol. Interestingly, several of these genes are overexpressed in human cancers, particularly those with a poor prognosis, and are promising targets for anti-neoplastic therapy. Exploration of the mechanism underlying the lipogenic effects of As revealed that As induce activation of Sterol Regulatory Element-Binding Proteins (SREBPs), proteolytically activated transcription factors that play a central role in the control of cellular lipid homeostasis. The exact mechanism by which As interfere with the SREBP pathway remains to be investigated. Herein, we show that A-activation of the SREBP pathway depends on the presence of the structural elements required also for its sterol-regulated activation. Detailed studies revealed that As substantially increase the expression of the transporter protein SCAP in 2 of the 3 cell lines tested. Co-transfection of an expression vector encoding SCAP lead to a marked increase in the transcription of a key gene involved in cholesterol synthesis. In support of the involvement of activated SREBPs in the observed effects, SCAP-mediated transcriptional activation of this lipogenic gene was dependent on the presence of intact SREBP binding sites in its promoter region, and was counteracted by addition of dominant-negative SREBP forms. These data suggest that As activate the SREBP pathway by inducing a change in the normal cellular balance between SCAP and its retention protein complex.

Introduction

Since the development of endocrine therapy for the treatment of PCA (1), androgens (As) have been known to play a major regulatory role in PCA cell biology. In a search for novel A-regulated genes, using the LNCaP PCA cell line as an experimental paradigm, As were shown to coordinately stimulate the expression of several genes involved in the synthesis, transport and metabolism of fatty acids and cholesterol (2). Interestingly, increased lipogenesis, characterized by over-expression of fatty acid synthase (FAS), a key enzyme in the biosynthesis of fatty acids, has been observed in many human cancers, including PCA (3). Increased expression of FAS is considered as one of the most common molecular alterations found in PCA (4). As interference with lipogenesis delays tumor growth in several experimental models, these findings may have a significant therapeutic potential (3). In view of these observations, efforts were directed towards identifying the mechanism underlying the lipogenic effects of As in LNCaP PCA cells. These studies revealed that As activate SREBPs (2).

SREBPs are a family of transcription factors involved in the maintenance of intracellular cholesterol homeostasis, the control of fatty acid synthesis and the differentiation of adipocytes. Synthesized as inactive 125-kDa precursor proteins, SREBPs are anchored into intracellular membranes where they form a complex with SREBP-cleavage activating protein (SCAP). SCAP in turn interacts with a retention protein complex. When cellular sterol levels are low, interaction between SCAP and its retention protein complex is lost, permitting the SREBP/SCAP complex to translocate to the Golgi apparatus. Within this cell organelle two proteases [site 1 protease (S1P) and site 2 protease (S2P)] act to release a 68-kDa aminoterminal SREBP fragment. This transcriptionally active SREBP fragment (nuclear SREBP, nSREBP) migrates to the nucleus, where it activates the transcription of a large set of sterol regulatory element (SRE)-containing genes belonging to the pathways of fatty acid and cholesterol synthesis (5-7) (Figure 1).

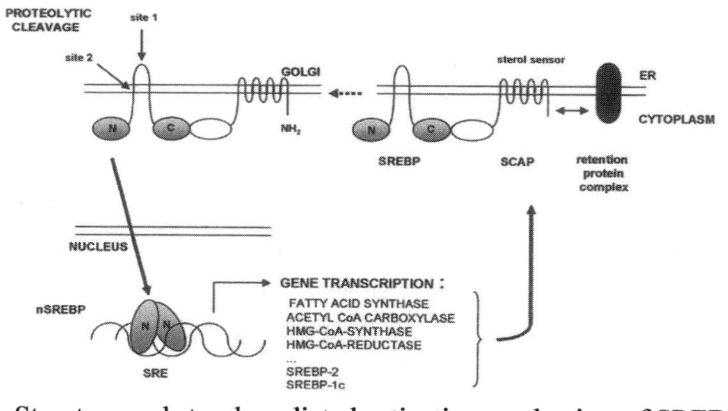

Figure 1. Structure and sterol-mediated activation mechanism of SREBPs.

Treatment of LNCaP cells was shown to lead to increased nuclear levels of mature active SREBPs. Moreover, the stimulatory effects of As on the expression of key lipogenic genes such as FAS and HMG-CoA-synthase (SYN) depended on the presence of intact SREs in the promoter regions of these genes and were counteracted by dominant-negative SREBP forms (2, 8). Recently, similar effects of As on SREBP-dependent transcription of lipogenic genes have been described also in MDA-PCa-2a (9) and PC346c (10) PCA cells (8). Taken together, these data provide evidence for the involvement of the SREBP pathway in the A-induction of the lipogenic program in PCA cells. The exact molecular mechanism(s) by which As affect the SREBP pathway remained to be investigated.

Results

Androgen Activation of the SREBP Pathway Relies on the Presence of the Structural Elements Required for its Sterol-Regulation. To unravel the mechanism(s) by which A treatment of PCA cells gives rise to an increase in the content of nSREBPs, we investigated whether the action of As involves the same segments of the SREBP precursor protein that are required for its sterol regulation (such as the cleavage sites for S1P and S2P and the SCAP-interacting carboxyterminal domain). By means of site-specific mutagenesis of an expression construct for SREBP-2 (pTK-HSV-BP2) expression constructs for SREBP-2 were generated that had undergone mutations (site 1 and site 2) or deletions (COOH terminus) at these crucial sites (Figure 2A). Recombinant SREBP was provided with an HSV epitope tag. Immunoblot analysis using an antibody directed against the HSV epitope was performed on total cell extracts (SREBP precursor protein expression), as well as on nuclear extracts (nSREBP expression) of LNCaP cells grown either in the presence or absence of As or in the presence or absence of sterols. As shown in Figure 2B, activation of SREBP was observed only in cells expressing wild type SREBP precursor proteins, indicating the importance of intact recognition sites for S1P and S2P and the need for interaction with SCAP for As to cause maturation of SREBP precursors (Figure 2B).

Androgen Treatment of LNCaP and MDA-PCA-2a Cells Markedly Upregulates the Expression of the Transporter Protein SCAP. The data shown in Figure 2) suggest that the structural elements involved in the A-activation of SREBP precursor proteins are identical to those required for sterol-regulated SREBP activation. Thus, we investigated whether As affect the expression of components essential for the sterol-regulation of the SREBP pathway. In the three cell lines studied, slight increases in the mRNA expression for SREBP-1c and SREBP-2, known nSREBP target genes (11, 12), and S2P were observed following A exposure. SREBP-1a and S1P were not affected (data not shown). In both LNCaP and MDA-PCa-2a cells, A treatment caused a more pronounced increase in the mRNA expression of SCAP. In PC346c cells such a stimulatory effect on the expression of SCAP was not observed (Figure 3).

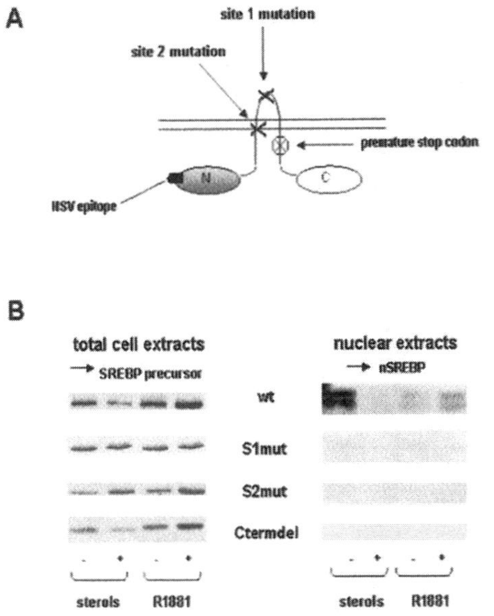

Figure 2. Mutation of sites required for sterol regulation of SREBPs prevents A-induced SREBP activation. Schematic representation of the locations of the mutations, as well as the HSV epitope tag (**A**). Total cell and nuclear extracts derived from LNCaP cells expressing either wild type SREBP-2 (wt) or SREBP-2 in which site 1 (S1mut) or 2 (S2mut) was mutated or in which the C-terminus was deleted (Ctermdel), incubated in the absence (-) or presence (+) of R1881, or in the +/- of sterols, were subjected to western blotting with an antibody directed against the HSV epitope (**B**). Results shown are representative of three independent experiments.

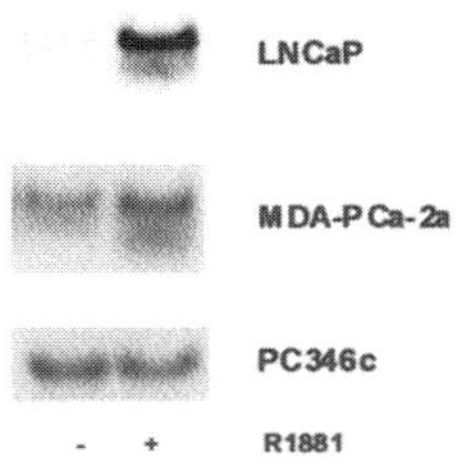

Figure 3. As markedly stimulate the expression of SCAP in LNCaP and MDA-PCa-2a cells. Cells were cultured in the presence (+) or absence (-) of 10^{-8}M R1881 for 2 (LNCaP and MDA-PCa-2a cells) or 3 (PC346c cells) days. Total RNA was isolated and northern blot analysis was performed with a SCAP probe. Data shown are representative of two independent experiments.

Overexpression of SCAP Leads to Activation of the SREBP Pathway. To explore whether increased SCAP expression is actively involved in the A-induced activation of the SREBP pathway, an expression vector encoding SCAP (pcDNA1.1-SCAP) was generated. COS-7 cells were transiently transfected with pcDNA1.1-SCAP and pTK-HSV-BP2. Two days later, nuclear extracts were prepared and immunoblot analysis was performed using an antibody directed against the HSV-epitope. Figure 4A shows that enhancing cellular SCAP levels does indeed result in a pronounced increase in mature nuclear SREBP. To investigate whether such a SCAP-induced elevation of nuclear SREBP levels is responsible for the observed increase in lipogenic gene expression, pcDNA1.1-SCAP was co-transfected into MDA-PCa-2a cells with a promoter-reporter construct derived from SYN. Co-transfection of pcDNA1.1-SCAP gave rise to a marked increase in SYN reporter activities. Mutation of SREs in this promoter fragment prevented SCAP-induced transcriptional activation of the reporter gene (Figure 4B). Co-transfection of MDA-PCa-2a cells with increasing amounts of an expression construct encoding a dominant-negative form of SREBP-1 (DN-SREBP) gradually prevented the stimulatory effect of enhanced SCAP expression on both FAS (data not shown) and SYN (Figure 4C) promoter-reporter constructs. Similar SCAP-induced stimulatory effects on the transcriptional activation of these genes were observed in LNCaP cells (data not shown) (8).

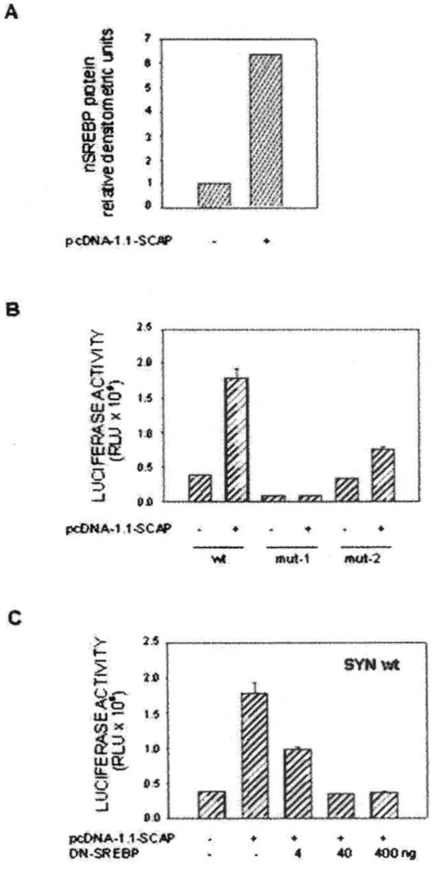

Figure 4. Effect of increased cellular SCAP levels on nSREBP-dependent lipogenic gene transcription. **(A)** COS-7 cells were transiently transfected with pTK-HSV-BP2 and 20 ng of an expression construct encoding human SCAP (pcDNA1.1-SCAP). Representative data from two independent experiments. **(B)** MDA-PCa-2a cells were transiently transfected with either a plasmid containing a luciferase reporter gene driven by a HMG-CoA-synthase (SYN) promoter fragment harboring two SREs (SRE-1 and SRE-2) (wt), or a similar construct in which either SRE-1 or SRE-2 was mutated (mut-1 and mut-2). Fourty ng of an expression construct encoding human SCAP (pcDNA1.1-SCAP) was added. **(C)** Increasing amounts of a plasmid encoding a dominant-negative form of SREBP (DN-SREBP) were added. Luciferase activity was expressed in relative luciferase units (RLU). The data represent the mean ± SEMs of incubations performed in triplicate.

Discussion

To investigate whether the molecular mechanism underlying the A activation of the SREBP pathway in PCA cells makes use of the same structural elements required for its sterol-regulated activation, site-specific mutations were made in the SREBP-2 precursor protein. These studies demonstrated the involvement of elements typically involved in the sterol-mediated activation of SREBPs such as the cleavage sites 1 and 2 and the SCAP-interacting SREBP-domain also in the A-induced maturation of SREBPs. These findings are in agreement with earlier observations that the size of the nSREBP fragments generated after A exposure is similar to the size of the fragments resulting from sterol-depletion (data not shown). Further studies investigating whether As affect the expression of elements involved in sterol-mediated activation of the SREBP pathway revealed that As induce an increase in the expression of SCAP in 2 of the 3 cell lines studied. Since alternative routes of SREBP activation, e.g. by cholesterol depletion, do not result in a similar stimulation of SCAP expression (data not shown), we concluded that the observed increase in SCAP expression is not an indirect effect of SREBP activation. In support of this conclusion is the finding that SCAP is up-regulated just before the accumulation of mature SREBP and well before the activation of lipogenic gene expression (data not shown) (8). The hypothesis of a direct link between the increase in SCAP expression and the stimulation of lipogenic gene expression by As was further corroborated by the finding that forced overexpression of SCAP led to enhanced nuclear accumulation of mature SREBP and to activation of lipogenic gene expression in an SRE- and SREBP-dependent manner in MDA-PCa-2a as well as LNCaP PCA cells. These data suggest that although A-activation of the SREBP pathway makes use of the same structural elements as does sterol-regulation, the mechanism by which As activate the SREBP pathway is different from the one observed under conditions of sterol depletion. Low intracellular concentrations of cholesterol are sensed by SCAP and reduce the interaction with its retention protein complex that holds the SREBP/SCAP complex in the endoplasmic reticulum (7). As a result the SREBP/SCAP complex is released from the endoplasmic reticulum and moves to the Golgi complex where the SREBP precursor is cleaved and activated. As might reach the same effect by promoting SCAP production and changing the balance between SCAP and its retention protein complex. Although A treatment of PC346c cells also gives rise to maturation of SREBP precursors and SREBP-dependent transcription of lipogenic genes, in this particular cell line the A-induced increase in the expression of SCAP could not be observed, suggesting that apart from up-regulation of SCAP other mechanisms may contribute to the A-induced activation of the SREBP pathway. In this respect it will be exciting to explore whether As also affect the expression of the recently identified proteins insig-1 and insig-2 (13, 14), components of the postulated SCAP retention complex.

Acknowledgments

The PC346c cell line was a kind gift from Dr. van Weerden (10). MDA-PCa-2a cells were kindly provided by Dr. Navone (9). This work was supported by 'Geconcerteerde Onderzoeksactie van de Vlaamse Gemeenschap', FWO, FB Insurance, VIVA, Belgische Federatie Tegen Kanker, IWT (H.H.) and Interuniversitary Poles of Attraction Programme-Belgian State.

References

1. Huggins C, Hodges CV (1941) Studies on prostatic cancer: effect of castration, of estrogen, and of androgen injection on serum phosphatases in metastatic carcinoma of the prostate. Cancer Res 1:293-297.

2. Swinnen JV, Ulrix W, Heyns W, et al (1997) Coordinate regulation of lipogenic gene expression by androgens: Evidence for a cascade mechanism involving sterol regulatory element binding proteins. Proc Natl Acad Sci USA 94:12975-12980.
3. Kuhajda FP (2000) Fatty-acid synthase and human cancer: new perspectives on its role in human biology. Nutrition 16:202-208.
4. Swinnen JV, Roskams T, Joniau S, et al (2002) Overexpression of fatty acid synthase is an early and common event in the development of prostate cancer. Int J Cancer 98:19-22.
5. Brown MS, Ye J, Rawson RB, et al (2000) Regulated intramembrane proteolysis: a control mechanism conserved from bacteria to humans. Cell 100:391-398.
6. Brown MS, Goldstein JL (1999) A proteolytic pathway that controls the cholesterol content of membranes, cells and blood. Proc Natl Acad Sci USA 96:11041-11048.
7. Yang T, Goldstein JL, Brown MS (2000) Overexpression of membrane domain of SCAP prevents sterols from inhibiting SCAP-SREBP exit from endoplasmic reticulum. J Biol Chem 275:29881-29886.
8. Heemers H, Maes B, Foufelle F, et al (2001) Androgens stimulate lipogenic gene expression in prostate cancer cells by activation of the sterol regulatory element-binding protein cleavage activating protein/sterol regulatory element-binding protein pathway. Mol Endocrinol 15:1817-1828.
9. Navone NM, Olive M, Ozen M, et al (1997) Establishment of two human prostate cancer cell lines derived from a single bone metastasis. Clin Cancer Res 3:2493-2500.
10. van Weerden WM, de Ridder CM, Verdaasdonk CL, et al (1996) Development of seven new human prostate tumor xenograft models and their histopathological characterization. Am J Pathol 149:1055-1062.
11. Sato R, Inoue J, Kawabe Y, et al (1996) Sterol-dependent transcriptional regulation of sterol regulatory element-binding protein-2. J Biol Chem 271:26461-26464.
12. Amemiya-Kudo M, Shimano H, Yoshikawa T, et al (2000) Promoter analysis of the mouse sterol regulatory element-binding protein-1c gene. J Biol Chem 275:31078-31085.
13. Yang T, Espenshade PJ, Wright ME, et al (2002) Crucial step in cholesterol homeostasis: sterols promote binding of SCAP to INSIG-1, a membrane protein that facilitates retention of SREBPs in ER. Cell 110:489-500.
14. Yabe D, Brown MS and Goldstein JL (2002) Insig-2, a second endoplasmic reticulum protein that binds SCAP and blocks export of sterol regulatory element-binding proteins. Proc Natl Acad Sci USA 99:12753-12758.

COMMUNICATIONS

Session II. Hormone Metabolism and Cell/Molecular Biology

Metabolism of 17β-Estradiol in ACI Rat Liver and Mammary Gland After Chronic Estradiol Treatment

Sonia Mesia-Vela, Rosa I Sanchez, Kenneth R. Reuhl, Allan H. Conney, and Frederick C. Kauffman

Summary

A comparative study of the effects of chronic 17β-estradiol (E_2) treatment on microsomal oxidation and conjugation of E_2 via Phase I and II enzymes in the ACI rat mammary gland (MG) and liver was performed. NADPH-dependent oxidation of E_2 was not detected in the MG, but was readily measured in the liver. Oxidation was not altered by chronic E_2 treatment. Ascorbic acid stimulated E_2 oxidation (non-enzymatically) in MG microsomes, but had no effect in the liver. Hepatic but not MG NADP(H):quinone oxidoreductase and glutathione *S*-transferase activities increased 4.0- and 2.0-fold, respectively, after 6 weeks (w) of treatment. MG catalase activity was decreased 64% after 28 w of E_2 treatment, when the rats had developed 100% incidence of MG adenocarcinomas. Moreover, the activities of phenolsulfotransferase SULT1A1and fatty acyl-CoA:E_2-acyltransferase ACO:E_2 decreased by 95 and 80%, respectively, in the MG but not in liver. Decreases in these enzymes were maximal after 6 w and preceded induction of MG tumors. Collectively, these data indicate that E_2 regulates the expression of antioxidant and E_2-metabolizing enzymes differentially in the ACI rat liver and MG. The decreased activities of SULT1A1 and ACO:E_2 may favor accumulation of E_2 available for receptor binding and conversion to catechol estrogens, both of which are implicated in E_2-induced mammary oncogenesis.

Introduction

Metabolism of E_2, the major endogenous estrogen (E), regulates both the level and activity of this hormone in MG and other target tissues. Alterations in E_2 metabolism have been implicated in the progression of hormone-dependent cancers because they may lead to the formation of abnormal amounts of reactive catechol estrogens (CE). Formation of CE in E-target organs may promote extensive oxidative stress and/or the generation of DNA mutations with the potential to initiate cancer (1-3). Formation of high amounts of 4-CE and their conjugates have been reported in human and rodent cancerous MGs (4-6); and 4-hydroxylase activity, an enzyme linked to CEs formation, has been observed in human breast tumor samples (7). Moreover, differential formation of 4- over 2-CE correlates with susceptibility to cancer in E-dependent organs such as the Syrian hamster kidney (8, 9), CD-1 mouse uterus (10), and rat pituitary (11). Thus, alterations of E_2-metabolism leading to the formation of high amounts of CE, and in the activity of enzymes that control CE availability, *i.e.,* SULTs and ACO:E_2, may influence the E_2-induction of MG tumors.

Herein, we evaluate the effect of chronic E_2 treatment on key oxidative and conjugative enzyme activities associated with E_2 metabolism in the MG and liver of ACI rats (12), a strain highly susceptible to the induction of mammary ductal adenocarcinomas (MDAs) after E exposure (12-15).

Materials and Methods

Animals and Treatments. Female ACI rats, 7-8 weeks of age, were treated with a single 20-mg pellet containing 3 mg of E_2 + 17 mg of cholesterol, implanted sub-pannicularly in the shoulder region, as previously described (16). Control animals received 20-mg cholesterol pellets. Animals were killed after 6, 12, or 28 weeks and tissues were removed for analyses.

Preparation of Sub-cellular Fractions. Cytosol and microsomal fractions were prepared by differential centrifugation as described previously (17), and stored at −80°C until used. Protein concentration was determined by the BCA™ protein assay kit.

Enzyme Assays. NAD(P)H quinone oxidoreductase (NQO1) was measured as reported by Jaiswal *et al.* (18). Cytosolic glutathione *S*-transferase (GST), glutathione peroxidase (GPx) and catalase (CAT) activities were measured as described earlier (19-21). Sulfotransferase (SULT) activity was determined as described previously (22) using PAP[^{35}S] as cofactor. NADPH- dependent oxidation of E_2 (CYP450 assay) was measured as described previously (23) using 5 μM (MG) or 50 μM (liver) [^{3}H]-E_2 at 37 °C, for 60 and 30 min, respectively. Blanks, heat-inactivated microsomes, were analyzed with and without NADPH or ascorbic acid. Glucuronidation of E_2 was assayed as described previously (24). ACO:E_2 activity was determined with 25 μM (MG) or 50 μM (liver) [^{3}H]-E_2 as described previously (25). Kinetic parameters were determined using six different concentrations of substrate.

Statistical Analysis. Differences among means were assessed by ANOVA followed by Bonferroni post hoc test for P values < 0.05. The kinetic parameters (*Km* and *Vmax*) of the enzymatic reactions were derived graphically by computer linear regression analysis using Khaleida Graph for Windows program.

Results

Enzyme Assays. Effects of MG and liver cytosol on antioxidant enzyme activities are depicted in Figure 1. In general, specific activities were notably higher in liver compared to MG ranging from ~ 6.0-fold higher for NQO1 to 100.0-fold higher for CAT. E_2-treatment for 6 and 12 w increased hepatic NQO1 activity by 4.1- and 3.7-fold, respectively (Figure 1B); while GST increased by 1.5- and 2.0-fold, respectively (Figure 1D). Both activities returned to control levels by week 28 (Figure 1B, D). Hepatic GPx activity did not change significantly over the course of the treatment (Figure 1, F); however, its activity increased significantly in MG tissue at 28 w (Figure 1E). CAT activity decreased in liver and MG tissue at 28 w (Figure 1G, H).

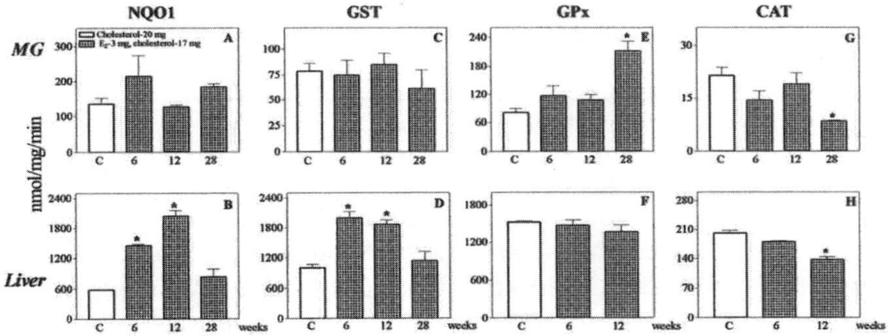

Figure 1. Antioxidant enzyme activities in MG and liver of E_2-treated ACI rats. Data represent the mean ± SE of 3 rats/group. Data from control animals (C) at different periods were similar and pooled together (n = 9). Statistically different from C.

SULT1A1 activity was reduced markedly (95%) after E_2 treatment (Figure 2), and preceded the appearance of E_2-induced MG tumors.

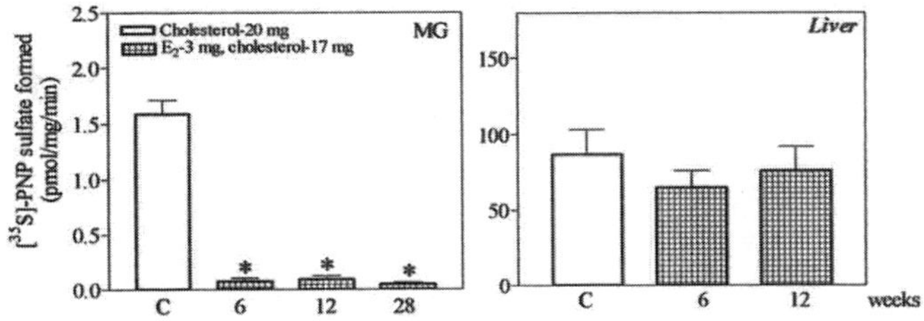

Figure 2. MG and liver SULT1A1 activity of E_2-treated ACI rats. Data represent the mean ± S.E. of 3rats/group. Data for control animals (C) at different periods were similar and pooled together (n = 9). *Statistically different from C.

SULT activity was measured using PNP, E_2, and DHEA as substrates targeting activities of SULT1A1, -E1, and -2A1, respectively. Kinetic parameters for liver and MG SULTs activities are described in Table 1.

Table 1. Kinetic Parameters of E_2-metabolizing Enzymes in MG and Liver of ACI Rat.

Enzyme System	Substrate	Tissue	Kinetic Parameters	
			Km (μM)	Vmax (pM/mg/min)
SULT1A1	PNP	Liver	2	6700
		MG	2.1	1.975
Sulfotransferase	E_2	Liver	10	5040
		MG	48	6.97
SULT2A1	DHEA	Liver	10.8	1860
		MG	Linear to 100 μM	nd

Liver microsomal NADPH-dependent oxidation of E_2 formed majoritary 2-OH-E_2 and estrone (E_1) and small amounts of 4-OH-E_2 and 2-OH-E_1 as reported previously (23, 26) (Table 2). E_2 exposure caused a modest increased in E_1 formation (~1.5-fold). In MG microsomes, several oxidized derivatives such as 6α-, 16α-, 16β-, and 2-OH-E_2, and 6-keto- and 2-OH-E_1, and E_1 were detected in small amounts (data not shown). Blanks without NADPH or with heat-inactivated protein formed similar amount of metabolites, however in the absence of ascorbic acid, they were not formed.

Table 2. Effect of E_2 on ACI hepatic NADPH-dependent Oxidation of E_2 [1]

Treatment	E_2 Metabolites formed (pM/mg/min)				E_2 Metabolized (pM/mg/min)
	4-OH-E_2	2-OH-E_2	2-OH-E_1	E_1	
6 Weeks					
Chol, 20 mg	9 ± 2	123 ± 11	24 ± 7	134 ± 13	392 ± 23 (17%)
E_2, 3 mg + Chol, 17 mg	11 ± 2	149 ± 10	32 ± 6	232 ±7[2]	530 ± 28 (25%)
12 Weeks					
Chol, 20 mg	7 ± 1	169 ± 12	41 ± 7	140 ± 28	473 ± 33 (24%)
E_2, 3 mg + Chol, 17 mg	11 ± 3	182 ± 36	102 ± 26	309 ± 42[2]	644 ± 64 (32%)

[1] Data represent the mean ± SE of 6 rats/group.
[2] Statistically different from control group ($p < 0.05$).

Control liver microsomes incubated with 100 μM E_2 and 2 mM UDPGA resulted in the formation of two metabolites: E_2-3β-glucuronide (128 ± 9 pmol/mg/min) and E_2-17β-glucuronide (87 ± 13 pmol/mg/min) respectively. The formation of these metabolites was not altered by chronic E_2 treatment at any of the time periods examined. MG microsomal formation of E_2 glucuronides was not detected under the same conditions used for liver microsome.

Liver or MG microsome incubations with [^{3}H]-E_2 in the presence of oleoyl-coenzyme A as a cofactor, followed by HPLC detection, revealed a single radioactive peak less polar than E_2 corresponding to E_2-17β-oleoyl ester as reported previously (25). Chronic E_2 treatment did not alter hepatic esterification, but was reduced by ~ 80 % of control values in the MG (Figure 3). This decrease in MG E_2 esterification occurred as early as 6 w. The Vmax but not the Km (8 μM) for this reaction was decreased markedly after E_2 treatment (from 6.017 to 2.725 pmol/mg/min).

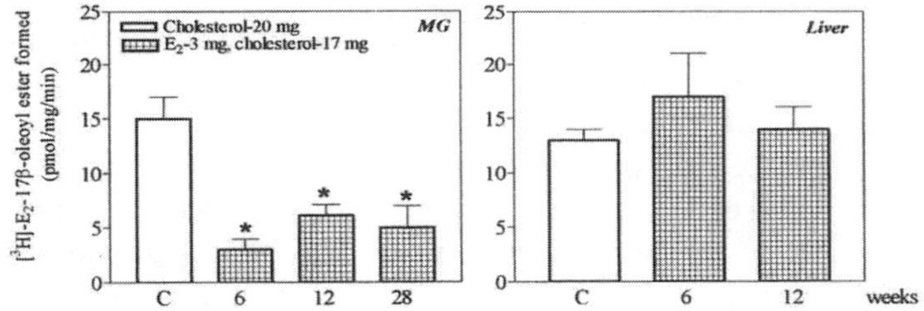

Figure 3. Effect of chronic E_2 treatment on ACO:E_2 activity in ACI rat MG and liver. Data represent the mean ± S.E. of 3 rats/group. Data from the control (C) group at different time periods were similar and pooled (n = 9). * Statistically different from C group, $p < 0.05$.

Discussion and Conclusions

Chronic, low dose E_2 treatment of ACI rats induced a marked decline in MG SULT1A1 and ACO:E_2, but not in the liver. Since both enzymes are involved in the inactivation of E_2, it is noteworthy that the marked decreases in the activity of both enzymes occurred before the appearance of the MG tumors. Sulfonation is an important pathway of E_2 inactivation in rodents and humans (27). Reduction in the E_2 sulfonation rate may prolong the availability of E_2 in target tissues, and enhance its interaction with estrogen receptors (ER). A possible link of impaired sulfonation to carcinogenesis is suggested by the recent finding that polyhalogenated aromatic hydrocarbons and their OH-metabolites are potent inhibitors of several SULTs including SULT1A1 (28).

Our finding that chronic E_2 treatment strongly reduced MG E_2 esterification rates are consistent with observations made in ERα$^+$ MCF-7 human breast cancer cells in which fewer E_2 esters were formed compared to those in ERα$^-$ MDA-MB-

231 and MDA-MB-330 human cells (29). Greater amounts of E_2 are present in $ER\alpha^+$ cells than in $ER\alpha^-$ cells (30) suggesting that the presence of unconjugated E_2 is of prime importance for initiation/development of MG tumorigenesis. Aside from storage, no other physiological function of esterified E_2 metabolites has been described. The possibility that reduction of E_2-esterification and sulfonation occurs after chronic exposure to E_2 leading to higher amounts of biologically active E_2 in the MG of ACI rats warrants further study.

Our data indicate that chronic E_2 treatment did not alter the specific activity of antioxidant enzymes in the MG as it did in the liver. This finding indicates that the regulation of expression of these enzymes by E_2 differs in liver and MG. The expression of NQO1 and GST is mediated, at least in part, through the antioxidant/electrophile response element (ARE/EphRE) (31) found in the regulatory regions of their genes. GPx is also induced in response to oxidative stress but through mechanisms that involve sp-1 and AP-2 regulatory sequences in the promoter region of the rat liver GPx gene which lacks the ARE/EphRE regulatory sequences (32). Also, the induction of cytosolic CAT activity suggests that oxidative stress after 28 weeks of treatment may be a consequence of the presence of MG tumors, and may not be a predominant factor during the initiation/development of these tumors in the rat.

In summary, the most prominent enzymatic changes noted in MG tissue of female ACI rats treated chronically with E_2 were marked decreases in sulfonation and esterification. These changes appeared specific for MG tissue and preceded the appearance of MG tumors. Also, MGs from E sensitive ACI rats readily formed several E_2 metabolites, including CEs in a non enzymatic *i.e.,* independent of NADPH and active microsomes way. Further studies are needed to determine whether these changes in conjugating enzymes are related to accumulation of E_2 available for $ER\alpha$ binding and promotion of MG tumors in ACI rats.

Acknowledgment

This research was funded, in part, by NIH Grant ES05022.

References

1. Stack D, Byun J, Gross ML, et al (1996) Molecular characteristics of catechol estrogen quinones in reactions with deoxyribonucleosides. Chem Res Toxicol 9:851-859.
2. Cavalieri EL, Stack DE, Devanesan PD, et al (1997) Molecular origin of cancer: catechol estrogen-3,4-quinones as endogenous tumor initiators. Proc Natl Acad Sci USA 99:10937-10942.
3. Cavalieri EL, Li K, Balu N, et al (2002) Catechol ortho-quinones: the electrophilic compounds that form depurinating DNA adducts and could initiate cancer and other diseases. Carcinogenesis 23:1071-1077.

4. Liehr JG and Ricci MJ (1996) 4-Hydroxylation of estrogens as marker of human mammary tumors. Proc Natl Acad Sci USA 93:3294-3296.
5. Castagnetta LA, Granata OM, Arcuri FP, et al (1992) Gas chromatography/mass spectrometry of catechol estrogens. Steroids 57:437-443.
6. Devanesan P, Santen RJ, Bocchinfuso WP, et al (2001) Catechol estrogen metabolites and conjugates in mammary tumors and hyperplastic tissue from estrogen receptor-alpha knock out (ERKO)/Wnt-1 mice: implications for initiation of mammary tumors. Carcinogenesis 22:1573-1576.
7. Abul-Hajj YJ (1982) Formation of estradiol 17 beta fatty acyl 17-esters in mammary tumors. Steroids 40:149-156.
8. Li JJ, Li SA (1987) Estrogen carcinogenesis in Syrian hamster tissue: role of metabolism. Fed Proc 46:1858-1863.
9. Liehr JG, Fang WF, Sirbasku DA, et al (1986) Carcinogenicity of catecholestrogens in Syrian hamsters. J Steroid Biochem 24: 353-356.
10. Newbold RR and Liehr JG (2000) Induction of uterine adenocarcinoma in CD-1 mice by catechol estrogens. Cancer Res 60:235-237.
11. Bui QD, Weisz J (1989) Monooxygenase mediating catecholestrogen formation by rat anterior pituitary is an estrogen-4-hydroxylase. Endocrinology 124:1085-1087.
12. Guad LR, Thomas RD, Green M. (2003) Diallyl sulfide inhibits diethylstilbestrol-induced lipid peroxidation in beast tissue of female ACI rats: implications in breast cancer prevention. Oncol Rep 10 (3):739-43.
13. Shull JD, Spady TJ, Snyder MC, et al (1997) Ovary-intact, but not ovariectomized female ACI rats treated with 17β-estradiol rapidly develop mammary carcinoma. Carcinogenesis 18:1595-1601.
14. Cutts JH, Noble RL (1964) Estrone-induced mammary tumors in the rat: I. Induction and behavior of tumors. Cancer Res 24:1116-1123.
15. Harvell DM, Strecker TM, Tochacek M, et al (2000) Rat strain-specific actions of 17β-estradiol in the mammary gland: correlation between estrogen-induced lobuloalveolar hyperplasia and susceptibility to estrogen-induced mammary cancers. Proc Natl Acad Sci USA 97:2779-2784.
16. Li SA, Weroha SJ, Tawfik O, et al (2002) Prevention of solely estrogen-induced mammary tumors by Tamoxifen: Evidence for estrogen receptor mediation. J Endocrinol 175:297-305.
17. Thomas PE, Reick LM, Ryan DE, et al (1983) Induction of two immunochemically related rat liver cytochrome P450 isozymes cytochromes P450c and P450d, by structurally diverse xenobiotics. J Biol Chem 258:4590-4598.
18. Jaiswal AK, Mcbride OW, Adesnik M, et al (1988) Human dioxin-inducible cytosolic NADP(H):menadione oxidoreductase. cDNA sequence and localization of gene to chromosome. J Biol Chem 263:13572-13578

19. Habig WH, Pabst MJ, Jakoby WB (1974) Glutathione S-transferases: The first enzymatic step in mercapturic acid formation. J Biol Chem 249:7130-7139
20. Flohe L, Gunzler WA(1984) Assay of glutathione peroxidase. Meth Enzymol 105:114-121
21. Aebi H (1984) Catalase in vitro. Methods Enzymol 105: 121-126
22. Foldes A and Meek JL (1973) Rat brain phenolsulfotransferase-partial purification and some properties. Biochem Biophys Acta 327:365-374.
23. Mesia-Vela S, Sanchez RI, Li JJ, et al (2002) Catechol estrogen formation in liver microsomes from female ACI and SD rats: comparison of 2- and 4-hydroxylation revisited. Carcinogenesis 23:1369-1372.
24. Sanchez RI, Mesia-Vela S, Kauffman FC (2003) Induction of NAD(P)H quinone oxidoreductase and Glutathione *S*-transferase activities in livers of female August-Copenhagen Irish Rats treated chronically with estradiol: Comparison to the Sprague-Dawley Rat. J Steroid Biochem Mol Biol 87 (2-3):199-206.
25. Xu S, Zhu BT and Conney AH (2001) Stimulatory effect of clofibrate and gemfibrozil administration on the formation of fatty acid esters of estradiol by rat liver microsomes. J. Pharmacol Exp Ther 296:188-197.
26. Zhu BT, Conney AH (1998) Functional role of estrogen metabolism in target cells: review and perspectives. Carcinogenesis 19:1-27
27. Glatt H (2000) Sulfotransferases in the bioactivation of xenobiotics. Chem Biol Interact 129:141-170.
28. Kester MH, Bulduk S, van Toor H, et al (2002) Potent inhibition of estrogen sulfotransferase by hydroxylated metabolites of polyhalogenated aromatic hydrocarbons reveals alternative mechanisms for estrogenic activity of endocrine disruptors. J Clin Endocrinol Metab 87:1142-1150
29. Martyn P, Smith DL, Adams JB (1987) Selective turnover of the essential fatty acid ester components of estradiol-17β lipoidal derivatives formed by human mammary cancer cells in culture. J Steroid Biochem 28:393-398.
30. Adams JB, Vrahimis R, Young CE (1991). Metabolism of lipoidal derivatives of estradiol 17β in human mammary cancer tissue and cell lines. J Steroid Biochem Mol Biol 39:751-758
31. Rushmore TH, Pickett CB (1990) Transcriptional regulation of the rat glutathione S-transferase Ya subunit gene. Characterization of a xenobiotic-responsive element controlling inducible expression by phenolic antioxidants. J Biol Chem 265:14648-14653.
32. Ho YS and Howard AJ (1992) Cloning and characterization of the rat glutathione peroxidase gene. FEBS Lett 301:5-9.

Modulation of Transforming and Clastogenic Activities of Catechol Estrogens by a Catechol-*O*-methyltransferase Inhibitor in Syrian Hamster Embryo Fibroblasts

Takeki Tsutsui, Takeo W. Tsutsui, Yukiko Tamura, and J. Carl Barrett

Summary

Catechol estrogens (CEs) are considered critical intermediates in estrogen (E)-induced carcinogenesis. Previously, we demonstrated that estradiol (E_2), estrone (E_1), and four of their catechol estrogens, 2- and 4-OHE_2 and 2- and 4-OHE_1 induced morphological transformation in Syrian hamster embryo (SHE) cells, and their transforming activities varied as follows: 4-OHE_1 > 2-OHE_1 > 4-OHE_2 > 2-OHE_2 ≥ E_1, E_2, which are consistent with the genetic effects, *i.e.*, chromosome aberrations and DNA adduct formation, of each E. To further elucidate the mechanism of hormonal carcinogenesis, we studied the effect of the catechol-*O*-methyltransferase (COMT) inhibitor Ro41-0960 on the transforming and clastogenic activities of the CEs using SHE cells. The frequencies of transformation and chromosome aberrations induced by 4-OHE_1 were not affected by co-treatment with Ro-41-0960, but those induced by 2-OHE_1 were markedly enhanced. The frequency of transformation induced by 4-OHE_1 was markedly decreased by E_2 in a concentration dependent manner, but this decrease was not inhibited by Ro41-0960. Cell treatment with E_2, 2-OHE_1, or 4-OHE_1 alone induced apoptosis as detected by the TUNEL method. Additive effect on the induction of apoptosis was observed in cells treated with E_2 + 2-OHE_1 or 4-OHE_1. The % apoptotic cells induced by E_2 and 4-OHE_1 decreased in the presence of Ro41-0960, while those induced by E_2 and 2-OHE_1 did not. These results suggest an important role of both the substrate specificity of COMT and the induction of apoptosis in CE-induced carcinogenesis.

Introduction

Es are carcinogenic in humans and rodents (1, 2). Accumulating evidence has suggested that estrogenic activity is necessary but not sufficient to explain the carcinogenic activity of Es *in-vivo* and *in-vitro* (3). Another mechanism related to genetic alterations has been suggested in studies of E-induced carcinogenesis (4, 5).

Endogenous Es are implicated as a possible etiological factor in the causation of certain types of human cancers such as breast, endometrium, ovary, prostate, and, possibly, brain cancers (6). E_2 undergoes oxidative metabolism at C-17 to yield E_1, which can be hydroxylated to yield 16α-OHE_1 and CEs (2-OHE_1 and 4-OHE_1). 16α-OHE_1 is metabolized to estriol by 17β-hydroxysteroid dehydrogenase. E_2 also can be hydroxylated to 2-OHE_2 or 4-OHE_2. These CEs are mainly inactivated by *O*-methylation, catalyzed by COMT, but conjugative metabolism by glucuronidation and/or sulfonation also plays a role in conversion of the CEs to hormonally inactive metabolites (6). CEs are considered critical intermediates in E-induced cancer (7-9). The oxidative metabolites of the CEs, *i.e.* E_2, E_1, 2, 3 or 3, 4-quinones, covalently bind to calf thymus DNA and form DNA adducts (10).

We have used SHE fibroblast cell cultures as a model system to study the ability of Es to transform cells directly (5). SHE cells do not express measurable levels of estrogen receptor (ERα), and E treatment is not mitogenic to the cells (5). Thus, estrogenic stimulation of cell proliferation can be excluded as the mechanism of action in this *in-vitro* assay. However, the cells do have the ability to metabolize Es (5). We have shown that treatment of SHE cells with E_2, E_1, or four of their CEs induces cellular transformation. The transforming abilities of the Es vary as follows: 4-OHE_1 > 2-OHE_1 > 4-OHE_2 > 2-OHE_2 ≥ E_1, E_2, which are consistent with the genetic effects, *i.e.*, chromosome aberrations and DNA adduct formation, of each E (11-13). To further elucidate the mechanism of hormonal carcinogenesis, herein, we studied the effect of a COMT inhibitor, Ro41-0960 on the transforming and clastogenic activities of the CEs using SHE cells.

Materials and Methods

Cells and Chemicals. SHE cell cultures were established from 13-day-gestation hamster fetuses and grown as previously described (11). E_2, 2-OHE_1, and 4-OHE_1 were purchased from Sigma (St. Louis, MO) and dissolved with DMSO at 3 mg/ml. DMSO was added to control cultures at a final concentration of 0.33%.

Cellular Transformation. Cells (2.5 x 10^5) were plated, incubated overnight, and treated with test Es for 48 h. After trypsinization, 2,000 cells were replated, and incubated for 7 days to form colonies. The cells were fixed with absolute methanol and stained with a 10% aqueous Giemsa solution. The number of surviving colonies with > 50 cells and morphologically transformed colonies were scored using previously established criteria (5, 11).

Chromosome Aberrations. SHE cells (5 x 10^6) were plated into 75-cm^2 flasks. After overnight incubation, the cells were treated with test Es for 24 h. Three hours before the end of the treatment time, Colcemid (GIBCO, Grand Island, NY) was administered at 0.2 μg/ml, and metaphase chromosomes were prepared. After trypsinization, cells were treated with 0.9% sodium citrate at room temperature for 13 min, fixed in Carnoy's solution (methanol:acetic acid, 3:1), and spread on glass slides using the air-drying method. Specimens were stained with a 3% Giemsa solution in 0.07 M phosphate buffer (pH 6.8) for 7 min. For determination of chromosome aberrations (gaps, breaks, exchanges, dicentric chromosomes, ring chromosomes, and fragmentations), 100 metaphases/experimental group were scored. Achromatic lesions greater than the width of the chromatid were scored as gaps, unless there was displacement of the broken piece of chromatid. If there was displacement, these were recorded as breaks.

Detection of Apoptotic Cells. SHE cells (10^6) were plated into 75-cm^2 flasks. After overnight incubation, cells were treated for 24 h with each test E. In some experiments, cells were pretreated with the caspase inhibitor Z-Asp-CH_2-DCB (Peptide Institute, Osaka, Japan) at 200 μM for 2 h. After harvesting with 0.1% trypsin, the cells were fixed in 4% neutral buffered formalin for 10 min, and a 50 μl volume of cell suspension at a density of 2 x 10^6 cells/ml was dropped on slide glasses and air-dried. To detect apoptotic cells, the cells on the slide glasses were stained using the TUNEL method and an apoptosis *in situ* detection kit (Wako Pure Chemical, Tokyo, Japan) according to the manufacturer's instructions. More than 1,500 cells were scored for each group, and the % apoptotic cells were determined from the number of apoptotic cells relative to the total number of cells x 100.

Results

Treatment of SHE cells with E_2, 2-OHE_1, or 4-OHE_1 induced morphological transformation in SHE cells. The transforming activities of these Es were ranked as follows: 4-OHE_1 > 2-OHE_1 > E_2. When SHE cells were co-treated with 2- or 4-OHE_1 together with the specific COMT inhibitor Ro41-0960 (3 μM), the transformation frequency induced by 4-OHE_1 was not affected, but that induced by 2-OHE_1 was markedly enhanced.

Chromosome aberrations in SHE cells were induced by treatment with 2- or 4-OHE_1, but not by E_2 treatment. The rank of the clastogenic activities of 2- or 4-OHE_1, determined by the induced frequencies of chromosome aberrations, was as follows: 4-OHE_1 > 2-OHE_1. The frequency of chromosome aberrations induced by 4-OHE_1 was not affected by co-treatment with Ro41-0960, but that induced by 2-OHE_1 was markedly enhanced.

The frequency of morphological transformation induced by 4-OHE_1 was markedly decreased by co-treatment with E_2 in a concentration dependent manner, but the decrease by E_2 was not inhibited by Ro41-0960. Cell treatment with E_2, 2-

OHE$_1$, or 4-OHE$_1$ alone induced apoptosis as detected by the TUNEL method. Additive effect on the induction of apoptosis was observed in SHE cells treated with E$_2$ in combination with 2-OHE$_1$ or 4-OHE$_1$. The induction of apoptosis by these Es was inhibited by the caspase inhibitor Z-Asp-CH$_2$-DCB. The % apoptotic cells induced by co-treatment with E$_2$ and 4-OHE$_1$ was decreased when Ro 41-0960 was administered to the cultures, while the % apoptotic cells induced by co-treatment with E$_2$ and 2-OHE$_1$ was not decreased by Ro41-0960.

Discussion

The effects of the COMT inhibitor Ro41-0960 on the transforming and clastogenic activities of CEs in SHE cells were examined. Because the transforming and clastogenic activities of E$_1$ CEs in SHE cells are higher than those of the E$_2$ CEs (12), we studied the effects of the COMT inhibitor on the transforming and clastogeniic activities induced by 2- or 4-OHE$_1$. The difference in the transforming activities between E$_1$ and E$_2$ CEs are consistent to the abilities of these CEs to induce chromosome aberrations, gene mutations at the Na^+/K^+ ATPase and/or HPRT loci, and DNA adduct formation in SHE cells (11-13), suggesting that genetic damage at the gene and/or chromosome levels acts as one of the causal mechanisms of CE-induced cellular transformation in SHE cells. In the hamster kidney tumor model, 4-OH CEs are carcinogenic, whereas 2-OH CEs are not (7, 8). Although the incidence is lower than that of 4-OHE$_2$, 2-OHE$_2$ induces uterine adenocarcinoma in mice (9). The hamster kidney contains high estrogen 4-hydroxylase activity (14) and the methylation of 4-OHE$_2$ by COMT is inhibited by 2-OHE$_2$ *in-vitro* (15). This may facilitate accumulation of 4-OHE$_2$ in the hamster kidney. However, few or low carcinogenic activity of 2-OHE$_2$ in hamsters and mice may also be due to rapid methylation and rapid metabolic clearance of 2-OHE$_2$ itself (16). The frequencies of transformation and chromosome aberrations induced by 4-OHE$_1$ were not affected by co-treatment with Ro41-0960, but those induced by 2-OHE$_1$ were markedly enhanced by the COMT inhibitor. The results suggest that SHE cells retain COMT activity which is more specific for 2-OHE$_1$ than for 4-OHE$_1$. Because SHE cells have endogenous metabolizing enzymes that exhibit oxidative and peroxidative activity (5), 2-OHE$_1$ could accumulate and undergo metabolic conversion in SHE cells to reactive intermediates such as quinone estrogen (CE-2, 3-quinone) that may enhance chromosome aberrations and cellular transformation. Studies on the substrate specificity of COMT in SHE cells are currently in progress.

The frequency of transformation induced by 4-OHE$_1$ was markedly decreased by co-treatment with E$_2$ in a concentration dependent manner, but the decrease by E$_2$ was not inhibited by Ro41-0960. This suggests that another mechanism rather than COMT may participate in the decrease in the E$_2$ induced transformation frequencies. Treatment of SHE cells with E$_2$ or 2- or 4-OHE$_1$ alone induced apoptosis which was enhanced by co-treatment with E$_2$ in combination with

2- or 4-OHE_1. The % apoptotic cells induced by co-treatment with E_2 and 4-OHE_1 were decreased by Ro41-0960, while that induced by co-treatment with E_2 and 2-OHE_1 was not. These findings also suggest that the substrate specificity of COMT affects the induction of apoptosis induced by Es, which leads modulation of the transformation frequencies induced by CEs in SHE cells.

In summary, CEs are considered critical intermediates in E-induced carcinogenesis. The transforming activity and genetic effects in SHE cells varied among CEs. The substrate specificity of COMT and the induction of apoptosis may modulate these activities.

References

1. IARC, IARC (1979) Monographs on the Evaluation of the Carcinogenic Risk of Chemicals to Humans, Sex Hormones (II). Lyon: International Agency for Research on Cancer Vol 21:139-362.
2. Yager JD Jr, Yager R (1980) Oral contraceptive steroids as promoters of hepatocarcinogenesis in female Sprague-Dawley rats. Cancer Res 40:3680-3685.
3. Huff J, Boyd J, Barrett JC (1996) Cellular and Molecular Mechanisms of Hormonal Carcinogenesis: Environmental Influences. New York: Wiley-Liss, 1996.
4. Liehr JG (1997) Dual role of oestrogens as hormones and pro-carcinogens: tumor initiation by metabolic activation of oestrogens. Eur J Cancer Prev 6:3-10.
5. Tsutsui T, Barrett JC (1997) Neoplastic transformation of cultured mammalian cells by estrogens and estrogenlike chemicals. Environ Health Perspect 105:619-624.
6. Zhu BT, Conney AH (1998) Functional role of estrogen metabolism in target cells: review and perspective. Carcinogenesis 19:1-27.
7. Liehr JG, Fang W-F, Sirbasku DA, et al (1986) Carcinogenicity of catechol estrogens in Syrian hamsters. J Steroid Biochem 24:353-356.
8. Li JJ, Li SA (1987) Estrogen carcinogenesis in Syrian hamster tissues: role of metabolism. Fed Proc 46: 1858-1863.
9. Newbold RR, Liehr JG (2000) Induction of uterine adenocarcinorma in CD-1 mice by catechol estrogens. Cancer Res 60:235-237.
10. Stack DE, Byun J, Gross ML, et al (1996) Molecular characteristics of catechol estrogen quinones in reactions with deoxyribonucleosides. Chem Res Toxicol 9:851-859.
11. Hayashi N, Hasegawa K, Komine A, et al (1996) Estrogen-induced cell transformation and DNA adduct formation in cultured Syrian hamster embryo cells. Mol Carcinogenesis 16:149-156.

12. Tsutsui T, Tamura Y, Yagi E, et al (2000) Involvement of genotoxic effects in the initiation of estrogen-induced cellular transformation: studies using Syrian hamster embryo cells treated with 17 β-estradiol and eight of its metabolites. Int J Cancer 86:8-14.
13. Yagi E, Barrett JC, Tsutsui T (2001) The ability of four catechol estrogens of 17 β-estradiol and estrone to induce DNA adducts in Syrian hamster embryo fibroblasts. Carcinogenesis 22:1505-1510.
14. Weisz J, Bui QD, Roy D, et al (1992) Elevated 4-hydroxylation of estradiol by hamster kidney microsomes: a potential pathway of metabolic activation of estrogens. Endocrinology 131:655-661.
15. Roy D, Weisz J, Liehr JG (1990) The *O*-methylation of 4-hydroxyestradiol is inhibited by 2-hydroxyestradiol: implications for estrogen-induced carcinogenesis. Carcinogenesis 11:459-462.
16. Roy D, Liehr JG (1999) Estrogen, DNA damage and mutations. Mutat Res 42:107-115.

Bi-directional Regulation of Human Progesterone Receptors and the Mitogen Activated Protein Kinase Pathway in Breast Cancer Cell Models

Emily Faivre, Ming Qiu, and Carol A. Lange

Summary

Breast cancers (BCs) often have increased mitogen-activated protein kinase (MAPK) activity. This pathway influences BC cell growth in part by targeting steroid hormone receptors. Activation of p42 and p44 MAPKs increases progesterone receptor (PR) transcriptional activity in the presence of progestins, and triggers their rapid down-regulation by the ubiquitin-proteasome pathway. In turn, progestins increase the expression of type I growth factor receptor tyrosine kinases that feed into MAPK activation. Recently, progestins have been shown to activate the p42/p44 MAPK module in a PR-dependent manner, but independently of their function as transcription factors. Mechanisms of bi-directional cross-talk between these two pathways are becoming well-documented. Herein, we provide an overview of the primary ways in which steroid hormone receptor and growth factor cross-talk occurs, using examples from our work with human PR as a model receptor; we demonstrate MAPK regulation of PR subcellular localization, transcriptional synergy, and regulation of cyclin D1 expression. Cross-talk between growth factor and PR-mediated signaling events is an important means by which growth regulatory genes are coordinately regulated, and may contribute to the growth of hormonally responsive normal breast tissue and to BC.

Introduction

Progestins are Mitogens in the Breast. BC is the most frequently diagnosed cancer in western women; over 180,000 new cases are identified annually in the

United States alone (American Cancer Society; www.cancer.org). Of those cases, approximately 60% of BC patients present with a hormone-dependent form of disease, wherein the proliferation of the cancer cells is under the control of the ovarian steroid hormones estrogen (E) and progesterone (P). Both E and P are important molecules implicated in normal mammary gland development as well as BC progression; however, several lines of evidence, including developmental studies conducted with E and PR-null mice (8,12), indicate that P is a primary mitogen in the breast epithelium (4, 22, 23). Further, recent clinical trials demonstrated a significantly greater BC risk for post-menopausal women receiving E and P hormone replacement therapy (HRT), as compared to E HRT alone (6), implicating P action in increased BC risk.

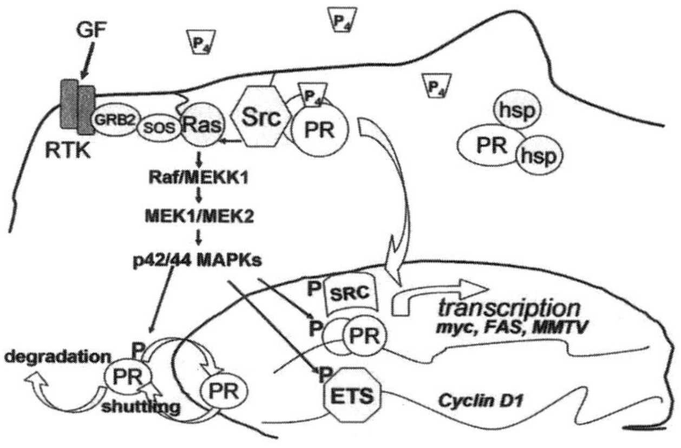

Figure 1. The role of MAPK activation in PR actions. PR and growth factors independently activate MAPKs. MAPKs positively regulate PR action via direct phosphorylation of PR or indirectly via phosphorylation of PR co-regulators. Direct phosphorylation of PR occurs in both the absence and presence of steroid hormone ligands, and plays a role in nuclear sequestration of unliganded PR (18). Phosphorylated (PO_4) PR may in turn act on PRE-containing gene promoters or other PR-regulated genes. Additionally, activation of MAPKs by PR provides for regulation of gene promoters independently of the transcriptional functions of PR (*i.e.*, via MAPK-regulated ETS factors).

The mitogenic actions of progestins may in partially explained by interactions between PR and canonical mitogenic signal transduction pathways. This article will focus on direct regulation of PR by phosphorylation, and the role of MAPK signaling in PR actions (11, 18, 21). We highlight recent findings: 1) PR and growth factors activate MAPKs independently, and this may result in positive regulation of PR action via "feed-back" regulation (*i.e.*, direct phosphorylation), occurring in both the absence and presence of steroid hormone ligands and on PRE-containing gene promoters, and 2) activation of MAPKs by PR provides for regulation of gene targets whose promoters do not contain PREs and are otherwise independent of PR-transcriptional activities. This important linkage provides for tight control of a large number of genes coordinately regulated in response to convergence of growth factor and PR-mediated signaling events (Figure 1).

Direct Phosphorylation of PR by MAPKs. Phosphorylation of the PR is an important mechanism of regulation, influencing subcellular localization (18), transcriptional activation (21), and degradation (11). PR Ser294 is a ligand-inducible PO_4 site, becoming rapidly PO_4 upon exposure to hormone (25). Ser294 is also a proline-directed (SP) or MAPK consensus site (PXXSP). Extensive studies have uncovered a role for MAPK PO_4 in the regulation of numerous PR functions (11, 18, 21). Ser294 PO_4 is required to sustain transcriptional synergy in the presence of progestins and growth factors (21), and to signal for ligand-induced receptor degradation (11). Additionally, Ser294 mediates nuclear translocation of PRs in the absence of ligand (18) (Figure 2).

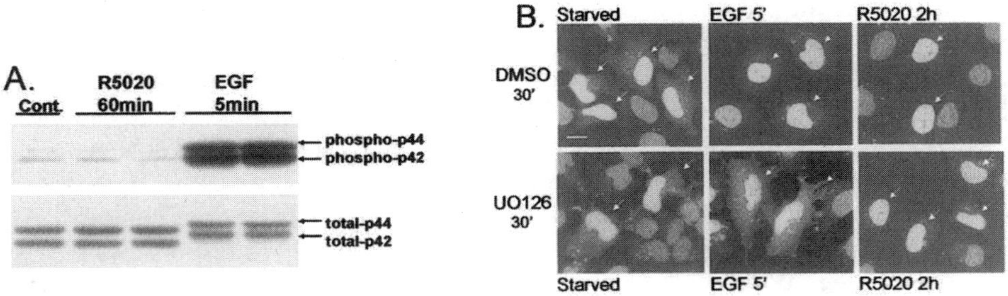

Figure 2. MAPK mediates nuclear localization of unliganded PR. **A.** MAPKs are - activated by growth factors, but not progestins. **B.** The MEK1/MEK2 inhibitor, U0126, blocks EGF- but not R5020-induced PR nuclear accumulation. Representative fields are shown from one of three independent experiments. Scale bar, 20 μM. Adapted with permission from Qiu, *et. al.* (18).

Growth factors, such as EGF, are strong activators of p42/p44 MAPKs (Figure 2A), and interestingly, within 5 min of EGF treatment in the absence of progestins, PR Ser 294 becomes PO_4, and a rapid nuclear accumulation of PR occurs, as measured by both fluorescence microscopy and cellular fractionation experiments (18). Mutation of the consensus MAPK site, Ser294 to Ala (S294A), abolished EGF-mediated translocation, however the ability of progestin (R5020) to induce nuclear localization of S294A PR was unaffected (18). To investigate the role of MAPK activation in regulation of both ligand-dependent and -independent PR localization (Figure 1), PRs were visualized by fluorescence microscopy under circumstances when MAPK activation is blocked. Hela cells transiently expressing GFP-PR fusion proteins were serum starved for 24 h, then pretreated with vehicle control or U0126, a small molecule inhibitor of MEK1 and MEK2, upstream kinases required for activation of p42/p44 MAPKs (Figure 2B). Cells were then treated with EGF for 5 minutes or the synthetic progestin, R5020, for 2 h; time points were determined to represent the peak of PR nuclear accumulation induced by either agent (18). A dual staining method was employed, wherein all nuclei were stained with propidium iodide, while GFP-PR in transfected cells were measured by direct fluorescence. Inhibition of p42/p44 MAPKs by pretreatment with U0126 blocked EGF-induced PR nuclear accumulation, but did not affect R5020-induced translocation (Figure 2B). These results suggest a mechanism for ligand-independent transcriptional activation of PR.

Role of MAPK Activation in PR Transcriptional Activity. We previously reported transcriptional synergy between progestins and constitutive active MEKK1, a strong activator of p42/p44 MAPKs, in both T47D and HeLa cervical carcinoma cells (21). This event occurred using wt PR-B, but not S294A PR-B. Interestingly, MAPK inhibition had little or no effect on PR transcriptional activity induced by ligand alone, but effectively blocked the large synergistic component of PR activity in the presence of both R5020 and MEKK1 (21). To directly determine the effects of MAPK activation on PR in the complete absence of ligand binding, we utilized N-terminal PR-B (NTB), a deletion mutant of PR lacking the C-terminal hormone-binding domain (24). Confocal microscopy using monoclonal antibodies to the PR N-terminus demonstrated that like wt-PR-B, NTB is found in both nuclear and cytoplasmic compartments of resting HeLa cells (Figure 3A). However, in contrast to wt-PR (Figure 2B), the cytoplasmic portion of NTB does not localize to the nucleus in response to progestin, but translocates only in response to EGF (5 min) treatment. We then compared NTB transcriptional activity to that of wt-PR-B ± R5020, and constitutively active MEKK1 in HeLa cells (Figure 3B). As previously reported (21), wt PR-B transcriptional activity was further increased by expression of constitutive active MEKK1 (Figure 3B), and increased in a concentration dependent manner over a 10-1000 ng range of NTB DNA (24). MEKK1 expression constitutively activated p42/p44 MAPKs in T47D cells (21) and HeLa cells (Figure 3B inset) and surprisingly, greatly increased NTB transcriptional activity. Again, addition of progestin had no effect on NTB transcriptional activity. To illustrate the MAPK-dependence of MEKK1 on PR activity, experiments were repeated with and without co-expression of dominant negative (dn) MEK-1 (13). Dn-MEK1 inhibited p42/p44 MAPK activity (Figure 3C inset) and blocked the effect of MEKK1 on NTB transcriptional hyper-activity (Figure 3C), thus indicating the participation of the p42/p44 MAPK signaling module. MEKK1 had no effect on a luciferase reporter gene lacking a PRE, or containing an ERE in place of the PRE, indicating that regulation is specific to transcriptional signaling by PR (not shown). These data demonstrate that ligand binding is not an absolute requirement for "hyper-activation" of NTB mutant PR by activated kinase pathways, and reveal a MAPK-dependent component of PR transcriptional activity that is separable from that induced by ligand alone.

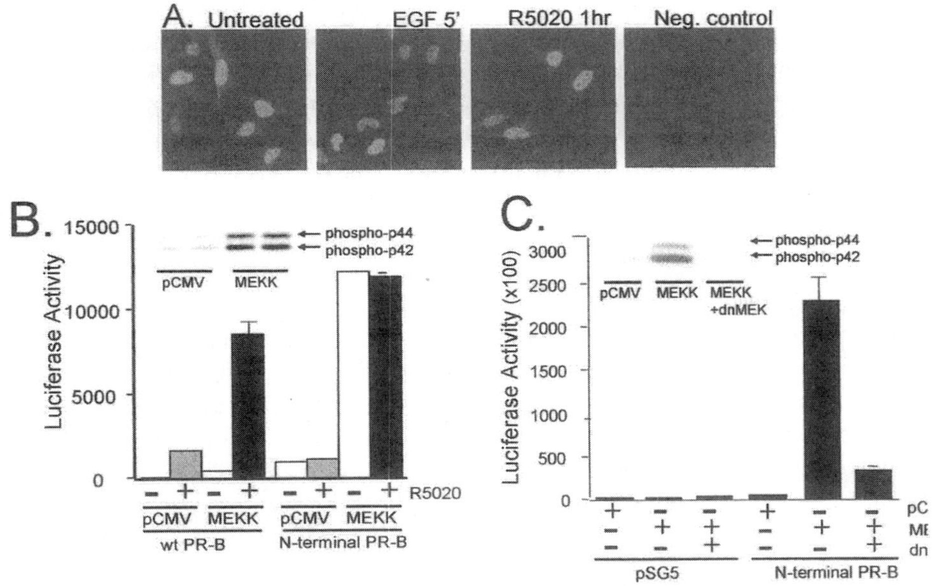

Figure 3. MAPK elevates N-terminal PR-B transcriptional activity. **A.** Nuclear translocation of NTB in response to EGF, but not R5020. HeLa cells were transiently transfected with N-terminal PR-B (24); NTB; lacking the C-terminal ligand binding domain). Cells were washed and placed in serum-free medium (SFM) for 24 h, then untreated (vehicle control) or treated with EGF (30 ng/mL) for 5 min, or R5020 (10 nM) for 2 h. NTB in transfected cells were visualized by IHC and direct fluorescence as described in the legend for Figure 2. Omission of the 1ary antibody indicated that staining was specific for NTB PR (Neg. Control). Scale bar, 20 μM. **B.** MEKK1 activates N-terminal PR-B. HeLa cells were transiently co-transfected with wt PR-B or NTB, and either pCMV control "empty" vector or MEKK1, and a PRE-driven luciferase reporter construct. Cells were treated with vehicle control (EtOH) or R5020 for 24 h. Cells were lysed and firefly luciferase/Renilla activity was measured using a Dual-luciferase assay kit. **(Inset:** MEKK1 activated p42/p44 MAPKs in HeLa cells, as measured using phospho-specific MAPK antibodies. **C.** Dominant negative dn-MEK blocks MEKK1 activation of NTB. Hela cells were transiently co-transfected with pSG5 control empty vector or NTB (in pSG5), and either pCMV control vector, MEKK1 (in pCMV), or both MEKK1 and dn-MEK. Cells were placed in SFM for an additional 24 hrs prior to lysis and luciferase/Renilla measurements were taken as above. **Inset:** MEKK1 activation of p42/p44 MAPKs is blocked by co-expression of dn-MEK1.

Activation of MAPK Modules by Liganded PR. In addition to the well-documented actions of PR as transcription factors, liganded PR induce a rapid activation of cytoplasmic signaling in a receptor-dependent manner, including that of c-Src^{p60} kinase (3, 14), PI3-K (5), and p42/p44 MAPKs (3, 14). P non-genomic effects, via the activation of analogous signaling molecules in fish and amphibian eggs is recognized as important for the oocyte maturation (15, 16, 19). In the mammalian system, however, the physiologic role(s) of activation of these signaling molecules is not yet clear, but could theoretically serve to potentiate several functions of the PR (Figure 1). One prospective role of this cytoplasmic activation is to provide a mechanism for rapid, direct PO_4 of PR coincident with ligand binding, thus allowing a form of PR-induced amplification of the signal or positive "feedback." We propose that this feedback has a dramatic influence on PR function (Figure 1).

Progestin stimulation of multiple cytoplasmic kinases in BC cells suggests that these pathways may contribute to the mitogenic actions of progestins (3). The activation of p42/p44 MAPKs is initiated by the upstream activation of Src^{p60} kinase via complex formation with the liganded PR (3). This complex formation is mediated by a direct interaction between the SH3 domain of Src^{p60} kinase and a polyproline region in the PR N-terminus. Other groups have reported that Src^{p60} activation is facilitated through a complex formed by the liganded PR, ERα, and Src^{p60} kinase (2,14).

To test the ability of PR to activate MAPKs, variant T47D-YB BC cells engineered to stably express only the full-length B isoform of the PR (20) were serum-starved for 24 h and treated for 1.5, 5, and 15 min with the synthetic progestin R5020 or vehicle control. Western blotting for the activated form of p42/p44 MAPKs using a phospho-specific antibody indicates a strong, rapid activation of p42/p44 MAPKs between 5 and 15 min of progestin treatment. At later time points, the activity was reduced to near basal levels by 60 min (data not shown). No activation of p42/p44 MAPKs was observed in PR null T47D-Y cells (20), demonstrating the PR dependence of this event (data not shown).

Another member of the MAPK family stimulated in response to growth factors and implicated in BC is p38 MAPK. The putative involvement of p38 MAPK in P signaling was suggested by the observation that a small molecule inhibitor of p38 (SB203580) could block transcriptional synergy of hormone activated PR with growth factors in a manner similar to MEK1 inhibitors (21). Therefore, we examined the activation of p38 MAPK by Western blotting with phospho-p38-specific antibodies. We found that p38 MAPK can be activated in response to progestin in a time course similar to that of p42/p44 MAPKs; the mechanism by which liganded-PR are able to activate p38 MAPK is not yet known.

PR Regulation of Cyclin D1 Expression Requires MAPK. As an alternative to feedback regulation of PR, rapid activation of cytoplasmic signaling cascades by liganded PR may provide a separate mechanism by which transcriptional regulation in the nucleus is achieved (Figure 1). For example, growth factor stimulation of the p42/p44 MAPK signaling module results in phosphorylation and activation of numerous transcription factors, including c-Ets-2, a member of the large family of

ETS transcription factors. ETS proteins promote G1 phase progression by the regulation of a number of genes involved in cell growth regulation, including the cell cycle regulator cyclin D1 (1).

The *cyclin D1* promoter is complex, containing multiple target elements for transcription factors inputs, but lacks a canonical PRE. Cyclin D1 protein levels increase in response to progestin treatments alone (7), and are synergistically regulated by progestins and growth factors in a MAPK-dependent manner (9, 10), a phenomenon that may partially explain the mitogenic actions of progestins. To demonstrate progestin-dependent regulation of cyclin D1 protein levels, T47D-YB BC cells were serum starved for 48 h, treated with or without the synthetic progestin R5020 for periods of 2, 6, 12, and 24 h. Cyclin D1 levels were examined by Western blotting (Figure 3). Within 2 h of R5020 treatment, cyclin D1 protein increased, peaking between 6-12 h, a time-course similar to that induced by growth factor treatment (17), by 24 h, cyclin D1 expression returned to basal levels.

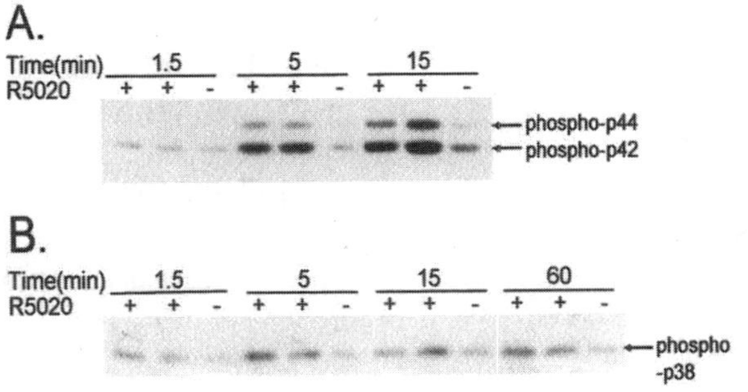

Figure 4. MAPK signaling is activated in response to R5020 treatment. Proteins were separated by SDS-PAGE, and activation of **(A)** p42/p44 MAPKs and **(B)** p38 MAPK was measured by Western blotting using phospho-specific MAPK antibodies.

To examine the MAPK-dependency of progestin-mediated regulation of cyclin D1 protein levels, T47D-YB cells were pre-treated for 30 min with or without U0126 (to inhibit MEK1/MEK2), or SB203580 (an inhibitor of p38 MAPK), followed by a 6 h treatment with or without R5020. Control experiments demonstrated the specificity of the inhibitors for their respective kinases (Figure 5B, bottom panel). Cyclin D1 protein levels were upregulated by R5020 in the absence of inhibitors. By comparison, R5020-dependent upregulation of cyclin D1 was significantly diminished in cultures pretreated with U0126 (Figure 5B). Inhibition of p42/p44 MAPKs also caused a reduction of basal cyclin D1 levels. These data suggest that p42/p44 MAPKs are required for both basal cyclin D1 expression levels, and efficient progestin-induced cyclin D1 upregulation. Inhibition of p38 had no effect on basal or R5020-induced cyclin D1 protein levels, demonstrating the specificity of progestin-activated p42/p44 MAPKs for cyclin D1 expression. Currently, we are testing the effects of progestins on the cyclin D1 promoter in reporter assays. PR-induced p38 MAPK may influence other actions of liganded PR (21) or the activity of additional MAPK-regulated transcription factors. Thus, the activation of cytoplasmic signaling pathways by liganded-PR provides both feedback onto the receptor for enhanced PR action at specific PR-regulated genes (Figure 3), and couples this to the regulation of additional growth regulatory genes (whose promoters function independently of PR but require PR-activated MAPKs).

Conclusion

Herein, we have demonstrated, using examples from our recent studies (18) that MAPKs can directly regulate PR subcellular localization (Figure 2). Previously, we showed that MAPK activation by growth factors induced rapid nuclear translocation of wt, but not S294A mutant PR; both receptors concentrated in the nucleus following progestin treatment (18). Notably, the MEK inhibitor, U0126, blocked EGF but not progestin-induced translocation of wt PR (Figure 2).

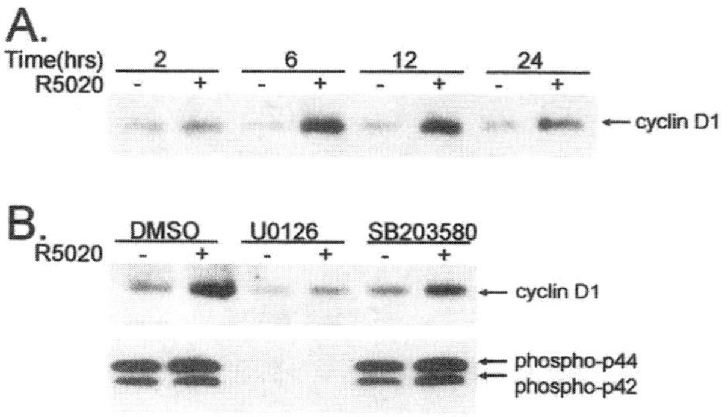

Figure 5. PR regulates cyclin D1 protein levels. **A.** Time course of cyclin D1 regulation by R5020. Western blot analysis of cyclin D1. **B.** Inhibition of MAPK reduced R5020-inducted cyclin D1 protein expression. Cyclin D1 protein was measured as in Figure 3A.

Also, MAPKs profoundly influence PR transcriptional activity (21). NTB-PR localized in the nucleus following growth factor treatment, but not progestin treatment, and exhibited greatly increased transcriptional activity that was entirely MAPK-dependent (Figure 3). Finally, progestins rapidly activated p42/p44 and p38 MAPKs (Figure 4) and increased the expression of the cell-cycle mediator, cyclin D1 (Figure 5). Changes in cyclin D1 expression required the activity of p42/p44 MAPKs, but not p38 MAPK.

In sum, upregulation of growth factor-mediated signaling pathways during BC progression is predicted to profoundly influence steroid hormone responsiveness via direct regulation of steroid hormone receptor nuclear localization and transcriptional activity of both PR and MAPK regulated gene targets. Cross-talk between MAPK pathways and steroid hormone receptors is emerging as an important regulatory paradigm that may contribute to ultimate endocrine failure in steroid receptor-driven cancers.

Acknowledgements

This work was supported by a training grant from the University of Minnesota Cancer Center (E. Faivre) and by NIH grant DK53825 (C.A. Lange). We are grateful to Kathryn B. Horwitz (University of Colorado Health Science Center, Denver) for her generous support and wonderful mentorship and for the use of the T47D-YB cell lines used in selected studies.

References

1. Albanese C, Johnson J, Watanabe G, et al (1995) Transforming p21ras mutants and c-Ets-2 activate the cyclin D1 promoter through distinguishable regions. J Biol Chem 270:23589-23597.
2. Ballare C, Uhrig M, Bechtold T, et al (2003) Two domains of the progesterone receptor interact with the estrogen receptor and are required for progesterone activation of the c-Src/Erk pathway in mammalian cells. Mol Cell Biol 23:1994-2008.
3. Boonyaratanakornkit V, Scott MP, Ribon V, et al (2001) Progesterone receptor contains a proline-rich motif that directly interacts with SH3 domains and activates c-Src family tyrosine kinases. Mol Cell 8:269-280.
4. Cardiff RD (1998) Are the TDLU of the human the same as the LA of mice? J Mammary Gland Biol Neoplasia 3:3-5.
5. Castoria G, Migliaccio A, Bilancio A, et al (2001) PI3-kinase in concert with Src promotes the S-phase entry of oestradiol-stimulated MCF-7 cells. Embo J 20:6050-6059.
6. Chlebowski RT, Hendrix SL, Langer RD, et al (2003) Influence of estrogen plus progestin on breast cancer and mammography in healthy postmenopausal women: the Women's Health Initiative Randomized Trial. Jama 289:3243-3253.
7. Groshong SD, Owen GI, Grimison B, et al (1997) Biphasic regulation of breast cancer cell growth by progesterone: role of the cyclin-dependent kinase inhibitors, p21 and p27(Kip1). Mol Endocrinol 11:1593-1607.
8. Korach KS (1994) Insights from the study of animals lacking functional estrogen receptor. Science 266:1524-1527.
9. Lange CA, Richer JK, Horwitz KB (1999) Hypothesis: Progesterone primes breast cancer cells for cross-talk with proliferative or antiproliferative signals. Mol Endocrinol 13:829-836.
10. Lange CA, Richer JK, Shen T, et al (1998) Convergence of progesterone and epidermal growth factor signaling in breast cancer. Potentiation of mitogen-activated protein kinase pathways. J Biol Chem 273:31308-31316.
11. Lange CA, Shen T, Horwitz KB (2000) Phosphorylation of human progesterone receptors at serine-294 by mitogen-activated protein kinase signals their degradation by the 26S proteasome. Proc Natl Acad Sci USA 97:1032-1037.
12. Lydon JP, DeMayo FJ, Funk CR, et al (1995) Mice lacking progesterone receptor exhibit pleiotropic reproductive abnormalities. Genes Dev 9:2266-2278.

13. Mansour SJ, Matten WT, Hermann AS, et al (1994) Transformation of mammalian cells by constitutively active MAP kinase kinase. Science 265:966-970.
14. Migliaccio A, Piccolo D, Castoria G, et al (1998) Activation of the Src/p21(ras)/Erk pathway by progesterone receptor via cross-talk with estrogen receptor [In Process Citation]. Embo J 17:2008-2018.
15. Muslin AJ, Klippel A, Williams LT (1993) Phosphatidylinositol 3-kinase activity is important for progesterone-induced Xenopus oocyte maturation. Mol Cell Biol 13:6661-6666.
16. Muslin AJ, MacNicol AM, Williams LT (1993) Raf-1 protein kinase is important for progesterone-induced Xenopus oocyte maturation and acts downstream of mos. Mol Cell Biol 13:4197-4202.
17. Neve RM, Holbro T, Hynes NE (2002) Distinct roles for phosphoinositide 3-kinase, mitogen-activated protein kinase and p38 MAPK in mediating cell cycle progression of breast cancer cells. Oncogene 21:4567-4576.
18. Qiu M, Olsen A, Faivre E, et al (2003) Mitogen activated protein kinase regulates nuclear association of human progesterone receptors. Mol Endocrinol 17:628-642.
19. Sagata N, Daar I, Oskarsson M, et al (1989) The product of the mos proto-oncogene as a candidate "initiator" for oocyte maturation. Science 245:643-646.
20. Sartorius CA, Groshong SD, Miller LA, et al (1994) New T47D breast cancer cell lines for the independent study of progesterone B- and A-receptors: only antiprogestin-occupied B- receptors are switched to transcriptional agonists by cAMP. Cancer Res 54:3868-3877.
21. Shen T, Horwitz KB, Lange CA (2001) Transcriptional hyperactivity of human progesterone receptors is coupled to their ligand-dependent down-regulation by mitogen-activated protein kinase-dependent phosphorylation of serine 294. Mol Cell Biol 21:6122-6131.
22. Shyamala G (1999) Progesterone signaling and mammary gland morphogenesis, J Mammary Gland Biol Neoplasia 4:89-104.
23. Shyamala G, Yang X, Cardiff RD, et al (2000) Impact of progesterone receptor on cell-fate decisions during mammary gland development. Proc Natl Acad Sci USA 97:3044-3049.
24. Tung L, Shen T, Abel MG, et al (2001) Mapping the unique activation function 3 in the progesterone B-receptor upstream segment. Two LXXLL motifs and a tryptophan residue are required for activity. J Biol Chem 276:39843-39851.
25. Zhang Y, Beck CA, Poletti A, et al (1995) Identification of a group of Ser-Pro motif hormone-inducible phosphorylation sites in the human progesterone receptor. Mol Endocrinol 9:1029-1040.

Abnormal Properties of Mutants in the Hinge Region of ERα: Implications in Breast Cancer

Carlos Martínez-Campa, Pedro Zuazua, Juana María García-Pedrero, Pedro Casado, Pedro Sánchez Lazo, and Sofía Ramos

Summary

In the search for differences between estrogen receptor (ER) α and ERβ, we proved that ERα but not ERβ directly interacts with calmodulin (CaM) through the hinge region. The transcriptional activity of a mutant unable to interact with CaM becomes insensitive to inhibition by CaM antagonists (W7). These residues are acetylated by p300 and substitution of lysine 302 and 303 with other residues enhance ERα–hormone sensitivity, suggesting that ERα acetylation normally suppresses ligand sensitivity. Also, the somatic mutation K303R has been identified in early pre-malignant breast lesions. ERα K303R normally binds E but shows increased E-induced transcriptional activation and increased proliferation in response to E when transfected into breast cancer cell (BC) lines. Herein, we show that mutations K303R and K303A, render an ERα unable to interact with CaM and therefore insensitive to W7. K303 homodimers and K303 mutant/wt heterodimers show increased sensitivity to E. Contrary to the wt ERα, AP1 transcriptional activity is inhibited by estradiol (E_2) and OH-Tamoxifen (OH-TAM) in both K303R and K303A mutants.

Introduction

The discovery of ERβ (1), made possible to show that ERα and β mediate different biological actions. ERβ appears to be the quantitatively dominant over ERα in bone, cardiovascular system, urogenital tract, kidney, central nervous system, immune system and lung, whereas ERα has a significant role in the reproductive system, both in males and females (2, 3). ERα is over-expressed in more than 60% of BCs. Activation of both receptors by E_2 induces ERα and β mediated transcription of genes regulated by E-responsive elements (ERE) (1). However, only ERα mediates E_2-dependent transcription of genes regulated by AP1 (4) and Sp1 (5). In contrast, ERβ specifically mediates JNK inhibition (6).

For all the reasons exposed above, the finding of selective inhibitors of ERα and β is of paramount importance, both to identify the roles of each receptor and to eliminate the deleterious effects of E therapy. Novel ligands that function as selective E or anti-estrogens for ERα or β have been reported (7, 8), and studies providing basis for some differential transcription activities between both receptors have also been published (9).

The interaction of CaM with ER has been demonstrated, and a CaM binding domain has been postulated (10). Since then, several reports have indicated that CaM was essential for the interaction of ERα with DNA, and for activation of responsive promoters (11). We have demonstrated that ERα is tightly bound to CaM. Its transactivation capacity is dependent on CaM, and therefore able to be inhibited by the CaM antagonist W7. In contrast, ERβ does not interact with CaM, and its transactivation capacity is not altered by W7. Furthermore, we converted the W7-sensitive ERα into a W7-insensitive ERα by mutating aminoacids located at the putative CaM binding site (12). Remarkably, these lysines are targets for acetylation by p300 (13) and, K303 is mutated to arginine in 34% of pre-malignant breast lesions (14). ERα (K303R) mutant homodimers, and interestingly, also K30R/wt and K303A/wt heterodimers show increased sensitivity to E. Moreover, contrary to the wt ERα, AP1 transcriptional activity mediated by both K303R and K303A mutants is inhibited by E_2 and OHTAM.

In summary, all these results strongly suggest that these residues (K302, K303) located in the hinge region might play a crucial role in the ERα regulation *in vivo* and open new ways of searching for ERα inhibitors which do not interfere ERβ-mediated functions.

Results

ERα but not β Interacts with Calmodulin. Full-length ERα and ERβ were [^{35}S]-labeled *in vitro* and immunoprecipitated with anti-ER or anti-CaM antibodies. ERα was co-immunoprecipitated with anti-CaM and polyclonal anti-ER's antibodies. In contrast, [^{35}S]-labeled ERβ was not co-immunoprecipitated with the anti-CaM antibodies, indicating that only ERα interacts with CaM. Additional evidence of direct CaM-ERα interaction was obtained using purified GST-ERs hybrid proteins and dansyl-CaM. Emission spectra of dansyl-CaM (λ_{333}nm) was determined before and after addition of the indicated GST- proteins or equivalent amounts of buffer in the controls. In the presence of GST-ERα (1-595) an enhancement of the fluorescence of dansyl-CaM was observed, as expected if the dansyl group bound to CaM reached a more hydrophobic environment as a consequence of CaM interaction with ERα. Moreover, the fluorescence of ERα-dansyl-CaM complex was decreased by 40% upon addition of 10^{-8} M E_2, indicating that the change induced by E_2 in ERα affects the ERα-CaM interaction pattern. However, GST-ERβ did not alter the fluorescence of dansyl-CaM independently of the presence of E_2. The data confirmed that of the two ERs only GST-ERα(1-595) interacts with CaM (12).

The Mutant ERα (K302, 303G) is Unable to Bind CaM. Sequence analysis of the postulated CaM binding site (10) in ERα (aa 298-310) revealed important differences between hERα and β. In particular, residues K302 and K303 of hERα are both substituted for glycine in hERβ. We tested whether these aminoacids were essential for CaM binding. For this purpose, K302 and K303 were substituted for G in ERα. Both wt ERα and the mutant ERα (K302,303G) were *in-vitro* labeled with $[^{35}S]$-methionine followed by immunoprecipitation with anti-CaM or anti-ER antibodies. The results obtained showed that 95% of wt ERα but only 18% of ERα (K302,303G) was immunoprecipitated with anti-CaM antibodies (12). These data strongly indicate that substitutions of K302 and K303 for G in ERα (K302,303G) renders a receptor with a notably reduced affinity for CaM.

Specific Inhibition of ERα-mediated Transcriptional Activation by W7. To further understand the different regulation of ERα and β by CaM, we transiently transfected HeLa cells with ERα, ERβ, or ERα K302,303G expression vectors, along with the 3x-ERE-TATA-Luc reporter plasmid. A wide range of W7 concentrations (10^{-11} to 10^{-7} M) specifically inhibit E_2-ERα-mediated transactivation while neither ERβ nor ERα K302,303G were affected (Table 1). These results indicate that W7 induces conformational changes in CaM which prevent ERα-dependent transcription probably by destabilization of the E_2-ERα-CaM-ERE complex (12). This destabilization neither occurs in the mutant receptor nor in ERβ due to its inability to bind CaM.

Table 1. Effect of W7 Over E_2-induced Transcriptional Activation[1].

Receptor	Control	$E_2(10^{-7}$ M)	$E_2(10^{-7}$ M) + W7(10^{-11} M)	$E_2(10^{-7}$ M) + W7(10^{-9} M)	$E_2(10^{-7}$ M) + W7(10^{-7} M)
ERα	1	6.2	6.0	3.9	2.1
ERβ	1	4.1	4.0	4.0	4.2
ERα K302,303G	1	7.0	6.8	6.7	6.7

[1] HeLa cells were transfected with 0.5 μg of 3x-ERE-TATA-Luc reporter plasmid, 0.1 μg of pcDNA-hERα (wt), pcDNA-hERβ (wt), or pcDNA-hERα K302,303G expression vectors and 50 ng of control plasmid pRL-TK. Cells were treated as indicated and luciferase activity was determined.

The Mutant ERα (K299A, R300G, K302A) is Sensitive to W7. The basic residues 299, 300, and 302 are dispensable for interaction with CaM. We transiently transfected HeLa cells with ERα, or ERα (K299A, R300G, K302A) expression vectors along with the 3x-ERE-TATA-Luc reporter plasmid. A wide range of W7 concentrations (10^{-11} to 10^{-7} M) specifically inhibit the transactivation mediated by both ERα (K299A, R300G, K302A) and wt ERα. These results indicate that all the basic residues that were modified are dispensable for interaction with CaM (Table 2).

The Mutant ERα (K303R) is Hypersensitive to E_2 and Does Not Respond to W7. We next substituted lysine 303 with arginine, exactly the same mutation described in pre-malignant breast lesions (14), maintaining intact the rest of basic residues in the CaM-binding site, and we assayed the response to low concentrations of E_2 and the effect of the CaM antagonist W7 on E_2-dependent transactivation of this mutant. It has been reported that ERα K303R is hypersensitive at low concentrations of E_2 (14). Indeed, in our particular culture conditions and using 3x-ERE-TATA-Luc as reporter we found that the mutant transactivates much more efficiently than the wt ERα at low concentrations of E_2 (Table 3).

Table 2. Effect of W7 Over E_2-induced Transcriptional Activation[1].

Receptor	Control	E_2 (10^{-7} M)	E_2(10^{-7} M) + W7(10^{-11} M)	+ W7(10^{-9} M)	+ W7(10^{-7} M)
ERα	1	6.3	6.2	3.8	2.3
ERα K299A, R300G, K302A	1	7.5	6.8	4.3	2.5

[1] HeLa cells were transfected with 0.5 μg of 3x-ERE-TATA-Luc reporter plasmid, 0.1 μg of pcDNA-hERα (wt), or pcDNA-hERα (K299A, R300G, K302A) expression vectors and 50 ng of control plasmid pRL-TK. Cells were treated as indicated and luciferase activity was determined.

Table 3. Activation of the wt ERα and ERα K303R by E_2 Concentrations and Effect of W7[1].

Receptor	Control	E_2 (M) (10^{-11})	(10^{-10})	(10^{-9})	(10^{-8})	(10^{-7})	E_2 (10^{-8} M) + W7(10^{-7} M)	+ Mel(10^{-7} M)
ERα	1	1.8	2.7	3.5	6.6	6.3	2.4	2.7
ERα K303R	1	3.0	7.1	16.5	12.7	10,1	12.7	12.9

[1] HeLa cells were transfected with 0.5 μg of 3x-ERE-TATA-Luc reporter plasmid, 0.1 μg of pcDNA-hERα, or pcDNA-hERα K303R expression vectors and 50 ng of pRL-TK. Cells were treated as indicated and luciferase activity was determined.

Contrary to the wt ERα, the mutant ERα K303R is not inhibited by CaM antagonists, strongly suggesting that the mutation impairs CaM binding to the ER. Similar results have been obtained when lysine 303 was substituted with alanine, excluding the possibility that the effect observed for K303R might be due to the presence of a positively charged arginine that can not be neutralised by acetylation. Therefore, it seems that lysine 303 is essential for the normal regulation of the wt ERα, and that mutation in this residue results in abnormal regulatory properties.

ERα/ERα K303R and ERα/ERα K303A Heterodimers are also Hypersensitive to E_2. Since ERα K303R transactivates more efficiently than the wt receptor, we next determined the activity of the heterodimers. To compare the activity of homodimers ERα/ERα with heterodimers ERα/ERα K303R and ERα/ERα K303A, we used a system designed by Tremblay, *et al.* (15). By mutating three residues located at the base of the first Zn finger module of the ERα DNA binding domain, the DNA-binding specifity of the ER can be made identical to that of the glucocorticoid receptor (GRE). The altered $ER\alpha_{GRE}$ is cotransfected with wt ERα, ERα K303R, or ERα K303A and a reporter plasmid containing a hybrid ERE-GRE that allows transcription to occur exclusively in the presence of the $ER\alpha_{GRE}$-ERα K303 heterodimers. K303 mutants/wt heterodimers (Table 4) transcribed much more efficiently than ERα homodimers when cells were treated with E_2 concentrations ranging from 10^{-9} to 10^{-7} M. Therefore, ERα K303 mutants are more sensitive to E_2 as homodimers, but also when they form heterodimers with wt ERα.

Table 4. Transcriptional Activity of Heterodimers[1].

		E_2 (M)				
Heterodimer	**Control**	**(10^{-11})**	**(10^{-10})**	**(10^{-9})**	**(10^{-8})**	**(10^{-7})**
$ER\alpha_{GRE}$/ERα	1	2.6	5.6	8.4	14.1	18.0
$ER\alpha_{GRE}$/ERαK303R	1	4.2	7.4	16.5	22.0	36.5
$ER\alpha_{GRE}$/ERαK303A	1	2.4	5.8	16.1	20.3	30.9

[1] HeLa cells were transfected with 500 ng of a reporter plasmid containing a hybrid ERE-GRE-Luc reporter, 50 ng of pcDNA-$hER\alpha_{GRE}$, 50 ng of pcDNA-hERα K303R or pcDNA-hERα K303A expression vectors and 50 ng of internal control plasmid pRL-TK. With this strategy, we can specifically measure the transcriptional properties of heterodimers ERα/ERαK303R or ERα/ERαK303R. Cells were treated with E_2 as indicated and luciferase activity determined.

ERα K303R Mediated Transactivation in AP1 is Inhibited by E_2 and OH-TAM. Next, we compared the effect of ERα K303R and ERα over AP1 responsive genes by transiently transfecting HeLa cells with Δcoll 73-Luc reporter plasmid. We found (as published) that AP1 activity was increased by EGF when HeLa cells were transfected with either ERα wt or ERα K303R. E_2 and OH-TAM significantly potentiated the activity of AP1 in ERα-transfected cells (12). These data agree with previous reports indicating that EGF synergize with E_2 but only in cells expressing ER. Similar results were also obtained with OHT (Table 5).

Importantly, the synergistic effect of EGF and either E_2 or OH-TAM observed for ERα was not observed when the cells were transfected with ERα K303R (Table 5). E_2 and OH-TAM treatment results in a strong inhibition of AP1, similarly to what happens for ERβ. We have previously described that CaM is a regulator of ERα/AP1 pathways since the transcription mediated by ERα is sensitive to CaM antagonists, and the inhibition by W7 was statistically significant (12). Here, we show that the mutation K303R that impairs CaM-binding to the ER results in abnormal properties of AP1 ERαK303R-mediated transactivation.

Table 5. Effect of E_2 and OH-TAM Over hERα–, hERβ– and hERα K303R-Dependent Transactivation at AP1 Promoters[1].

Receptor	Control	EGF	EGF + E_2 (10^{-7} M)	EGF + OH-TAM(10^{-6} M)
ERα	1	2.3	5.2	5.4
ERαK303R	1	2.6	1.5	1.2
ERβ	1	2.7	1.6	1.4

[1] HeLa cells were transfected with either 0.1 μg of the ERα, ERβ, or hERα K303R expression vector as indicated, 50 ng of internal control plasmid pRL-TK and 0.5 μg of AP1-containing reporter plasmid (Δ coll. 73-Luc). Cultures were stimulated for 48 h with 1 μg/ml EGF, or 1 μg/ml EGF + either 10^{-7} M E_2 or 10^{-6} M OH-TAM as indicated. Luciferase activity was determined.

Conclusions

ERα but not ERβ directly interacts with CaM through the hinge region. Substitutions (K302,303G), (K303R), or (K303A) but not (K299A, R300G, K302A) of ERα render a mutant whose transcriptional activity becomes insensitive to inhibition by CaM antagonists. ERα K303R is hypersensitive to E_2 as homodimer but also as heterodimer with wt ERα. Contrary to wt ERα, ERα K303R-mediated transactivation at AP1 promoters is inhibited by E_2 and OH-TAM.

References

1. Kuiper GG, Enmark E, Pelto-Huikko M, et al (1996) Cloning of a novel receptor expressed in rat prostate and ovary. Proc Natl Acad Sci USA 93:5925-5930.
2. Weihua Z, Saji S, Makinen S, et al (2000) Estrogen receptor (ER) beta, a modulator of ERalpha in the uterus. Proc Natl Acad Sci USA 97:5936-5941.
3. Gustafsson JA (1999) Novel mechanism of estrogen action. In: Paoleti R, Crosignani PG, Kenemans P, Wenger NK and Jackson AS (eds) Women's Health and Menopause. Risk Reduction Strategies-Improved Quality of Health, London: Kluver Academ Publishers, 33-38.
4. Paech K, Webb P, Kuiper GG, et al (1997) Differential ligand activation of estrogen receptors ERalpha and ERbeta at AP1 sites. Science 277:1508-1510.

5. Saville B, Wormke M, Wang F, et al (2000) Ligand-cell and estrogen receptor subtype (alpha/beta)-dependent activation at GC-rich (Sp1) promoter elements. J Biol Chem 275:5379-3587.
6. Srivastava S, Weitzmann MN, Cenci S, et al (1999) Estrogen decreases TNF gene expression by blocking JNK activity and the resulting production of c-Jun and JunD. J Clin Invest 104:503-513.
7. Sun J, Meyers MJ, Fink BE, et al (1999) Novel ligands that function as selective estrogens or antiestrogens for estrogen receptor-alpha or estrogen receptor-beta. Endocrinology 140:800-804.
8. Gaido KW, Maness SC, McDonnell DP, et al (2000) Interaction of methoxychlor and related compounds with estrogen receptor alpha and beta, and androgen receptor: structure-activity studies. Mol Pharmacol 58:852-858.
9. Patrone C, Pollio G, Vegeto E, et al (2000) Estradiol induces differential neuronal phenotypes by activating estrogen receptor alpha or beta. Endocrinology 141:1839-1845.
10. Castoria G, Migliaccio A, Nola E, et al (1988) *In vitro* interaction of estradiol receptor with Ca2+-calmodulin. Mol. Endocrinol 2:167-174.
11. Biswas DK, Reddy PV, Pickard M, et al (1998) Calmodulin is essential for estrogen receptor interaction with its motif and activation of responsive promoter. J Biol Chem 273:33817-33824.
12. García Pedrero JM, del Rio B, Martínez-Campa C, et al (2002) Calmodulin is a selective modulator of estrogen receptors. Mol Endocrinology 16:947-960.
13. Wang C, Fu M, Angeletti RH, et al (2001) Direct acetylation of the estrogen receptor alpha hinge region by p300 regulates transactivation and hormone sensitivity. J Biol Chem 276:18375-18383.
14. Fuqua SA, Wiltschke C, Zhang QX, et al (2000) A hypersensitive estrogen receptor-alpha mutation in premalignant breast lesions. Cancer Res 60:4026-4029.
15. Tremblay GB, Tremblay A, Labrie F, et al (1999) Dominant activity of activation function 1 (AF-1) and differential stoichiometric requirements for AF-1 and -2 in the estrogen receptor alpha-beta heterodimeric complex. Mol Cell Biol 19:1919-1927.

The Effects of Eicosapentaenoic Acid Upon Proliferation and 17β–Dehydrogenase Activity in MCF-7 Breast Carcinoma Cells

Alison S. Whitehouse, Eftychia Oikonomou, and Eric F Adams

Summary

Estradiol 17-β dehydrogenase (OE_2DH) is an enzyme involved in both the activation and the inactivation of estrogen, which plays an important role in breast cancer (BC). Polyunsaturated fatty acids (PUFA's) are also implicated in the development of BC, although the mechanisms are not fully understood. The aim of this investigation was to examine the growth effects of the n-3 PUFA-eicosapentaenoic acid (EPA) on the MCF7 human BC cell line in parallel with its effects upon OE_2DH. A bell shaped growth inhibition of MCF7 cells was observed at around 50 μM. A single dose of EPA significantly inhibited cell proliferation at every time point measured over a twelve-day period. In the presence of 17β–estradiol (E_2) alone, cell proliferation was increased peaking at 1nM E_2. The effect was completely abolished by co-incubation with 50 μM EPA. OE_2DH activity was measured in both the reductive [estrone (E_1) → E_2] and the oxidative [E_2 → E_1] directions. These results show that whilst 50 μM EPA weakly stimulates OE_2DH activity in the reductive direction, it strongly drives the reaction in the oxidative direction. This effect was specific to EPA as the structurally related n-6 fatty acid arachidonate, had no effect. These results suggest that EPA might, in part, exert its effects in BC through driving OE_2DH activity towards the production of the less biologically potent E_1 as well as having a direct inhibitory effect on activity.

Introduction

It is well established that PUFAs have a protective effect in BC. The aim of this research was to investigate the effect the EPA on the OE_2DH system. OE_2DH is responsible for the interconversion of E_1 and E_2. Of the three endogenous estrogens, E_2 is the most potent because it has the highest affinity for estrogen receptors (ER). E_1, on the other hand has very little and, at times, no significant biological activity. Several studies have indicated increased levels of E_2 in malignant breast tissue when compared to normal breast tissue (1-4).

Whilst it has been shown that progestins and estrogens directly stimulate reductive OE_2DH activity (5), there are relatively few other studies examining the control of this complex. To the best of our knowledge, only one publication has considered the effects of fatty acids upon the activity of this enzyme. Blomquist, *et al.* (6) showed that *isolated* human placental OE_2DH activity was inactivated by several unsaturated fatty acids (oleic, arachidonic, linoleic, linolenic), and postulated that the soluble enzyme may contain a binding site at or near pyridine nucleotide binding sites. The work presented herein is novel since it is the first to consider the effects of fatty acids upon the OE_2DH system within a functioning cell, and to consider the significance of this process in cancer.

Blomquist, *et al.* suggested that fatty acid manipulation of the OE_2DH system was possible. Adams, *et al.* (5) showed that altering the OE_2DH activity had a profound influence upon the growth of MCF7 cells in culture. Taken together these observations (6, 5) suggest a potential mechanism through which dietary PUFA's might influence cell growth in malignant breast tissue. The aim of this report is to investigate whether EPA might in part exert its anti-tumour effects through manipulation of E_2DH activity.

Methods

MCF7 cells (ECACC) were maintained in a pre-confluent state in Eagle's MEM supplemented with 1% penicillin/streptomycin, 1% glutamine, non essential amino acid solution, and pH'd to 7.2-7.4. Cells were seeded at $1x10^6$ in T25 flasks and allowed to attach overnight. The cells were subsequently grown in phenol red free medium and the fetal calf serum used was stripped with dextran coated activated charcoal (8). Cis-5,8,11,14,17 eicosapentaenoic acid was prepared in an equimolar $NaHCO_3$ and equal mass bovine serum albumin in PBS and stored under N_2. Cells were treated with up to 100 μM EPA (9), repeat doses were applied every three days unless otherwise stated.

The measurement of OE_2DH activity was carried out according to the methods of Gompel, *et al.* (10) and Adams, *et al.* (11). To measure reductive and oxidative enzyme activities, cells were incubated in 2 nM/l (40 Ci/mM) $[^3H]$- E_1 or $[^3H]$- E_2 respectively, for four h at 37°C . Reaction blanks in which cells were omitted were run in parallel. After the incubation, 2 ml of the media was removed and added to $[^{14}C]$- E_2 or $[^{14}C]$- E_1 (5000 cpm) recovery labels. Steroids were then extracted into 4ml ether and dried at 40°C under N. Thin layer chromatography using 4:1 v/v dichloromethane:ethyl acetate solvent, separated the steroids and both product and recovery activity determined by scintillation counting according to James and Newton (12). Results are expressed as % control fmol product formed/cell and are a function of recovery and cell number. An n of 3 was used for each variable and each experiment was repeated at least twice. In all experiments cells were counted by measuring nuclei number as described by Butler, *et al.* (13). Statistical significance was determined with one-way ANOVA and Tuckey's post test as appropriate.

Results

The effects of a single 50 μM dose of EPA upon MCF7 proliferation in vitro, is shown in Figure 1. The increase in cell numbers evident after three days was significantly reduced by incubation in EPA and continued at days 5 and day 7. Following 7 days, cell numbers in both treatment and control groups fell, and although EPA treated groups continued to demonstrate significant reductions in cell number, it is to be anticipated that the highly confluent state of both groups resulted in significant amounts of non-specific cell death.

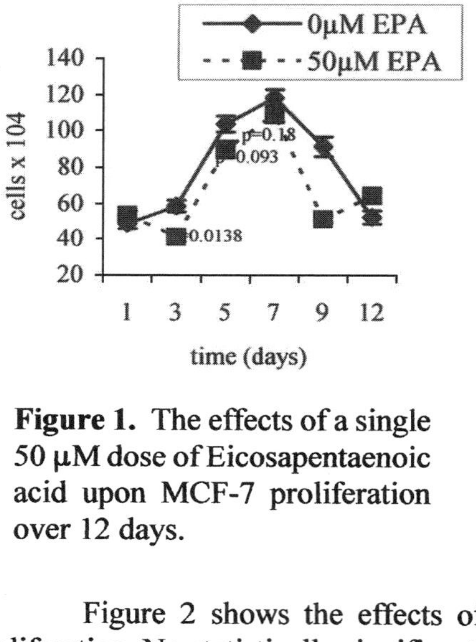

Figure 1. The effects of a single 50 μM dose of Eicosapentaenoic acid upon MCF-7 proliferation over 12 days.

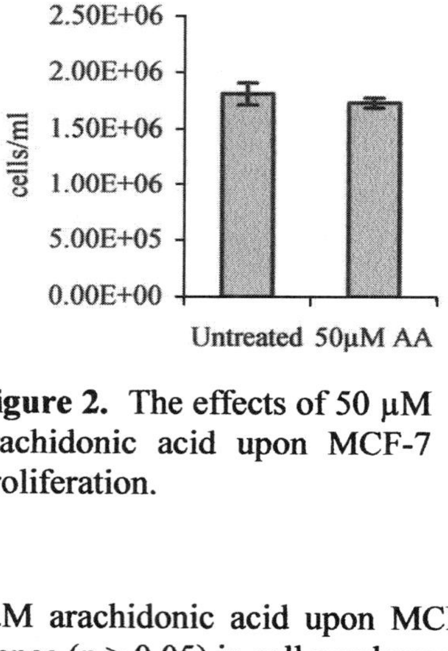

Figure 2. The effects of 50 μM arachidonic acid upon MCF-7 proliferation.

Figure 2 shows the effects of 50 μM arachidonic acid upon MCF-7 proliferation. No statistically significant difference ($p > 0.05$) in cell numbers was observed with this structurally similar fatty acid, suggesting that the effect was specific to EPA. A dose response of 5-100 μM EPA (Figure 3) stimulated both the reductive and oxidative actions of OE_2DH activity in MCF7 cells. However the effect upon the reductive activity did not reach statistical significance when compared to respective control. Oxidative activity was much more profoundly affected with a peak of stimulation at 20-50 μM EPA. Multiple dosing of the structurally similar n-6 arachidonic acid had no effect ($p > 0.05$) upon the reductive (Figure 4) or oxidative (Figure 5) action of OE_2DH activity in MCF7cells.

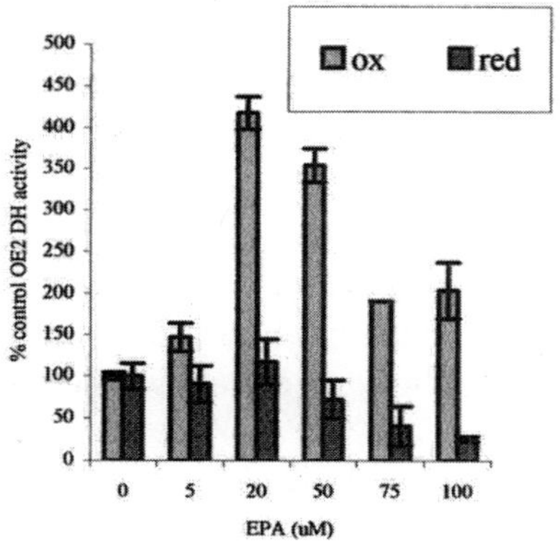

Figure 3. The effects of Eicosapentaenoic acid upon the reductive ($E_1 \rightarrow E_2$) and oxidative ($E_2 \rightarrow E_2$) action of OE_2DH.

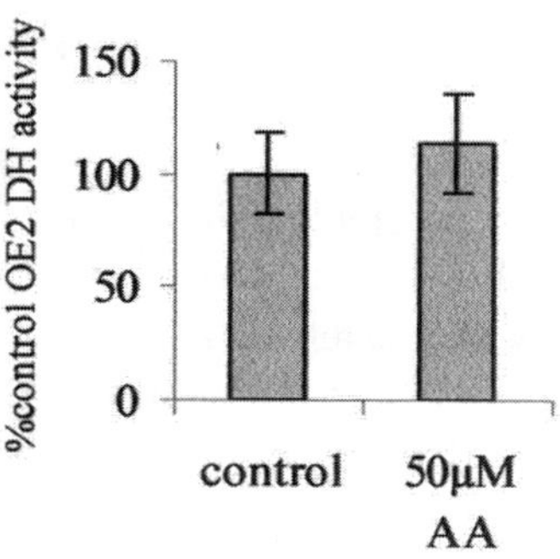

Figure 4. Effects of arachidonic acid upon the reductive ($E_1 \rightarrow E_2$) action of OE_2DH.

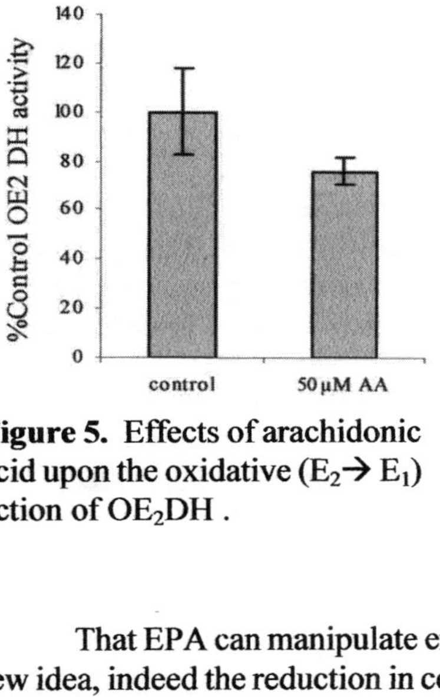

Figure 5. Effects of arachidonic acid upon the oxidative ($E_2 \rightarrow E_1$) action of OE_2DH .

Conclusion

We have presented data here that demonstrates that the polyunsaturated fatty acid EPA significantly inhibits MCF7 proliferation with a single dose for up to 7 days. The effect is dose dependent and specific to EPA and not the n-6 AA. We have also presented the novel finding that EPA can preferentially drive the OE_2DH system (the enzyme complex responsible for altering the bioavailability of estrogen) towards the production of the less biologically potent E_1. This effect was dose dependent and showed a highly statistically significant correlation with a fall in cell numbers.

That EPA can manipulate enzyme systems in malignant breast tissue is not a new idea, indeed the reduction in cell numbers observed following EPA treatment of breast tumours is often ascribed to the inhibition of the PUFA metabolizing enzyme – (cyclo-oxygenase) COX2 which promotes mammary carcinogenesis. It is unlikely however, and many reports support the idea (14) that suppression of tumour growth is entirely mediated through COX dependent pathways.

The question that arises then is 'how can fatty acids influence enzyme activity?' One possible explanation is that as PUFA's compete for the same desaturases and as one fatty acid could inhibit the conversion of another, EPA for example could be incorporated into the biologically labile fraction of phospholipids at the expense of the n-6 arachidonic acid. This could alter the eicosanoid axis to generate an entirely different set of products which tend to suppress oxidative OE_2DH activity rather than augment it. There is no documented evidence that n-6 fatty acids or their metabolites have a reductive effect upon OE_2DH activity and so this hypothesis remains to be tested. Although a loss of Δ-6 desaturase activity has recently been reported in MCF7 cells (15), suggesting that the ability to elongate and Δ -6 desaturate the parent n-3 and n-6 PUFAs may have been compromised, EPA is post Δ-5 and so it is still feasible that metabolic competition with some n-6s may explain our findings.

Furthermore, EPA might act through the generation of reactive oxygen species (ROS) generated during fatty acid peroxidation. It is well established that ROS act as subcellular messengers and although the molecular mechanisms are not fully understood it is possible that oxidation or reduction of protein sulfhydrils in intracellular proteins leads to conformational changes which can promote protein complex formations and/or the release of inhibitory subunits resulting in altered functioning (16). In addition, Stoll (17) demonstrated that n-3 fatty acids increase apoptosis in mammary tumours in a manner that is associated with increased lipid peroxidation and could be reversed by anti-oxidants. Although whether this is related to steroid hormone responsiveness is questionable as Gonzales, *et al.* (18) showed that anti-oxidants can increase lipid peroxidation and suppress human breast carcinoma growth in the ERα negative MDA-MB-321 cell line. Similarly Chamras, *et al.* (9) showed that the inhibitory effects of EPA upon MCF7 growth was not due to lipid oxidation or reversed by vitamin E.

Fatty acids can also modify signal transduction by direct binding to a peptide (19, 20). It seems that they can be linked to an amino acid residue directly or indirectly as a component of a phosphatidyl moiety attached to COOH terminal through an intervening glycan structure, reviewed in Muszbeck and Laposata, (20). It is possible that EPA might exert its oxidative effects upon OE_2DH activity via a direct linkage to the complex or as recent studies have shown that the overall OE_2DH activity of human breast cells is due to the action of at least three different enzymes, two of which catalyse the oxidative direction whilst the third catalyses both oxidative and reductive directions (21), EPA might differentially alter the activity of these enzymes in favour of the oxidative activity.

Another tempting yet speculative option is that transcription of the OE_2DH complex itself may have been altered. EPA can influence transcription in MCF7 cells, Tiwari, *et al.* (22) showed that both EPA and linoleic acid could induce expression (genes 1-8) in MCF7 cells and induced second messenger signals similar to the ones generated by interferons. Also, fatty acids can both inhibit (androgen) and stimulate (estrogen) steroid hormone receptor binding patterns (23), demonstrating that EPA works at least in part by altering ligand binding capacities. Thus it seems likely that the effects of EPA are multi-factorial and modulation of the OE_2DH system represents another facet of a concerted attack on malignant cells.

In conclusion we have demonstrated that OE_2DH activity can be profoundly influenced by the fatty acid EPA, particularly in the oxidative direction, which may in turn influence tissue availability of E_2 and mammary tumor progression.

Acknowledgements

We thank the Biochemical and Biophysical Sciences Research Council for financial support.

References

1. Fishman J, Nisselbaum JS, Menendez-Botet C, et al (1997) Estrone and estradiol content in human breast tumours: relationship to estradiol receptors. J Steroid Biochem 8:893-896.
2. Van Landeghem AAJ, Poortman J, Nabuurs M, et al (1985) Endogenous concentration and subcellular distribution of estrogens in normal and malignant human breast tissue. Cancer Res 45:2900-2906.
3. Bonney RC, Reed MJ Baranek PA, et al (1986) Metabolism of $[H^3]$oestradiol in vitro by normal breast and tumour tissue in postmenopausal women. J Steroid Biochem 24:361-364.
4. McNeill JM, Reed MJ, Beranek PA, et al (1986) The effect of epidermal growth factor, transforming growth factor and breast tumour homogenates on the activity of oestradiol 17β-hydroxysteroid dehydrogenase in cultured adipose tissue. Cancer Lett 31:213-219.
5. Adams EF, Coldham NG, James VHT (1988) Steroidal regulation of oestradiol-17β dehydrogenase activity of the human breast cancer cell line MCF-7. J Endocrinol 118:149-154.
6. Blomquist CH, Lindemann NJ, Hakanson EY (1985) Inactivation of soluble 17β-hydroxysteroid dehydrogenase of human placenta by fatty acids. J Steroid Biochem 23:357-363.
7. Bertois Y, Katzenellenbogen JA, Katzenellenbogen BS (1986) Phenol red in tissue culture media is a weak estrogen: implications concerning the study of estrogen-responsive cells in culture. PNAS USA 83:2496-2500.
8. Braunsberg H, Coldham NG, Leake RE, et al (1987) Action of progestogen on human breast cancer cells: Mechanisms of stimulation and inhibition. Eur J Cancer Clin Oncol 23:563-571.
9. Chamras H, Ardashian A, Heber D, et al (2002) Fatty acid modulation of MCF7 human breast cancer cell proliferation, apoptosis and differentiation. J Nutr Biochem 711-716.
10. Gompel A, Malet C, Spritzer P, et al (1986) Progestin effect on cell proliferation and 17β hydroxysteroid dehydrogenase activity in normal human breast cells in culture. J Clin Endocrinol Metab 63:1174-1180.

11. Adams EF, Coldham NG, James VHT (1988) Steroidal regulation of oestradiol-17β dehydrogenase activity of the human breast cancer cell line MCF-7. J Endocrinol 118:149-154.
12. Newton CJ, James VHT (1985) 17β-Hydroxysteroid dehydrogenase activity in leiomyoma and myometrium and its relationship to concentrations of oestrone, oestradiol and progesterone throughout the menstrual cycle. J Steroid Biochemistry 22:487-493.
13. Butler WB, Kelsey WH, Goran N (1981) Effects of serum and insulin on the sensitivity of the human breast cancer cell line MCF-7 to estrogens and anti-estrogens. Cancer Res 41:81-88.
14. Boudreau MD, Sohn KH, Rhee SH, et al (2001) Suppression of tumour cell growth both in nude mice and in culture by n-3 polyunsaturated fatty acids: Mediation through cyclooxygenase-independent pathways. Cancer Res 61:1386-1391.
15. Grammatikos SI, Subbaiah PV, Victor TA, et al (1984) N-3 and n-6 fatty acid processing and growth effects in neoplastic non-cancerous human mammary epithelial cell lines. Br J Cancer 70:219-227.
16. Allen RG, Tresini M (2000) Oxidative stress and gene regulation by fatty acids in the adipose cell. Free Rad Biol Med 28:463-469.
17. Stoll BA (2002) N-3 fatty acids and lipid peroxidation in breast cancer inhibition. Br J Nutr 87:193-198.
18. Gonzalez MJ, Schemmel RA, Dugan L Jr, et al (1993) Dietary fish oil inhibits human breast carcinoma growth: a function of increased lipid peroxidation. Lipids 28:827-832.
19. Nunez AE (1995) Fatty acids and cell signalling. Prostaglandins, Leukotrienes Essential Fatty Acids 52:117-120.
20. Muszbeck L, Laposata L (1993) Covalent modification of proteins by arachidonate and eicosapentaenoate. J Biol Chem 268:18243-18248.
21. Tait GH, Newton CJ, Reed MJ, et al (1987) Multiple forms of oestradiol 17β hydroxysteroid dehydrogenase in human breast tissues. J Endocrinol 115 (suppl.): Abstract no 85.
22. Tiwari RK, Mukhopadyhay NT, Telang NT, et al (1991) Modulation of gene espression by selected fatty acids in human breast cancer cells. Anti Cancer Res 11:1383-1388.
23. Prinsloo SE, Answegen CHV (2002) Effect of fatty acids on estradiol and testosterone to whole DU-145 prostate cells. Prostaglandins, Leukotrienes, Essential Fatty Acids 66:419-425.

Effect of Dietary Genistein on Estradiol-induced Mammary Carcinogenesis in the ACI Rat

Valerie K. Turan, Kenneth R. Reuhl, and Paul E. Thomas

Summary

We evaluated the possible effectiveness of genistein (GEN) as an inhibitor of mammary cancer in the ACI rat, a strain genetically susceptible to estrogen-induced mammary carcinogenesis. In this study, three-week old ACI rats were fed a phytoestrogen-free diet or a diet containing 250 ppm GEN. At nine weeks of age, the rats were implanted with 20 mg pellets containing either 2 mg 17β-estradiol (E_2)/18 mg cholesterol or cholesterol alone. The appearance of palpable mammary tumors (MTs) was recorded. Our results show that E_2-treated rats fed the GEN diet both prepubertally and as adults had the highest incidence of palpable MTs. E_2-treated rats fed the GEN diet only prepubertally had delayed appearance of palpable MTs but the incidence quickly approached that of the rats fed GEN for the whole treatment period. Dietary GEN fed to adult E_2-treated rats had the lowest incidence of MT. Rats implanted with the cholesterol pellet and fed either diet did not develop MTs.

Introduction

In the year 2000, breast cancer (BC) was the leading cancer in women, and the one with the highest mortality worldwide (1). The American Cancer Society estimates that BC will have the highest number of new cases, and the second highest number of new deaths in women, after lung cancer, in the USA in the year 2002 (ACS webpage, cancer.org). Epidemiological studies indicate that BC risk includes factors related to reproductive history and non-hormonal factors. Risk factors relating to reproductive history may reflect longer lifetime exposures to endogenous ovarian hormones. Factors that increase the number of ovulatory cycles increase the risk, while reducing the number of cycles may be protective against BC (2).

For this study, we examined mammary carcinogenesis in the ACI rat. This strain is highly susceptible to mammary carcinogenesis when continuously exposed estrogens, but expresses a low rate of spontaneous mammary tumors (MTs) (3). Continuous treatment with E_2 induces MTs in ovary-intact but not ovariectomized adult female ACI rats (4), and when hormone treatment was ended by removing the E_2 implants, all of the tumors regressed, demonstrating that the MTs are dependent upon exogenous E_2 (5).

Fisher 344 (F344) rats that are continuously exposed to E_2 rapidly developed prolactin-producing pituitary tumors (6). Ovariectomized F344 rats fed an energy-restricted diet demonstrated an inhibition of prolactin-producing pituitary tumors from continuously administered E_2 (7), while ovariectomized ACI rats administered continuous E_2 and fed the energy-restricted diet did not demonstrate an inhibition of prolactin-producing pituitary tumors (8). Together, Spady, *et al.* concluded that the effects of energy restriction are strain-specific and are strongly affected by genetic background (7, 8). Ovary-intact ACI rats subjected to dietary energy restriction demonstrated inhibition of MT development from continuously administered E_2 (9). Prolactin-producing pituitary tumors induced in ovary-intact ACI rats by E_2 were not inhibited by dietary energy restriction, indicating that the inhibitory effects of energy restriction on MT development is tissue specific and is independent of circulating levels of prolactin and E_2 (10).

Li, *et al.* have studied the ACI rat mammary carcinogenesis model using cholesterol pellets as the vehicle E_2 (11). They evaluated the serum levels of E_2 in ACI rats implanted with 2 mg E_2/18 mg cholesterol pellets over a 6.0-mo period. The serum levels, which plateaued around 2 mo, ranged from 52 to 89 pg/ml (11), which are within normal range for cycling Sprague-Dawley (SD) rats (12). Concomitant treatment of ACI rats with pellets containing tamoxifen, a known anti-estrogen, prevented induction of the E_2-induced MTs (11). The ACI rat E_2-induced mammary carcinogenesis model has some critical distinctive differences from other chemical carcinogen-induced mammary carcinogenesis models. Early genomic destabilization (aneuploidy) and *c-myc* amplification, features commonly found in invasive human ductal carcinoma *in-situ* and ductal BC, were reported in E_2-induced ACI rat MTs but not in chemical carcinogen-induced MTs from other rat strains (13).

Epidemiological evidence suggests that Asian women consuming a diet high in soy are less susceptible to BC than women consuming low levels of soy (14-16), yet when they emigrate to the USA, only the first generation maintains this protection (17). Since the protection is not conferred from one generation to the next, the protection has been attributed to dietary or environmental factors as opposed to genetically conferred protection. It has been suggested that this protection can be attributed to consumption of the phytoestrogens present in soy (18). In a BC case-control study of patients in Singapore, premenopausal women consuming modest amounts of soy had ~50% reduction in BC (16), while a meta-analysis of case-control and cohort studies indicates that in postmenopausal women high soy intake did not reduce cancer risk (19). Together, it suggests that the timing of the exposure to genistein (GEN), the primary phytoestrogen in soy, may be an important factor to consider with regard to its chemopreventative capabilities.

Considerable debate exists regarding the effect of dietary GEN on mammary carcinogenesis in rodent models. In a series of studies, female SD rats were fed GEN in their diets at various time points in their development to evaluate how the timing of exposure affected DMBA-induced MTs (20-22). Female SD rat pups fed dietary GEN on days 0-20 postpartum and injected with DMBA developed fewer MTs than did rats fed a diet lacking GEN, in a dose/dependent manner (20). Since DMBA exposure occurred after GEN treatment had ceased, the authors concluded that the prevention of chemically induced MTs resulted from events following early GEN treatment, rather than a direct GEN action during MT formation and development. In a related study, female SD rats were fed a diet containing GEN through breeding and pregnancy, and were switched to a diet without GEN following parturition. Female pups were injected with DMBA and monitored for MTs. Prenatal GEN exposure did not affect DMBA-induced mammary carcinogenesis, suggesting that the critical time for GEN chemoprevention of DMBA-induced MTs is the postnatal to perinatal period (22). The authors demonstrated that the greatest protection from DMBA-induced MTs occurred with GEN exposure prepubertally and as adults. GEN, fed from breeding to weaning or from parturition to weaning, reduced the number of MTs/rat, yet when fed from breeding to parturition there was no effect (20, 21). The authors concluded that in this model, GEN exposure must occur prepubertally to exert a chemopreventative effect (20).

Gallo, *et al.* fed SD rats an extract of soy and evaluated its effect on DMBA-induced mammary carcinogenesis. Female SD rats were fed a soy-supplemented diet from weaning until the end of the study. DMBA was injected to induce MTs. There was an increase in the MT latency with the dietary soy, but no difference in final incidence or multiplicity (23, 24). Appelt and Reicks and Constantinou, *et al.* studied the effects of dietary soy or isoflavones on DMBA-induced MTs in adult SD rats. Appelt and Reicks observed nonsignificant decreases in MT incidence and burden (25), while Constantinou found nonsignificant decreases in MT multiplicity, but no effect on incidence (26).

Gotoh, *et al.* studied the effect of dietary soy, miso, or biochanin A, a precursor of GEN, on NMU-induced mammary carcinogenesis. Adult female SD rats were injected with NMU, and subsequently fed a supplemented diet. There was no significant difference in the incidence of MTs, however MT multiplicity was decreased in all three supplemented diet groups (27). Cohen *et al.* studied MTs induced by NMU in adult female F344 rats. One week before NMU injection, rats were fed diets supplemented with soy protein. They found that soy protein had no inhibitory effect on MTs (28).

The E_2-induced MT of the ACI rat is a unique model in that does not require an exogenous carcinogen, and the rats are ovary-intact and continue progressing through the estrous cycle. In this study, 3-week old female ACI rats were fed diets with or without GEN. At 9 weeks of age, the rats were implanted with pellets, composed of cholesterol with or without E_2, and fed the control or

GEN diet. The results of this study will allow us to determine not only whether GEN can modulate the ability of E_2 to induce MTs, but also whether the timing of GEN exposure is an important factor in the E_2 induction of MTs in the ACI rat.

Materials and Methods

Three-week old female ACI rats were obtained from Harlan Sprague-Dawley and immediately allowed free access to the AIN-76A diet (AIN) or AIN supplemented with 250 ppm GEN, their prepubertal diet. The AIN diet is a casein-based, open formula purified diet that contains nondetectable levels of phytoestrogens. At 9 weeks of age, they were subcutaneously implanted in the shoulder region with a 20 mg pellet containing either cholesterol or 2 mg E_2 + 18 mg cholesterol as previously described (11). The rats remained either on AIN or were switched to the GEN-containing diet. Treatment protocols are summarized in Table 1. Body weights were recorded weekly. Rats were palpated twice weekly for MTs, with the date of appearance and location of each MT recorded. When the MT reached ~ 3 cm^2 in size or when rats showed clinical evidence of discomfort the rats were euthanized.

Results and Conclusions

Table 1 summarizes the results after 36 weeks of treatment. The cumulative MT incidence and the % rats/treatment group with palpable MT show that rats fed GEN prepubertally, GEN/E_2/AIN and GEN/E_2/GEN, are more susceptible to E_2-induced mammary carcinogenesis. There were not significant differences in the mean MT latency period between E_2-treated rats regardless of the diet regimen. The rats fed GEN prepubertally, GEN/E_2/AIN and GEN/E_2/GEN, had greater MT burden, but the difference was not significant.

Table 1. Incidence of Palpable MTs after 36 Weeks of Treatment.

Treatment Group n = 15-16	**% Incidence**	**Mean Latency[1]**	**Number Necropsied[2]**	**Tumor Burden**	**Necropsy Week**
AIN/Chol[3]/AIN	0	-	0	-	-
AIN/Chol/GEN	0	-	0	-	-
GEN/Chol/AIN	0	-	0	-	-
GEN/Chol/GEN	0	-	0	-	-
AIN/E_2[4]/AIN	87	23.2	8	2.3	28.4
AIN/E_2/GEN	75	25.8	5	2.6	31.2
GEN/E_2/AIN	94	25.5	4	3.0	30.3
GEN/E_2/GEN	100	21.9	8	4.0	29.8

[1] Mean MT latency and autopsy week presented in weeks after pellet implantation.
[2] Number autopsied does not include rats alive after 36 weeks of treatment.
[3] Chol, 20 mg cholesterol
[4] E_2, 18 mg Chol + 2 mg E_2

Dietary consumption of 250 ppm GEN had no effect on body weight or general health when consumed by prepubertal or adult rats (Figure 1). Rats implanted with E_2 pellets weighed significantly more than those treated with cholesterol alone at 15 and 21 weeks after treatment, regardless of diet regimen, though these body weight differences disappeared after 27 weeks post implantation.

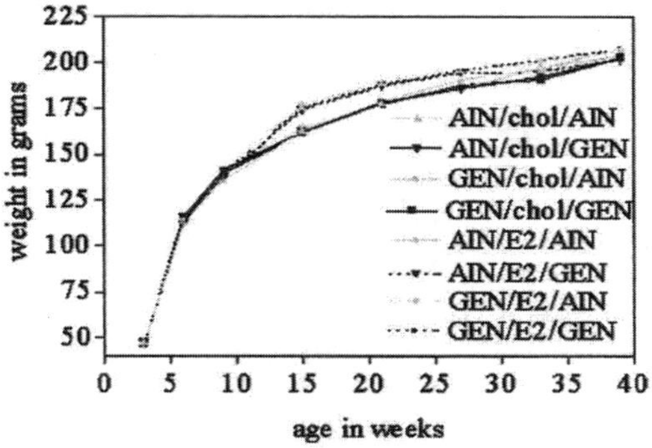

Figure 1. Variations in body weight after different treatments. (▲) AIN/Chol/AIN, (▼) AIN/Chol/GEN, (●) GEN/Chol/AIN, (■) GEN/Chol/GEN, solid lines; (▲) AIN/E_2/AIN, (▼) AIN/E_2/GEN, (●) GEN/E_2/AIN, (■) GEN/E_2/GEN, dotted lines.

Preliminary results show that at 36 weeks post pellet implantation, E_2-treated rats fed the GEN diet prepubertally and as adults, GEN/E_2/GEN, had the highest incidence and lowest latency of palpable MTs (Figure 2). E_2-treated rats fed the GEN diet only prepubertally, GEN/E_2/AIN, had delayed appearance of palpable MTs, but the MT incidence quickly approached that of the rats fed GEN throughout the treatment period. Dietary GEN fed to adult E_2-treated rats, AIN/E_2/GEN, resulted in the lowest incidence of MTs. Rats implanted with cholesterol alone, fed either diet, did not develop MTs.

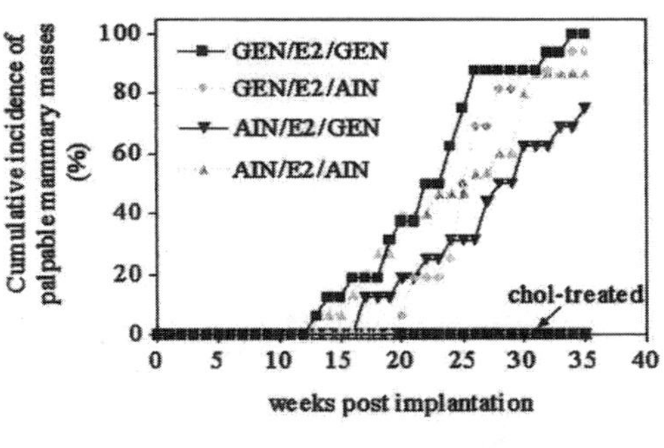

Figure 2. Cumulative incidence of palpable MTs, starting after nine weeks of age. (▲) AIN/E_2/AIN, (▼) AIN/E_2/GEN, (●) GEN/E_2/AIN, (■) GEN/E_2/GEN, (■) Chol-treated, all diets.

At 36 weeks of treatment, the results observed with prepubertal GEN exposure in the ACI rat E_2-induced carcinogenesis model do not correspond to those reported for the SD rat DMBA-induced MT model, which suggests that dietary exposure to GEN prepubertally is chemopreventative (27-29). This discrepancy may be due in part to the differences between E_2-induced MTs in the ACI rat and chemically induced MTs in other strains that were described by Li, *et al.* (20).

Acknowledgement

Portions of this work were supported by ES 05022, GM 44982, and a Collaborated Research Award from the Cancer Institute of New Jersey.

References

1. Parkin DM, Bray FI, Devesa SS (2001) Cancer burden in the year 2000. The global picture. Eur J Cancer 37:S4-66.
2. Hilakivi-Clarke L, Forsen T, Eriksson JG, et al (2001) Tallness and overweight during childhood have opposing effects on breast cancer risk. Br J Cancer 85:1680-1684.
3. Maekawa A Odashima S (1975) Spontaneous tumors in ACI/N rats. J Natl Cancer Inst 55:1437-1445.
4. Shull JD, Spady TJ, Snyder MC, et al (1997) Ovary-intact, but not ovariectomized female ACI rats treated with 17beta-estradiol rapidly develop mammary carcinoma. Carcinogenesis 18:1595-1601.
5. Harvell DM, Strecker TE, Tochacek M, et al (2000) Rat strain-specific actions of 17beta-estradiol in the mammary gland: correlation between estrogen-induced lobuloalveolar hyperplasia and susceptibility to estrogen-induced mammary cancers. Proc Natl Acad Sci USA 97:2779-2784.
6. Wiklund J, Wertz N, Gorski J (1981) A comparison of estrogen effects on uterine and pituitary growth and prolactin synthesis in F344 and Holtzman rats. Endocrinology 109:1700-1707.
7. Spady TJ, Lemus-Wilson AM, Pennington KL, et al (1998) Dietary energy restriction abolishes development of prolactin-producing pituitary tumors in Fischer 344 rats treated with 17beta-estradiol. Mol Carcinog 23:86-95.
8. Spady TJ, Pennington KL, McComb RD, et al (1999) Estrogen-induced pituitary tumor development in the ACI rat not inhibited by dietary energy restriction. Mol Carcinog 26:239-253.
9. Harvell DM, Strecker TE, Xie B, et al (2001) Diet-gene interactions in estrogen-induced mammary carcinogenesis in the ACI rat. J Nutr 131:3087S-3091S.
10. Harvell DM, Strecker TE, Xie B, et al (2002) Dietary energy restriction inhibits estrogen-induced mammary, but not pituitary, tumorigenesis in the ACI rat. Carcinogenesis 23:161-169.
11. Li SA, Weroha SJ, Tawfik O, et al (2002) Prevention of solely estrogen-induced mammary tumors in female ACI rats by tamoxifen: evidence for estrogen receptor mediation. J Endocrinol 175:297-305.
12. Butcher RL, Collins WE, Fugo NW (1974) Plasma concentration of LH, FSH, prolactin, progesterone and estradiol- 17beta throughout the 4-day estrous cycle of the rat. Endocrinology 94:1704-1708.

13. Li JJ, Papa D, Davis MF, et al (2002) Ploidy differences between hormone- and chemical carcinogen-induced rat mammary neoplasms: comparison to invasive human ductal breast cancer. Mol Carcinog 33:56-65.
14. Wu AH, Ziegler RG, Horn-Ross PL, et al (1996) Tofu and risk of breast cancer in Asian-Americans. Cancer Epidemiol Biomarkers Prev 5:901-906.
15. Adlercreutz CH, Goldin BR, Gorbach SL, et al (1995) Soybean phytoestrogen intake and cancer risk. J Nutr 125:757S-770S.
16. Lee HP, Gourley L, Duffy SW, et al (1991) Dietary effects on breast-cancer risk in Singapore. Lancet 337:1197-1200.
17. Ziegler RG, Hoover RN, Pike MC, et al (1993) Migration patterns and breast cancer risk in Asian-American women. J Natl Cancer Inst 85:1819-1827.
18. Messina MJ, Persky V, Setchell KD, et al (1994) Soy intake and cancer risk: a review of the in vitro and in vivo data. Nutr Cancer 21:113-131.
19. Trock BJ, White BL, Clarke R, et al (2000) Meta-analysis of soy intake and breast cancer risk. J Nutr 130:653S-680S.
20. Fritz WA, Coward L, Wang J, et al (1998) Dietary genistein: perinatal mammary cancer prevention, bioavailability and toxicity testing in the rat. Carcinogenesis 19:2151-2158.
21. Lamartiniere CA (2002) Timing of exposure and mammary cancer risk. J Mammary Gland Biol Neoplasia 7:67-76.
22. Lamartiniere CA, Cotroneo MS, Fritz WA, et al (2002) Genistein chemoprevention: timing and mechanisms of action in murine mammary and prostate. J Nutr 132:552S-558S.
23. Gallo D, Giacomelli S, Cantelmo F, et al (2001) Chemoprevention of DMBA-induced mammary cancer in rats by dietary soy. Breast Cancer Res Treat 69:153-164.
24. Gallo D, Ferrandina G, Giacomelli S, et al (2002) Dietary soy modulation of biochemical parameters in DMBA-induced mammary tumors. Cancer Lett 186:43-48.
25. Appelt LC, Reicks MM (1999) Soy induces phase II enzymes but does not inhibit dimethylbenz[a]anthracene-induced carcinogenesis in female rats. J Nutr 129:1820-1826.
26. Constantinou AI, Lantvit D, Hawthorne M, et al (2001) Chemopreventive effects of soy protein and purified soy isoflavones on DMBA-induced mammary tumors in female Sprague-Dawley rats. Nutr Cancer 41:75-81.
27. Gotoh T, Yamada K, Yin H et al (1998) Chemoprevention of N-nitroso-N-methylurea-induced rat mammary carcinogenesis by soy foods or biochanin A. Jpn J Cancer Res 89:137-142.
28. Cohen LA, Zhao Z, Pittman B, et al (2000) Effect of intact and isoflavone-depleted soy protein on NMU-induced rat mammary tumorigenesis. Carcinogenesis 21:929-935.

Cytogenetic Analysis of Genomic Destabilization in Solely Estrogen-Induced Female ACI Rat Mammary Neoplasms

Dan Papa, Jonathan J. Li, and Sara Antonia Li

Summary

A high frequency of aneuploid cells in carcinoma *in-situ* (CIS) and in small and large mammary tumors of female ACI rats are induced by estrogen (E) treatment alone. Employing G-banding karyotype analysis, consistent numerical chromosome alterations were detected as gains in chromosome 7, 11, 13, 19, and 20, and loss of chromosome 12, when the data were analyzed by binomial distribution for a frequency of occurrence ≥ 30.0%. Recurrent chromosome gains were also detected in chromosomes 3, 4, 6, 10, 14, and 15, whereas infrequent losses were observed in chromosome 5. Regional genomic alterations were detected by comparative genomic hybridization (CGH) in eleven individual E-induced primary mammary gland tumors (MGTs) obtained from an equal number of female ACI rats. CGH analysis revealed consistent regional gains at *1p11, 1q11-q22, 4q41-q44, 7q31-q33, 11p12-q11, 13p*, and *20p12*.

Introduction

Sporadic breast cancer (BC) comprises ~85-90% of all BC cases among women. Consequently, an understanding of the underlying mechanisms of BC causation is of utmost public health importance (1). Evidence has accumulated that estrogens (E), either endogenous or exogenously ingested, play a crucial role in the causation and development of human BC (2, 3). Probably, the most compelling argument for a crucial role for Es in BC causation has come from the Breast Cancer Prevention Trial in which Tamoxifen, an anti-estrogen, was shown to decrease the BC risk in women considered at increased risk for the disease (4). Recently, solely E-induced primary MGT in female ACI rats, as well as androgen-promoted MGTs in male Noble rats exhibited high aneuploid frequencies (5). In addition, in human ductal carcinoma *in-situ* (DCIS), and in primary breast neoplasms, high frequency of chromosomal instability and aneuploidy were reported, 65-78% and 85-92%, respectively (6, 7).

Taken together, these data clearly demonstrate that genomic destabilization is a distinctive event that occurs early in E-induced oncogenic processes, both in E-induced animal tumor models and in human sporadic ductal BC. The findings presented herein extend our previous aneuploidy studies in primary MGTs comparing numerical chromosomal changes determined by karyotype analysis to regional chromosomal alterations determined by *in-situ* comparative genomic hybridization (CGH).

Materials and Methods

Cell Culture of Untreated Spleen Cells from ACI Rats. Spleens from untreated, female ACI rats were removed immediately following decapitation, and ground between two autoclaved microscope glass slides in RPMI 1640 with L-glutamine. After centrifugation, the pellet was resuspended in culture medium supplemented with fetal bovine serum, and penicillin/streptomycin. Cell culture suspensions were incubated at 37°C for 4 to 5 days, with Concanavalin A and β-mercaptoethanol. After colcemid treatment, slides were prepared for CGH analysis.

Chromosome Preparation, Karyotyping, and Aneuploidy in E-induced Primary ACI Mammary Tumors. Female ACI rats bearing MGTs were used for the cytogenetic studies. Metaphase spreads were prepared from MGT according to a protocol described by us (8). An ACI rat G-banded chromosome classifier, developed in our lab, based on a system previously described (9), and implemented into QUIPS software was used to identify ACI rat metaphase chromosomes, and to determine consistent/nonrandom whole chromosome gains and losses.

Chromosome Preparation for CGH. ACI rat metaphase chromosome spreads were prepared from control untreated spleen tissue cultures according to standard procedures (10).

DNA Extraction. Control untreated ACI rat MGs and MGTs were quickly frozen in liquid nitrogen, and the DNA extracted by the LiCl protocol (11).

DNA Labeling and CGH. A CGH nick translation kit was used for direct DNA labeling for CGH, following the manufacture's recommendations. The probe preparation, hybridization, and post-hybridization steps were carried out according to the University of Colorado Health Science Center, Cancer Center Cytogenetics Core FISH Protocol 12 *'Comparative Genomic Hybridization with Directly Labeled Probes'*.

Image Acquisition and CGH Evaluation. Image analysis of CGH metaphase spreads was performed using a Photometrics cooled charge-coupled device camera, mounted on a Nikon Eclipse E400 fluorescence microscope with filter equipment for

FITC-, Texas Red-, and DAPI-fluorescence. Chromosomes were identified using digitally inverted images of DAPI-banded metaphases and an ACI rat idiogram implemented in the software. Thresholds of 1.20 were used for over-representation, and 0.80 for under-representation. CGH hybridizations were verified by exchanging fluorescent labels as previously described (12, 13).

Statistical Analysis. The data, both for karyotype and CGH analysis, were analyzed by the exact binomial distribution test that determines the occurrence or non-occurrence of an event. A criterion of ≥ 30% frequency of occurrence within any given MGT was considered as non-random/consistent chromosome number change, whereas frequencies of 21 to 30% were classified as recurrent changes.

Results

A total of 76 metaphase spreads (9-12/MGT), obtained from 8 individual E-induced primary MGT from 8 female ACI rats were subjected to G-banding karyotype analysis. A representative karyotype from a primary MGT is shown in Figure 1. Note trisomies in chromosomes 6, 8, 10, 13, and 19 and tetrasomies in 7, 11, and 15. The chromosomal alterations seen in 3, 4, and 20 were random. The total gain frequency of chromosomes 7 and 20 were 100% (8/8); for 19, 87.5% (7/8); and for 11 and 13, 62.5% (5/8). Only chromosome 12 was lost with a frequency of 62.5% (5/8) (Table 1). Recurrent chromosome gains were only detected in chromosomes 3, 4, 6, 10, 14, and 15 (data not shown). Generally, trisomies were observed with a 3.0- to 4.0-fold higher frequency in chromosomes 13, 19, and 20.

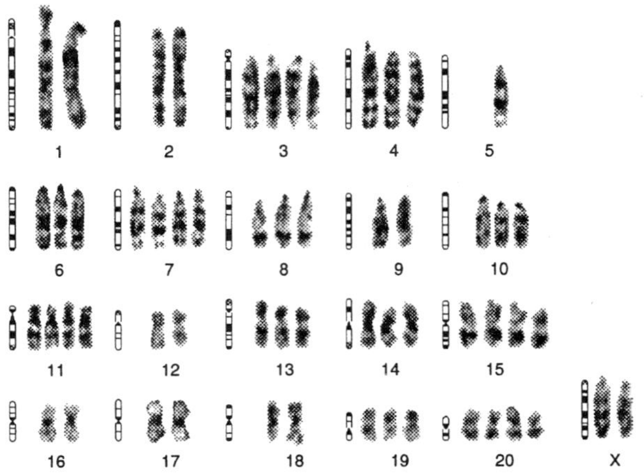

Figure 1. Representative Giemsa-banded karyotype from a 5.5-mo E-induced MGT Note trisomies in chromosomes 6, 8, 10, 13, and 19; and tetrasomies in 7, 11, and 15. The chromosomal alterations seen in 3, 4, and 20 were random.

Table 1. Non-random Numerical Chromosomal Alterations in E-induced ACI Rat Primary Mammary Gland Tumors

Chromosome	Gains %			Losses %		
	Ratio[1]	(%) 100.0 62.5	p[2] value	Ratio[1]	(%)	p[2] value
7	(8/8)	100.0	0.0000			
11	(5/8)	62.5	0.0104			
12				(5/8)	62.5	0.0104
13	(5/8)	62.5	0.0104			
19	(7/8)	87.5	0.0001			
20	(8/8)	100.0	0.0000			

1. Number of MGTs showing either gains or losses for a certain chromosome at a frequency of ≥30%/total number of MGTs
2. p value for binomial distribution

Employing CGH analyses, profound regional genomic alterations were detected in all E-induced primary MGTs examined. Negative control CGH analyses were based on successfully hybridized metaphases with two differentially labeled controls, untreated female reference spleen DNA (Figures 2, 3). Moreover, CGH analysis revealed nonrandom amplified regions in chromosome 1 (two distinct regions), 4, 7, 9, 11, 13, 19, and 20, when a stringent criterion of ≥ 30% CGH frequency of occurrence was employed within any given MGT (Table 2). The overall consistently amplified regions among 11 MGTs ranged between 54.5 to 81.9%.

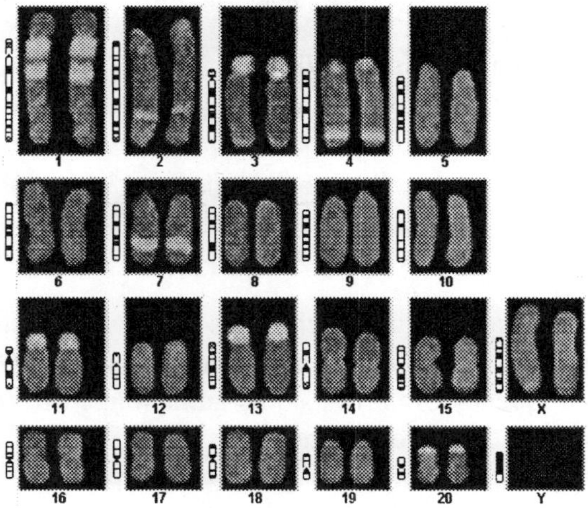

Figure 2. Representative CGH karyotype from a female ACI rat E-induced MGT. Tumor DNA was detected in green (SpectrumGreen FITC) and normal DNA in red (SpectrumRed TRITC). Chromosomes were identified using digitally inverted images of DAPI-banded metaphase chromosomes. A rat G-banded chromosome classifier was implemented in the VYSIS image analysis software. Note regional gains on chromosomes 1, 3, 4, 7, 11, 13, and 20.

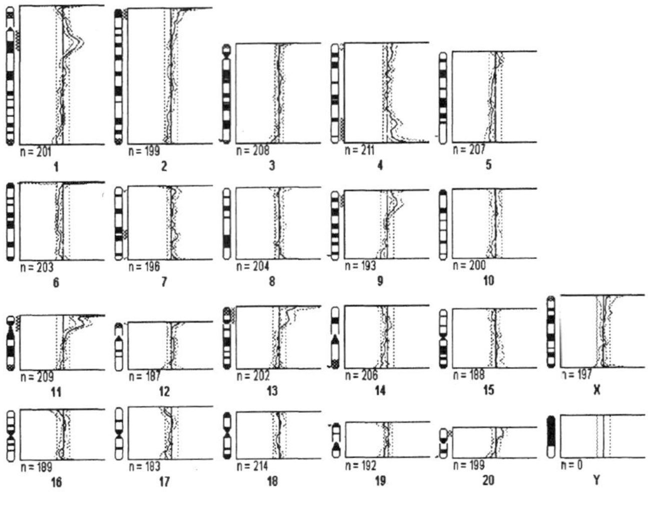

Figure 3. Ratio profiles of gre
red fluorescence intensities after
from a female ACI rat E-ind
MGT. Profiles were normalized
average green/red ratio of 1.0.
and left lines depict the uppe
lower threshold of 1.2 for over
0.80 for under-represent
respectively. Ratio profiles
displayed along with rat idiogra

Table 2. Frequency of genomic alterations determined by CGH analysis in ACI rat E-induced MGT[1].

Chromosome	Genomic alteration	Region[2]	Locus[2]	Ratio[3]	% frequency	p-value[4]
1	*p11*			6/11	54.5	0.056
1	*q11-q22*	*1q21-22*	*Ccne1*	7/11	63.6	0.017
4[5]	*q41-q44*	*4q42-44*	*Ccnd2*	7/11	63.6	0.005
7[5]	*q31-q33*	*7q33*	*c-myc*	9/11	81.9	0.005
9	*q11-q13*			6/11	54.5	0.056
11[5]	*p-q11*			6/11	54.5	0.056
13[5]	*p*			7/11	63.6	0.017
20[5]	*p12*			9/11	81.9	0.005

[1]Eleven individual MGTs were examined with an average of 10 metaphases/MGT
[2]From Entrez Genome, Ratus norvegicus Map View (www.ncbi.nlm.nih.gov/mapview/maps.cgi).
[3]Frequencies ≥ 30%.
[4]Analyzed by binomial distribution.
[5]Chromosomal alterations identified by karyotype analysis

Conclusions

Employing karyotype analysis, consistent whole chromosome alterations (gains/losses) were detected in E-induced MGTs, when a stringent frequency (≥ 30%) of occurrence was used. The frequency of chromosome gains ranged from 100% (chromosomes 7 and 20) to 62.5% (chromosomes 11 and 13). Only chromosome 12 was lost with a frequency of 62.5%. These chromosomal alterations were accompanied by numerous structural alterations of the mitotic apparatus, anaphase chromosome bridges, lagging chromosomes, and frequent multipolar nuclear spindles (data not shown).

Our findings are in agreement with previous studies showing that chromosomal instability (CIN) may contribute to the oncogenic process by eliciting gains of chromosomes containing oncogenes, and losses of chromosomes containing tumor suppressor genes (14, 15, 16, 17). The highly consistent gain of chromosome 7 represents an example of such event. Chromosome 7, where *c-myc* resides at *7q33* (18) was gained in 100% of the E-induced primary MGTs, and previous FISH performed in our lab using a fluorescent-labeled *c-myc* cDNA probe on MGT metaphase spreads (5) revealed a specific and consistent fluorescent signal on all three homologues of a trisomic chromosome 7. In addition, the *7q33* region was amplified in 81.9% of similar primary MGT examined by CGH. Interestingly, in E-induced primary ectopic uterine stem cell tumors in the kidney of the male Syrian hamster, a similar consistent chromosome gain frequency of 88.0% was seen in chromosome 6, where *c-myc* resides at *6qb3* (19). Thus, two solely E-induced and dependent tumors, in two different E-target tissues and species, exhibit similar frequencies of *c-myc* amplification in their tumors. The results of the CGH analysis and the frequency of occurrence of *c-myc* amplification in female ACI rat MGT is similar to that reported in human BC samples using CGH analysis (20, 21).

Acknowledgements

This investigation was supported by NIH Grant CA33519 (SAL).

References

1. King M-C, Rowell S, Love SM (1993) Inherited breast and ovarian cancer. What are the risks? What are the choices? J Am Med Assoc 269:1975-1980.
2. Chen C-L, Weiss NS, Newcomb P, et al (2002) Hormone replacement therapy in relation to breast cancer. JAMA 287:734-741.
3. Nelson HD, Humphrey LL, Nygren P, et al (2002) Postmenopausal hormone replacement therapy. Scientific Reviews. JAMA 288:872-881.
4. Fisher B, Constantino JP, Wickerham DL, et al (1998) Tamoxifen for prevention of breast cancer: Report of the National Surgical Adjuvant Breast and Bowel Project P-1 Study. J Nat Cancer Inst 90:1371-1388.
5. Li JJ, Papa D, Davis MF, et al (2002) Ploidy differences between hormone- and chemical carcinogen-induced rat mammary neoplasms: Comparison to invasive human ductal breast cancer. Mol Carcinogen 33:56-65.
6. Peiro G, Lemma A, Clement MA, et al (1997) Prognostic value of S-phase fraction in lymph node-negative breast cancer by image and flow cytometry analysis. Mod Pathol 10:216-222.
7. Arnerlov C, Endin SO, Cajander S, et al (2001) Intratumoral variation in DNA ploidy and S-phase fraction in human breast cancer. Anal Cell Pathol 23:21-28.

8. Li SA, Liao D-Z, Li JJ (2000) Estrogen-induced breast cancer in female ACI rats. In: Li JJ, JR Daling, SA Li (editors) Hormonal Carcinogenesis III. New York: Springer-Verlag, 178-188.
9. Levan G (1974) Nomenclature for G-bands in rat chromosomes. Hereditas 77:37-51.
10. Islam, MQ, Levan G (1987) A new fixation procedure for quality G-bands in routine cytogenetic work. Hereditas 107:127-30.
11. Gemmel NJ, Akyiama S (1996) An efficient method for the extraction of DNA from vertebrate tissues. Trends Genet 12:338-339.
12. Kallioniemi A, Kallioniemi O-P, Piper J (1994) Optimizing comparative genomic hybridization for analysis of DNA copy number changes in solid tumors. Genes Chromosomes Cancer 10:231-243.
13. Mohaptra G, Kim DH, Feuerstein BG (1995) Detection of multiple gains and losses of genetic material in ten glioma cell lines by comparative genomic hybridization. Genes Chromosomes Cancer 13:86-93.
14. Pihan, GA, Doxey SJ (1999) The mitotic machinery as a source of genetic instability. Seminars Cancer Biol 9:289-302.
15. Duesberg P, Rausch C, Rasnik D, et al (1998) Genetic instability of cancer cells is proportional to their degree of aneuploidy. Proc Nat Acad Sci USA 95:13692-13697.
16. Pihan GA, Purohit A, Wallace J, et al (1998) Centrosome defects and genetic instability in malignant tumors. Cancer Res 58:3974-3985.
17. Lengauer C, Kinzler KW, Vogelstein B (1997) Genetic instability in colorectal cancers. Nature 401:297-300.
18. Tsichlis PN, Lohse MA, Szpirer C, et al (1985) Cellular DNA regions involved in the induction of rat thymic lymphomas (Mlvi-1, Mlvi-2, Mlvi-3, and c-myc) represent independent loci as determined by their chromosomal map location in the rat. J Virol 56:938-942.
19. Papa D, Li SA, Li JJ (2003) Comparative genomic hybridization of estrogen-induced ectopic uterine-like stem cell neoplasms in the hamster kidney: Non-random chromosomal alterations. Mol Carcinogen 38:97-105.
20. Kreipe H, Feist H, Fischer L, et al (1993) Amplification of *c-myc* but not of *c-erb-B-2*, is associated with high proliferative capacity in breast cancer. Cancer Res 53: 1956-1961.
21. Courjal F, Theillet C (1997) Comparative genomic hybridization analysis of breast tumors with predetermined profiles of DNA amplification. Cancer Res 57:4368-4377.

Microarray Analysis of Estrogen-induced Protection Against Breast Cancer

Lakshmanaswamy Rajkumar, Demi-Nhung Dang, Mark D. Hartnett, David L. Hirschberg, Kenneth C. Loh, Raphael C. Guzman, Gudmundur Thordarson, and Satyabrata Nandi

Summary

Pregnancy early in life reduces the risk of breast cancer (BC) in women and this effect is universal. This phenomenon is also observed in rodents. We have shown that a treatment with high pregnancy levels of 17β-estradiol (E_2) for 7 to 21 days is effective in conferring protection against mammary carcinogenesis. We determined the difference in gene expression between hormone-protected rats and unprotected rats. Nine weeks old female Lewis rats were treated with 10 microgram (unprotected), or 200 microgram (protected) of E_2 in silastic capsules for 3 weeks. Control rats received silastic capsules with no hormone. The rats were terminated 8 weeks after the removal hormone treatment. Mammary RNA was used for microarray analysis. Using Agilent Rat cDNA Microarrays, with 14,815 unique clones, we have analyzed genes from *Rattus norvegicus* and *Rattus rattus* which are annotated as "mRNA" or "gene EST" in GenBank. The genes involved in growth promotion like interleukin 18, kit oncogene, thyrotropin stimulating hormone receptor, cyclin dependent kinases etc. were down regulated in the protected group compared to the unprotected groups. In contrast, genes involved in growth inhibition like early growth response 1, insulin-like growth factor binding proteins and genes involve in apoptosis and DNA repair like T-cell death activated gene, CD47, histone acetyltransferase were up regulated in the protected animals compared to the unprotected. These findings define a pattern of gene expression that could serve to determine the efficacy of protective hormone treatments and help identify potential biomarkers for prevention of mammary cancers.

Introduction

BC is the most common cancer in women worldwide. Attempts at prevention of BC are important areas of clinical and experimental investigations and many different approaches are being used for such studies (1-4). A full term pregnancy before the age of 20 is the only natural phenomenon known that can drastically reduce the risk of BC in women of all ethnic backgrounds worldwide (5-8). This universal protective effect of early pregnancy is clearly of major consideration in devising strategies for BC prevention. Rats that have been exposed to chemical carcinogens before or after undergoing a full-term pregnancy are protected from mammary carcinogenesis (9-15). Mice that undergo a full term pregnancy also have a greatly reduced susceptibility to chemical carcinogen induced mammary carcinogenesis compared to nulliparous animals (16-18).

Hormonal prevention strategies have used exogenous hormonal treatment to mimic the protective effect of pregnancy against BC. High levels of E_2 and progesterone (P) given for 30 days beginning 15 days after carcinogen administration inhibited mammary carcinogenesis in Sprague Dawley rats treated with the carcinogens (19, 20). Medina (21) also showed that treatment with E_2 and P, prior to initiation of cancer cells with a carcinogen was highly protective. Human chorionic gonadotropin treatment either before or after carcinogen exposure induces protection from mammary carcinogenesis (13). Short-term (7-21 days) treatment with pregnancy level of E_2 with or without P is highly effective in conferring protection against chemical carcinogen induced carcinogenesis in rats (22, 23).

Despite having strong epidemiological, clinical, and experimental evidences suggesting parity-related or hormone-induced refractoriness against mammary carcinogenesis, very little is known about the physiological and molecular mechanisms governing this parity-induced protection. To understand the molecular bases of the protective effect of hormonal treatment, we analyzed the difference in gene expression levels between the protected and the susceptible states using cDNA microarrays. We used a global approach to identify differences in gene expression patterns that may be useful to establish the efficacy of protective hormone treatment, and as biomarkers for the success or failure of hormone treatments.

Materials and Methods

Animals. Virgin Lewis rats were from Charles River Laboratories (Wilmington, MA), housed in a temperature-controlled room with 12-h light/dark schedule, fed food (Teklad 8640; Teklad, Madison, WI), and water ad libitum. All the procedures followed University of California Animal Care and Use Committee guidelines.

Estradiol Treatment. All doses of E_2 were packed in individual silastic capsules (size 0.078 inch i.d. x 0.125 inch o.d., 2 cm in length; Dow Corning Corporation, Midland, MI) in a cellulose matrix. Control animals received silastic capsules containing only cellulose. All silastic capsules were dorsally implanted s.c., and primed before implantation by soaking in media 199 (GIBCO) overnight at 37°C.

Persistent Effect of Different Doses of Estradiol on Gene Expressions in the Mammary Gland. At 9 week old rats were divided into 3 groups, each consisting of 3 rats, and receiving one of the following treatments: (i) Control, (ii) 10 μg E_2 (non-protective), and (iii) 200 μg E_2 (protective). Each treatment was continued for 3 weeks. At the end of the treatment, the silastic capsules were removed. The rats were terminated, 8 weeks after the removal of the hormone treatment. Mammary glands were removed, immediately frozen in liquid N, and stored at -80°C.

RNA Isolation. RNA was isolated from frozen samples using Trizol (Life Tech). Total RNA was subjected to DNAse treatment, and further purified with RNeasy columns (Qiagen). The quality of RNA was analyzed using the Agilent 2100 Bioanalyzer, and quantified using a Hitachi UV spectrophotometer.

Microarray Analysis. Using Agilent Rat cDNA Microarrays, with 14,815 unique clones, we analyzed genes from *Rattus norvegicus* and *Rattus rattus* which are annotated as "mRNA" or "gene EST" in GenBank. Experiments were performed in replicates. Samples were labeled and reverse labeled with Cy3 and Cy5. The combined plots show high reproducibility of the experimental set. Bioinformatic analysis was carried out using Rosetta Resolver software.

Results

Microarray Analysis of the Short-Term Estradiol-induced Changes in Gene Expression. The objective of this study was to define a pattern of gene expression that would serve as a signature profile for the efficacy of preventive hormone treatments. This approach may identify genes which otherwise would not have been considered in the targeted approach. We used RNA from whole mammary gland for the microarray analysis, since stromal-epithelial interactions have been implicated in the normal and cancer development of the mammary gland. The use of a low dose of E_2 that does not confer protection should permit the discrimination of genes induced by E_2 treatment, but not associated with protection. We identified ~ 20 genes differentially expressed between the protected and unprotected groups. These genes are listed in Tables 1 and 2.

Genes Up-regulated in the Protected Group. Several genes involved in growth inhibition like early growth response 1, insulin-like growth factor binding proteins were up-regulated in the animals which received the protective E_2 treatment compared to the non-protective E_2 dose treatment and controls. CD47, T-cell death activated gene, double stranded RNA dependent activator protein kinase and p53 apoptosis associated target are involved in apoptosis, and these genes are up-regulated in the protected group compared to the unprotected groups. Serine protease inhibitor, which influences apoptosis and inhibits adhesion, migration and invasion, is highly expressed in the 200 μg E_2 treated animals compared to the rats which were treated with 10 μg E_2 and the controls. Histone acetyltransferase, DNA repair gene, was up-regulated in the protected animals compared to the unprotected.

Table 1. Genes Up-regulated in the Protected Group (200 μg E_2).

Gene	Gene Symbol	Accession Number	Fold Change	Function
Early Growth Response1	Egr1	M18416	3.5	Growth regulation
T-cell death associated gene	Tdag	AF192802	2.5	Pro-apoptotic
Cluster of Differentiation 47	CD47	D87659	2.1	Pro-apoptotic
Double stranded RNA dependent activator protein kinase	Prkra	AF083032	1.9	Pro-apoptotic
p53 apoptosis- associated target	Perp	AF249870	1.8	Pro-apoptotic
Insulin-like growth factor binding protein 5	Igfbp5	AF139830	2.1	Binding protein
Insulin-like growth factor binding protein 2	Igfbp2	M31672	2.1	Binding protein
Retinoblastoma binding protein 2 homolog	Rbp2	AAD16061	1.7	Binding protein
Serine protease Inhibitor, Kunitz Type 2	Spint2	AF099020	2.2	Protease inhibitor
Histone acetyl-transferase 1	Hat1	AAC02425	1.9	DNA-repair

Genes Down-regulated in the Protected Group. There were consistent decreases in growth promoting genes and genes involved in cell cycle regulation in the mammary glands of protected animals in comparison to the unprotected ones. These include thrombospondin, thyroid-stimulating hormone receptor, interleukins and cyclin-dependent kinases. Genes encoding for superoxide dismutase and myeloperoxidase were also lowly expressed in the protective hormone treatment group. Further, genes involved in anti-apoptosis (metallothionein-1), neo-angiogenesis (matrix metalloproteinase 11/stromelysin 3), and oncogenesis (kit) were down-regulated in the 200 μg E_2 treated rats compared to 10 μg E_2-treated and control rats.

Table 2. Genes down-regulated in the protected group (200 μg E_2).

Gene	Gene Symbol	Accession Number	Fold Change	Function
Thrombospondin	THBS	M63470	3.2	Growth regulation
Thyroid-stimulating hormone receptor	Tshr	M34842	1.9	Growth regulation
Interleukin 18	Il18	U77776	1.9	Growth regulation
Cyclin-dependent Kinase 2	Cdk2	D28754	1.8	Cell cycle
Cyclin-dependent Kinase 4	Cdk4	L11007	1.4	Cell cycle
Kit	Kit	X65997	1.7	Oncogene
Metallothionein-1	Mt1a	M11794	3.9	Anti-apoptotic
Superoxide dismutase 3	Sod3	Z24721	2.6	Oxidoreductase
Myeloperoxidase	Mpo	X15313	1.8	Peroxidase
Matrix metallo-proteinase 11	Mmp11	U46034	1.8	Neo-angiogenesis

Conclusion

The data from the present investigation demonstrates that short-term treatment with 200 μg of E_2, which results in high pregnancy levels of E_2 in circulation, persistently alters the gene expression patterns. Although pregnancy levels of E_2 with or without P conferred protection against mammary carcinogenesis, the mechanism of this protection is not clear. Our findings illustrate that there may be several different molecular pathways which may be altered in the protected animals. Recently, persistently altered expression of genes in the mammary glands of parous rats was reported (24). In that report, certain genes involved in growth promotion were persistently down regulated in parous mammary glands; while TGF-β3 and several of its transcriptional targets were up-regulated in parous mammary glands. Other investigators studying protection induced by $E_2 + P$ treatment found persistent alterations in the expression of several known genes. Specifically, persistent changes in estrogen receptor (ERα), p53, and retinoblastoma binding protein (RbAp46) a gene implicated in the regulation of cell proliferation and differentiation following hormone treatment (25, 26) were identified.

Short-term treatment with high pregnancy levels of E_2 resulted in persistent down-regulation of genes involved in growth promotion, cell cycle, anti-apoptosis, angiogenesis, and oncogenesis. On the other hand, the protective treatment persistently up-regulated genes involved in growth inhibition, apoptosis and DNA-repair. The data obtained could be helpful to understand the mechanisms involved in estrogen–induced protection and will also facilitate the identification of biomarkers associated with protection.

Acknowledgement

Supported by California Breast Cancer Research Program 8PB-0132.

References

1. Jordan VC (2000) Progress in the prevention of breast cancer: concept to reality. J Steroid Biochem Mol Biol 74:269-277.
2. RosenbaumSmith SM, Osborne MP (2000) Breast cancer chemoprevention. Am J Surg 180:249-251.
3. Brown PH, Lippman SM (2000) Chemoprevention of breast cancer. Breast Cancer Res Treat 62:1-17.
4. Leris C, Mokbel K(2001) The prevention of breast cancer: an overview. Curr Med Res Opin 16: 252-257.
5. MacMahon B, Cole P, Lin TM, et al (1970) Age at first birth and breast cancer risk. Bull World Health Organ 43:209-221.
6. Henderson BE, Powell D, Rosario I, et al (1974) An epidemiologic study of breast cancer. J Natl Cancer Institute 53:609-614.
7. Dupont WD, Page DL (1987) Breast cancer risk associated with proliferative disease, age at first birth, and a family history of breast cancer. Am J Epidemiol 125:769-779
8. Kelsey JL (1993) Breast cancer epidemiology: summary and future directions. Epidemiol Rev 15:256-263.
9. Welsch CW (1985) Host factors affecting the growth of carcinogen-induced rat mammary carcinomas: a review and tribute to Charles Brenton Huggins. Cancer Res 45:3415-3443.
10. Sinha DK, Pazik JE, Dao TL (1988) Prevention of mammary carcinogenesis in rats by pregnancy: effect of full-term and interrupted pregnancy. Br J Cancer 57:390-394.
11. Moon RC (1969) Relationship between previous reproductive history and chemically induced mammary cancer in rats. Int J Cancer 4:312-317.
12. Dao TL (1971) Inhibition of tumor induction in chemical carcinogenesis in the mammary gland. Prog ExpTumor Res 14:59-88.
13. Russo J, Russo IH (1994) Toward a physiological approach to breast cancer prevention. Cancer Epidemiol Biomarkers Prev 3:353-364.
14. Thordarson G, Jin E, Guzman RC, et al (1995) Refractoriness to mammary tumorigenesis in parous rats: is it caused by persistent changes in the hormonal environment or permanent biochemical alterations in the mammary epithelia? Carcinogenesis 16:2847-2853.

15. Grubbs CJ, Juliana MM, Hill DL, et al (1986) Suppression by pregnancy of chemically induced preneoplastic cells of the rat mammary gland. Anticancer Res 6:1395-1400.
16. Marchant J (1955) Influence of pregnancy and lactation on the incidence of mammary carcinoma induced with methylcholanthrene in female mice of the IF strain. J Pathol Bacteriol 70:415-418.
17. Swanson SM, Guzman RC, Collins G, et al (1995) Refractoriness to mammary carcinogenesis in the parous mouse is reversible by hormonal stimulation induced by pituitary isografts. Cancer Lett 90:171-181.
18. Medina D, Smith GH (1999) Chemical carcinogen-induced tumorigenesis in parous, involuted mouse mammary glands. J Natl Cancer Inst 91:967-969.
19. Huggins C, Moon RC, Mori S (1962) Extinction of experimental mammary cancer. I. Estradiol-17β and progesterone. Proc Natl Acad Sci USA 48:379-386.
20. Grubbs CJ, Juliana MM, Whitaker LM (1988) Short-term hormone treatment as a chemopreventive method against mammary cancer initiation in rats. Anticancer Res 8:113-117.
21. Sivaraman L, Stephens LC, Markaverich BM, et al (1998) Hormone-induced refractoriness to mammary carcinogenesis in Wistar-Furth rats. Carcinogenesis 19:1573-1581.
22. Guzman RC, Yang J, Rajkumar L, et al (1999) Hormonal prevention of breast cancer: mimicking the protective effect of pregnancy. Proc Natl Acad Sci USA 96:2520-2525.
23. Rajkumar L, Guzman RC, Yang J, et al (2001) Short-term exposure to pregnancy levels of estrogen prevents mammary carcinogenesis, Proc Natl Acad Sci USA 98:11755-11759.
24. D'Cruz CM, Moody SE, Master SR, et al (2002) Persistent parity-induced changes in growth factors, TGF-beta3, and differentiation in the rodent mammary gland. Mol Endocrinol 16:2034-2051.
25. Ginger MR, Gonzalez-Rimbau MF, Gay, JP, et al (2001) Persistent changes in gene expression induced by estrogen and progesterone in the rat mammary gland. Mol Endocrinol 15:1993-2009.
26. Medina D, Sivaraman L, Hilsenbeck SG, et al (2001) Mechanisms of hormonal prevention of breast cancer. Ann NY Acad Sci 952:23-35.

Pregnancy Level of Estrogen Prevents Mammary Cancers

Raphael C. Guzman, Lakshmanaswamy Rajkumar, Gudmundur Thordarson, and Satyabrata Nandi

Summary

A full-term pregnancy early in life reduces the risk of breast cancer (BC) in women. Parity protection against mammary carcinogenesis is also observed in rats and mice. Administration of high pregnancy levels of ovarian hormones to nulliparious rats induces protection from N-methyl-N-nitrosourea (MNU)-induced mammary cancers. We have demonstrated that 17β-estradiol (E_2) alone or in combination with progesterone (P) are effective in preventing mammary cancer while P alone is not. We have now determined the actual daily amount of E_2 required to confer protection against mammary cancer. Rats were injected with MNU at 7 weeks of age. Two weeks later the rats were treated with 10 nanograms, 100 nanograms, or 1 μgram of E_2 per day for 3 weeks in mini-osmotic pumps. All the animals also received a 30 milligram P silastic capsule for 3 weeks. Serum levels of E_2 immediately after treatments were 6, 30, and 118 pg/ml, respectively. Control rats had a 90% mammary cancer incidence while rats treated with 1 μicrogram of E_2 per day had no mammary cancers, nine months after MNU treatment. Treatments with 10 nanograms or 100 nanograms of E_2 per day did not significantly decrease the mammary cancer incidence compared to the controls; they had a cancer incidence of 78% and 65%, respectively. The blood hormone data indicates that treatment with 1 microgram of E_2 per day results in high late pregnancy levels of E_2 and this level of E_2 is effective in conferring protection against mammary carcinogenesis. This short-term chemoprevention treatment is as effective as full-term pregnancy, ovariectomy or long-term tamoxifen treatment and there is no permanent loss of ovarian function. This treatment can be used as a paradigm for developing strategies for human BC prevention.

Introduction

A full term pregnancy before the age of 20 is the only natural phenomenon known that drastically reduces the risk of BC in women of all ethnic backgrounds worldwide (1). This universal protective effect of early pregnancy is clearly of major consideration in devising strategies for the prevention of BC. Rats (2-10) and mice (11-13) that undergo a full-term pregnancy also have a greatly reduced susceptibility to chemical carcinogen induced mammary carcinogenesis compared to nulliparous animals. The protective effects of pregnancy have been reported in rats undergoing pregnancy before or subsequent to exposure to chemical carcinogens (14, 15). The physiological and molecular mechanisms for the protective effect of pregnancy have not been defined, but are thought to be related to increases in hormones during pregnancy causing persistent alterations in the systemic hormonal milieu and/or target cells of the parous females.

We have demonstrated that E_2 alone or in combination with P is effective in preventing mammary cancer, while P alone is not (16). We determined the lowest dosage of E_2 given alone that would be effective in preventing mammary cancer. Rats were injected with MNU at 7 weeks of age. Two weeks later the rats were treated with 30 mg, 200 μg, 100 μg or 20 μg of E_2 in silastic capsules for three weeks (17). Serum levels of E_2 immediately after treatments were 144, 95, 68, and 50 pg/ml, respectively. The serum levels of E_2 from the 30 mg, 200 μg, and 100 μg treated rats were in the range found during pregnancy. Control rats had a 100% mammary cancer incidence, 9.0 mo after treatment, while rats treated with 30 mg or 200 μg or 100 μg of E_2 had a cancer incidence ranging from 15 to 20%. The cancer incidence was not different from controls at the lowest dosage of E_2, but the multiplicity was reduced. The multiplicity of mammary cancers was 0.2 (30 mg), 0.3 (200 μg), 0.6 (100 μg), 1.2 (20 μg) and 3.0 (controls). These results demonstrate that short-term treatments with doses of E_2 equivalent to levels during pregnancy are highly effective in preventing mammary cancer. Treatment with E_2 equivalent to pregnancy levels does not induce full lobulo-alveolar differentiation yet it is highly effective in conferring protection. Additional studies have also shown that 1 or 2 weeks treatment with pregnancy levels of E_2 (200 μg) and P (30 mg) in silastic capsules is highly effective in reducing the mammary cancer incidence in MNU treated rats (16). Based on these studies, we wanted to determine the daily amount of E_2 necessary to induce protection from mammary carcinogenesis.

Materials and Methods

Determination of Pregnancy Levels of Estradiol in Lewis Rats. Seven-week-old female Lewis rats were mated. The appearance of a vaginal plug was designated as day 0 of pregnancy. Serum samples were collected a days 1, 5, 10, 15, and 20 of pregnancy. E_2 levels were determined by a specific radioimmunoassay for E_2 (16).

Diffusion of Estradiol from Silastic Capsules. Silastic capsules were prepared containing 30 mg of E_2 or 200 μg E_2 + 30 mg cellulose. Individual silastic capsule were incubated at 37 C in 1 ml of isotonic saline. The saline was collected at 24 h intervals for three weeks. The levels of E_2 in each sample were determined by a radioimmunoassay for E_2 (16, 17).

Treatment of Rats with Different Amounts of Estradiol per Day. Lewis rats were treated with MNU at 7 weeks of age (16). Rats received no hormone treatment or were implanted with mini-osmotic pumps to deliver 10 ng, 100 ng, or 1 μg E_2 in sesame oil/day, beginning at 9 weeks of age, and continuing for 3 more weeks. Rats were palpated weekly for overt mammary cancers and terminated at 44 weeks of age. Serum samples were collected at the end of the treatment to assay E_2 levels.

Results

Serum Levels of Estradiol During Pregnancy in Lewis Rats. Serum levels in pregnant Lewis rats did not reach maximal levels until the third week of pregnancy (90-140 pg/ml) (Table 1).

Table 1. E_2 Serum Levels During Pegnancy in the Lewis Rat.

Day of Pregnancy	Estradiol pg/ml
Day 1	8.0 $\pm$ 1.9
Day 5	9.2 $\pm$ 1.0
Day 10	21.0 $\pm$ 1.7
Day 15	137.1 $\pm$ 5.4
Day 20	89.6 $\pm$ 9.8

Diffusion of E_2 from Silastic Capsule. Silastic capsules containing 30 mg of E_2 released approximately 1 μg of E_2 per day aftter incubation in saline. Capsules containing 200 μg of E_2 released ~ 100 ng of E_2 per day. Capsules filled with these amounts of E_2 have been highly effective in conferring protection from mammary carcinogenesis. We hypothesized that daily sustained treatment with a total of 100 ng to 1 μg of E_2 should result in high pregnancy levels of E_2 and confer protection from mammary carcinogenesis.

Effects of the Administration of Different Amounts of E_2 in Mini-osmotic Pumps. Only treatment with 1 μg of E_2 per day resulted in high pregnancy levels of E_2 (120 pg/ml). The other doses resulted in lower levels found during the early part of pregnancy. Only the 1 μg per day dose was able to confer significant protection in terms of mammary cancer incidence (0%) compared to controls (92%). The 100 ng (78%) and 10 ng (65%) mammary cancer incidences were not significantly different from controls. The 1 μg dose (0), 100 ng dose (1.25), and 10 ng dose all reduced the multiplicity of mammary cancers compared to controls (2.5) (Table 2).

Table 2. Serum Levels of E_2, Mammary Cancer Incidence, and Multiplicity in E_2-Rats.

Treatment	E_2 pg/ml	Mammary Cancer % Incidence	Mammary Cancer Multiplicity
Control	8.0 ± 1.9	92	2.50
10 ng E_2	5.8 ± 1.0	78	1.40
100 ng E_2	30.1 ± 5.0	65	1.25
1 μg estradiol	118.0 ± 9.8	0	0

Conclusions

Our results indicate that short-term treatment with any dose of E_2 that results in serum levels equivalent to the high levels of E_2 during the third week of pregnancy are able to confer persistent protection from chemical carcinogen induced mammary carcinogenesis in rats.

Our goal is to use the rodent model system to develop a safe, efficient mechanism based, short-term hormonal intervention for protection from mammary carcinogenesis that can serve as an experimental paradigm for developing BC prevention strategies for women. We hypothesize that short-term treatment with high pregnancy levels of E_2 or parity confers protection from mammary carcinogenesis by causing persistent alterations in the hypothalmo-pituitary axis resulting in a decrease in the blood levels of the mammogenic hormones, prolactin and growth hormone, causing a reduction in the promotion of carcinogen initiated mammary cells to frank mammary cancers

References

1. MacMahon B, Cole P, Lin TM, et al (1970) Age at first birth and breast cancer risk. Bull World Health Organ 43:209-221.
2. Leris C, Mokbel K (2001) The prevention of breast cancer: an overview. Curr Med Res Opin 16:252-257.
3. Moon RC (1969) Relationship between previous reproductive history and chemically induced mammary cancer in rats. Int J Cancer 4:312-317.
4. Dao TL (1971) Inhibition of tumor induction in chemical carcinogenesis in the mammary gland. Prog ExpTumor Res 14:59-88.
5. Russo J, Russo IH (1994) Toward a physiological approach to breast cancer prevention. Cancer Epidemiol Biomarkers Prev 3:353-364.
6. Russo IH, Russo J (1993) Physiological bases of breast cancer prevention. Eur J Cancer Prev 2 Suppl 3:101-111.
7. Russo J, Rivera R, Russo IH (1992) Influence of age and parity on the development of the human breast. Breast Cancer Res Treat 23:211-218.

8. Thordarson G, Jin E, Guzman RC, et al (1995) Refractoriness to mammary tumorigenesis in parous rats: is it caused by persistent changes in the hormonal environment or permanent biochemical alterations in the mammary epithelia? Carcinogenesis 16:2847-2853.
9. Grubbs CJ, Juliana MM, Hill DL, et al (1986) Suppression by pregnancy of chemically induced preneoplastic cells of the rat mammary gland. Anticancer Res 6:1395-1400.
10. Grubbs CJ, Hill DL, McDonough KC, et al (1983) N-nitroso-N-methylurea-induced mammary carcinogenesis: effect of pregnancy on preneoplastic cells. J Natl Cancer Inst 71:625-628.
11. Marchant J (1955) Influence of pregnancy and lactation on the incidence of mammary carcinoma induced with methylcholanthrene in female mice of the IF strain. J Pathol Bacteriol 70:415-418.
12. Swanson SM, Guzman RC, Collins G, et al (1995) Refractoriness to mammary carcinogenesis in the parous mouse is reversible by hormonal stimulation induced by pituitary isografts. Cancer Lett 90:171-181.
13. Medina D, Smith GH (1999) Chemical carcinogen-induced tumorigenesis in parous, involuted mouse mammary glands. J Natl Cancer Inst 91:967-969.
14. Welsch CW (1985) Host factors affecting the growth of carcinogen-induced rat mammary carcinomas: a review and tribute to Charles Brenton Huggins. Cancer Res 45:3415-3443.
15. Sinha DK, Pazik JE, Dao TL (1988) Prevention of mammary carcinogenesis in rats by pregnancy: effect of full-term and interrupted pregnancy. Br J Cancer 57:390-394.
16. Guzman RC, Yang J, Rajkumar L, et al (1999) Hormonal prevention of breast cancer: mimicking the protective effect of pregnancy. Proc Natl Acad Sci USA 96:2520-2525.
17 Rajkumar L, Guzman RC, Yang J, et al (2001) Short-term exposure to pregnancy levels of estrogen prevents mammary carcinogenesis. Proc Natl Acad Sci USA 98:11755-11759.

Hormonal Dependence of Mammary Premalignant Progression

Daniel Medina, Frances S. Kittrell, Anne Shepard, Jamel Hill, and Powel Brown

Summary

The p53 null mammary epithelium transplant model has been extensively characterized at the genetic, hormonal, and biological levels. Tumors progress from ductal hyperplasias and ductal carcinoma *in-situ* (DCIS) and are aneuploid, metastatic and approximately 80% are estrogen receptor (ERα) negative. The normal and premalignant mammary stages of development are ERα positive and hormone responsive. Continuous estrogen (E) or progesterone (P) treatment markedly enhances tumorigenesis in p53-null mammary epithelium. Blocking E signaling by tamoxifen or blocking P signaling by deleting the PR or blocking E and P signaling by ovariectomy markedly decreases hormone-induced mammary tumorigenesis in p53-null epithelium. The tumors that do arise are ERα negative. These results suggest that this model is appropriate to examine issues regarding the timing and duration of tamoxifen treatment on pre-malignant progression and the use of combined prevention strategies to delay the occurrence of invasive breast cancer (BC).

Introduction

There are numerous models of mouse mammary tumorigenesis (1, 2). These include traditional models in which the oncogenic stimulus is the murine mammary tumor virus (MMTV), a chemical carcinogen [*i.e.,* 7,12-dimethyl[a]benzanthracene (DMBA)], radiation or reproductive hormones (2), or the numerous transgenic and gene-deletion models developed in the past 15 years (1). Each model provides unique and specific information on genes and signal transduction pathways that can induce and/or strongly promote murine mammary tumorigenesis and theoretically provide information relevant to subsets of human BC. The exact relevance to human BC remains to be determined for the vast majority of these models. A model characterized recently is the BALB/c p53-null mammary epithelium, where deletion of the tumor suppressor gene p53 results in enhanced tumorigenic risk (3, 4). The p53-null mammary epithelium progresses trough ductal hyperplasia and DCIS prior to becoming invasive BC (5). The ductal hyperplasias are immortal, exhibit high telomerase activity, and are aneuploid and ERα positive. These properties are different from the biological and cellular properties of the traditional alveolar hyperplasias found in the MMTV, chemical carcinogen, and spontaneous models of mammary tumorigenesis (6). Interestingly, the tumors arising from both types of hyperplasias are predominantly ERα negative. Because the p53-null mammary epithelial cell possesses an ERα pattern that mimics the expression patterns found in subsets of human ductal hyperplasia, DCIS, and invasive BCs, we investigated the hormone dependence of normal and hyperplastic stages, and the tumorigenic response to the SERM, tamoxifen.

Materials and Methods

BALB/c (both p53-wt and p53-null) and FVB mice were bred and maintained at 70°F the room temperature in a closed conventional mouse colony at the Baylor College of Medicine with food and water provided *ad libitum*. The animal facility is AALAC accredited. The basic experimental protocol was as previously published (3). Mammary duct samples from 8-10 week-old BALB/c p53-null female mice were transplanted into the cleared mammary fat pads of 3.0-week-old BALB/c p53-wt or (BALB/c x FVB)F_1 mice. The transplanted ducts grow and fill the mammary fat pads in 6-8 weeks. The recipient mice were left untreated or received a pituitary isograft under the kidney capsule at 5 weeks of age. The pituitary isograft provides a continuous increased circulating level of prolactin (PLN) and P that result in differentiation of the mammary epithelium and sustained elevation in cell proliferation (7). A tamoxifen pellet (5 mg) was implanted subcutaneously in the upper back. The pellets were replaced at 3.0-mo intervals. Mice were palpated weekly over a 55-60 week study period. The tumor incidences were evaluated statistically by Fisher's exact test.

Results

Ovarian-dependent Growth of p53-null Normal Mammary Epithelium. The growth of the p53-null mammary epithelium was absolutely dependent on ovarian hormones as the cells did not exhibit expansive growth in ovariectomized (ovx) mice (Figure 1). The growth of p53-null mammary epithelium in ovx mice was inhibited to the same extent as the growth of p53-wt epithelium (84% vs. 77%, respectively; $P > 0.05$). Thus, in these assays and those previously reported, the p53-null normal mammary epithelium behaved similarly to the p53-wt normal mammary epithelium. The p53-null mammary epithelium at 8-12 weeks post transplantation exhibits a normal distribution and cellular localization of ERα and PR. At 8 weeks post transplantation, the ERα score for wt and p53-null epithelium was 5-6 on a scale of 8, using the system described in (8).

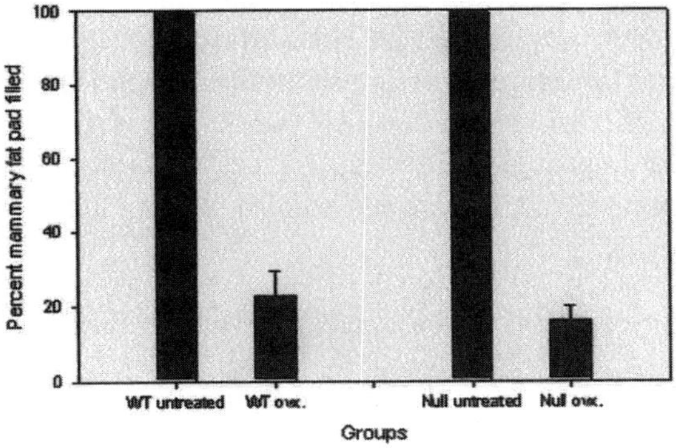

Figure 1. Growth of normal mammary duct transplants in ovx mice. There were 8 transplants/group. Mice were ovx at 5 weeks of age and mammary transplants were analyzed as whole mounts at 11 weeks of age. (Reproduced with permission from Cancer Research.)

Hormone-induced Tumorigenesis in p53-null Mammary Epithelium. Previous results demonstrated that hormones provided by a pituitary isograft (P and PLN) induce normal morphogenesis in p53-null mammary epithelium (3). The results in Table 1 show that chronic levels of E_2 and P, administered singly, markedly enhanced tumorigenesis compared to the untreated group. There were no significant differences in tumor incidences or tumor latencies among the different hormone treatment groups ($P > 0.05$). In contrast, ovariectomy at 5 weeks of age almost totally blocked tumorigenesis (Table 1) and resulted in a tumor incidence less than that in untreated mice ($P < 0.05$).

Table 1. Hormone-induced Tumorigenesis in p53-null Mammary Epithelium.

Group	No. of Tumors/No. of Transplants (%)	TE_{50} (Weeks)
Intact	11/26 (42)	>62
Ovariectomized	2/26 (7.7)	>62
Estrogen	17/26 (65.4)	38
Progesterone	24/28 (85.7)	37

Tumorigenesis in Progesterone Receptor-deficient Mammary Epithelium. The marked dependence on P-mediated signaling for tumorigenesis was tested directly by cross-breeding the p53-null mice with FVB PRKO mice to generate p53 –/–, PR –/– mammary ducts. The results in Table 2 show that mammary tumorigenesis in hormone-stimulated p53-null epithelial cells was markedly reduced in the absence of P signaling (*i.e.,* when the PR was deleted) from 84% to 32%. In these mice, P levels are high due to the PLN secreted by the pituitary isograft.

Tamoxifen Inhibition of p53-null Mammary Tumorigenesis. The tumorigenic response of the p53-null mammary epithelium is shown in Figure 2. The presence of a pituitary isograft increased the tumorigenic response ($p \leq 0.05$) compared to that of untreated epithelium [26 of 28, 93%, TE_{50} (50% tumor endpoint) = 36 weeks vs. 10 of 28, 36%, $TE_{50} > 58$ weeks, respectively]. The addition of tamoxifen completely eliminated the enhanced tumorigenic response induced by the chronic hormone stimulation (Tam. = 11 of 28, 39%, $TE_{50} > 58$ weeks), so this group effectively behaved like the untreated group. Of interest, 30% of the tumors were ERα+ in the untreated group, while no ERα+ tumors arose in the tamoxifen treated group.

Table 2. Tumorigenesis in p53-null and PR-deficient Mammary Epithelium.

Group	No. of Tumors/No. of Transplants (%)	TE_{50} (weeks)
p53 –/–, PR +/+, pit[1]	21/25 (84)	41
p53 –/–, PR –/–, pit	8/25 (32)	>52

[1] pit, pituitary isograft.

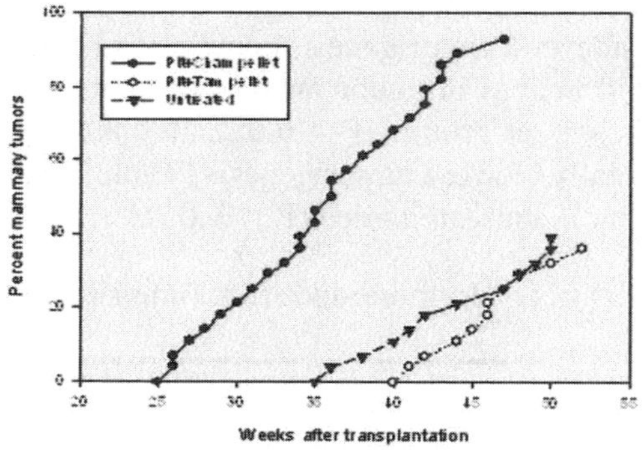

Figure 2. Effect of tamoxifen on tumorigenesis in BALB/c p53-null mammary epithelium.

Conclusions

The p53-null mammary epithelium model has been characterized at the biological and genetic level in detail. The tumors are aneuploid, metastatic, and primarily ERα-negative (about 20% of the primary tumors are ERα-positive). However, the p53-null normal mammary epithelium is ERα-positive and absolutely dependent on ovarian steroid hormones for growth, functional differentiation and tumorigenesis. Thus, this model is the best characterized transgenic model that develops both ERα-positive and ERα-negative mammary tumors. Hormonal stimulation by E and/or P or PLN/P markedly enhances tumorigenesis. Blocking estrogen signaling by either ovariectomy or tamoxifen or blocking P signaling by knocking out the PR greatly reduces tumorigenic capability of the p53-null mammary epithelium. Given the marked hormone dependence of the p53-null normal mammary epithelium, and the strong response to prolonged hormone stimulation, we tested the ability of the SERM, tamoxifen, to block or delay tumorigenesis in this model. Continuous tamoxifen exposure markedly delayed the development of ERα-negative tumors arising in p53-null mammary cells.

The p53-null mammary epithelium model used in this study appears to be an appropriate model as it mimics several of the important characteristics found in human BC development; *i.e.* E/P-dependence for tumorigenesis; genomic instability at the premalignant stage, and premalignant progression through DCIS. The presence of both ERα-positive and ERα-negative mammary tumorigenesis offers an opportunity to study molecular characteristics of these two subtypes of breast cancer in the same animal model. This model may be particularly useful to address the effect of duration and timing of tamoxifen exposure on ERα-positive and ERα-negative mammary tumorigenesis. In addition, this model would be appropriate to test the combination of anti-estrogens and other cancer preventive agents.

References

1. Cardiff RD, Moghanaki D, Jensen RA (2000) Genetically engineered mouse models of mammary intraepithelial neoplasia. J Mammary Gland Biol Neoplasia 5:421-437.
2. Medina D (2000) Mouse models for mammary cancer. In: Ip MM, Asch BB (editors), Methods in Mammary Gland Biology and Breast Cancer Research. New York: Kluwer Academic Press, 3-17.
3. Jeryr DJ, Kittrell FS, Kuperwasser C, et al (2000) A mammary-specific model demonstrates the role of the *p53* tumor suppressor gene in tumor development. Oncogene 19:1052-1058.
4. Kuperwasser C, Hurlburt GD, Kittrell FS, et al (2000) Development of spontaneous mammary tumors in BALB/c p53 heterozygous mice. Am J Pathology 157:2151-2159.
5. Medina D, Kittrell FS, Shepard A (2002) Biological and genetic properties of the p53 null preneoplastic mammary epithelium. FASEB J 16:881-883.
6. Medina D (2002) Biological and molecular characteristics of the premalignant mouse mammary gland. BBA Reviews on Cancer 1603:1-9.
7. Christov K, Swanson SM, Guzman RC, et al (1993) Kinetics of mammary epithelial cell proliferation in pituitary isografted BALB/c mice. Carcinogenesis 14:2019-2025.
8. Harvey JM, Clark GM, Osborne CK, et al (1999) Estrogen receptor status by immunohistochemistry is superior to the ligand-binding assay for predicting response to adjuvant endocrine therapy in breast cancer. J Clin Oncol 17:1474-1481.

The Possible Role of IGF-I and Androgens in the Development of Canine Inflammatory Mammary Carcinoma

Juan Carlos Illera, Gema Silván, M. Dolores Pérez-Alenza, Ana R. Sánchez-Archidona, Ana Nieto, and Laura Peña

Summary

Inflammatory mammary cancer is the most aggressive spontaneous type of mammary cancer in women and dogs. Our results indicate that a special endocrine mechanism could be involved in its pathogenesis with relevant action of IGF-1 and androgens.

Introduction

Inflammatory mammary cancer is the most aggressive spontaneous type of mammary cancer both in women (inflammatory breast carcinoma, IBC) (1) and dogs (inflammatory mammary carcinoma, IMC) (2). The clinical presentation is sudden and resembles an inflammatory process (dermatitis or mastitis) (1). The signs include diffuse erythema, edema, induration, and others, with or without the existence of mammary nodule (1, 2). The only distinctive feature for the histological diagnosis is the observation of the massive invasion of dermal lymphatics by neoplastic cells (1). Fortunately, it is a rare type of mammary cancer (1%-6% in women), although its prevalence seems to be increased in the later years in both species, probably under the influence of the same environmental carcinogens (3). Although, an animal model of IBC has been established in scid/nude mice (4), the dog is the unique animal specie in which spontaneous IMC has been reported. Several similarities have been found between human and canine IMC concerning epidemiology, histopathology, and clinical characteristics (2, 3). Canine IMC has been proposed as a spontaneous model for human IBC, being especially useful in studies on new therapeutics approaches (3).

In the last decades hormonal dependency of breast cancer (BC) has been extensively studied, overall concerning the implications of estrogen receptor (ERα) and progesterone receptor (PR) in carcinogenesis, prognosis and therapy. However, the role of androgens in normal and neoplastic mammary growth is less known.

Different studies indicate androgens can stimulate or inhibit the growth of mammary cells depending on the cell line and clone under study (5, 6). Normal and neoplastic breast tissues contain and produce several forms of androgens (5). Finally, androgens mediate their function by binding to the androgen receptor (AR) which has been detected in normal and neoplastic human and rat mammary tissues (5, 7, 8).

The mammalian family of insulin-like growth factors (IGFs) consists of two related peptides with structural homology to pro-insulin (IGF-I and IGF-II), two known receptors (type 1 and type 2; IGF-1R and IGF-2R), and at least 10 bindings proteins (IGFBPs). The role played by the IGF system on cancer has focused attention of researchers for many years. The IGF family plays a role in the growth of normal and neoplastic cells since they can have mitogenic, transforming and antiapoptotic actions (9). Also, some studies indicate a role of IGF-I in cell-cell adhesion and cell migration (10).

Regarding IBC, two genes have been identified to be differently expressed after differential display: overexpression of RhoC GTPase and the specific loss of a novel gene lost in inflammatory breast carcinoma (LIBC). Interestingly, LIBC seems to be a member of the low-affinity IGF-BP family (11). Previous results in the dog indicating an especial endocrine mechanism involved in IMC have been reported (3, 12). The purpose of this study was to investigate the possible role of IGF-I and androgens in the development of canine IMC by determining dehydroepiandrosterone (DHEA), androstenedione, testosterone, and IGF-I profiles in tissue homogenates and serum of IMC dogs, as well as the immunolocalization of androgen and IGF-I receptors. These results were compared to mammary dysplasias, benign mammary tumors, and other non-IMC malignant mammary tumors.

Materials and Methods

Animals and Samples. Serum and tissue samples of 30 animals presented at Veterinary Teaching Hospital of Veterinary School (Madrid), with at least one mammary tumor, were prospectively collected. The animals were physically and radiographically examined. Surgical excision was the treatment of choice except in IMC cases, in which surgery is not the first choice recommended. Samples were obtained by "tru-cut" biopsy or necropsy. For this study, samples were divided in two adjacent fragments, one was fixed in 10% neutral formalin (for histological and immunohistochemical study), and the other was frozen and stored for hormonal detection. Five control animals without history of any disease were included , serum and samples of two different mammary glands were obtained. To determine the esrus cycle at sampling, vaginal smears were performed.

Hormonal Study. From 35 female dogs, a total of eighty-six mammary samples were collected: 10 normal mammary gland (NMG), 21 dysplasias (DYSP), 26 benign (BMT), 22 malignant (MMT), and 7 IMC.

Plasma and Homogenate IGF-I Ensymeimmunoassay (EIA). IGF-I levels both in plasma and tissue homogenates were determined by a commercial assay kit (Non-extraction IGF-I ELISA, DSL-10-2800), validated for species and tissues.

Plasma and Homogenate DHEA, Androstenedione, and Testosterone EIA. Plasma and tumor homogenate DHEA, androstenedione and testosterone concentrations were assayed by amplified EIA previously validated for this species and tissues (12).

Histopathology. Immunohistochemistry of IGF-1R and AR. The samples for histopathology and immunohistochemistry were fixed in 10% buffered formalin; paraffin embedded and cut in 4 μm sections, following routine methods. Histopathological diagnosis was done on H&E sections following the WHO's classification for canine mammary tumors and dysplasias .

For immunohistochemistry, a selection of each tissue group was used: NMG (n = 5), BMT and DYSP (n = 10), MMT-non-CI (n = 20), MMTCI (n = 20) and 13 achieved CI cases were also included. Immunohistochemistry was performed on deparaffined sections using the streptavidin-biotin-complex peroxidase method after a high temperature unmasking protocol. The primary antibodies used were a mouse monoclonal antibody anti-IGF-1R α subunit (clone 24-31, Neomarkers Int.; dilution 1/50; incubation overnight at 4°C), and a polyclonal rabbit anti-AR (clon 2F12, Novocastra dilution 1/10; incubation overnight at 4°C). Known positive and negative control samples were used. The intensity of the staining was evaluated in all sections luated by two observers simultaneuously according to the following scale: Negative = 0, Low = +, Moderate = ++, and Intense = +++.

Statistical Analysis. The Biomedical Data Program (BMDP), Statistical Software Inc. (Los Angeles, CA, USA), was used for statistical analysis. In all statistical comparisons, $P < 0.05$ was accepted as denoting significant differences.

Results

The serum and tissue hormonal levels are shown in Figures 1 and 2, respectively. The serum levels of all the androgens studied were significantly higher levels in IMC animals than in controls, or animals bearing BMTs, MMT, and non-IMC ($p < 0.01$) (DHEA 11.82 ± 3.56 ng/ml, A4 3.81 ± 0.69 ng/ml, T4 35.43 ± 2.27 ng/ml). In contrast, levels of serum IGF were considered normal in IMC dogs (IMC 7.08 ± 4.38 ng/ml, NMG 13.68 ± 2.17 ng/ml) and increased in dogs with BMT (127.22 ± 50.41 ng/ml) and dogs with MMTnon-IMC (49.15 ± 19.14 ng/ml). In tissue homogenates, IGF-1 values of CI were not significantly different than in other MMTnon-IMC. Tissue androgens were dramatically increased in IMCs when compared to the rest of samples studied ($p < 0.001$) (DHEA 702.22 ± 89.93 ng/mg, A4 631.73 ± 70.73 ng/mg, T 287.43 ± 6.89 ng/mg, IGF-1 13.68 ± 2.17 ng/mg).

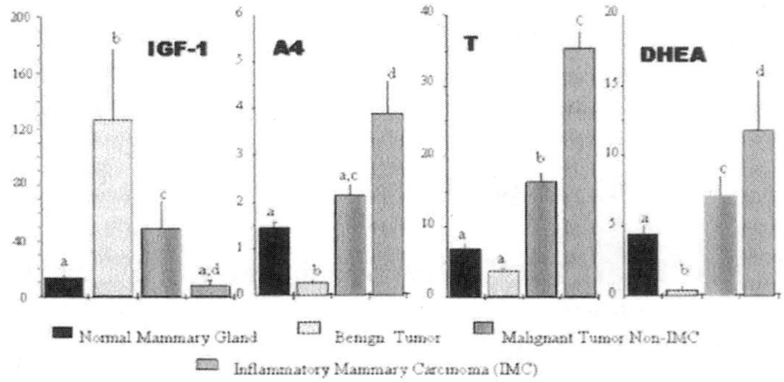

Figure 1. Serum concentrations of IGF-1, A4, T and DHEA in normal mammary gland, benign tumor, malignant tumor, and inflammatory mammaty carcinoma. Different superscript denoting statistical differences ($p < 0.01$).

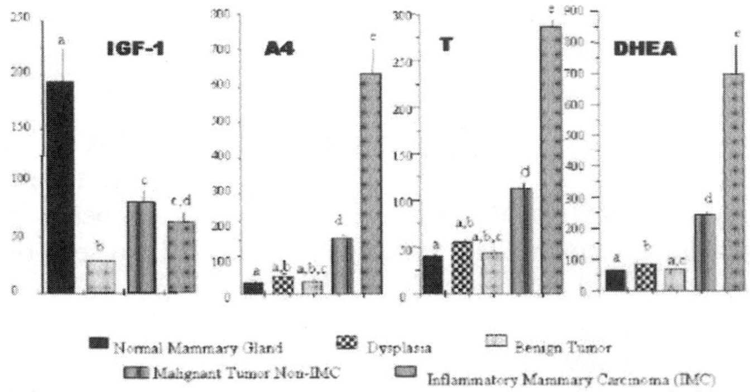

Figure 2. Tissue homogenates concentrations of IGF-1, A4, T, and DHEA in normal mammary gland, dysplasias, benign tumor, malignant tumor, and inflammatory mammary carcinoma. Different superscript denoting statistical differences ($p < 0.01$).

IGF-1R immunostaining was found in epithelial and myoepithelial cells. In positive mixed tumors, cartilaginous and osseous cells h also displayed IGF-1R expression. Occasionally, some scattered stromal cells, endothelial cells and fibroblasts, showed immunostaining. In most of the positive cases, the staining was present only in the cytoplasm, although a nuclear staining was also seen in some cases. NMGs ($n = 5$) had very low expression of IGF-1R immunostaining and they were considered negative. DMGs and BMT ($n = 10$) were also mostly negative, although some cases showed low to moderate levels of IGF-1R expression. None of the MMTnon-IMC studied were negative, most of them had low or moderate level of IGF-1R expression. Half of the IMC tumors (10/20) had a high level of IGF-1R labeling.

AR expression was present in epithelial and myoepithelial cells, mainly in the nucleus, but some cytoplasmic staining was also found. NMGs and DMGs stained for AR with low to moderate intensity. BMTs had also a low-moderate AR-expression, but few cases (6.3%) were AR-negative. Most of the MMT were AR-positive. In IMC cases, a high level of AR positive staining (+++) was found, only 5.0% of the cases were AR-negative. Considering all the samples studied, the expression of IGF-1R and AR was inversely correlated ($p < 0.05$).

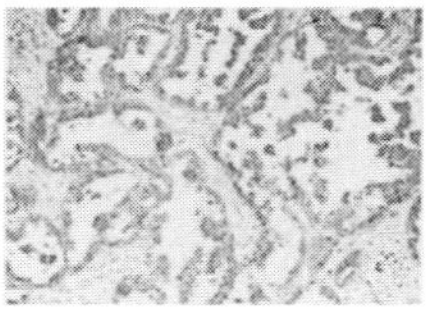

Figure 3. IMC. Tubulo-papillary carcinoma. Streptavidin biotin-peroxidase anti-IGF-1R. IGF-1R is strongly expressed in epithelial cells.

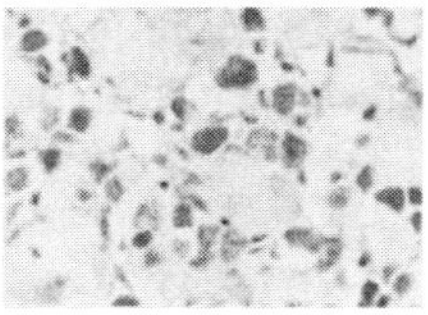

Figure 4. IMC. Streptavidin biotin-peroxidase anti AR. Infiltrating cells are AR-positive.

Discussion

Little is known about the etiology and pathogenesis of IMC. Lately, some mechanisms have been indicated to be especially involved in this type of mammary cancer and not in other malignant non-inflammatory mammary neoplasms (2, 3, 11, 12). In dogs, some histopathologic evidences indicate a lipid secretion by IMC, possibly steroids, and a marked expression of the first-step steroidogenic enzyme P450scc (3, 12). In this study, serum and tissue levels of DHEA, A_4, and T were significantly elevated in IMC with respect to other malignant non-IMC mammary tumors and to NMG. Also a strong AR immunoexpression in IMC cases was observed. DHEA, DHEA sulphate (DHEAS), and Adiol (5-androstene-3ß, 17ß-diol) of adrenal origin have been detected at high amounts in normal and cancerous breast tissues (5). DHEA may affect breast cancer cells by *in-situ* estrogen production, or by the activation of the ERα, although in a previous study on canine IMC (3), none of the cases studied were ERα-positive. In our opinion, further studies on the expression of ERβ receptor in IMC are necessary.

Studies using [^{3}H] androstenedione demonstrated that this substrate can be aromatized to estrone in breast carcinoma homogenates (13). The adrenal cortex secretion of DHEA and A4 in dogs remains controversial (14). In our opinion, a high proportion of the androgens found in the canine mammary tissues may be attributed to a local synthesis. Our findings indicate a possible different role for these hormones in the pathogenesis of IMC, acting in an autocrine/paracrine manner.

This is the first report regarding AR in canine mammary tissues and the first on IMC. Literature regarding immunohistochemistry of AR in human breast cancer offers contradictory results on the extent of its expression, relation with histological type, prognosis, and correlation with other steroid receptors (7, 8). Several studies have indicated a positive correlation among AR content, PR, and the absence of correlation with ERα (7). Interestingly, canine IMC has also a high content of PR (71.4 % of positive IMC cases) (3).

The IGF family can have a role in cancer by mitogenic, transforming, and anti-apoptotic actions (9). Also, some studies indicate a role of IGF-I in cell-cell adhesion and cell migration (10). Two previous studies suggested the possible participation of the IGF system on canine IMC: its typical occurrence in lutheal phase (2) and the specific loss of a low-affinity insulin-like growth factor-binding protein gene which has been described in human IBC (11). The IGF-1 expression in breast cancer by tumor and stromal cells supports the theory of a the local production of IGF-1. Our study is the first on IGF system in canine mammary tumors and the first on IMC. According to our results, the local content of IGF-1 is significantly increased in all malignant mammary tumors of the dog (non-inflammatory and inflammatory cases) respect to the normal levels. In contrast, serum contents of IGF-1 are significantly reduced in IMC respect to non-IMC malignant tumors, and are similar to those found in normal bitches. This fact could indicate a higher local consume of IGF-1 in IMC. This hypothesis is supported by the greater expression of IGF-1R found in IMC tissues. In the present study, there was an increase of IGF-1R expression with malignancy, according to previous studies on human breast cancer (15). A function of the IGF-1R in E-cadherin-dependent adhesion has been suggested, although the mechanism of IGF-1R regulation of cell-cell adhesion is not clear (10). On the other hand IGF-1R has been shown to mediate motility in breast cancer cells *in-vitro* (10). It is interesting to note that enhanced E-cadherin expression has been related to IBC, suggesting a different pathogenetic mechanism of human IBC compared with other non-invasive mammary carcinomas (16).

Our results indicate an inverse correlation between AR and IGF-1R immunostaining which could be explained by the antagonist action of androgens on estrogen metabolism. It has been reported a co-regulation of estrogens and IGFs in breast tumors where high expression of IGF-1R correlated to ER expression (10).

In conclusion, IGF-1 and androgens have a remarkable local action in inflammatory breast carcinoma. This fact can be useful in the development of new therapeutic approaches especially design for inflammatory carcinoma of the breast.

Acknowledgements

We thank Pedro Aranda and Mario Hernando for their technical assistance.

References

1. Tavassoli FA (1999) Inflammatory Carcinoma. Infiltrating carcinomas:special types. In: Pathology of the Breast, 2nd ed., New York: McGraw Hill, 538-541.

2. Perez-Alenza MD, Tabanera E, Peña L (2001) Inflammatory mammary carcinoma in dogs: 33 cases (1995-1999). J Am Vet Med Assoc 219:1110-4.
3. Peña L, Perez-Alenza MD, Rodriguez-Bertos A, et al (2003) Canine inflammatory mammary carcinoma: histopathology, immunohistochemistry and clinical implications of 21 cases. Breast Cancer Res Treat 78:141-148.
4. Alpaugh ML, Tomlinson JS, Shao ZM, et al (1999) A novel human xenograft model of inflammatory breast cancer. Cancer Res 59:5079-5084.
5. Liao DJ, Dickson RB (2002) Roles of androgens in the development, growth, and carcinogenesis of the mammary gland. J Steroid Biochem Mol Biol 80:175-189.
6. Birrell SN, Bentel JM, Hickey TE, et al (1995) Androgens induce divergent proliferative responses in human breast cancer cell lines. J Steroid Biochem Mol Biol 52:459-467.
7. Langer M, Kubista E, Schemper M, et al (1990) Androgen receptors, serum androgen levels and survival of breast cancer patients. Arch Gynecol Obstet 247:203-209.
8. Kuenen-Boumeester V, Van der Kwast TH, van Putten WL, et al (1992) Immunohistochemical determination of androgen receptors in relation to oestrogen and progesterone receptors in female breast cancer. Int J Cancer 52:581-584.
9. Rubin R, Baserga R (1995) Insulin-like growth factor-I receptor. Its role in cell proliferation, apoptosis, and tumorigenicity. Lab Invest 73:311-331.
10. Surmacz E, Guvakova MA, Nolan MK et al (1998) Type I insulin-like growth factor receptor function in breast cancer. Breast Cancer Res Treat 47:255-267.
11. Van Golen KL, Davies S, Whu ZF, et al (1999) A novel putative low-affinity insulin-like growth factor-binding protein, LIBC (lost in inflammatory breast cancer), and RhoC GTPase correlate with the inflammatory breast cancer phenotype. Clin Cancer Res 5:2511-2519.
12. Peña L, Silvan G, Perez-Alenza MD, et al (2003) Steroid hormone profile of canine inflammatory mammary carcinoma: a preliminary study J Steroid Biochem Mol Biol 84:211-216.
13. Killinger DV, Perel, Daniilescu D, et al (1987) Aromatase activity in the breast and other peripheral tissues and its therapeutic regulation. Steroids 50:523-536.
14. Labrie F, Luu-The V, Labrie C, et al (2001) DHEA and its transformation into androgens and estrogens in peripheral target tissues: intracrinology. Frontiers Neuroendocrinol 22:185-212.
15. Pezzino V, Papa V, Milazzo G et al (1996) Insulin-like growth factor-I (IGF-I) receptors in breast cancer. Ann N Y Acad Sci 784:189-201.
16. Kleer CG, van Golen KL, Braun T, et al (2001) Persistent E-cadherin expression in inflammatory breast cancer. Mod Pathol 14:458-464.

Hormonal Activation of the Gab-1 Docking Protein in Uterine Cells

Diane M. Klotz, Jon A. Proctor, David K. Walmer, R. Gregg Richards, and Richard P. DiAugustine

Summary

Docking or adapter proteins link signals emanating from receptors to other downstream intracellular molecules and amplify the initial signal. Therefore, the activation of a docking protein can indicate the prior activation of a receptor, such as a receptor tyrosine kinase (RTK). The Gab-1 (for Grb-2-associated binder-1) docking protein is a substrate for many different RTKs. As part of our effort to define the signaling pathways stimulated by 17β-estradiol (E_2) and progesterone (P) in the uterus, we investigated the activation of uterine Gab-1 in response to these hormones. In this study, Gab-1 was immunoprecipitated from CD-1 mouse uterine extracts, followed by immunoblotting with appropriate antibodies. In cycling mice, maximal uterine Gab-1 tyrosine phosphorylation (Gab-1-pY) was observed during late diestrus/early proestrus. To determine the role for specific ovarian hormones in uterine Gab-1 activation, ovariectomized mice were treated with vehicle, E_2, P, or E_2 + P, and 12 hours later uteri were analyzed. Consistent with the data from cycling mice, either E_2 or E_2 + P stimulated tyrosine phosphorylation of Gab-1. Uterine Gab-1 from E_2- and E_2 + P-treated mice coimmunoprecipitated with p85, the regulatory subunit of phosphatidylinositol 3-kinase; uterine Gab-1 from E_2 + P-treated mice also coimmunoprecipitated with SHP-2, a tyrosine phosphatase. These data suggest that E_2 and/or P activated an RTK signaling pathway(s) that utilized Gab-1. To identify this RTK, mice were administered IGF-1 or EGF. Uterine Gab-1 was tyrosine phosphorylated by 5 minutes post-treatment in EGF-, but not IGF-1-, treated mice, suggesting that EGFR may mediate steroid hormone-induced uterine Gab-1 activation.

Introduction

Docking, or adapter, proteins are intracellular proteins that often link signals emanating from receptor tyrosine kinases (RTKs) in the cell membrane to downstream cytoplasmic signaling molecules, such as Src homology 2 (SH2)

domain-containing proteins. Docking proteins can function to serve a number of different signaling purposes including the amplification of receptor signals and the dissociation of the intracellular signal from endocytosis of the activated RTK. Additionally, the presence of multiple docking proteins in a signaling complex provides an expanded choice of signaling pathways that can be regulated by a stimulated receptor, or a single docking protein can mediate the integration of multiple RTK signals into a common downstream pathway (reviewed in 1). Based on these functions, activation of a docking protein may serve as an indication of RTK activation.

Grb-2-associated binder-1 (Gab-1) is an epithelial docking protein that has been localized to the cell membrane in areas of cell-cell contact (2) and shares amino acid homology with the insulin receptor substrate (IRS) adapter protein family (reviewed in 3). In addition, Gab-1 contains multiple tyrosine-containing motifs that serve as binding sites for SH2 domain-containing proteins. Gab-1 is a substrate for many different RTKs, including epidermal growth factor receptor (EGFR), insulin-like growth factor-1 receptor (IGF-1R), insulin receptor, and the hepatocyte growth factor (HGF) receptor Met (4-6). Upon tyrosine phosphorylation by RTKs, Gab-1 can function as a docking protein for SH2 domain-containing proteins and other signaling molecules such as Grb-2, SHP-2, and the p85 subunit of phosphatidylinositol 3 (PI3)-kinase (reviewed in 3).

As part of our continuing effort to understand the growth factor signaling pathways regulated by E_2 and progesterone (P) in the female reproductive tract, we investigated the activation of uterine Gab-1 in response to these steroid hormones.

Materials and Methods

Animals. All animals were handled in accordance with an approved NIEHS/NIH animal study protocol. Cycling adult female CD-1 mice were used, or adult female CD-1 mice were ovariectomized, held for 10-14 days to clear endogenous ovarian hormones, and then used in studies. Uteri from cycling mice were removed, and prior to homogenization a section of one uterine horn and vaginal tissue were retained for histological evaluation of the stage of estrus. For hormone treatments ovariectomized animals were given an intraperitoneal (IP) injection of either sesame oil or the following hormones in sesame oil: 250 ng E_2, 1 mg P, or 250 ng E_2 + 1 mg P; uteri were removed 12 h post-injection. For growth factor treatments ovariectomized animals were given an IP injection of either 200 μg long R^3-IGF-1 or 100 μg recombinant human EGF in phosphate buffered saline containing 0.1% bovine serum albumin (BSA); uteri were removed 5 minutes post-injection and processed as described.

Gab-1 Immunoprecipitation. Uterine horns were homogenized on ice in homogenization buffer. Homogenates were pre-cleared by incubation with protein A-Sepharose (PAS) and then subjected to immunoprecipitation with 0.1 μg/ml anti-Gab-1 polyclonal antibody with an equal volume of immunoprecipitation buffer for 2 h at 4° C. Antigen-antibody complexes were captured with PAS for 3 h at 4° C and subjected to 3 washes in immunoprecipitation buffer. Precipitated antigen was eluted from the protein A-Sepharose by resuspending the pellets in Laemmli sample buffer and boiling for 5 min.

Immunoblots. Immunoprecipitated proteins were resolved by SDS-PAGE and transferred to PVDF membranes. Membranes were blocked in Tris-buffered saline with 0.1% Tween-20 (TBS-T) with 5% BSA and subsequently probed with anti-phosphotyrosine, anti-Gab-1, anti-EGFR, anti-p85, or anti-SHP-2 antibodies for 2 h at room temperature. After washing with TBS-T, blots were incubated with horseradish peroxidase-conjugated secondary antibodies for 1 h at room temperature, and Enhanced Chemiluminescence was used for detection of immunoreactive proteins.

Results

Uterine Gab-1 Activation During Various Stages of the Estrous Cycle. To examine if there may be a physiological function for Gab-1 in uterine signaling, uteri were collected from adult ovary-intact, cycling mice. As demonstrated in Figure 1, activation of uterine Gab-1 (~115 kD), as identified through Gab-1 tyrosine phosphorylation, varied throughout the estrous cycle. Uterine Gab-1 activation was greatest during the late diestrus/early proestrus stages of the cycle. During this time serum P levels are declining and E_2 levels are rising (7), implicating E_2, P, or an appropriate ratio of the two hormones in the activation of uterine Gab-1. Protein levels of Gab-1 remained constant across the different stages of estrus.

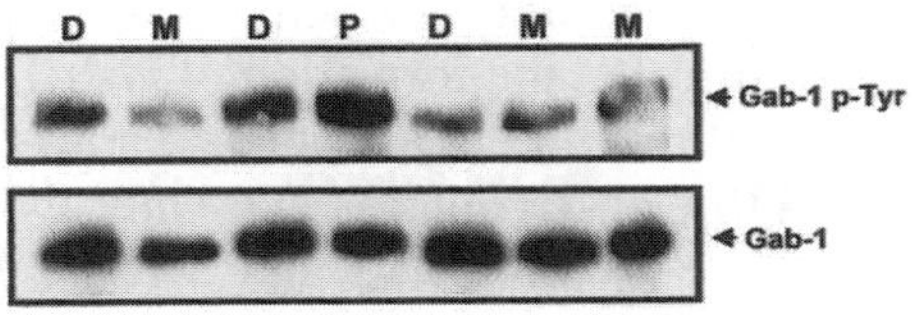

Figure 1. Tyrosine phosphorylation of uterine Gab-1 (Gab-1 p-Tyr) varies throughout the mouse estrous cycle indicating that optimal Gab-1 stimulation may have a specific hormone requirement. D, diestrus; M, metestrus; P, proestrus.

Ovarian Steroid Hormone Activation of Uterine Gab-1. To determine a role for the ovarian steroid hormones in uterine Gab-1 activation, ovariectomized mice were treated with vehicle, E_2, P, or E_2 + P. Figure 2 shows that there was minimal Gab-1-pY in samples from vehicle-treated animals. Gab-1 phosphorylation was increased by all steroid hormone treatments according to the following hierarchy: $E_2 + P > E_2 > P$; treatment with the combination of E_2 +

P resulted in the greatest level of uterine Gab-1-pY/activation. Hormone treatment did not alter Gab-1 protein levels.

The observation that each steroid hormone alone activated uterine Gab-1 to a differing degree may indicate that different cellular pools of Gab-1 are activated by the individual hormones. Alternatively, as Gab-1 has multiple tyrosine residues available for phosphorylation (reviewed in 3), the individual hormones may phosphorylate the same Gab-1 molecule but on tyrosine residues that are distinct for that hormone. For example, E and P may stimulate different RTKs that phosphorylate the same Gab-1 molecule on different tyrosine residues, and, therefore, the extent of phosphorylation of the Gab-1 molecule may vary with the nature of hormone treatment. Additionally, while E or P alone may be able to stimulate the production of a ligand necessary for RTK-stimulation of Gab-1, an appropriate ratio of the two hormones together may be necessary for optimal production of such a ligand.

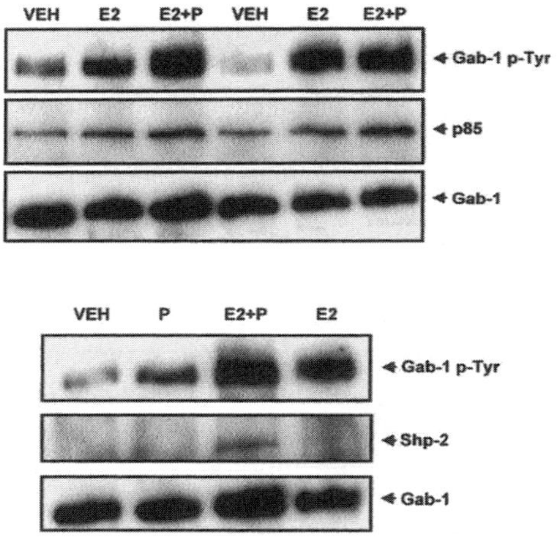

Figure 2. Stimulation of uterine Gab-1-pY and signaling complexes by ovarian steroid hormones. A. Stimulation of Gab-1-pY (Gab-1-pY, top panel) and increase in the amount of the p85 subunit of PI3-kinase bound to Gab-1 (middle panel); bottom panel demonstrates the levels of Gab-1 protein. B. Stimulation of Gab-1-pY (top panel) and increase in the amount of SHP2 bound to Gab-1 (middle panel); bottom panel demonstrates the levels of Gab-1 protein.

In addition to activation of uterine Gab-1, ovarian steroid hormones also stimulated the formation of Gab-1 signaling complexes. Both E_2 and E_2 + P treatment resulted in an increase in the amount of the p85 subunit of PI3-kinase bound to Gab-1 (Figure 2A). Only E_2 + P treatment stimulated formation of a complex between Gab-1 and SHP-2 (Figure 2B). The formation of Gab-1/p85 and Gab-1/SHP-2 signaling complexes in response to steroid hormones indicates that Gab-1 is actively participating in a signaling cascade in response to the ovarian steroids, and that the signaling cascades involve PI3-kinase and SHP2. This observation may be a further indication that multiple Gab-1-mediated signals are elicited in the uterus by different ovarian steroids or that a combination of the two steroids is obligatory for an optimal Gab-1-mediated signal. The interaction between Gab-1 and p85 is necessary for activation of the PI3-kinase/Akt signaling pathway in response to many different stimuli, and the production of PIP3 by activated PI3-kinase may function to create a positive feedback loop between Gab-1 and PI3-kinase activation, since PIP3 can bind to Gab-1 to promote further PI3-kinase activation (reviewed in 3). Studies have shown that the Gab-1/SHP2 interaction is important for activation of SHP2 phosphatase activity, which can lead to activation of MAP kinase and subsequent MAP kinase-regulated cellular responses (reviewed in 3).

Other studies suggest that SHP-2 regulates the strength and duration of Gab-1-associated PI3-kinase activity by dephosphorylating the p85 binding sites of Gab-1 after EGF stimulation, preventing p85 from further binding to Gab-1 (8). Thus, another hypothesis is that E and P regulate competing signals, one that stimulates a Gab-1/p85 interaction and another that serves to promote the SHP-2 interaction and inhibit signaling from the Gab-1/p85 complex.

Growth Factor Activation of Gab-1. Levels of IGF-1 and several EGFR ligands are increased in the uterus by E_2 and/or P (9, reviewed in 10, 11). To investigate whether EGF or IGF-1 stimulates uterine Gab-1-pY, ovariectomized adult mice were administered vehicle, EGF, or IGF-1. Figure 3 demonstrates that EGF, but not IGF-1, stimulated uterine Gab-1-pY. This suggests that in the uterus signaling through EGFR may contribute to the activation of Gab-1 and formation of Gab-1 signaling complexes in response to ovarian steroids. This particular experiment further indicates that IGF-1R signaling is not involved in uterine Gab-1 signaling after ovarian steroid hormone action. Both EGFR and IGF-1R were tyrosine phosphorylated after treatment with the respective ligands, demonstrating that both growth factors appropriately stimulated their cognate receptors (data not shown).

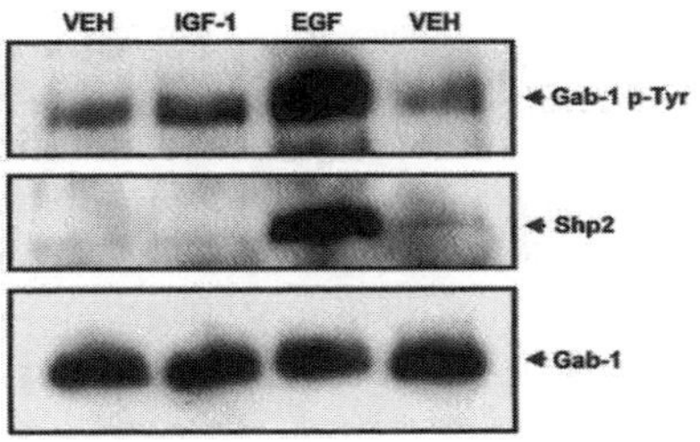

Figure 3. Growth factor-stimulated Gab-1-pY. EGF but not IGF-1 stimulated uterine Gab-1-pY(Gab-1 pY, top panel); Gab-1 protein levels were not altered by growth factor treatment (bottom panel).

Immunohistochemistry confirmed localization of Gab-1 in the uterine luminal and glandular epithelium (Figure 4); Gab-1 immunoreactivity, however, was not observed in the stroma or myometrium. When extracts of uterine luminal epithelial cells were analyzed, an increase in EGFR-pY and Gab-1-pY was observed in EGF-treated mice when compared to that of vehicle-treated controls (data not shown). This suggests that Gab-1 is a substrate for activated EGFR in the epithelium. In these and other experiments of the present study, Gab-1 was not detected in EGFR immunoprecipitates, nor was EGFR detected in Gab-1 immunoprecipitates.

The EGFR ligands EGF and TGF-α are increased in the uterus by E_2 (10-12), and amphiregulin mRNA is increased in the uterus by P (9). A role in

E_2-induced uterine cell proliferation has been suggested for EGF (13), while it has been hypothesized that TGF-α and amphiregulin function during implantation (9 and references therein). Furthermore, heparin binding-EGF (HB-EGF) is another EGFR ligand that is regulated in the uterus by the ovarian steroid hormones; treatment with E + P increases the level of HB-EGF in rodent uterine stromal cells (14). The number of different EGFR ligands that are stimulated in the uterus by ovarian steroids makes EGFR a good candidate for investigation of an RTK pathway that mediates hormone-stimulated Gab-1 signaling this organ. Indeed, in many other cell systems EGFR has been demonstrated to stimulate Gab-1 phosphorylation (3 and references therein). Thus, as a specific role for Gab-1 in the uterus remains undefined, EGFR continues to be a valid target for investigation in the activation of uterine Gab-1.

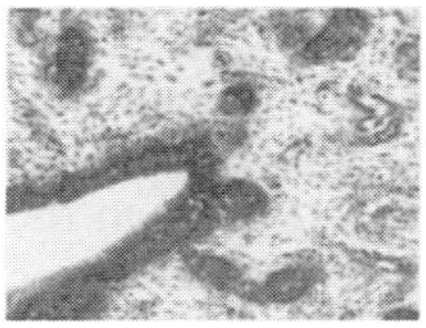

Figure 4. Immunolocalization of Gab-1 in the uterus. Gab-1 immunoreactivity is observed in the luminal and glandular epithelium.

Conclusions

The extent of uterine Gab-1-pY is dependent on the stage of the mouse estrous cycle, with the greatest level of tyrosine phosphorylation observed during late diestrus/early proestrus, suggesting that a kinase acting on the Gab-1 substrate is regulated by ovarian steroid hormones. Consistent with such a role for ovarian steroid hormones, E_2 alone or E_2 + P treatment of ovariectomized adult mice stimulated uterine Gab-1-pY. Additionally, ovarian steroid treatment resulted in the formation of Gab-1 signaling complexes containing p85 and/or SHP2, indicating that E_2 and P stimulate active signaling cascades that utilize Gab-1. Furthermore, EGF treatment resulted in Gab-1-pY, which may be indicative of a possible role for EGFR signaling in uterine Gab-1 activation. The phosphorylation of uterine Gab-1 in response to steroid hormones and EGF was recapitulated *in vitro* in uterine endometrial cancer cells by EGFR ligands, as well as by HGF (data not shown). HGF/Met signaling has been demonstrated to promote proliferation, migration, and lumen formation of human endometrial cells *in vitro* (15), providing another suitable candidate for a mediator of E_2- and/or P-stimulated Gab-1 activation. Elucidation of the different signaling pathways through which the ovarian steroid hormones exert their effects on the uterus will serve to provide investigators with new pathways to target in the development of endocrine and anti-hormone therapies.

References

1. White MF (1998) The IRS-signalling system: A network of docking proteins that mediate insulin action. Mol Cell Biochem 182:3-11.

2. Maroun CR, Holgado-Madruga M, Royal I, et al (1999) The Gab-1 PH domain is required for localization of Gab1 at sites of cell-cell contact and epithelial morphogenesis downstream from the Met receptor tyrosine kinase. Mol Cell Biol 19:1784-1799.
3. Liu Y and Rohrschneider LR (2002) The gift of Gab. FEBS Letters 515:1-7.
4. Holgado-Madruga M, Emlet DR, Moscatello DK, et al (1996) A Grb2-associated docking protein in EGF- and insulin-receptor signaling. Nature 379:560-564.
5. Weidner KM, DiCesare S, Sachs M, et al (1996) Interaction between Gab1 and the c-Met receptor tyrosine kinase is responsible for epithelial morphogenesis. Nature 384:173-176.
6. Winnay JN, Bruning JC, Burks DJ, et al (2000) Gab-1-mediated IGF-1 signaling in IRS-1-deficient 3T3 fibroblasts. J Biol Chem 275:10545-10550.
7. Walmer DK, Wrona MA, Hughes CL, et al (1992) Lactoferrin expression in the mouse reproductive tract during the natural estrous cycle: correlation with circulating estradiol and progesterone. Endocrinology 131:1458-1466.
8. Zhang SQ, Tsiaras WG, Araki T, et al (2002) Receptor-specific regulation of phosphatidylinositol 3'-kinase activation by the protein tyrosine phosphatase Shp2. Mol Cell Biol 22:4062-4072.
9. Das SK, Chakraborty I, Paria BC, et al (1995) Amphiregulin is an implantation-specific and progesterone-regulated gene in the mouse uterus. Mol Endocrinol 9:691-705.
10. Murphy LJ, Murphy LC (1994) Steroid hormone induction of growth factor and oncogene expression in the uterus. In: Khan SH, Stancel GM, editors. Protooncogenes and growth factors in steroid hormone induced growth and proliferation. Boca Raton: CRC Press, 31-45.
11. Murphy LJ, Gong Y, Murphy LC (1991) Growth factors in normal and malignant uterine tissue. Ann NY Acad Sci 622:383-391.
12. DiAugustine RP, Petrusz P, Bell GI, et al (1988) Influence of estrogens on mouse uterine epidermal growth factor precursor protein and messenger ribonucleic acid. Endocrinology 122:2355-2363.
13. Hom YK, Young P, Wiesen JF, et al (1998) Uterine and vaginal organ growth requires epidermal growth factor receptor signaling from stroma. Endocrinology 139:913-921.
14. Zhang Z, Funk C, Roy D, et al (1994) Heparin-binding epidermal growth factor-like growth factor is differentially regulated by progesterone and estradiol in rat uterine epithelial and stromal cells. Endocrinology 134:1089-1094.
15. Sugawara J, Fukaya T, Murakami T, et al (1997) Hepatocyte growth factor stimulated proliferation, migration, and lumen formation of human endometrial epithelial cells in vitro. Biol Reprod 57:936-942.

Endometrial Adenocarcinoma in Syrian Hamsters Treated with Diethylstilbestrol, Tamoxifen and N-Ethyl-Nitrosourea

Jaime Ferrer, Faustino Pérez-Mínguez, Antonio Leal, Amando Peydró, and Antonio Llombart-Bosch

Summary

The synthetic estrogen diethylstilbestrol (DES) causes marked abnormalities in the female hamster genital tract, after either prenatal or postnatal exposure, leading to endometrial hyperplasia and carcinoma. Acting as an initiating event, DES altering uterine development may facilitate the abnormal response of promoting agents. Tamoxifen (TAM) is an antiestrogen that competes for central and peripheral estrogen receptor (ERα). TAM exerts agonistic effects on E-dependent endometrial proliferation. N-ethyl-N-nitrosourea (ENU), a potent mutagenic agent, induces tumors in a variety of organs, predominantly in the peripheral nervous system. To test whether ENU and TAM treatment in a model of hyperestrogenism, the histopathologic endometrial alterations following exposure to DES, ENU and TAM (alone or in combination) were analyzed in ovariectomized female Syrian hamsters. Herein, the incidence and characteristics of the lesions found in the endometrium, and the progressive transformation from early alterations to malignant processes, are presented and discussed.

Introduction

The synthetic estrogen DES is a endocrine disruptor that causes marked abnormalities in the female genital tract of the Syrian golden hamster (SGH), after either prenatal or postnatal exposure (1, 2), leading to endometrial hyperplasia and carcinoma (3). Acting as an initiating event, DES alters uterine development and may facilitate its abnormal response to promoting agents (4). DES-induced epithelial changes have been linked to cell-specific imbalances in the expression of cell proliferation oncogenes (*c-jun*, *c-fos*, *c-myc*) and apoptosis (*bax*, *bcl-2*, *bcl-x*) (5).

TAM is an antiestrogen that competes for central and peripheral ERα. Used mainly as treatment of hormone-dependent human breast carcinomas, TAM has agonistic effect on E-dependent endometrial proliferation (6). ENU, a potent mutagenic agent, induces tumors in a variety of organs, predominantly in the peripheral nervous system (7, 8). The aim of the present study was to evaluate the effect of TAM and ENU in a DES-induced model of endometrial carcinoma.

Materials and Methods

Seven groups (n = 40, each) of 3.0-mo-old, ovariectomized female SGH, were treated with DES, TAM, and ENU, either alone or in combination (Table 1). DES pellets (25 mg) were implanted subcutaneously, every 3.0 mo. In a single dose, TAM was administered orally at 4.0 mo (0.1 mg/8 g food); and ENU ip (20 mg/kg) at 5.0 mo. Two groups, female intact and ovariectomized (ovx), of untreated, age-matched SGH served as controls.

Table 1. Groups (40 each) of 3.0-mo-old, ovariectomized female SGH, were treated as follows: DES, 25 mg pellet subcutaneously/every 3.0 mo; and single doses of TAM orally, 0.1 mg/ 8g food, at the 4.0 mo; ENU, ip, 20 mg/kg, at the 5.0 mo. Two groups of untreated, age-matched female SGH (intact and ovx) served as controls.

I	II	III	IV	V	VI	VII
DES	TAM	ENU	DES + TAM	DES + ENU	TAM + ENU	DES + TAM + ENU

SGHs were euthanized at 7.0, 8.0, and 9.0 mo after treatment, and subjected to macroscopic pathological examination. Tissues were harvested and processed as previously described (9). Dissected and sectioned organs were fixed in 10% buffered formalin, embedded in paraffin, and 3 to 5 μm sections prepared and stained with hematoxylin-eosin (HE). Uterine horns and cervix were also serially sectioned. Statistical analysis was performed by means of χ^2 test.

Results

Endometrial hyperplasia was observed in all SHGs treated with DES. It was more frequent in SHGs treated with DES alone (group I, 69.2%) than in those receiving DES + TAM (group IV, 55%), or DES + TAM + ENU (group VII, 23.8%). Macroscopically, it consisted of swelling of one or both uterine horns, together with an enlargement of the endometrium, that displayed a fleshy and cystic appearance. Microscopically, hyperplasia disclosed an increased glandular/stromal ratio, with medium- and large-sized glands with pseudostratified epithelium. Some apoptotic figures were present, but the epithelial cells did not have nuclear atypia. Hyperplasia can be subdivided into two types: simple, with a stromal enlargement that parallels that of glandular parenchyma, and cystic, in which large cysts and papillary foldings are formed in endometrial glands, causing a marked thinning of the intervening stroma (Figure 1). In the latter, microgland formation was visible within the epithelium.

Simple hyperplasia appeared as an earlier lesion, since it was more frequently detected in 7.0-mo treated SGHs. After later treatments, there were many instances of cystic hyperplasia. Both hyperplasias were observed in pre and and malignant lesions (see below).

Complex (atypical) hyperplasia was more frequent after TAM + ENU (group VI) treatment than after DES alone (group 1), 25% vs. 17.8%, respectively, although this difference was not statistically significant. The lowest frequency of complex hyperplasia was found in after DES + TAM + ENU treatment (group VII). Most of these lesions were diagnosed as truly malignant.

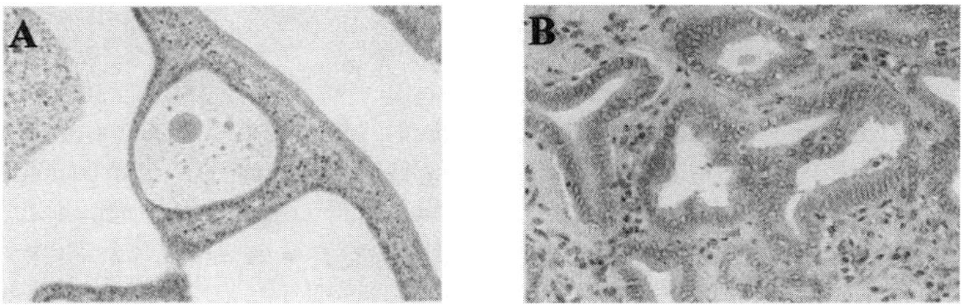

Figure 1. **A.** Endometrial cystic hyperplasia with dilated glands and thinned stroma. Nuclear atypia was absent (HE, 100 x). **B.** Complex hyperplasia with architectural disorganization and nuclear atypia. Proliferating glands did not invade the surrounding stroma (HE, 400 x).

Complex hyperplasia was defined by the presence of architectural and cytological atypia (Figure 1). It appeared in the context of cystic hyperplasia, where some foci of increased glandular population were seen. These glands revealed complex architectural features, with branching units, crowding and a relative loss of their spatial orientation. The surrounding stroma was markedly diminished. The hyperplastic glands were formed by a single layer of cells with enlarged nuclei, with frequent loss of polarization and prominent nucleoli. A low number of mitotic figures, some of them atypical, were observed. No stromal invasion was associated with this kind of lesion. The stroma showed the usual cellular appearance of the hyperplastic endometrium.

The highest frequency, 55%, of endometrial adenocarcinoma was detected after DES + TAM + ENU treatment (group VII). This difference was statistically significant compared to DES alone (group I, 12.5%), and DES + TAM (group IV, 18%) treatment groups. Endometrial adenocarcinoma appeared as unilateral or bilateral enlargement of the uterus, which was more evident than that of the hyperplasia. Carcinomas were fleshy, soft, reddish masses, often with hemorrhage

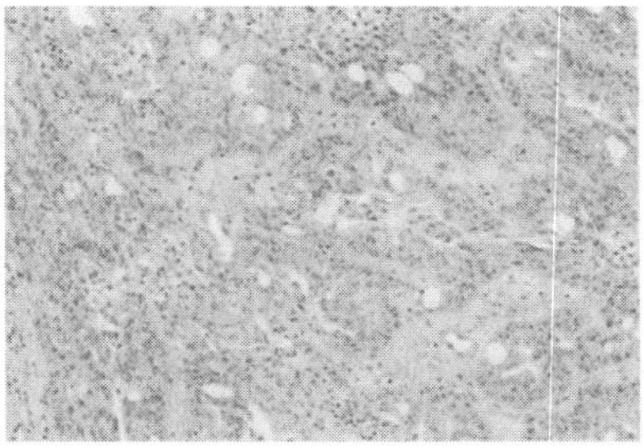

Figure 2. Irregular masses of atypical, infiltrating glands characteristic of endometrial adenocarcinoma (HE, 200 x).

and necrosis. Histologically, most of these tumors were of endometrioid type, *i.e.,* they aberrantly replicate the endometrial architecture, showing irregular glands, cribiform areas, solid nests and chords that infiltrated the surrounding stroma in a diffuse manner (Figure 2). A minor proportion of the tumors were predominantly of serous (papillary) type, a pattern that in a small number of tumors merged focally with that of the endometrioid type. Squamous cell areas were also seen in some tumors. Neoplastic cells disclosed a variable degree of nuclear atypia, ranging from round, non-polarized, relatively uniform nuclei with small nucleoli, to pleomorphic, irregular nuclei with clumped chromatin and prominent nucleoli. A high number of atypical mitoses were noticed. Focally, some of the less well-differentiated tumors lost their glandular configuration, and were arranged in a fascicular pattern with spindle cells closely resembling those of a sarcoma. Glandular hyperplasia, whether cystic or complex (atypical), was seen in the vicinity of some carcinomas. Occasionally, atypical hyperplasia showed a direct transition to a more invasive, malignant lesion.

Conclusions

Sustained exposure to DES induces endometrial malignant transformation in ovx female SGHs. Sequential histopathological events, from glandular hyperplasia and cystic changes leading to atypical hyperplasia and adenocarcinoma, were observed in this study. Combined treatment TAM + DES caused an increased incidence of atypical hyperplasia and adenocarcinoma, although not statistically significant. Simultaneous exposure to DES, TAM, and ENU resulted in the highest number of malignancies, may be due to the carcinogenic potency of ENU.

Acknowledgment

Supported by grant FIS 01/702, Ministry of Health, Madrid, Spain.

References

1. Leavitt WW, Evans RW, Hendry WJ 3rd (1981) Etiology of DES-induced uterine tumors in the Syrian hamster. Adv Exp Med Biol 138:63-86.

2. Hendry WJ 3rd, DeBrot BL, Zheng X, et al (1999) Differential activity of diethylstilbestrol versus estradiol as neonatal endocrine disruptors in the female hamster (*Mesocricetus auratus*) reproductive tract. Biol Reprod 61:91-100.
3. Gilloteaux J, Paul RJ, Steggles AW (1982) Upper genital tract abnormalities in the Syrian hamster as a result of in utero exposure to diethylstilbestrol. I. Uterine cystadenomatous papilloma and hypoplasia. Virchows Arch A Pathol Anat Histopathol 398:163-183.
4. Hendry WJ 3rd, Zheng X, Leavitt WW, et al (1997) Endometrial hyperplasia and apoptosis following neonatal diethylstilbestrol exposure and subsequent estrogen stimulation in both host and transplanted hamster uteri. Cancer Res 57:1903-1908.
5. Zheng X, Hendry WJ 3rd (1997) Neonatal diethylstilbestrol treatment alters the estrogen-regulated expression of both cell proliferation and apoptosis-related proto-oncogenes (c-jun, c-fos, c-myc, bax, bcl-2, and bcl-x) in the hamster uterus. Cell Growth Differ 8:425-434.
6. Wade GN, Powers JB (1993) Tamoxifen antagonizes the effects of estradiol on energy balance and estrous behavior in Syrian hamsters. Am J Physiol 265:R559-R562.
7. Diwan BA, Rehm S, Rice JM (1996) Age- and dose-dependent transplacental carcinogenesis by N-nitrosoethylurea in Syrian golden hamster. J Cancer Res Clin Oncol 122:643-652.
8. Shibuya T, Morimoto K (1993) A review of the genotoxicity of 1-ethyl-1-nitrosourea. Mutat Res 297:3-38.
9. Llombart-Bosch A, Peydro-Olaya A, Cremades A, et al (1996) Interstitial cell origin of estrogen-induced kidney tumors in the Syrian golden hamster. In: Li JJ, Li SA, Gustafsson JA, Nandi S, Sekely LJ, editors. Hormonal Carcinogenesis II. New York: Springer Verlag, 185-192.

Hormonal Regulation of ZEB-1 and Implication for Progression of Human Reproductive Cancers

Bynthia M. Anose, Michael P. Linnes, and Michel M. Sanders

Summary

The human Zinc finger E-box Binding (ZEB)-1 protein belongs to a family of transcription factors involved in critical developmental processes. Yet little is known of the mechanisms by which ZEB-1 is regulated. Our lab has recently demonstrated that the expression of ZEB-1 is induced by estrogen (E) in the ovarian cancer cell line Ov266 and that it is regulated by dihydrotestosterone (DHT) in the human PC-3/AR prostate carcinoma cell line. Interestingly, a dose-response assay indicates the expression of ZEB-1 is induced by 5 nM DHT and repressed at higher dosages. Cloning and analysis of approximately 1000 bp upstream of the translation start site of hZEB-1 revealed a number of putative E and androgen (A) response elements. Transient transfection assays indicate that this region is sufficient to confer responsiveness to both steroids. To determine whether expression of ZEB-1 could serve as a marker of tumor progression, real-time PCR (rtPCR) assays were performed on staged human reproductive carcinomas. Preliminary results indicate that expression of ZEB-1 increases as the normal ovary transforms to a primary carcinoma and continues to increase as the cancer progresses to an invasive and finally a metastatic state. There is an approximate 12-fold elevation in the expression of ZEB-1 in metastatic ovarian carcinoma relative to its expression in *in situ* cancers. rtPCR is currently being utilized to investigate the potential changes in ZEB-1 expression in breast and prostate cancer during the progression of these diseases. These data raise the possibility that overexpression of ZEB-1 contributes to the progression of reproductive carcinomas.

Introduction

Nuclear receptors (NR) comprise a large family of ligand-activated transcription factors known as the steroid hormone receptor superfamily; this is the largest eukaryotic family of transcription factors (1). The ability of these factors to influence gene transcription and other downstream events relies on the binding of their cognate ligands, the steroid hormones (2). Steroids facilitate the changes in conformation and dissociation of chaperone proteins from the NR that allow it to bind its respective hormone response element (HRE). The resultant changes in gene expression lead to varied and important physiological consequences.

Although considerable information is known about how NR regulate transcription, relatively little is known about the transcriptional cascades that they activate in vertebrates. Interestingly, we have found that E (3) and A (B.M. Anose and M.M. Sanders, unpublished) induce the ZEB-1/δEF1 (ZEB-1) transcription factor. This establishes a novel signaling cascade for these steroids. ZEB-1 is ubiquitously expressed and is required for mesodermal development (4, 5). Homologues of this important transcription factor have been cloned in many species: mouse, *Drosophila*, rat, hamster (6-10). Of particular interest is the observation that it can both repress (11-16) and activate transcription (3, 17). Understanding the actions of ZEB-1 has been complicated by the presence of a second family member, SIP1 or ZEB-2 (18 and references therein).

ZEB-1 is 1114 amino acids long, and it contains numerous functional domains that are reminiscent of classical transcriptional regulators (Fig. 1). The 7 zinc fingers are of particular note because of their potential roles in determining whether ZEB-1 activates or represses transcription on a specific target gene.

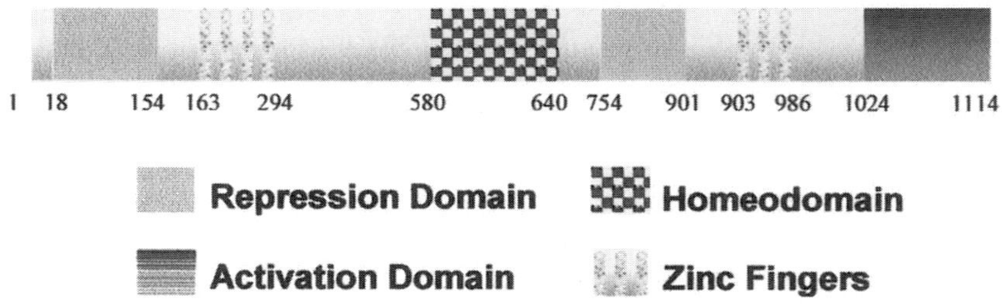

Figure 1. Human Zinc finger E-box Binding (ZEB)-1 protein contains 1114 amino acids. It is a typical zinc finger homeodomain protein containing two sets of Krüppel-type zinc fingers and a central homeodomain.

Materials and Methods

Transient Transfections. The PC-3/AR prostatic carcinoma cell line (kindly provided by K. Burnstein) was cultured as described (19). Transfections were carried out with Lipofectamine 2000 reagent (Invitrogen 2003) per the manufacturer's protocol. 6 hours post-transfection, steroid treatments were administered. The ovarian carcinoma cell line Ov266 (kindly provided by K. Kalli) was maintained as described (20). Transfections were carried out by electroporation in phenol red-free media containing the appropriate steroids. After 24 h, both cell lines were harvested. β-galactosidase assays were performed on total cell lysates with detection by luminometer, measuring 1sec/tube.

Real-Time RT PCR. Snap-frozen biopsy tissue samples were obtained through the University of Minnesota Cancer Center Tissue Procurement Facility (TPF) and the Cooperative Human Tissue Network (CHTN). Total RNA was harvested from each sample using the RNeAsy Extraction kit (Qiagen 2003) per the manufacturer's protocol. 2 μg RNA aliquots were reverse transcribed to create template cDNAs. 5 μL of each reaction was transferred to a SYBR Green rtPCR reaction and run on the *iCycler* machine (BioRad) using primers specific to human ZEB-1.

Results

Regulation of the hZEB-1 Promoter by Sex Steroids. Approximately 1 kb of the human ZEB-1 5'-flanking region was cloned by us from human genomic DNA by PCR and subcloned into the pBlueTOPO vector (Invitrogen). This construct was named pBlueZEB974. To test the responsiveness of this DNA region to A, transient transfections were carried out in the PC-3/AR prostate carcinoma cell line, and the cells were treated with 100 nM DHT for 24 h prior to harvesting for analysis of β-galactosidase expression. The high, constitutive expression of β-galactosidase from the control pCMV*SPORTβGal plasmid confirms that the transfections were performed successfully. The highly inducible expression of the pMMTV-LacZ plasmid confirms the cells are A-responsive. Interestingly, 100 nM DHT represses expression of the ZEB-1 construct by 10-fold.

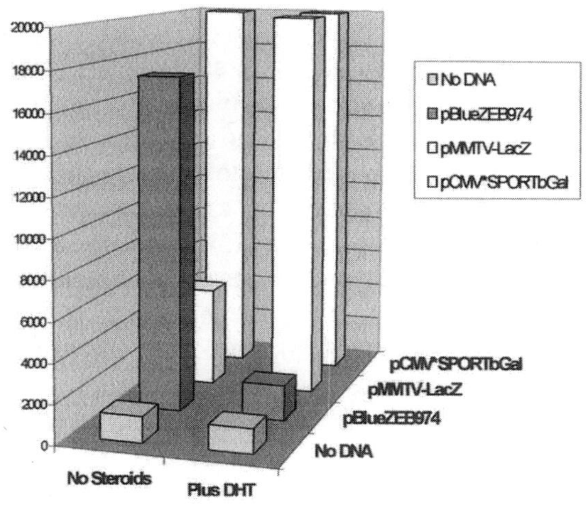

Figure 2. The PC-3/AR A-responsive prostate cell line was transiently transfected with 3 *LacZ* reporter plasmids: pBlueZEB974, which contains 974 bp of the hZEB-1 5'-flanking region, pMMTV-LacZ which is an A-responsive control, or with pCMV*SPORTβGal, which is a transfection control. The cells were then treated with 100 nM DHT for 24 h.

Similarly, the E-responsive Ov266 ovarian carcinoma cell line was transfected with pBlueZEB974, pCMV*SPORTβGal and an estrogen-inducible reporter construct (Fig. 3). The cells were treated with 100 nM 17β-estradiol (E_2) for 24 h and processed as the PC-3/AR cells. In contrast to the effects of A in the PC-3/AR cells, pBlueZEB974 was induced about 5-fold by E in the Ov266 cells.

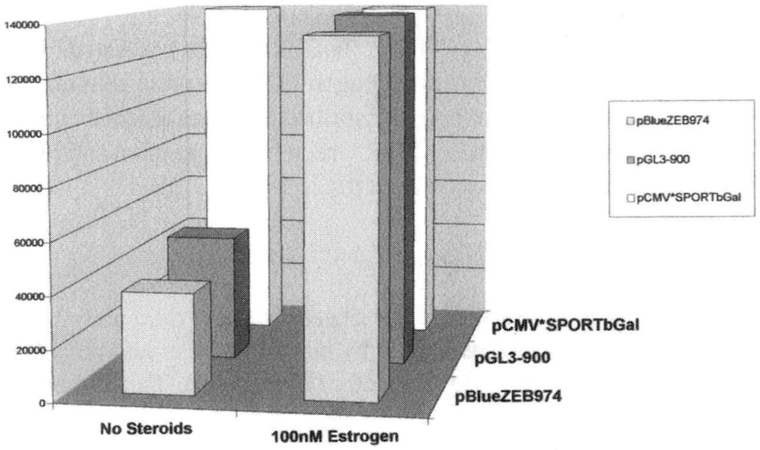

Figure 3. Ov266 cells were transiently transfected with pBlueZEB974, pGL3-900 (an E-responsive positive control), or pCMV*SPORTβGal. The cells were treated with 100 nM E_2 for 24 h.

Expression of hZEB-1 in Human Reproductive Carcinomas. To investigate the possibility that hZEB-1 may be involved in the etiology of cancers in sex steroid-responsive organs, the expression patterns of ZEB-1 in staged human ovarian and prostate carcinomas was studied. RNA was harvested from human biopsy samples and subjected to rtPCR. Based on the threshold crossing values (C_T), it appears that hZEB-1 expression does indeed increase during ovarian cancer progression (Fig. 4). Specifically, its expression in invasive carcinoma is 2-fold higher than its expression in *in situ* carcinoma, and its expression in metastatic carcinoma is 6-fold higher than in invasive carcinoma. Therefore, ZEB-1 expression overall increases approximately 12-fold as a primary ovarian cancer progresses to a metastatic state.

ZEB-1 resides at chromosomal location 10p11.2-p12, which is associated with some prostate cancers. Additionally, human RNA master blots reveal high expression of ZEB-1 in the prostate gland. This suggests, in conjunction with our observation that ZEB-1 is regulated by A, a potential role for ZEB-1 in prostate cancer. To study this possibility, human prostate samples were obtained from the CHTN. Thus, far only 2 samples have been obtained and neither contains carcinoma. Analysis of these samples indicates that the expression of hZEB-1 is very similar in the normal prostate and in benign prostatic hyperplasia.

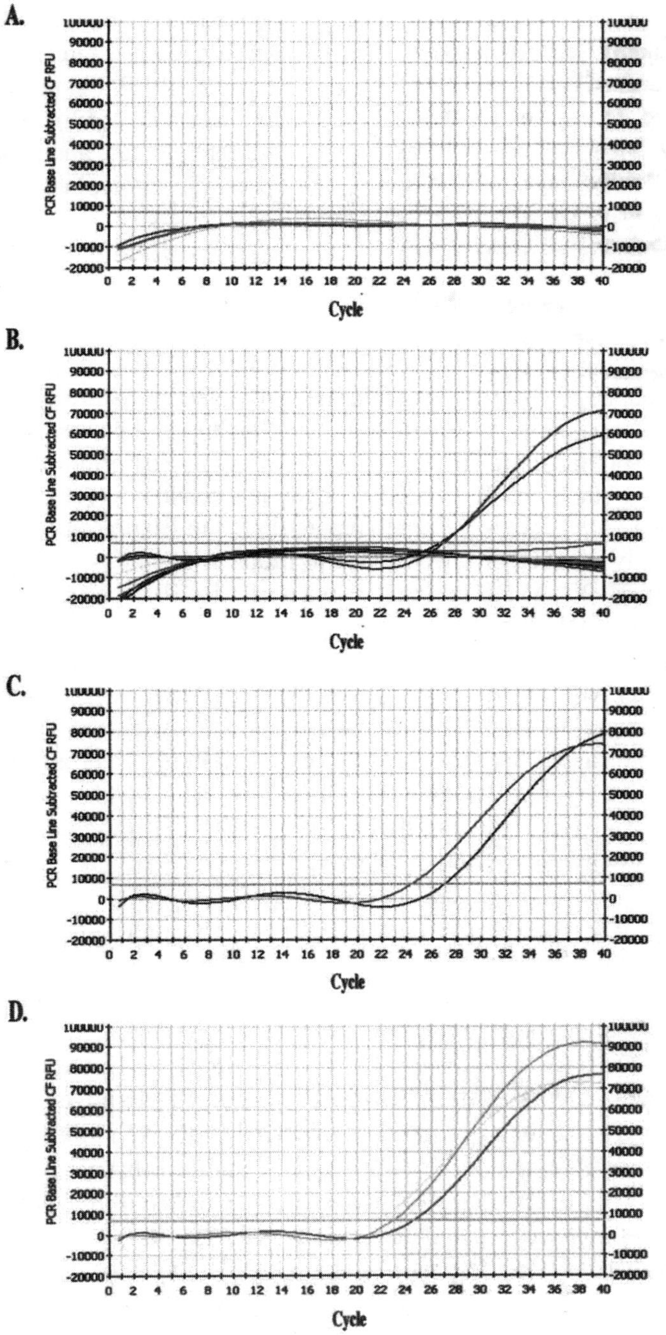

Figure 4. rtPCR was used to sensitively measure expression levels of hZEB-1 mRNA in ovarian biopsy tissues. **A.** In normal ovary, hZEB-1 expression levels are low, nearly undetectable N = 4 **B.** In primary *in situ* ovarian carcinoma, hZEB-1 expression is heterogeneous, with some samples having high and some very low levels. N = 12 **C.** In invasive ovarian carcinoma, hZEB-1 expression levels are approximately 2-fold higher than those in the *in situ* carcinomas. N = 2 **D.** In metastatic ovarian carcinoma, hZEB-1 expression levels are approximately 6-fold higher than those in invasive carcinomas, and 12-fold higher than those in the primary *in situ* carcinomas. N=3

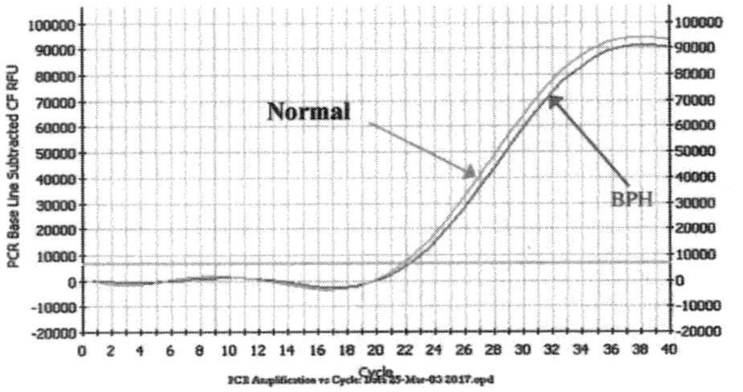

Figure 5. rtPCR was used to sensitively measure expression of hZEB-1 mRNA in prostate biopsy tissues. hZEB-1 expression is nearly identical in normal prostate and benign prostatic hyperplasia. N = 2

Discussion

Steroid hormones influence or control a large number of molecular pathways, and they induce an array of physiological responses in both healthy and disease states. The data presented here demonstrate that human ZEB-1 is a steroid hormone target gene. It is regulated by both A and E, in a dose-dependent manner. Recent dose-response assays (data not shown) demonstrate that the expression from pBlueZEB974 is actually induced at lower concentrations of DHT (1-5 nM) and repressed at doses from 7-100 nM DHT. Thus, it appears the hZEB-1 promoter may be exquisitely regulated by DHT in a highly sensitive manner.

Interestingly, expression of ZEB-1 appears to increase as the normal ovary transforms to a primary carcinoma, and continues to increase as the cancer progresses to an invasive and finally a metastatic state. Its expression is unchanged in early stages of prostate cancer but has recently been correlated with progression of these cancers to the metastatic state (B.M. Anose and M.M. Sanders, unpublished). Thus, ZEB-1 is a molecular link between steroid hormones and the progression of various human reproductive cancers.

References

1. Evans R (1988) The steroid and thyroid hormone receptor superfamily. Science 240:889-895.
2. Tsai MJ, O'Malley B (1994) Molecular mechanisms of action of steroid/thyroid receptor superfamily members. Ann Rev Biochem 63:451-486
3. Chamberlain EM, Sanders MM (1999) Identification of the novel player δEF1 in estrogen transcriptional cascades. Mol Cell Biol 19(5):3600-3606.
4. Funahashi J, Sekido R, Murai K, et al (1993) δ-crystallin enhancer binding protein δEF1 is a zinc finger-homeodomain protein implicated in postgastrulation embryogenesis. Development 119:433-446.
5. Watanabe Y, Kawakami K, Hirayama Y, et al (1993) Transcription factors positively and negatively regulating the Na,K-ATPase α1 subunit gene. J Biochem 114:849-855.

6. Genetta T, Kadesch T (1996) Cloning of a cDNA encoding a mouse transcriptional repressor displaying striking sequence conservation across vertebrates. Gene 169:289-290.
7. Franklin A, Jetton T, Shelton K, et al (1994) BZP, a novel serum-responsive zinc finger protein that inhibits gene transcription. Mol Cell Biol 14:6773-6788.
8. Cababillas A, Darling D (1996) Alternative splicing gives rise to two isoforms of Zfhep, a zinc finger/homeodomain protein that binds T3-response elements. DNA Cell Biol 15:643-651.
9. Fortini M, Lai Z, Rubin G (1991) The *Drosophila zfh-1* and *zfh-2* genes encode novel proteins containing both zinc-finger and homeodomain motifs. Mech Development 34:113-122.
10. Lai Z, Fortini M, Rubin G (1991) The embryonic expression patterns of *zfh-1* and *zfh-2*, two *Drosophila* genes encoding novel zinc-finger homeodomain proteins. Mech Development 34:123-134.
11. Sekido R, Kasumi M, Yusuke K, et al (1997) Two mechanisms in the action of repressor δEF1: binding site competition with an activator and active repression. Genes Cells 2:771-783.
12. Postigo A, Dean DC (1999) ZEB represses transcription through interaction with the corepressor CtBP. Proc Natl Acad Sci 96:6683-6688.
13. Sekido R, Murai K, Funahashi J-I, et al (1994) The δ-crystallin enhancer-binding protein δEF1 is a repressor of E2-box-mediated gene activation. Mol Cell Biol 14:5692-5700.
14. Ikeda K, Halle J-P, G. Stelzer, et al (1998) Involvement of negative cofactor NC2 in active repression by zinc finger-homeodomain transcription factor AREB6. Mol Cell Biol 18:10-18.
15. Yasui DH, Genetta T, Kadesch T, et al (1998) Transcriptional repression of the IL-2 gene in Th cells by ZEB. J Immunol 160:4433-4440.
16. Postigo A, Dean DC (1997) ZEB, a vertebrate homolog of *Drosophila* Zfh-1, is a negative regulator of muscle differentiation. EMBO J 16:3935-3943.
17. Lazarova DL, Bordonaro M, Sartorelli AC (2001) Transcriptional regulation of the vitamin D3 receptor gene by ZEB. Cell Growth Diff 12:319-326.
18. Van de Putte T, Maruhashi M, Francis A, et al (2003) Mice lacking *Zfhx1b*, the gene that codes for Smad-Interacting Protein-1, reveal a role for multiple neural crest cell defects in the etiology of Hirschsprung Disease-Mental Retardation Syndrome. Am J Hum Genet 72:465-470.
19. Dai JL, Maiorino CA, Gkonos PJ, et al (1996) Androgenic up-regulation of androgen receptor cDNA expression in androgen-independent prostate cancer cells. Steroids 61:531-539.
20. Conover CA, Hartmann LC, Bradley S, et al (1998) Biological characterization of human epithelial ovarian carcinoma cells in primary culture: the insulin-like growth factor system. Exp Cell Res 238:439–449.

Evaluation of Messenger RNA of Pituitary Tumour-transforming Gene-1 (PTTG1) as a Molecular Marker for Micrometastasis

Manuel Valladares Ayerbes, Lourdes Calvo, Guillermo Alonso, Pilar Iglesias, Maria J Lorenzo, Inmaculada Brandón, Mar Haz, Marga Reboredo, Silvia Antolín, and Luis Antón Aparicio

Summary

Pituitary tumor-transforming gene-1 (PTTG1) has been implicated as an inhibitor of chromatid separation; its overexpression may lead to chromosomal instability. PTTG1 functions through SH3-mediated signal transduction pathways and activation of growth factors, including basic fibroblast growth factor (bFGF) mediated-angiogenesis. In order to detect micrometastasis in solid tumor patients, we developed a model system based on reverse transcriptase (RT)-PCR amplification of PTTG1 mRNA. We analyzed PTTG1 expression, using RT-PCR, in a panel of human tumors cell-lines (hTCL), including breast, gastrointestinal, small-cell lung cancer, and haematopoietic neoplasms. In addition, to newly hTCL established in our laboratory were investigated: Pancreatic, MBQ-OJC1. Small cell lung carcinoma, JCA-OJC3. Colon, JJPF-OJC4. Specific amplicon for PTTG1 was confirmed by RT-PCR in all the hTCL tested. In breast cell lines expression for PTTG1-mRNA was not related to invasiveness, tumorogenicity, or estrogen receptor (ERα) status. PTTG1-mRNA was examined by PCR amplification of cDNA obtained from normal lymph nodes, bone marrow, and peripheral blood (cellular and plasmatic mRNA). Using a hot-start PCR and different amounts of input RNA, PTTG1 transcript was detected in control bone marrow but not on lymph nodes. PTTG1-mRNA was detected in peripheral blood from 4/14 healthy donors analyzed. Our data confirms that PTTG1 is abundantly expressed in human tumors of different histologic types. PTTG1 qualitative mRNA detection may be useful as a surrogate molecular marker for angiogenic phenotype and invasive tumor, however, it is not useful as a micrometastasis marker due to its transcription in normal bone marrow and peripheral blood.

Introduction

Tumor-cell dissemination and metastasis are complex processes whose outcome depends upon cancer cells properties and host-tumor interactions. Metastatic phenotype included activation of growth factors signaling pathways, differentiation, de-regulated adhesion, epithelial-mesenchymal transition, invasiveness, and angiogenesis. Experimental studies suggest that few isolated cancer cells detected in blood and distant sites may begin to grow, and only scarce micrometastasis may persist to form clinical significant metastases (1). Markers that are associated with metastatic phenotype would be useful for detection and characterization of isolated micrometastatic carcinoma cells. These biological markers may turn to be reliable predictors of subsequent metastasis development.

PTTG1 (also known as securin and tumor transforming protein 1 gene) has been implicated as an inhibitor of chromatid separation; its overexpression may lead to chromosomal instability. PTTG1 functions through SH3-mediated signal transduction pathway. Northern blot analysis from various normal human tissues revealed no mRNA PTTG1 expression in most of them, including peripheral blood leukocytes (2). PTTG induces an angiogenic phenotype in both *in-vitro* and *in-vivo* models. Also, a high PTTG mRNA expression is associated with an angiogenic phenotype in human tumors (3). These actions may be mediated through bFGF.

In order to detect micrometastasis in solid tumor patients, we have developed a model system based on RT-PCR amplification of PTTG1 mRNA. A panel of hTCL, as a surrogate model, were used, including three new hTCL developed in our laboratory. Specificity of the PTTG1-mRNA assay was also examined by PCR amplification of cDNA obtained from normal lymph nodes, bone marrow (BM), and peripheral blood (PB).

Materials and Methods

Cell Lines. A total of 20 cancer cell lines were used in this study, including: Breast cancer, MCF-7, MDA-MB468, T47D, BT-549, PM1. Colorectal cancer, Gp5d, LoVo, DLD1, LS513, HT29, JJPF-OJC4. Gastroesophageal adenocarcinoma, OE19. Pancreatic carcinoma, MBQ-OJC1. Small cell lung cancer, H146, DMS92, JCA-OJC3. Hepatocellular cancer, HepG2. Haematopoietic neoplasm, KG-1, K562, Jurkat. All cell lines were maintained in RPMI-1640 medium supplemented with fetal calf serum, L-glutamine, penicillin, streptomycin and amphotericin at 37°C in 5% CO_2. Cells from adherent cultures were recovered with trypsin-EDTA or non-enzymatic cell dissociating reagent. Cell pellets from suspension cultures were obtained. Characteristics of the hTCL MBQ-OJC1, JCA-OJC3, and JJPF-OJC4 developed in our laboratory were shown (Table 1).

RNA Extraction and RT-PCR. Purification of RNA from cell cultures was performed with High Pure RNA Isolation Kit (Roche) as suggested by the manufacturer. From each cell line, at 50-70 % confluence, 10^6cells were obtained for RNA isolation. Total RNA was treated with DNase I. The RT was performed from 1-μl aliquots of total cellular RNA using 1st strand cDNA synthesis kit (Roche). cDNA synthesis was carried out in a final volume of 20μl with random hexamer, dNTPs (20nM each), buffer, 25 units RNAse inhibitor, and 40 units Avian myeloblastosis virus reverse transcriptase (AMV), at 42°C for 60 min. Aliquots (5 μl) of the 1st strand cDNA were amplified using PCR master (Roche) in a total volume of 50 μl. Termocycling profile: Denaturation 94°C, 2 min, and 35 cycles of 30 s at 94°C, 60 s at 60°C, 45 s at 72°C, and 7 min at 72°C as final elongation. PCR products were electrophoresed through 2% agarose gel and stained with ethidium bromide. β-2 microglobulin served as positive control target. Negative controls were included in each experiment. Specific primers are also shown (Table 2).

Table 1. Characteristics of the Cell Lines Developed in our Laboratory Used to PTTG1 mRNA Assay.

Cell Line	Pathology	Specimen Origin	Primary Site	*In Vitro* Growth	Marker Gene Expression	Other Features
MBQ-OJC1	Adeno-carcinoma	Ascites	Pancreas	Epithelial, Monolayer	Cytokeratins Ep-CAM, EGFR	Mutation K-ras and p53
JCA-OJC3	Small-cell carcinoma	Pleural effusion	Lung	Aggregates, suspension	Ep-CAM, NCAM	Syntaxina positive
JJPF-OJC4	Adeno-carcinoma	Ascites	Colon	Epithelial-like monolayer	Cytokeratins, CK20, Ep-CAM, EGFR	3D growth

Analysis of PTTG1 Expression in Normal Human Tissues. Nucleic acids isolated from human normal BM and lymph nodes were purchased from BD Biosciences–Clontech and examined by RT-PCR for PTTG1 mRNA expression. Total RNA was derived from pooled Normal Human BM. cDNA synthesis was carried out as described using different amounts of RNA (up to 2 μg). cDNA were synthesized from Poly-A^+ RNA from human normal lump nodes pooled from 34 male/female Caucasians. PCR amplification of cDNA (up to 0.8 μg for lymp nodes) was analyzed as described. Blood samples (3-5 ml) were collected from healthy donor volunteers (n = 14) in EDTA tubes. Samples were lized in less than 1 h in stabilization reagent (Roche) without cell and plasma separation. The mixtures were stored at –20°C until mRNA extraction. Isolation reagent for blood and BM (Roche) was used for cellular and plasmatic mRNA extraction. Complementary DNA was generated with AMV according 1st strand cDNA synthesis kit. PCR amplification was performed using "hot start" PCR (5) with AmpliWax (Perkin-Elmer) and Taq DNA Polymerase (Roche).

Table 2. Primers Characteristics (References 2, 4, respectively).

Marker	Gene Map Locus	Primers	Length (bp)	Temp Annealing
PTTG 1	5q33	5'-CGA TGC CCC AC CAG CCT TAC C-3' 5'-CAAGCTCTC TCT CCT CGT CAA GG-3'	316	60
β-2 Micro-globulin	15q21-q22	5'-CGA GCA GAG AAT GGA AAGTC-3' 5'-GAT GCT GCT TAC ATG TCTCG-3'	268	60

DNA Sequencing. PCR products were used as template DNA. Products were purified by enzymatic method (ExoSAP-It, Amersham USB). DNA sequencing was performed in a reference facility on ABI 3700 (Applied Biosystems) using Big Dye Terminators. Sense human PTTG1-specific primer (5'-CGATGCCCCACCAGCCTTACC-3') was used.

Results

RT-PCR revealed PTTG1 expression in both hormone-related and unrelated cancer models. A specific 316-bp product for PTTG1 was demonstrated in all the human tumor cell lines tested (Figure 1). In breast cancer cell lines (6) expression for PTTG1-mRNA was not related to highly invasive ability (BT-549), tumorogenicity (T47D, MDA-468), or positive ERα status (MCF-7, T47D, PM1).

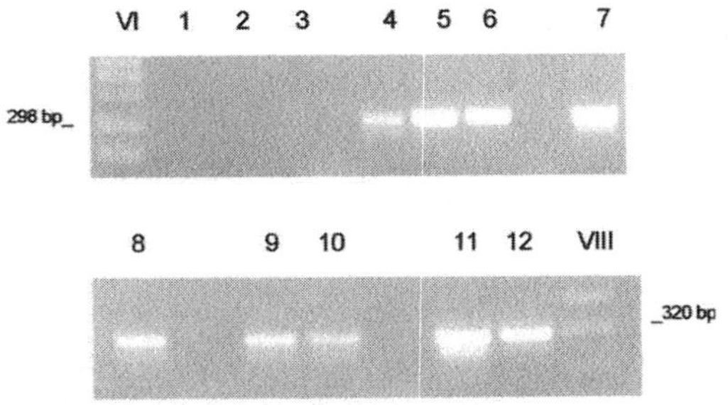

Figure 1. PTTG1 mRNA expression in human tumor cell lines. Molecular weight markers VI and VIII (Roche). Lanes 1 to 3, negative controls; Lane 4, KG1; Lane 5, K562; Lane 6, Jurkat; Lane 7, DLD1; Lane 8, LS513; Lane 9, MBQ-OJC1; Lane 10, HepG2; Lane 11, BT549; Lane 12, MCF7.

DNA sequencing of PCR products obtained from cell lines JJPF-OJC4 and JCA-OJC3 were performed. A BLAST search confirmed a 98 and 99 % identity between the input sequence from JJPF-OJC4 and JCA-OJC3, respectively, and the data base sequence for human PTTG-1. Specificity (Table 3) of the PTTG1-mRNA assay was examined by PCR amplification of cDNA obtained from normal lymph nodes, BM, and peripheral blood (cellular and plasmatic mRNA).The PTTG1 transcript was not detected in control lymph nodes (n = 34), using 500 and 800 ng of LN-derived cDNA for PCR amplification.

Table 3. Specificity of the PTTG1-mRNA RT-PCR Assay.

BM mRNA (n = 83)	PTTG1 Expression	BM mRNA (n = 36)	PTTG1 Expression
Total RNA (ng)		Total cDNA (ng)	
2000	+	2000	-
1000	+	1000	-
700	+	800	-
600	+	600	-
Blood Healthy Donors mRNA (n = 14)	4 (26.6 %)		

A semi-quantitative approach to evaluate the expression for PTTG-1 in BM was used. Total RNA was derived from two lots of pooled normal human blood marrow (male/female; Lot 1, n = 83 and Lot 2, n = 36). PCR amplification was performed using Ampliwax and "hot start" PCR. Different amounts of input RNA from Lot 1 were used for reverse transcription and a constant fraction (25%) of the cDNA was amplified subsequently in the PCR. Expression for PTTG1 was demonstrated with input RNA ranged from 600 to 2000 nanograms.

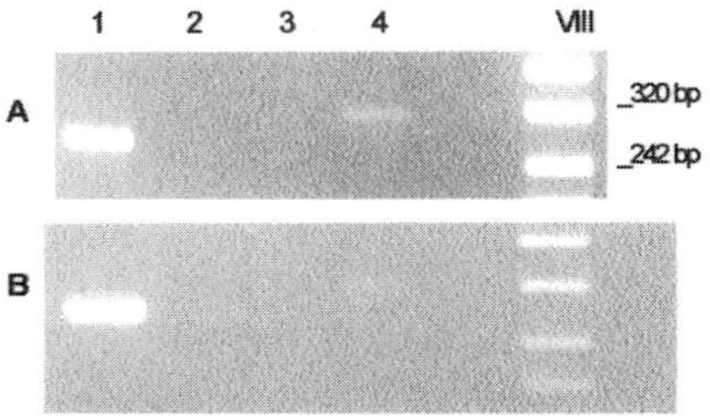

Figure 2. Specificity of PTTG1 mRNA analysis. Low level of PTTG1 mRNA expression (Lane 4) in blood from two healthy controls (A and B). Lane 1, β2-microglobulin; Lanes 2 and 3, negative results for epidermal growth factor receptor mRNA and cytokeratin 20 mRNA, respectively. MW Marker VIII (Roche).

PTTG1-mRNA was detected in peripheral blood in 4/14 healthy donors analyzed. Mean mRNA concentration was similar (165 ng ± 111.8, sd) in both positive and negative samples (p = NS, parametric and nonparametric test) (Figure 2). PCR and PTTG1-specific primers amplified different amounts of cDNA obtained after reverse transcription of pooled BM-derived RNA of lot 2. No PTTG1 expression was detected with cDNA ranged from 600 to 2000 ng (Figure 3).

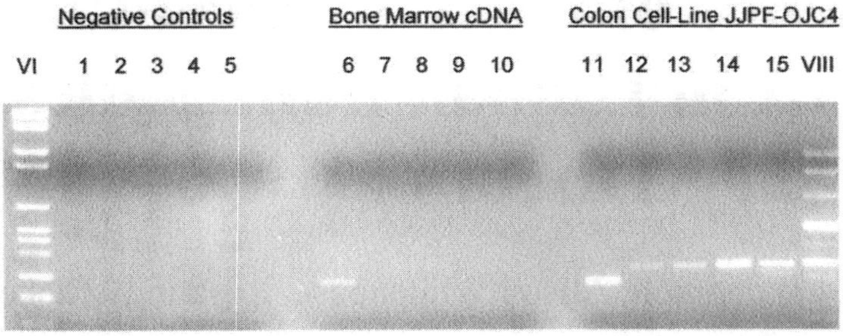

Figure 3. PCR and PTTG1-specific primers amplified different amounts of cDNA of pooled BM-derived RNA (n = 36). Lanes 1-5: Negative controls. Lanes 6 and 11: β2-microglobulin (cDNA = 600 ng). Lanes 7-10: No PTTG1 expression was detected in BM, cDNA ranged from 600 ng. (Lane 7, 800 ng; Lane 8, 1 μg; Lane 9-10, 2 μg). Lanes 12-15: PTTG1 expression in colon cancer cell-line JJPF-OJC4 with cDNA. (Lane 12, 600 ng; Lane 13, 800 ng; Lane 14, 1μg; Lane 15, 2 μg MW Marker VI/VIII (Roche).

Acknowledgments

Supported by a grant of the Universidade da Coruña 2000-5435256011.

References

1. Chambers AF, Naumov GN, Vantyghem SA, et al (2000) Molecular biology of breast cancer metastasis. Clinical implications of experimental studies on metastatic inefficiency. Breast Cancer Res 2:400-407.
2. Zhang X, Horwitz GA, Heaney AP, et al (1999) Pituitary tumor transforming gene (PTTG) expression in pituitary adenomas. J Clin Endocrinol Metab 84:761-767.
3. Ishikawa H, Heaney AP, Yu R, et al (2001) Human pituitary tumor-transforming gene induces angiogenesis. J Clin Endocrinol Metab 86:867-874.
4. Mapara MY, Korner IJ, Hildebrandt M, et al (1997) Monitoring of tumor cell purging after highly efficient immunomagnetic selection of CD34 cells from leukapheresis products in breast cancer patients: comparison of immunocytochemical tumor cell staining and reverse transcriptase-polymerase chain reaction. Blood 89:337-344.
5. Chou Q, Russell M, Birch DE, et al (1992) Prevention of pre-PCR mis-priming and primer dimerization improves low-copy-number amplifications. Nucleic Acid Res 20:1717-1723.
6. Zajchowski DA, Bartholdi MF, Gong Y, et al (2001) Identification of gene expression profiles that predict the aggressive behavior of breast cancer cells. Cancer Res 61:5168-5178.

Presence of CCK-B Receptor mRNA in Human Functionless Pituitary Tumors

Eftychia Oikonomou, Alison Whitehouse, Rosalind Mitchell, and Eric F. Adams

Summary

Cholecystokinin (CCK) is a peptide found widely distributed throughout the gut and nervous systems. There is growing evidence that it affects growth and differentiation of several types of cell, including anterior pituitary cells, via two receptor subtypes, A and B. An autocrine/paracrine role for CCK in GH_3 rat pituitary tumor cells has been reported. Herein, we demonstrate that functionless human pituitary tumors also possess CCK-B receptors. Total RNA was extracted from 30 human functionless pituitary tumors, reverse transcribed into cDNA, and subjected to PCR using primers specific for CCK-A and CCK-B receptors. The primers were targeted against different exons of the receptor genes, allowing identification of cDNA amplification. PCR bands of the predicted length were observed in all tumors using CCK-B receptor primers. Restriction digestion and direct sequence analysis of the products provided further evidence that they represented CCK-B receptor mRNA. CCK-A receptor mRNA was not detected by this RT-PCR analysis. In culture, CCK and CCK octapeptide sulphate powerfully stimulated phosphatidylinositol hydrolysis, providing evidence for functional activity of the CCK-B receptors. Similar analyses on pituitary-derived fibroblasts yielded negative results, indicating that the CCK-B receptor mRNA was expressed by tumors cells. These results suggest a role for CCK in the development of human functionless pituitary tumors.

Introduction

CCK is a member of the brain-gut peptide family. In the central nervous system (CNS), CCK is an abundant neurotransmitter, especially in the cerebral cortex and hippocampus. However, its functional significance is partially understood. The different molecular forms of amidated CCK found in tissues are different on the species and the tissue specific heterogeneity. The brain specializes in producing the small forms, such as CCK-8, while the intestine produces larger forms like CCK-58, 33, and 22. Even though all these forms contain CCK-8 amide and share the same biological activity, they differ in their amino terminal extension (1).

Molecular studies have identified two members of the CCK receptor family, the peripheral CCK-A type and the central CCK-B/gastrin type receptors (classification decreed by the IUPHAR Committee on Receptor Nomenclature and Drug Classification X11lth International Congress of Pharmacology, Munich, July 1998). Signal transduction at CCK-A receptors has been well characterized and receptor activation is coupled via guanine nucleotide binding protein to intracellular Ca^{2+} mobilization, and phosphatidylinositol (PI) hydrolysis, which may ultimately result in secretion such as the amylase release observed in pancreatic acinar cells (2). In addition, CCK-A receptors are known to undergo desensitization in response to agonists. Desensitization of G-protein coupled receptors seems to be associated with phosphorylation of the receptor at serine and threonine residues (3).

In contrast, CCK-B receptor have not been as widely characterized, and the transduction is less well-established, although a number of recent reports suggest activation of the phospholipase C specific for polyphosphoinositides, thereby generating inositol 1,4,5-trisphosphate (IP_3), as above, which in turn releases intracellular Ca^{2+} and diacylglycerol. Because diacylglycerol activates protein kinase C (PKC), this signal transduction process might be responsible for desensitization of the function of CCK-B receptors (3). More recent results reported on CCK-B-mediated intracellular Ca^{2+} increases in GH_3 cells (2). Past studies, where the CCK-B receptor was cloned from human brain, revealed a one potential site for PKC phosphorylation in the intracellular loop of its amino acid sequence (3). In addition, GTP and its analogues inhibited radiolabeled CCK-8 binding to Gh3 cell membranes, suggesting a coupling of CCK-B receptor to a G-protein (4).

CCK has been identified in both pituitary and hypothalamus where a modulator role of CCK is implicated in control of pituitary hormone secretion, with CCK stimulating secretion of a variety of hormones in both *in-vitro* and *in-vivo* studies (2). It has been previously shown that the GH_3 pituitary tumor cell line expresses a single population of functionally coupled CCK-B receptors, which stimulate intracellular Ca^{2+} mobilization and PI turnover (5, 3). Based on these data, studies regarding the proliferative effects of these peptides on mitogenesis in the GH_3 pituitary tumor cell line demonstrated that CCK acts as proliferative agent for pituitary GH_3 cells. The proliferative potencies were consistent with their CCK-B receptor binding affinities, with CCK octapeptide sulphate (CCK-8s) being the most potent agonist, indicating that the effects were mediated via G-protein coupled CCK-B receptors (5). Furthermore, there is some evidence that it may be involved in the hypothalamic-pituitary-adrenal (HPA) axis (6). The present studies were therefore designed to investigate whether human functionless pituitary tumours may also possess CCK-R and, if so, whether CCK peptides can influence IP_3.

Materials and Methods

Reverse Transcription Polymerase Chain Reaction (RT-PCR). An autocrine/paracrine role for CCK in GH_3 rat pituitary tumor cells has been reported via the CCK-B receptor subtype. To examine the suggested presence of the CCK-B and the possible involvement of the CCK-A receptor subtypes on human pituitary adenomas, total RNA was extracted from 30 human functionless pituitary tumours (specimens originated from Queen Elizabeth Medical Centre, University of Birmingham, UK), reverse transcribed into cDNA, and subjected to PCR using primers specific for CCK-A and CCK-B receptors. The primers were targeted against different exons of the receptor genes, allowing identification of cDNA amplification. The primers for CCK-A were designed to yield a PCR product of 278 bp in length, which was cut into two fragments of 46 and 232 bp by restriction digestion with Hinfl. The primers for CCK-B were designed to yield a PCR product of 309 bp in length, which was cut by PstI into three fragments of 12, 129, and 168 bp. For the full analysis, RT-PCR products were salt-ethanol precipitated and run on 1% agarose gels from which the DNA bands were excised and purified with Qiaex gel extraction kit. Purified PCR DNAs were digested overnight at 37°C with an excess of respective enzymes, and the cut and uncut PCR DNAs were resolved on a 2% agarose gel. DNA bands were visualized under UV light.

Cell Culture: Phosphatidylinositol (PI) Turnover. At surgery, a portion of each pituitary adenoma was placed into culture medium, transported to the laboratory, and processed for cell culture. After dispersion of the pituitary tissue with collagenase, the cells were seeded into glass culture tubes. The pituitary cells were allowed to equilibrate and attach during the following 24 h, after which analyses were conducted. PPI hydrolysis was assessed by measuring accumulation of [^{3}H] IPs in the presence of 10 mM LiCl. After 24 h seeding, cells were incubated in the presence of [^{3}H] *myo*-inositiol for 24 h at 37°C. The radiolabel was added at a concentration of 5 μCi/well in 1ml of fresh medium. After 24 h, cells were washed twice with 2 ml of serum free medium (SFM), incubated at 37°C for 2 h in SFM containing cold inositol, LiCl, and the agonists. Both CCK-33 and CCK-8s peptides were tested. Following aspiration of test medium, release of accumulated IPs was achieved by cell incubation of the cells with 1 ml of 3.3% perchloric acid at 4°C for 20 min. The perchloric acid extracts were neutralized with 80 μl 10 M KOH, and the aqueous phase chromatographed on 1 ml Dowex anionic exchange columns suspended in water. Columns were washed in a stepwise manner, first with 0.1M formic acid to remove contaminants, then with 1 M formic acid, IP_1, IP_2, and IP_3. Aliquots (0.4 ml) were mixed with 15 ml of scintillation fluid and counted. The cell membranes were dissolved in 1 M NaOH, and subsequently neutralized with 1 M HCl. Aliquots (0.4 ml) were mixed with 15 ml of scintillation fluid and counted. Results represent the amount of radioactivity in the aqueous phase as a % of total radioactivity (membranes plus aqueous).

Results

RT-PCR/Restriction Digestion. PCR bands of the predicted length were observed in all tumors using CCK-B receptor primers (Figure 1A). After digestion with PstI, the predicted band pattern was obtained, providing strong evidence that the RT-PCR DNAs represented CCK-B receptor (Figure 1B). Similar analyses on pituitary-derived fibroblasts yielded negative results, indicating that the CCK-B receptor mRNA was only expressed by tumors cells (data not shown).

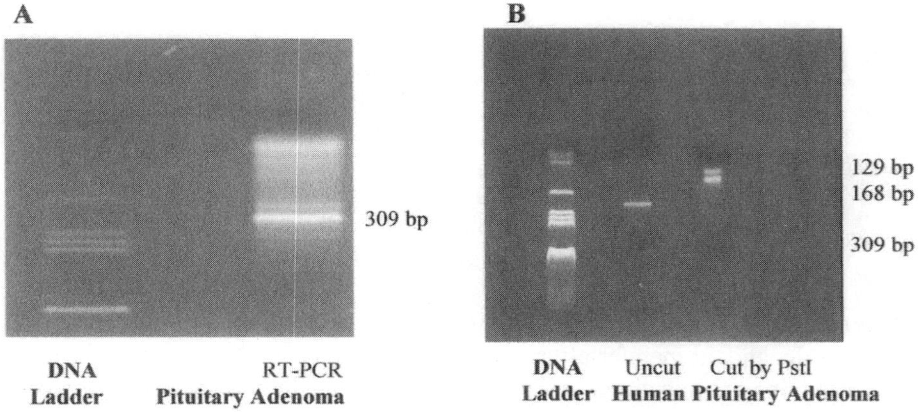

Figure 1. A. An agarose gel electrophoresis separation of PCR product of CCK-B receptor mRNA in human pituitary adenomas. The size of the amplified PCR band was 309 bp. **B**: An agarose gel electrophoresis separation of uncut and cut PCR DNAs of CCK-B receptor mRNA in human pituitary adenomas by PstI. PstI cuts twice at position 12 and 141 (CTGCA/G) and gave rise to three pieces of 12, 129, and 168 bp.

Cell Culture: Phosphatidylinositol (PI) Hydrolysis. The responses to two CCK receptor agonists were examined. Both non-selective agonists CCK-8s (1 nm to 100 nM) and CCK-33 (2 to 200 nM) produced dose/dependent increase in PPI turnover, providing evidence for functional activity of the CCK-B receptors (Figure 2).

Discussion

Previous studies have indicated that activation of the CCK-B receptor on GH_3 rat pituitary tumor cells leads to Ca^{2+} mobilization in association with stimulation of PI turnover and IP_3 production (7). The present findings are consistent with these studies and show that human functionless pituitary tumors cells also possess biologically active CCK-B receptors. Moreover, using reverse transcription followed by restriction digestion, we have demonstrated that functionless human pituitary tumors express CCK-B receptor mRNA but not CCK-A receptor. Taken together, these data correlate with previous findings that suggest GH_3 cells express a single class of high-affinity binding sites for radiolabeled CCK-8 with a Kd of 48 pM. The binding sites had high affinity for CCK-8 and CCK-B antagonists, but low affinity for CCK-A antagonists, indicating that the binding sites are for CCK-B receptors (4).

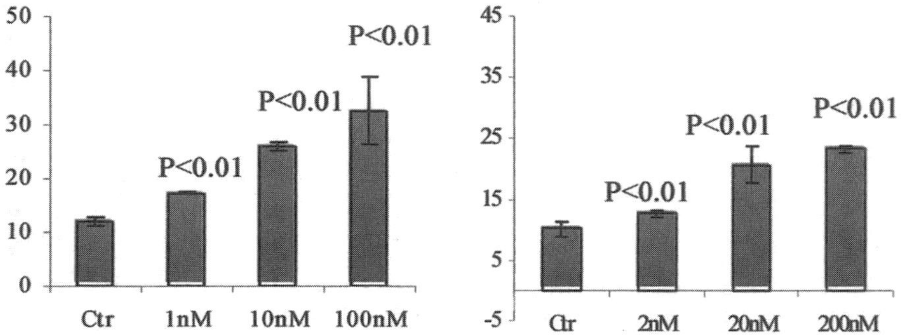

Figure 2. Agonist induced increase in PPI turnover in human pituitary tumor cells in a dose dependant manner after 6.0-h incubations. Mean values of all triplicate results obtained for each dosage are presented. Results are shown as number of cells per flask x10^{-5}. The bars represent the mean ± SD.

The role of IPs as intracellular second messengers is well established, with activation of cell surface receptors, promoting hydrolysis of membrane-bound PI 4,5-bisphosphate to generate IP_3 and diacylglycerol. The former mobilizes Ca^{2+} from intracellular stores, and the latter can activates the PKC enzyme family, bringing about protein phosphorylation. Many other IPs have been identified although their precise roles are still under investigation. In the present study, we have examined CCK-B receptor signal transduction. The results suggest that PI turnover is not a signaling mechanism exclusive to CCK-A receptors and demonstrate that CCK-B receptors on cells extracted from human pituitary adenomas are part of the family of receptors which employ IPs in cell signaling (2).

With respect to desensitization of CCK-B receptors in GH_3 cells, it has been reported that they show both homologous and heterologous desensitization. The initial exposure to CCK-8 reduced the responsiveness to the second exposure to CCK-8, leading to the conclusion that desensitization of the CCK-B receptors is mediated by some intracellular messengers (3). The fact that a supramaximal decline in response was consistently observed in agonist dose/response curves at high concentrations suggests that it may be desensitization. Some recent studies have shown evidence that PI hydrolysis is biphasic with a faster initial rate during the first 30 sec of agonist exposure, and with a subsequent rate reduction afterwards, indicating a possible mediation by rapid receptor phosphorylation. Evidence suggests that CCK receptors undergo rapid but partial desensitization through phosphorylation in a concentration–dependent manner in response to agonists (8). In addition, Li suppresses the supply of free inositol for PI resynthesis, which may result in a stimulus-dependent desensitization of the agonist effects (8). It is possible that similar desensitization mechanisms are present in human functionless pituitary tumor cells, although further confirmation studies are required.

The role of CCK-B receptors in pituitary cell function remains to be determined. However, there is considerable evidence that CCK receptors are coupled to a mitogenic response (5). In addition, PI hydrolysis is coupled to hormone secretion (9, 10), and thus it is conceivable that CCK may modulate the effects of hypothalamic releasing and inhibiting factors. These two observations together with the present findings raise the possibility that CCK may play a role in human pituitary tumorigenesis.

Conclusions

The data presented demonstrate that human functionless pituitary tumors express biologically active CCK-B receptors which are coupled to the PPI second messenger system as shown by stimulation by CCK peptides of IP_3 production. As such, these results suggest that there may be a role for CCK in the development of human functionless pituitary tumors. However, the type and size of stimulus required to initiate mitogenesis in a particular cell strongly depends on the combination of other factors, which may be present in the extracellular environment. Thus, further work is required to confirm the mitogenic effects of CCK peptides on human pituitary cells and to determine which type of interaction CCK may have with other factors (*i.e.*, hypothalamic factors) involved in pituitary cell function.

Acknowledgements

The authors would like to thank British Association for Cancer Research as well as the Lennox Children's Cancer Fund for financial support.

References

1. Mergery C Beinfeld, Wenger W (2002) CCK processing by pituitary GH3 cells, human teratocarcinoma cells NT2 and hNT differentiated human neuronal cells Evidence for a differentiation-induced change in enzyme expression and pro CCK processing. Life Sciences 70:1251-1258.
2. Smith AJ, Freedman SB (1996) CCK-B receptor-mediated stimulation of polyphosphoinositide turnover in GH3 pituitary cells in response to cholecystokinin and pentagastrin. Life Sciences 58:883-895.

3. Shinohara S, Kawasaki K (1994) Desensitization of cholecystokinin B receptors in GH3 çells. J Neurochemistry 62:1352-1356.
4. Kuwahara T, Takamiya M, Nagase H, et al (1993) GH3 cells, an anterior pituitary cell line, express CCK-B receptors. Peptides 14:647-649.
5. Smith AJ, McKernan RM (1999) Proliferative effects of cholecystokinin in GH3 pituitary cells mediated by CCK2 receptors and potentiated by insulin. Br J Pharmacology 126:79-86.
6. Abelson JL, Young EA (2003) Hypothalamic-pituitary adrenal response to cholecystokinin-B receptor agonism is resistant to cortisol feedback inhibition. Psychoneuroendocrinology 28:169-180.
7. Kuwahara T, Nagase H, Takamiya M, et al (1993) Activation of CCK-B receptors elevate cytosolic Ca2+ levels in a pituitary cell line. Peptides 14:801-805.
8. Roettger BF, Ghanekar D, Rao R, et al. (1997) Antagonist-stimulated internalization of the G protein-coupled cholecystokinin receptor. Mol Pharmacol 3:357-362.
9. Lei T, Buchfelder M, Fahlbusch R, et al (1995) Growth hormone releasing peptide (GHRP-6) stimulates phosphatidylinositol (PI) turnover in human pituitary somatotroph cells. J Mol Endocrinology 14:135-138.
10. Petersen B, Buchfelder R, Adams EF (1996) Pituitary adenylate cyclase-activating polypeptide directly stimulates LH and FSH secretion by human pituitary gonadotrophinomas. Exp Clin Endocrinol Diabetes 104:250-255.

Estrogen-Induced Mutations and Its Role in the Development of Tumorigenesis

Kamleshwar P. Singh, Jose Antonio López-Guerrero, Antonio Llombart-Bosch, and Deodutta Roy

Summary

Using RAPD-PCR fingerprinting, the detection of mutations induced by estrogen (E) exposure was studied in nonmalignant- and malignant cells. Cells exposed to 17α-and 17β-estradiol, diethylstilbestrol, bisphenol A, and α-zearalanol induced mutations in different regions of the genome. They consisted of insertion/deletion as a result of point or length mutations and quantitative changes as a result of hypoploidy or polyploidy. We also detected several mutated loci in tumor-free tissues, adjacent to tumors, and DES-induced tumors. The data suggest that these mutations occurred in early stages of stilbene E-induced renal carcinogenesis, and that they may be induced as a result of stilbene E treatment.

Introduction

Chronic exposure to estrogen (E) induces carcinogenesos in some animal systems (1-3). Target candidate regions of the genomes susceptible to mutations in response to 17β-estradiol (E_2) or diethylstilbestrol (DES) exposure are yet to be identified. Several PCR-based methods, Random Amplified Polymorphic DNA (RAPD), multiplex PCR, Single Strand Confirmation Polymorphism (SSCP), and Short Tandem Repeats (STR), have been used to detect mutations. These methods, except for AP-PCR/ RAPD, detect mutations using a known gene sequences as primers for amplification. Morover, they exclude the possibility of detecting mutations in unknown/novel genes(s) involved in tumorigenesis. RAPD uses an arbitrarily chosen oligonucleotide primer for the amplification of genomic DNA or cDNA fragments (4). In this study, we used RAPD/AP-PCR fingerprinting to detect mutations induced by E exposure in normal and malignant cells.

Materials and Methods

Syrian hamsters (6-8 weeks old) were divided into two groups. The control group (n = 6) received SC implants of 25 mg cholesterol pellets, and the treated group (n = 19) DES pellets every 3.0 mo (2, 3) in combination with ethylnitrosorea (ENU)

(3). For screening the mutagenic potential of Es, the mouse Leydig TM3 cells (ATCC) were treated with 100 ng/ml of 17α- and β-E_2, DES, bisphenol A, α-zearalanol or the vehicle (ethanol). After 72 hs of treatment, the cells were collected and genomic DNA was extracted. Sets of twenty OPA, OPC, OPK, OPE, OPAA and OP-26 oligonucleotide random 10-mer primers were from Operon Technologies, Alameda, CA. RAPD-PCR was performed as previously described by us (4).

Results and Discussion

Representative RAPD-PCR DNA fingerprints for TM3 cells treated with 17α- and β E_2, DES, bisphenol A, α-zearalanol or vehicle (ethanol) are shown in Figure 1.

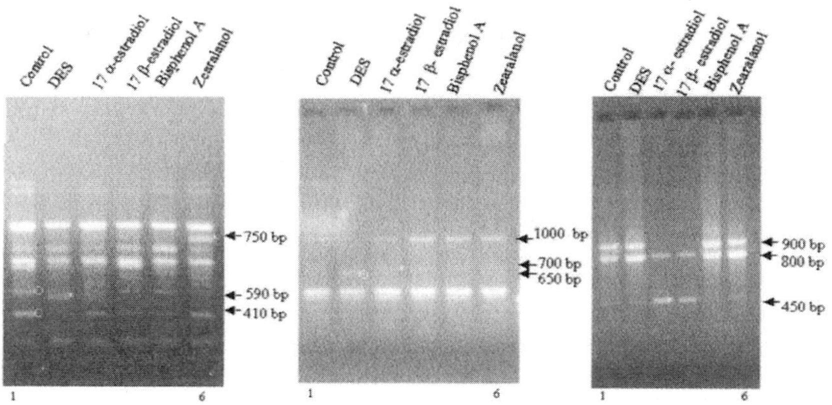

Figure 1. Representative RAPD fingerprints generated by primers OPA17, OPC 14, and OPE 26-2 depicting mutations induced by Es: DES, 17α- and β-E_2, bisphenolA, and zearalanol) in the genome of Leydig cells. Arrows indicate the insertion, deletion, or intensity differences in amplification products and their sizes.

Insertions of a 590 bp DNA fragment amplified with primer OPA17 (asterisk in left panel of Figures 1, lane 2) and a 650 bp amplified DNA fragment with primer OPC 14 (asterisk in middle panel of Figure 1, lane 2) were observed in the RAPD fingerprints of DES treated cells compared to untreated control cells. Similarly, deletions of 600 bp and 410 bp DNA fragments amplified with primer OPA17 (left panel of Figure 1, lane 2), and 1000 bp and 700 bp DNA fragments amplified with primer OPC14 (middle panel of Figure 1, lane 2) were observed in the RAPD fingerprints of DES treated cells compared to the DNA from untreated control cells. Cells exposed to either 17α- or 17β-E_2 exhibited a deletion of 900 bp, an insertion of 450 bp, and a loss of intensity of 800 bp DNA fragments amplified by primer OPE 26-2 (Right panel of Figure 1, lanes 3 and 4). Cell exposed to zearalanol exhibited an insertion of 750 bp, a deletion of 590 bp, and a loss of intensity of 700 bp DNA fragments amplified with primer OPA17 (Left panel of Figure 1, lane 6). A loss of intensity of a 410 bp product was observed in the RAPD fingerprint of bisphenol A treated cells as compared to control cell DNA (Left panel of Figure 1, lane 5). Thus, cells exposed to E exhibited mutations in different regions of the genome, and were both qualitative (insertion/deletion as a result of point or length mutations) and quantitative (as a result of hypo or polyploidy) in nature (a schematic diagram is shown in Figure 2).

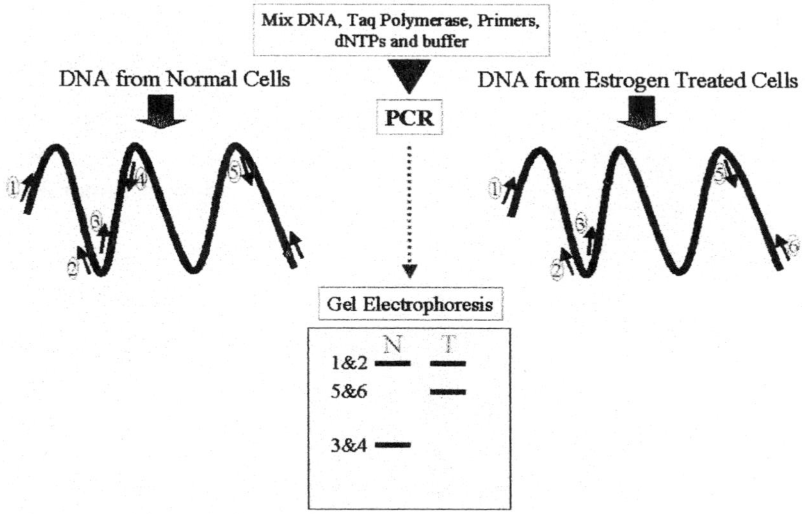

Figure 2. Diagram of the principle for the detection of E-induced mutation(s). Mutation at the primer binding sites may result in the loss of a primer binding site (indicated by a red dot in right panel, site number 4 present in untreated normal cells), or it may generate a new primer binding site (site number 6 in right panel). This loss/gain of primer binding sites may result in the deletion (generated by priming site 3 and 4), or the insertion of an amplification product (generated by priming site 5 and 6) in the RAPD fingerprints.

We have reported that the frequency of DES-induced kidney tumors in Syrian hamsters is around 40%, both in males and females. However, significant differences were observed between DES + ENU treated gonadectomized females and males, with frequencies of 91% and 53%, respectively (Figure 3). DES-induced tumor DNA was subjected to RAPD-PCR analyses after DNA digestion with *EcoRI, BamHI, Alu1* and Msp*I*. Tumor and adjacent non tumor tissue DNA digested with *Alu1* exhibited two mutated loci after OPCO3 amplification (an insertion of 498 bp indicated by asterisk in lanes 9 and 10 and a deletion of 700 bp indicated by an open circle in lane 8 of Figure 4).

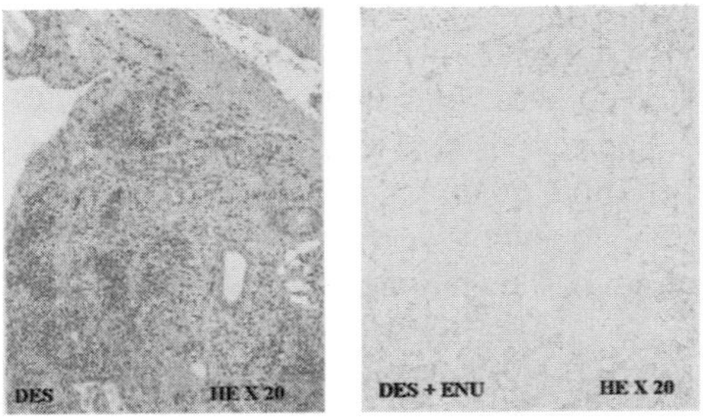

Figure 3. Representative photographs of DES- and ENU + DES-induced renal adenocarcinomas in Syrian hamsters.

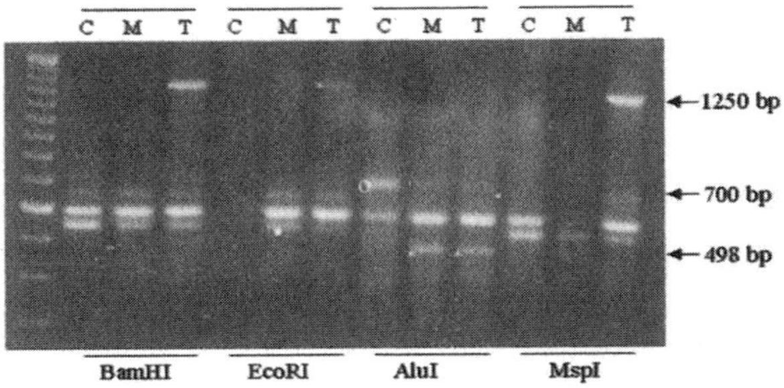

Figure 4. Representative RAPD fingerprints from a DES-induced hamster kidney tumors (T), adjacent tumor free tissue (M), and kidneys from age-matched controls (C). Arrows indicate the insertion/deletion and their base pair size.

The data presented suggest that the mutations observed occurred in early stages of stilbene E-induced renal carcinogenesis. These mutations were induced as a result of E treatment. In BamHI, EcoRI, and MspI digested DES-induced tumor DNA, we observed an insertion of 1250 bp DNA fragment amplified by OPCO3 primer (Figure 4, indicated by asterisk in lanes 4, 7, and 13). In addition, co-treatment with ENU influences mutations in the *Alu* repeat sequences in the genome of stilbene E-induced hamster kidney tumors (data not shown). This finding agrees with our previous report (3). The present study revealed that the RAPD/ AP-PCR fingerprinting is a useful tool in the identification and characterization of the regions of the genome of E-exposed nonmalignant and malignant cells that underwent various changes compared to the normal untreated cells.

We have previously reported that Es or their metabolites are genotoxic (5-7). DES and E_2 (after conversion to catechol E_2) are oxidized to quinone primarily by cytochrome P450 1A1 (8). DES or E_2 semiquinones are also formed in the process of oxidation of stilbene E or catechol Es. The semiquinones react with molecular oxygen and generate superoxide, which is further reduced to the hydroxyl radical. Various types of DNA damage are produced by either E metabolites or free radicals binding covalently to DNA (5-7).

Acknowledgment

This work was supported by a NIH grant ES10851 to DR and a FIS 01/702 grant, Madrid, Spain to ALB.

References

1. National Toxicology Program (NTP) (2002) Federal report on carcinogens. Federal Register 67:77283-77285.
2. Llombart-Bosch A, Peydro-Olaya A (1997) Urinary system. Monographs on Pathology of Laboratory Animals II SI. New York: Springer Verlag, 141-152.
3. Singh, KP, Lopex-Guerrero JA, Llombart-Bosch A, Roy D (2004) Age, sex and co-exposure to N-ethyl-N-nitrosourea influence mutations in the Alu repeat sequences in diethylstilbestrol-induced kidney tumors in Syrian hamsters. Mutagenesis 19:67-73.
4. Singh KP, Roy D (1999) Detection of mutation(s) or polymorphic loci in the genome of experimental animal and human cancer tissues by RAPD/AP-PCR depend on DNA polymerase. Int. J. Oncol 14:753-758.
5. Roy D, Qui Q (2002) Estrogen, immunoactivation, gene damages and development of breast, endometrial, ovarian prostate, and testicular cancers. Recent Res Devel Steroid Biochem Mol Biol 3:1-32.
6. Cavalieri E, Frenkel K, Rogan E, et al (2000) Estrogens as endogenous genotoxic agents--DNA adducts and mutations. JNCI Monograph 27:75-94.
7. Roy D, Singh KP (2004) Estrogen-induced genetic alterations and its role in carcinogenicity. Current Genomics, in press.
8. Roy D, Bernhardt A, Strobel HW, Liehr JG (1992) Catalysis of the oxidation of steroid and stilbene estrogens to estrogen quinone metabolites by the beta-naphthoflavone-inducible cytochrome P450 IA family. Arch Biochem Biophys 296:450-456.

Englitazone Delays Fetal Growth in Late Gestation in the Rat

Julio Sevillano, Inmaculada C. López-Pérez, Emilio Herrera, M. Pilar Ramos, and Carlos Bocos

Summary

The mechanisms regulating hepatocyte proliferation are relevant to liver development, carcinogenesis, and regeneration. Studies of hepatocyte proliferation control during late foetal and postnatal development have been used as a model to understand such mechanisms. Since peroxisome proliferator-activated receptor gamma (PPARγ) ligands have been implicated in the inhibition of growth and differentiation of certain human cancers, in the present study, we have investigated the effect of englitazone (EG), a PPARγ ligand, on foetal and postnatal development. Our results indicate that, EG administered semi-chronically to pregnant rats, produced a body weight reduction on the progeny. This effect may be related to the diminished level of plasma IGF-I found in the neonates from treated-mothers. Surprisingly, despite receiving an anti-diabetic drug, foetus and neonates showed high levels of insulin, and were hyperglycemic. The plasma levels of leptin, other putative mitogenic factor, were not affected by the treatment. In the liver of neonates from mothers receiving EG, the expression of PPARα, IR, PI3K and IRS-1 was unchanged, as was the phosphorylation of MAPK. Nevertheless, an increase on Akt phosphorylation was observed on liver of neonates from treated-mothers, confirming a remarkable change on the mitogenic insulin/IGF-I pathway. In conclusion, the growth inhibitory effect reported herein may be associated with the ability of PPARγ ligands to reduce IGF-I concentrations and produce an insulin resistance state on foetus/neonates. These data strengthen the idea that PPARγ ligands have potential benefits on cancer treatment.

Introduction

The mechanisms regulating hepatocyte proliferation are relevant to liver development, carcinogenesis, and regeneration. Hepatocyte proliferation during late foetal and postnatal development has been used as a model to understand the mitogenic signalling pathways involved in such mechanisms (1). PPARγ is a member of the steroid nuclear receptor superfamily, a large class of ligand-activated transcription factors regulating gene expression. Their activated receptors regulate the expression of target genes after binding to peroxisome proliferator responsive elements (PPRE). Initially, PPARγ was known for its regulatory roles in insulin sensitization and adipocyte differentiation. More recent studies have shown that PPARγ has an important role in cell proliferation and cancer (2). Thus, thiazolidinediones (TZD), anti-diabetic drugs that activate PPARγ, have been implicated in the growth inhibition and differentiation of certain cancers, *i.e.,* human hepatocellular carcinoma (2). Therefore, in the present study, the effect of EG (a TZD that function as a PPARγ ligand) on foetal and postnatal development was investigated.

Materials and Methods

Animals, Drug Administration, and Collection of the Samples. The experimental protocol was approved by the Animal Research Committee of the University San Pablo-CEU in Madrid, Spain. Female SD rats weighing 180-210 g were mated. Pregnancy day 0 was determined when spermatozoids were found in vaginal smears. From day 16 of gestation, rats were given a daily dose of 50 mg of EG/kg body weight, suspended in 2% Tween-80, by oral gavage, at 9.00 AM. On the morning of the 20th day of pregnancy, corresponding to 4 days of treatment, the rats were decapitated. The conceptus was dissected, weighed, and the fetuses counted and weighed. Fetuses were decapitated, blood was collected from all pups of the same mother, pooled into receptacles containing Na_2-EDTA for immediate plasma separation at 4°C. A different set of pregnant rats received the same EG treatment from day 16 of gestation for 5 days. The rats were allowed to deliver. On the day of birth, neonates were decapitated, and blood was collected as described above. Neonate livers were dissected, and those from the same mother were pooled, frozen in liquid N, and stored at -80°C until further analysis.

Analyses. Plasma aliquots, kept at –30° C, were used to measure glucose by an enzymatic colorimetric test (Glucose oxidase, GOD/PAP method, Roche Diagnostics, Barcelona, Spain) (3). Plasma insulin was determined using a specific ELISA kit for rats (Mercodia, Uppsala, Denmark) with a detection range of the 0.07-5.5 μg insulin/ml (1.8% intra-assay variation, 3.8% inter-assay variation). Plasma leptin was assayed by ELISA using a commercially available kit specific for rat leptin (Assay Designs, Inc., Ann Arbor, MI) with a detection range of 0.06-3.6 ng leptin/ml (11.6% intra-assay variation, 11.0% inter-assay variation). IGF-I was measured by a competitive binding enzyme immunoassay using a rat IGF-1 EIA kit (DRG Diagnostics) with a detection limit of 30 ng/ml (7.7% intra-assay variation, 9.4% inter-assay variation).

Protein Extraction and Immunoblotting. Aliquots (50 mg) of frozen liver were powdered in liquid N in a precooled mortar, and lysed in ice-cold 30 mM HEPES buffer pH 7.4, containing 5 mM EDTA; 1% Triton X-100, 0.5% sodium deoxycholate, and 2 mM protease inhibitor for 30 min. Cellular debris was pelleted and discarded by centrifugation at 17,000 g for 30 min at 4 °C. Supernatant fractions were collected and their protein concentration determined by the BCA protein assay. Aliquots (25 μg of protein) from each experimental condition were subjected to 7.5% SDS-PAGE, and electrophoretically transferred to PVDF membranes. The blots were probed with primary antibodies: Anti-insulin receptor -GLUT-2, -IRS-1, -IRS-2; -PI3K, -Akt1, -p Akt1, -MAPK1/2, PPAR-α, and -β-actin , followed by corresponding secondary antibodies conjugated with horseradish peroxidase. Immunoreactive bands were visualized using the enhanced chemiluminescence system (ECL), and quantified by densitometry. The intensity of the proteins studied was corrected by the values obtained from the immunodetection of β-actin.

Statistical Evaluation. The results are expressed as the mean ± SEM of 4-10 animals/group. The data were analyzed for homogeneity of variance with Levene test. Values were log-transformed to equalize the variance between conditions. Statistical comparisons between two groups were made using the Student *t* test with 95% confidence limits (4) using the SPSS program (version 9.0.1).

Results

EG semi-chronically administered to pregnant rats induced a significant decrease in neonatal body weight (Figure 1). Retarded fetal development and impaired postnatal growth in rats have been described for pioglitazone and rosiglitazone (5), using doses similar to the one used herein. Since it has been postulated that IGF-I promotes growth and differentiation in a variety of tissues (6), we determine whether the decreased body weight observed in neonates (Figure 1) may be related to a decreased IGF-I plasma levels in fetuses and neonates.

As shown in Figure 2, foetal plasma IGF-I from EG-treated mothers showed a decrease trend that was confirmed in the EG-treated neonates. Plasma IGF-I was significantly lower in the neonates from mothers treated for five days with EG, than in pups from control rats (Figure 2). Several authors have previously reported the importance of plasma IGF-I in body size at birth (7, and references therein).

Surprisingly, as shown Table 1, despite treatment with an antidiabetic drug, fetuses and neonates from mothers receiving EG showed higher levels of plasma insulin than their respective controls, and the neonates were hyperglycemic. These data suggests that pups from EG-treated mothers have developed an insulin resistant condition. In a recent report, the specific deletion of the gene encoding IGF-I in murine liver (8), produced transgenic mice with a marked reduction in circulating IGF-I levels, insulin resistant, and hyperinsulinaemic (7).

Leptin is a putative mitogenic factor that regulates intrauterine growth (9). However, in the present study fetus and neonate plasma leptin levels were not altered by EG treatment (Table 1), although the neonates from treated-mothers showed significantly lower body weight (Figure 1).

Figure 1. Average body weight of fetuses and neonates from mothers receiving medium (Control) or englitazone (EZ) for four (fetuses) or five days (neonates). Values represent the mean ± SEM, n = 80-100. Statistical comparisons between groups receiving different treatment are shown by asterisks (***, $p < 0.001$).

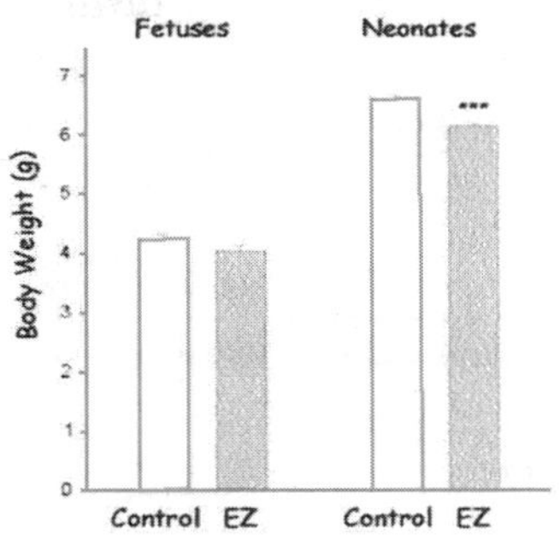

Figure 2. Plasma IGF-I levels in fetuses and neonates from mothers receiving medium (Control) or englitazone (EZ) for four (fetuses) or five days (neonates). Values represent the mean ± SEM, n = 4-8. Statistical comparisons between groups receiving different treatment are shown by an asterisk (*, $p < 0.05$).

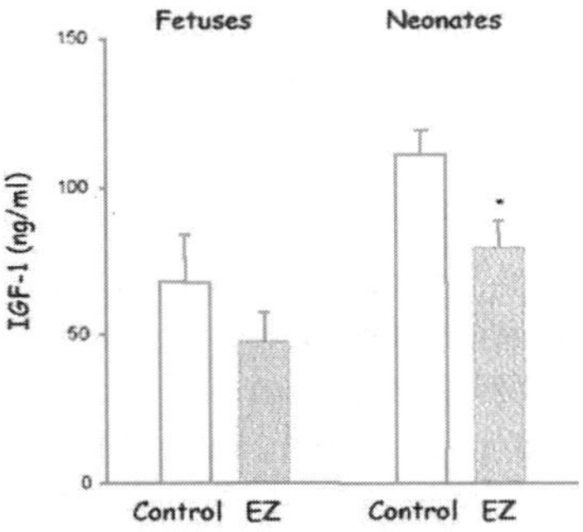

Table 1. Plasma Glucose, Insulin and Leptin Levels in Fetuses and Neonates from Mothers Receiving Medium (Control) or Englitazone for Four (Fetuses) or Five Days (Neonates)[1].

	FETUSES		NEONATES	
	Control	**Englitazone**	**Control**	**Englitazone**
Glucose (mg/dl)	59.45 ± 9.66	60.53 ± 7.18	88.33 ± 7.8	174.02 ± 15.66[2]
Insulin (μg/L)	1.53 ± 0.10	2.16 ± 0.101[2]	0.44 ± 0.07	1.65 ± 0.26[3]
Leptin (ng/ml)	5.34 ± 0.56	5.03 ± 0.59	2.71 ± 0.54	2.78 ± 0.64

[1] Values represent the mean ± SEM, n = 4-10
[2] Statistically significant, $p < 0.001$ [3] Statistically significant, $p < 0.002$

In order to understand the molecular events inducing the insulin resistant state observed in the neonates from EG-treated mothers, we considered the possibility that it may be influenced by a mitogenic signaling pathway, thus, we examined the hepatic expression of IR, PI3K, IRS-1, IRS-2 and MAPK (ERK-1 and ERK-2). As shown in Figure 3, none of the levels of the proteins examined were modified by EG treatment, neither was the phosphorylation of MAPK (data not shown). Furthermore, the amount of GLUT-2, an important glucose transporter in the liver, was also unaltered by the EG treatment (Figure 3). In addition, we analyzed the hepatic expression of PPARα, a transcription factor related to mitogenic processes (1), on neonates from mothers receiving EG. No significant changes were observed (Figure 3).

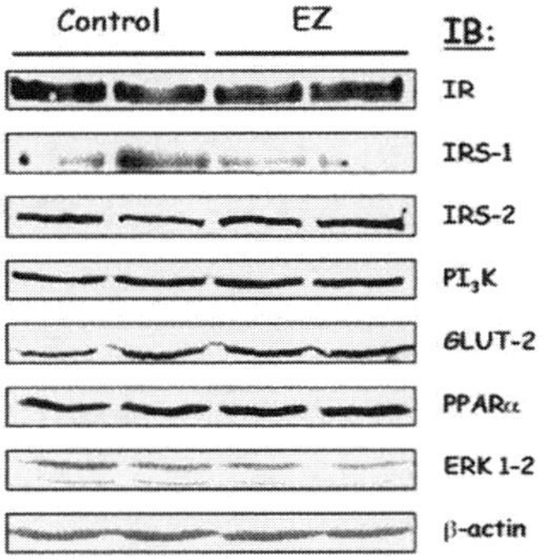

Figure 3. Different insulin- and mitogenic-signaling proteins in liver of neonates from mothers receiving medium (Control) or englitazone (EZ). PPARα, IR, PI3K, IRS-1, IRS-2, MAPK (ERK-1 and ERK-2), and GLUT-2 proteins were determined by Western immunoblotting (IB), as described in Materials and Methods. Detection of β-actin was used as a control. Representative autoradiographs from seven animals per group.

Interestingly, as shown in Figure 4, an increase on hepatic Akt phosphorylation in neonates from EG-treated mothers was observed, confirming a remarkable change on the mitogenic insulin/IGF-I pathway. This constitutively active form of Akt found in the liver of neonates from mothers receiving EG may be related to their hyperinsulinemic state (Table 1). This basal phosphorylation of Akt may prevent further phosphorylation of the enzyme in response to insulin, thus acting as a negative control mechanism. In fact, although these animals have elevated insulin levels, they were hyperglycemic.

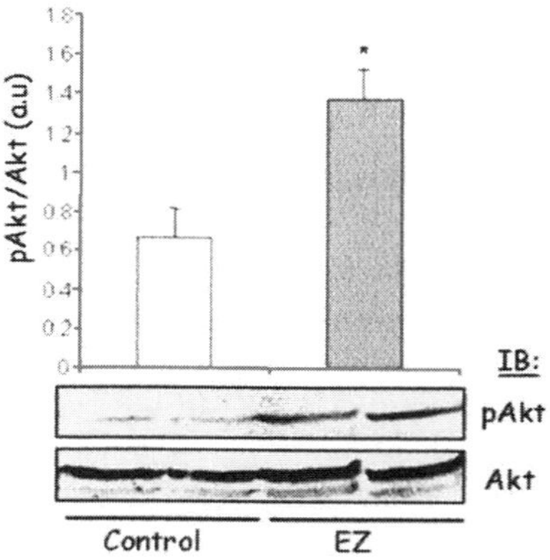

Figure 4. Basal Akt phosphorylation in liver of neonates from mothers receiving medium (Control) or englitazone (EZ). The autoradiographs identify the hepatic Akt protein (lower lane) and the corresponding (Ser473)-phosphorylated Akt (upper lane). The data represent the signal due to the phosphorylation corrected by the Akt protein determined by Western immunoblotting. Data represent the mean ± SEM of four animals per group. Statistical comparisons between groups receiving different treatment are shown by asterisks (*, $p < 0.05$).

Conclusions

The growth inhibition reported herein in the pups from mothers receiving EG during late gestation may be associated with their reduced IGF-1 levels. Actually, the capability of PPARγ ligands to reduce IGF-1 concentration, has been suggested by Stoll (10). Furthermore, since in the late foetal and postnatal development, the cellular proliferation take great importance, a direct role of PPARγ, which activation has been shown to inhibit cell proliferation and induce differentiation (11), cannot be discarded. Thus, these results may be related to the fact recently reported that acromegalic patients with reduced expression of PPARγ in the colonic mucosa and increased serum IGF-1 levels, present an increased prevalence of colonic polyps (12).

Nevertheless, the retarded growth of the pups from the mothers treated with EG, might be also related to their insulin resistance state, since it is well known that insulin is an important mitogenic factor in the foetal growth (13). This possibility, along with the highest Akt phosphorylation observed in liver of neonates from treated mothers, deserve further study.

Altogether, these data strengthen the idea that PPARγ ligands have potential benefits on cancer treatment, and represent an example of how PPAR γ and IGF-1 are involved in processes related to cell proliferation and cancer (11, 12).

Acknowledgements

The present study was carried out with grants from the Fondo de Investigación Sanitaria, Instituto de Salud Carlos III (01/0524), Ministerio de Ciencia y Tecnología (PM99/011), and Universidad San Pablo-CEU (10/99-00). Julio Sevillano is a recipient of a pre-doctoral fellowship from the Fundación Ramón Areces, Spain. The authors thank Jose M. Garrido for his help in handling the animals.

References

1. Gruppuso PA, Boylan JM, Vaslet CA (2000) Identification of candidate growth-regulating genes that are overexpressed in late gestation fetal liver in the rat. Biochim Biophys Acta 1494:242-247.
2. Rumi MAK, Sato H, Ishihara S, et al (2001) Peroxisome proliferator-activated receptor γ ligand-induced growth inhibition of human hepatocellular carcinoma. Br J Cancer 84:1640-1647.
3. Barham D, Trinder P (1972) An improved coluor reagent for the determination of blood glucose by the oxidase system. Analyst 97:142-145.
4. Rosner B (1994) Fundamentals of Biostatistics. USA: Duxbury Press.
5. O'Moore-Sullivan TM, Prins JB (2002) Thiazolidinediones and type 2 diabetes: new drugs for an old disease. Med J Aust 176:381-386.
6. Di Cola G, Cool MH, Accili D (1997) Hypoglycemic effect of insulin-like growth factor-1 mice lacking insulin receptors. J Clin Invest 99:2538-2544.
7. Holt RIG (2002) Fetal programming of the growth hormone-insulin-like growth factor axis. Trends Endocrinol Metab 13:392-397.

8. Sjogren K, Liu J-L, Blad K, et al (1999) Liver-derived insulin-like growth factor I (IGF-I) is the principal source of IGF-I in blood but is not required for postnatal body growth in mice. Proc Natl Acad Sci USA 96:7088-7092.
9. Hoggard N, Hunter L, Duncan J, et al (1997) Leptin and leptin receptor mRNA and protein expression in the murine fetus and placenta. PNAS 94:11073-11078.
10. Stoll BA (2002) Linkage between retinoid and fatty acid receptors: implications for breast cancer prevention. Eur J Cancer Prev 11:319-325.
11. Sarraf P, Mueller E, Jones D, et al (1998) Differentiation and reversal of malignant changes in colon cancer through PPARγ. Nat Med 4:1046-1052.
12. Bogazzi F, Ultimieri F, Raggi F, et al (2002) Peroxisome proliferator activated receptor γ expression is reduced in the colonic mucosa of acromegalic patients. J Clin Endocrinol Metab 87:2403-2406.
13. Molina-Font J (1998) Nutrition and fetal growth. Early Hum Dev 5:S51-S60.

The Tumor Suppressor Gene PTEN Plays a Role in Cell Cycle Regulation and Apoptosis in Prostate Cancer Cell Lines

Alice Hlobilková, Michaela Šváchová, Jana Knillová, Eva Pimrová, Petra Říháková, Per Guldberg, and Zdeněk Kolář

Summary

Phosphatase and tension homology deleted on chromosome ten (PTEN) is a tumor suppressor gene with protein and lipid phosphatase activity frequently altered in several types of human cancers. Herein, we demonstrate the effect of transfected wild type (wt) PTEN and its mutants (mt) L57W, H123Y, G129R that have lost lipid and protein phosphatase activity; and G129E characterized by loss of only lipid phosphatase activity, during the cell cycle and on apoptosis. We characterized the expression of important proteins regulating the cell cycle and the PI3-K/PKB/Akt pathway, analyzed PTEN cellular localization, and performed PTEN mutation analysis in DU-145 and PC-3 cells. The transfection of wt PTEN and its mt with defective lipid phosphatase activity (G129E) inhibited S-phase entry of DU-145 and PC-3 cells. In DU-145 cells, transfection of wt PTEN induced apoptosis. An inverse expression of PTEN and phosphorylated PKB/Akt was observed. We did not find any mutations of the PTEN gene in either cell line. PTEN influenced the cell cycle of tested cells in a p53 and pRb independent manner, and the effect was cell type specific.

Introduction

PTEN/MMAC (mutated in multiple advanced cancers) is a tumor suppressor gene localized on chromosome 10q23 (1, 2). PTEN is one of the most common targets of mutation in human cancer, with a mutation frequency approaching that of p53. Germline mutations of PTEN have been reported in three inherited syndromes: Cowden´s, Lhermitte-Duclos, and Bannayan-Zonana syndromes. The common feature of these diseases is a predisposition to hamartomas, benign tumors containing differentiated but disorganized cells (3-5). PTEN mutations have been found in sporadic glioblastomas, malignant melanomas, carcinomas of the prostate, breast, kidney, urinary bladder, uterus, and other organs, mostly in advanced stages of tumor progression (6-10). The protein product of PTEN (55 kD) possesses phosphatase activity for lipid and protein substrates, and it plays a role in the PI3-K/PKB/Akt pathway (11). PTEN catalyzes dephosphorylation of Ptd-Ins-(3,4,5)-P_3 at position 3 on the inositol ring, and regulates the level of Ptd-Ins-(3,4,5)-P_3 *in vivo* (12). PTEN causes inhibition of the G_1-S phase transition in the cell cycle by increasing the expression of cdk inhibitor $p27^{Kip1}$ (13). PTEN expression has been described mainly in the cytoplasm of tested cells, although it has been detected in the nucleus as well (7, 14, 15). Finally, the importance of PTEN increases with evolutionary level (16).

Earlier experiments revealed inhibition of the S-phase entry in breast cancer cell lines following transfection of PTEN with an active protein phosphatase domain, whereas the lipid phosphatase domain was unessential. The cell cycle inhibition was detected only in cell lines carrying the wt tumor suppressor genes p53 and Rb. In this study, we have analyzed the expression of the proteins regulating the cell cycle and the PI3-K/PKB/Akt pathway, as well as, the influence of overexpressed PTEN and its mutants on inhibition of G_1-S transition and apoptosis induction.

Material and Methods

Mutation Analysis of PTEN. Mutation analysis was performed by denaturating gradient gel electrophoresis and PCR described by Guldberg, *et al.* (17).

Plasmids and Mutagenesis. Human wt PTEN DNA, kindly provided by Dr. Steck (University of Texas, Houston), was subcloned into the pxmyc vector according to Lukas, *et al.* (18). PTENpxmyc mutagenesis was performed as previously described (19).

Cell Culture and Transfection. The human androgen insensitive prostate cancer cell lines DU-145 and PC-3 were cultured in medium DMEM (Invitrogen), supplemented with 10% foetal bovine serum (Invitrogen), 2 mM glutamine (Invitrogen), and antibiotics (penicillin 1 U/ml and streptomycin 1 μg/ml, Invitrogen). Gene transfer experiments were done on exponentially growing cell populations using an electroporation (transfection efficiency was 10-20%) (20). Purified DNA of the control empty vector, pxmyc, PTENpxmyc and its mutants L57Wpxmyc, H123Ypxmyc, G129Rpxmyc, and G129Epxmyc were transfected in combination with a pCMVCD-20 plasmid encoding the CD-20 surface antigen to facilitate identification of cells expressing transgenes.

Analysis of Cell Cycle Effects and Apoptosis. Transfected cells were cultured on glass coverslips for 36 or 48 h, and 12 h before cell fixation, 0.1 mM bromdeoxyuridine (BrDU) was added to the culture medium. The cells were stained with mouse monoclonal CD-20 antibody (Becton Dickinson), fixed in a cold mixture of methanol and acetone (1:1), incubated with rabbit anti-mouse biotinylated antibody (Amersham) and with fluorescent conjugate streptavidine-Texas red (Amersham). Cell DNA was denaturated with HCl (1:5). The cells were stained with mouse monoclonal BrDU antibody (DAKO) and FITC-conjugated secondary antibody (DAKO). Analyses of nuclei for apoptotic morphological changes were performed by staining with bisbenzimide–Hoechst 33258 (1 μg/ml, Sigma). The percentage of S-phase inhibition was determined as previously determined (19).

Immunocytochemical Analysis of Endogenous PTEN. Exponentially growing DU-145 and PC-3 cells cultured on glass coverslips were fixed and stained with mouse monoclonal antibody against PTEN (26H9, Cell Signaling).

Table 1. Summary of the Antibodies Used for the Detection of Endogenous Proteins Regulating the Cell Cycle and PI3-K/PKB/Akt Pathway.

Antibody	Species	Clone	Company/Source
$p21^{Waf1/Cip1}$	Mouse	DCS-61, DCS-62	Danish Cancer Society, DK
$p27^{Kip1}$	Mouse	Ab-3	NeoMaerkers, USA
Cyclin $D_{1,2,}$	Mouse	5D-4	Danish Cancer Society, DK
Cyclin D_3	Mouse	DCS-22	Danish Cancer Society, DK
Cyclin E	Mouse	HE-12	Danish Cancer Society, DK
Cyclin A	Mouse	Ab-3	Oncogene, USA
Cyclin B1	Mouse	GSN1	Santa Cruz, USA
Cdk2	Mouse	D-12	Santa Cruz, USA
PTEN	Mouse	26H9	Cell Signaling, USA
Phospho PKB/Akt Ser473	Mouse	4E2	Cell Signaling, USA
p53	Mouse	DO-1	Danish Cancer Society, DK
Phospho pRb Ser 795	Mouse	Rb-10	Danish Cancer Society, DK
Total pRb	Mouse	14001A	Pharmingen, USA
Mcm-7	Mouse	DCS-141.1	Danish Cancer Society, DK
Cdk4	Rabbit	Polyclonal, C-22	Santa Cruz, USA
Cdk6	Rabbit	Polyclonal, C-21	Santa Cruz, USA
PI3-K-85α	Rabbit	Polyclonal, Z8	Santa Cruz, USA
Total PKB/Akt	Rabbit	Polyclonal	Santa Cruz, USA

Electrophoresis and Western Blot Analysis. DU-145 and PC-3 cells were harvested and lysed in cold protein extraction buffer. The proteins (20-50 μg/sample) were electrophoretically separated on polyacrylamide gels in the presence of SDS, and transferred into nitrocellulose membranes by semi-dry method (21). Detection of the endogenous proteins regulating the cell cycle and PI3-K/PKB/Akt pathway was made using the antibodies summarized in the Table 1.

Results

Detection of PTEN Mutations and Immunocytochemical Analysis of Endogenous PTEN. Mutation analysis failed to confirm mutations of the PTEN gene in DU-145 and PC-3 cells. The immunocytochemical detection of endogenous PTEN revealed cytoplasmic staining of a granular character in DU-145 cells. PTEN protein was not detected in PC-3 cells (Figure 1).

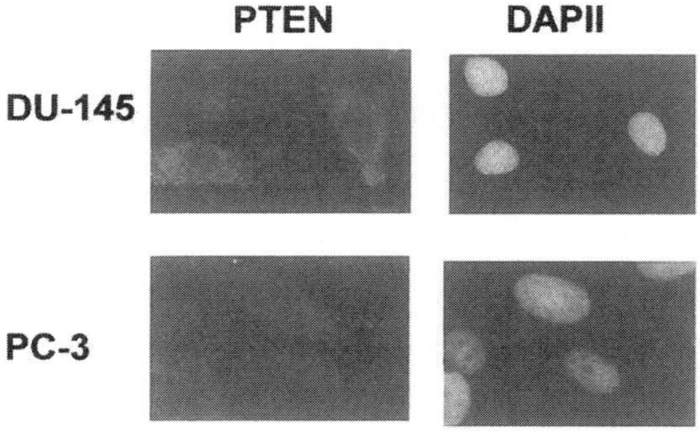

Figure 1. Immunofluorescence detection of endogenous PTEN using anti-PTEN antibody, clone 26H9, magnification 1000×.

Analysis of Protein Expression Regulating the Cell Cycle and PI3-K/PKB/Akt Pathway in DU-145 and PC-3 Cells. Western blot analysis of PTEN expression was not detected in PC-3 cells. This finding may be explained by either very low expression levels or changes at the post-transcription level. An inverse correlation between the expression of PTEN and phosphorylated PKB/Akt in both prostate cancer cell lines was detected. The total PKB/Akt was unchanged. DU-145 cells did not express pRb, but overexpressed a mt form of p53. In PC-3 cells, p53 was not detected, while pRb and its phosphorylated form were expressed at high levels (Figure 2).

Effect of Overexpressed PTEN on the Cell Cycle and Apoptosis. Overexpression of wt PTEN and its mutant with defective lipid phosphatase activity (G129E) led to inhibition of G_1-S transition in both cell lines, DU-145 with defective expression of p53 and pRb, and PC-3 with defective p53. Overexpression of PTEN mutants with defective lipid and protein phosphatase activity (L57W, H123Y and G129R) had no cell cycle effect (Figure 3). Overexpression of wt PTEN induced apoptosis in DU-145 cells (increase about 69% against the control empty vector). The influence of overexperessed PTEN on apoptosis in PC-3 cells was not shown.

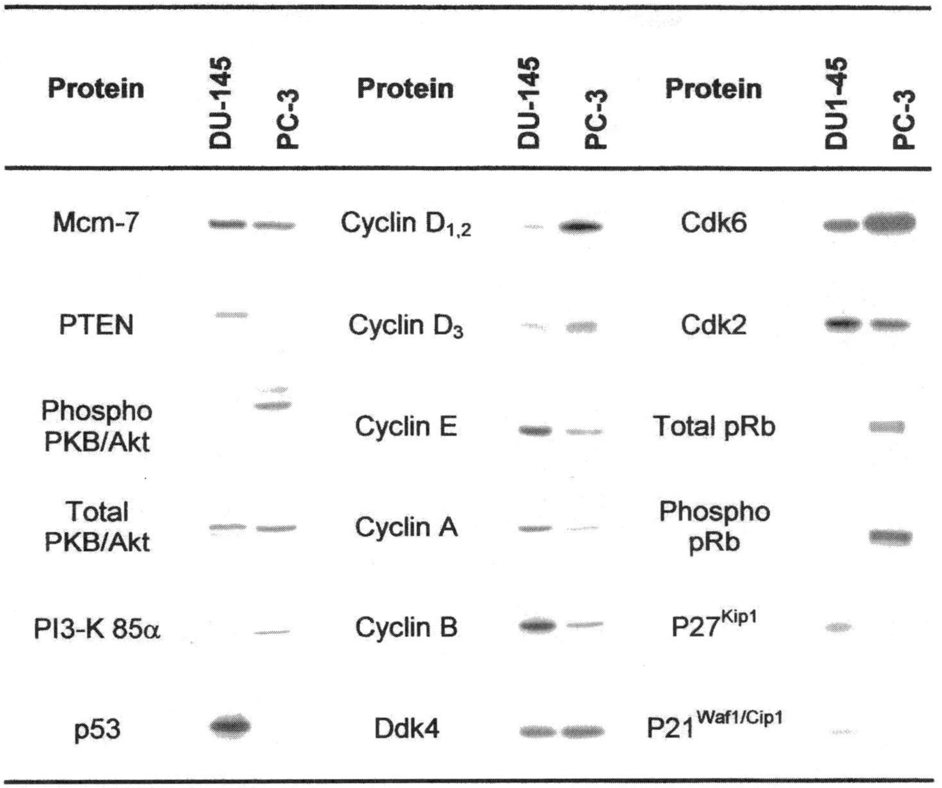

Figure 2. Western blot analysis of protein expression involved in cell cycle and PI3-K/PKB/Akt pathway regulation in exponentially growing DU-145 and PC-3 cells

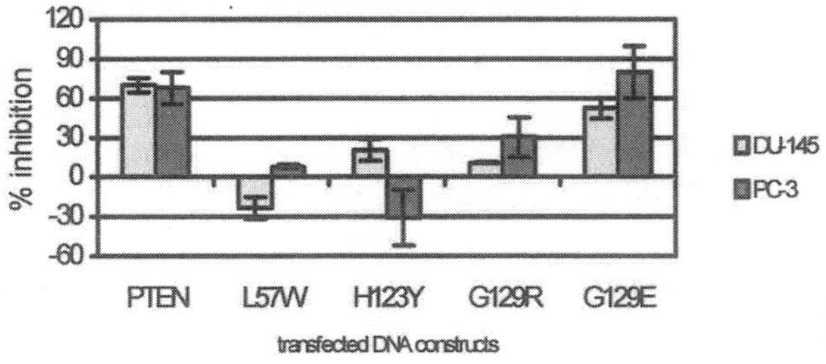

Figure 3. S phase inhibition after overexpression of PTEN and its mts in DU-145 and PC-3 cells. S-phase inhibition of the control vector was set at 0%.

Conclusions

In tested prostate cancer cell lines PTEN blocked the cell cycle G_1-S transition in a p53 and pRb independent manner. These data are not in agreement with already published mechanisms in some breast cancer cells. Protein phosphatase activity is essential for the PTEN cell cycle effect in both prostate cancer cell lines. The protein and lipid phosphatase activity are essential for apoptosis induction in DU-145 cells. The inverse correlation between PTEN and phosphorylated PKB/Akt supports their opposite role in the PI3-K/PKB/Akt pathway.

Acknowledgements

This work was supported by grant IGA MZ CR NC/6779-3 and MSM 151100001.

References

1. Li J, Yen C, Liaw D, et al (1997) PTEN, a putative protein tyrosine phosphatase gene mutated in human brain, breast, and prostate cancer. Science 275:1943-1947.
2. Steck PA, Perhouse MA, Jasser SA, et al (1997) Identification of a candidate tumour suppressor gene, MMAC1, at chromosome 10q23.3 that is mutated in multiple advanced cancer. Nat Genet 15:356-362.
3. Liaw D, Marsh DJ, Li J, et al (1997) Germline mutations of the PTEN gene in Cowden disease, an inherited breast and thyroid cancer syndrome. Nat Genet 16:64-67.
4. Padberg GW, Schot JDL, Vielvoye GJ et al (1991) Lhermitte-Duclos disease and Cowden syndrom: a single phakomatosis. Ann Neurol 29:517-523.
5. Marsh DJ, Dahia PL, Zheng Z, et al (1997) Germline mutations in PTEN are present in Bannayan-Zonana syndrome. Nat Genet 16:333-334.
6. Steck PA, Lin H, Langford AL, et al (1999) Functional and molecular analyses of 10q deletions in human gliomas. Genes Chromosomes Cancer 24:135-143.
7. McMenamin ME, Soung P, Perera S,et al (1999) Loss of PTEN expression in paraffin-embedded primary prostate cancer correlates with high Gleason score and advanced stage. Cancer Res 59:4291-4296.
8. Tsao H, Zhang X, Benoit E, et al (1998) Identification of PTEN/MMAC1 alterations in uncultured melanomas and melanoma cell lines. Oncogene 16:3397-3402.
9. Gordon BM, Lu Y, Fang X, et al (2001) The role of genetic abnormalities of PTEN and the phosphatidylinositol 3-kinase pathway in breast and ovarian tumorigenesis, prognosis, and therapy. Seminars Oncology 16:125-141.
10. Kondo K, Yao M, Kobayashi K, et al (2001) PTEN/MMAC1/TEP1 mutations in human primary renal-cell carcinomas and renal carcinoma cell lines. Int J Cancer 91:219-224.
11. Besson A, Robbins SM, Yong VW (1999) PTEN/MMAC1/TEP1 in signal transduction and tumorigenesis. Eur J Biochem 263:605-611.
12. Maehama T, Dixon J E (1998) The tumor suppressor. PTEN/MMAC1, dephosphorylates the lipid second messenger, phosphatidylinositol 3,4,5-trisphosphate. J Biol Chem 273:13375-13378.

13. Cheney IW, Neuteboom STC, Vaillancourt M-T,et al (1999) Adenovirus-mediated gene transfer of MMAC1/PTEN to glioblastoma cells inhibits S phase entry by the recruitment of p27Kip1 into cyclin E/CDK2 complexes. Cancer Res 59:2318-2323.
14. Sano T, Lin H, Chen X, et al (1999) Differential expression of MMAC/PTEN in glioblastoma multiforme: relationship to localization and prognosis. Cancer Res 59:1820-1824.
15. Whang YE, Wu X, Suzuki H, et al (1998) Inactivation of the tumor suppressor PTEN/MMAC1 in advanced human prostate cancer through loss of expression. PNAS USA 95:5246-5250.
16. Huang H, Potter CJ, Tao W, et al (1999) PTEN affects cell size, cell proliferation and apoptosis during Drosophila eye development. Development 126:5365-5372.
17. Guldberg P, Straten P, Birk A, et al (1997) Disruption of the MMAC1/PTEN gene by deletion or mutation is a frequent event in malignant melanoma. Cancer Res 57:3660-3663.
18. Lukas J, Parry D, Aagaard L, et al (1995) Retinoblastoma –protein-dependent cell-cycle inhibition by the tumour suppressor p16. Nature 375:503-506.
19. Hlobilkova A, Guldberg P, Thullberg M, et al (2000) Cell cycle arrest by the PTEN tumour suppressor is target cell specific and may require protein phosphatase activity. Exp Cell Res 256:571-577.
20. Lukas J, Bartek J, Strauss M (1994) Efficient transfer of antibodies into mammalian cells by electroporation. J Immunol Methods 170:255-259.
21. Sambrook J, Russel DW (2001) Commonly used techniques in molecular cloning. In: Molecular Cloning, A Laboratory Manual. New York: Cold Spring Harbor Laboratory Press, A8.40-A8.55.

The Coactivators CBP and p300 in Androgen Independent Prostate Cancer

Jose D. Debes, Zoran Culig, and Donald J. Tindall

Summary

Prostate cancer (PCA) is the most frequently diagnosed malignancy and the second leading cause of death as a result of cancer in men in the USA and many other western countries. In the year 2003 there will be approximately 220,900 cases diagnosed and 28,900 deaths due to PCA. Steroid hormones, particularly androgens (As), play a major role in PCA, but their precise role is not clear. The majority of PCA responds to A ablation therapy by temporary remission. However, most tumors will eventually relapse in an A-refractory state. A recent study from our laboratory suggests that androgen receptor (AR) function may play a major role in the proliferation of A-refractory PCA cells. Several reports suggest that the AR co-activators are involved in A-dependent and A-independent PCA. We have shown that p300 and CBP are involved in A-independent transactivation of the AR.

Introduction

Several coactivators interact with the AR and enhance its activity. The members of the 160-kd nuclear receptor coactivator family (p160) have been shown to interact with the AR and enhance its transcriptional activity. Some members of the p160 family include the glucocorticoid receptor interacting-protein 1, the steroid receptor coactivator-1 and the receptor-associated coactivator 3 (1). These proteins bind to the N-terminal activation domain of the AR and can enhance ligand-dependent activation (2, 3). It has been suggested that alterations in the AR coactivators or their interaction can allow AR activity to be enhanced by non-androgenic molecules. In this regard, it has been shown that p160 coactivators interact with BRCA1 and the AR. Indeed, BRCA1 enhances AR transactivation activity (4). A second group of coactivators are known to alter the ligand-dependent activation of the AR. These include the AR-associated proteins ARA54, ARA55 and ARA70. Activation of the AR by 17β-estradiol is enhanced by ARA55 and ARA70 (5-7). In addition, it has been shown that anti-As like hydroxyflutamide (HFM) and bicalutamide (Bic), among others, can promote the interaction between the AR and ARA70. On the other hand the anti-A HFM enhances the AR activity in the presence of ARA55 (6, 8). In LNCaP cells the coactivator ARA54 enhances transcriptional activity of the AR (which contains the T877A mutation) by both 17β-estradiol and HFM in LNCaP cells (which presents the T877A mutation) (9). Indeed, increased expression of coactivators PIAS1 (33% of 43 tumors) and Ran/ARA24 (81% in 43 tumors) have been reported in primary PCA, whereas decreased expression of ELE1/ARA70, ARA54, and no change in ARA54, SRC1 and TMF1/ARA160 have been found (10). In addition, LNCaP cells transfected with ARA70 cDNA retarded proliferation. Thus, co-activators for the AR may also act as tumor suppressors. Recently p300, as well and its binding protein PCAF, have been shown to enhance AR ligand-dependent activity on PCA cells (11), and data from our laboratory indicates that p300 produces the same effect in the absence of As (12). The p300 homologous CREB-binding protein (CBP) has been found to be expressed in PCA and to enhance AR transcriptional activity by A and anti-A.

Thus, there are several ways by which coactivators may be involved in the A-independent progression of PCA through the AR. Overexpression of coactivators may allow the activation of the AR by non-steroidal molecules as well as growth factors or cytokines, leading ultimately to progression of PCA. Herein, we present some of our published data about the role of p300 in interleuking-6 (IL-6)-mediated transactivation of the AR and CBP in the transactivation of the AR by anti-A.

Materials and Methods

Transfections and Luciferase Assays. LNCaP cells (2.8 X 10^5 in 6 well plates) were plated and 24 h, transfected using the Gene Porter Transfection System (Gene Therapy Systems, Inc., San Diego, CA) with plasmids containing full-length p300 (pCI.p300) or its mutant acetyltransferase negative derivative (pCI.p300-HAT), both previously described in the figure legends. Plasmid pCMVE1A12S was used at 10 ng per plate. Luciferase reporter containing a PSA promoter (2ug per plate) was used to measure AR transactivation activity in all transfections. Twenty-four hours post-transfection cells received fresh media containing 9% charcoal-stripped FBS (CSS) either with or without 50ng/ml IL-6 (R & D Systems, Mpls, MN). Dual Luciferase assays were performed according to the manufacturer's instructions (Promega, Madison, WI). Transfection efficiency was monitored by co-transfection with pGFP (1 μg per plate, Promega) and visualized with a Zeiss fluorescent microscope at 488 nm. Routinely, transfection efficiencies of 45-50% were obtained. More details are described elsewhere (12). Details on luciferase assays for CBP experiments are also described elsewhere (24).

Results

p300 and PCA. p300 has been shown to interact with the AR during its ligand-dependent activation in PCA cells (11). The same study shows that a point mutation in the acetylation site of the AR abrogated the A-dependent transactivation of the receptor. However, p300 role in A-independent transactivation of the AR remains to be elucidated.

One growth factor involved in PCA progression is, as be mentioned above, IL-6. This cytokine has been found to transactivate AR in the presence and absence of As, indicating that it might be involved in the A-independent progression of PCA. Many points in the pathway by which IL-6 regulates the AR remain undefined. Thus, further studies are needed.

IL-6 has been shown to transactivate the AR through the MAPK pathway (13). We assessed this effect and found that IL-6 induced inhibition of the MAPK pathway with a MEK-1 inhibitor significantly reduced the IL-6-mediated transactivation of the AR, and the phosphorylation of ERK-1 by IL-6. However, we found that increasing amounts of p300 overcame the inhibition of the MAPK pathway leading to transactivation of the AR (12). These data suggested that the effects of p300 on AR may be downstream of the MAPK pathway, or that IL-6 could induce AR activation through a different pathway involving p300.

It is known that the adenoviral oncoprotein E1A binds and sequesters p300, thereby inhibiting its HAT activity (14). When we transfected cells with EIA IL-6 could no longer transactivate the AR indicating that p300 sequestration by E1A impeded IL-6 effects on the AR (12). E1A also bind to other factors, like CBP. Nonetheless, when increasing amounts of p300 were transfected into cells, E1A-mediated inhibition was released (Figure 1). In these experiments AR protein was not affected by either IL-6 treatment, or by E1A or p300 transfection, indicating that these events may be taking place at the level of transcription (12). It has been proposed that acetylation alters the positive charge of histones, thus destabilizing the nucleosome bed and leaving DNA susceptible to regulation by different transcription factors (14). A mutant p300 lacking histone acetyltransferase activity (p300-HAT) failed to reverse E1A-mediated repression of AR activity, indicating the importance of the histone acetyltransferase activity of p300 for its interaction with AR in A-independent transactivation (12). Inducing p300 degradation or limiting its RNA production should have similar effects on AR transactivation in PCA cells. Thus, we used synthetic duplex RNAs (~25 nucleotides in length) specific for p300 that induce specific cleavage of p300 mRNA, thus knocking down its expression. When we transfected siRNA specific for we found that cells showed no p300 protein expression at 48 h (12). Since PSA is a known AR gene target and is dependent on AR activity, we evaluated PSA production after siRNA-p300 in PCA cells. We found that IL-6 treatment increased PSA expression, however, in cells that were transfected with siRNA-p300, IL-6 had no positive effect on PSA expression. AR protein expression was not altered during these experiments. These results confirmed that p300 plays a direct role in the transactivation of the AR by IL-6 (12). Our experiments do not preclude the possibility that other histone acetylators may also regulate the activity of the AR in the absence of As. This is the case of CBP, which is known to interact with the AR and induce its transactivation.

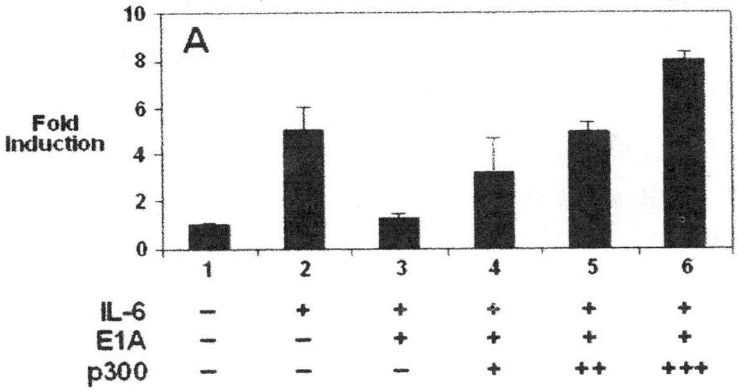

Figure 1. Inhibition of p300 abrogates IL-6-mediated transactivation of the AR. LNCaP cells were transfected with PSA-LUC reporter (2 μg), E1A 12S (10 ng), and increasing amounts of p300 (+, 0.25 μg; ++, 0.5 μg; +++, 1 μg). Twenty-four h after transfection, cells received fresh medium containing either 50 ng/ml of IL-6 or vehicle alone. Cells were lysed 12 h later and luciferase activity measured. [Reprinted with permission from Cancer Research (35)]

CBP and PCA. CBP was originally identified as a co-activator of CREB (cAMP response element binding protein) (15). In addition to CREB. CBP binds to several other specific factors like c-Fos and c-Myb (16-17), and enhances the activity of different nuclear receptors like the thyroid hormone, glucocorticoid, and estrogen receptors (18-19).

It has been demonstrated that CBP is an important coactivator for the AR during A-dependent transactivation (15). Therefore, we investigated the effects of CBP in A-independent transactivation of the AR. Acquisition of agonistic properties of AR antagonists has been reported previously, and is of interest because of the use of these drugs in endocrine therapy. Several studies have reported that HFM induces agonist properties in mutated ARs (20, 21). The mechanism by which HFM induces these effects in human PCA is not well established.

Since alterations in AR co-activators can induce a functional change in the receptor response, we decided to assess the effect of CBP over-expression in PCA cells. We found that HFM induced a greater response of wild-type AR activity in the presence of over-expressed CBP in DU-145 cells (22) (Figure 2). Moreover, when the mutated AR 730 val-met (that has increased activation by metabolic products of HFM) was used, HFM induce more activity than with the wild-type receptor (22). When the anti-A Bic was used, the levels of induction of AR activity were lower than with HFM, nonetheless, CBP over-expression enhanced the agonist activity of Bic. When LNCaP cells, which posses endogenous AR, and LNCaP-abl cells, a derivative with long-term steroid deprivation, were transfected with CBP and treated with HFM. AR activities induced by HFM were higher both in the absence and presence of over-expressed CBP in LNCaP-abl cells when compared to parental LNCaP (22). AR protein expression was not altered by CBP overexpression indicating the activity was at the level of transcription (22). In addition, CBP was found to be expressed in the nuclei of all the cell lines used for this study, as well as in PCA tissue specimens (22).

In this study, it was shown that agonistic/antagonistic balance of non-steroidal anti-As, mainly HFM, is influenced by the transcriptional coactivator CBP. Because of its association with a number of different factors, CBP might be involved in the ligand-independent activation of the AR and could be considered as a potential target for PCA therapy.

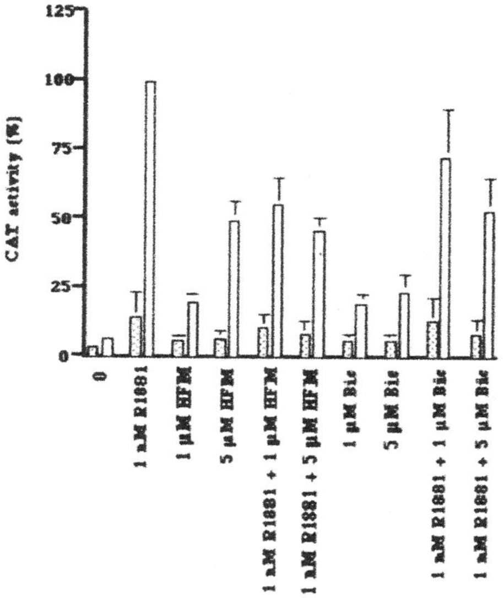

Figure 2. AR activity in the absence (dotted bars) or presence (open bars) of over-expressed CBP. DU-145 cells were transfected with the reporter gene ARE_2TATA-CAT, wild-type AR, and CBP cDNA. Cells were treated with R1881 and or hydroxyflutamide (HFM) or bicalutamide (Bic) and CAT activity was measured. [Reprinted with permission from the American Journal of Pathology (57)].

Conclusions

In the last few years much attention has been given to AR co-activators and their role in A-independent PCA. It is clear that many of these co-activators play a major role in transactivation of the AR in the absence of As. Considering the paucity of treatment options available to patients with A-independent PCA, the discovery of a new group of potential targets represented by AR co-activators is encouraging.

We have shown that both p300 and CBP play an important role in the A-independent transactivation of the AR in PCA (Figure 3). Both co-activators have the potential of becoming targets for treatment of PCA. Moreover, there is still a question of whether these co-activators could serve as markers for A independent progression of this neoplasm. Further studies are needed regarding the relation of p300 and CBP in PCA. Indeed, the role of these coactivators in A-independent transactivation of the AR needs to be investigated in depth considering the potential of both proteins.

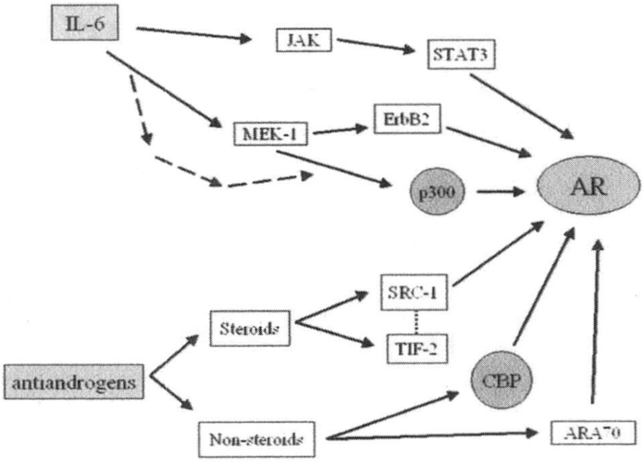

Figure 3. AR transactivation pathways induced by IL-6 or anti-As in PCA.

References

1. Chen D, Ma H, Hong H, et al (1999) Regulation of transcription by a protein methyltransferase. Science 284:2174-2177.
2. Alen P, Claessens F, Verhoeven G, et al (1999) The androgen receptor amino-terminal domain plays a key role in p160 coactivator-stimulated gene transcription. Mol Cell Biol 19:6085-6097.
3. Ma H, Hong H, Huang SM, et al (1999) Multiple signal input and output domains of the 160-kilodalton nuclear receptor coactivator proteins. Mol Cell Biol 19:6164-6173.
4. Park JJ, Irvine RA, Buchanan G, et al (2000) Breast cancer susceptibility gene 1 (BRCAI) is a coactivator of the androgen receptor. Cancer Res 60:5946-5949.
5. Fujimoto N, Yeh S, Kang HY, et al (1999) Cloning and characterization of androgen receptor coactivator, ARA55, in human prostate. J Biol Chem 274:8316-8321.
6. Kang HY, Yeh S, Fujimoto N, et al (1999) Cloning and characterization of human prostate coactivator ARA54, a novel protein that associates with the androgen receptor. J Biol Chem 274:8570-8576.
7. Miyamoto H, Yeh S, Wilding G, et al (1998) Promotion of agonist activity of antiandrogens by the androgen receptor coactivator, ARA70, in human prostate cancer DU145 cells. Proc Natl Acad Sci USA 95:7379-7384.
8. Miyamoto H, Yeh S, Lardy H, et al (1998) Delta5-androstenediol is a natural hormone with androgenic activity in human prostate cancer cells. Proc Natl Acad Sci USA 95:11083-11088.
9. Grossmann ME, Huang H, Tindall DJ (2001) Androgen receptor signaling in androgen-refractory prostate cancer. J Natl Cancer Inst 93:1687-1697.

10. Li P, Yu X, Ge K, et al (2002) Heterogeneous expression and functions of androgen receptor co-factors in primary prostate cancer. Am J Pathol 161:1467-1474.
11. Fu M, Wang C, Reutens AT, et al (2000) p300 and p300/cAMP-response element-binding protein-associated factor acetylate the androgen receptor at sites governing hormone-dependent transactivation. J Biol Chem 275:20853-20860.
12. Debes JD, Schmidt LJ, Huang H, et al (2002) P300 mediates androgen-independent transactivation of the androgen receptor by interleukin 6. Cancer Res 62:5632-5636.
13. Ueda T, Bruchovsky N, Sadar MD (2002) Activation of the androgen receptor N-terminal domain by interleukin-6 via MAPK and STAT3 signal transduction pathways. J Biol Chem 277:7076-7085.
14. Giordano A, Avantaggiati ML (1999) p300 and CBP: partners for life and death. J Cell Physiol 181:218-230.
15. Aarnisalo P, Palvimo JJ, Janne OA (1998) CREB-binding protein in androgen receptor-mediated signaling. Proc Natl Acad Sci USA 95:2122-2127.
16. Bannister AJ, Oehler T, Wilhelm D, et al (1995) Stimulation of c-Jun activity by CBP: c-Jun residues Ser63/73 are required for CBP induced stimulation in vivo and CBP binding in vitro. Oncogene 11:2509-2514.
17. Dai P, Akimaru H, Tanaka Y, et al (1996) CBP as a transcriptional coactivator of c-Myb. Genes Dev 10:528-540.
18. Chakravarti D, LaMorte VJ, Nelson MC, et al (1996) Role of CBP/P300 in nuclear receptor signalling. Nature 383:99-103.
19. Hanstein B, Eckner R, DiRenzo J, et al (1996) p300 is a component of an estrogen receptor coactivator complex. Proc Natl Acad Sci USA 93:11540-11545.
20. Fenton MA, Shuster TD, Fertig AM, et al (1997) Functional characterization of mutant androgen receptors from androgen- independent prostate cancer. Clin Cancer Res 3:1383-1388.
21. Tan J, Sharief Y, Hamil KG, et al (1997) Dehydroepiandrosterone activates mutant androgen receptors expressed in the androgen-dependent human prostate cancer xenograft CWR22 and LNCaP cells. Mol Endocrinol 11:450-459.
22. Comuzzi B, Lambrinidis L, Rogatsch H, et al (2003) The transcriptional co-activator cAMP response element-binding protein- binding protein is expressed in prostate cancer and enhances androgen- and anti-androgen-induced androgen receptor function. Am J Pathol 162:233-241.

Expression Study of Estrogen Receptor-related Receptors and Steroid Hormone Receptors in Human Prostatic Cells

C.P. Cheung, Lung-Wai Chan, Ki Lui, Uwe Borgmeyer, Shiuan Chen, and Franky L. Chan

Summary

Estrogen receptor-related receptors (ERRs) are closely related to the estrogen receptors (ERs) in their protein structures and consist of 3 closely related isoforms: *ERRα, ERRβ and ERRγ.* However, ERRs do not bind to estrogens (Es) or any other known physiological ligands, and thus they are classified as "*orphan nuclear receptors*". Studies on the expression pattern of ERRs in prostatic cells and prostate is scarce. As the first step towards a better understanding of the significance of ERRs in prostate, we investigated their expression profiles in human prostatic cell lines, two primary prostatic carcinoma xenografts, and prostatic tissues by RT-PCR, immunohistochemistry, and Western blot analysis. The results of RT-PCR showed that ERRα transcripts were widely detected in all tested cell lines and xenografts; ERRβ transcripts were either not detected or weakly expressed in prostatic cell lines and xenografts, while ERRγ transcripts were moderately expressed in C4-2B and PC-3 cells, and two xenografts. Similar expression patterns were also shown in human hyperplastic and neoplastic prostatic tissues. ERRγ immunohistochemistry showed that immunosignals were localized to the prostatic epithelial cells. We also compared mRNA expression of ERRs with that of ERs in prostatic cell lines. The results showed that ERRs and ERs were co-expressed in most cell lines and xenografts. Based on these observations, we speculate that the two members of nuclear receptor subfamily, ERRs and ERs, may control overlapping regulatory pathways in the prostatic cells.

Introduction

In addition to androgens, Es have been suggested to play certain roles in the normal development and neoplastic growths of prostate gland, (1, 2). However, the exact functional roles of Es and their receptors in the prostate are still unclear and remain controversial. The effects of Es on the prostate are complex and involve both direct and indirect actions.

The direct effects of Es on the prostate or their roles in prostate cancer and benign prostatic hyperplasia are believed to be mediated through ERs, which are present in two subtypes, ERα and ERβ. Both ER subtypes are expressed in human and rodent prostates, and exhibit distinct expression pattern in different tissue components. The expression of both ERα and ERβ in prostate has developed into a view that the two ERs might play different functional roles in the prostate, and Es and their antagonists exert different actions on prostatic cells. However, studies from ER knock-out mouse models show that targeted disruption of either ER (α or β) or both ERs does not affect the prostatic phenotype and function (3), suggesting that other ER-related signaling pathways could exist and be involved in the prostatic growth and functions.

ERRs belong to the nuclear receptor subfamily III and consist of three closely related isoforms: α, β and γ. All three ERR isoforms are highly homologous to ERs in their protein structures. However, they do not bind to Es or any other known physiological ligands, and thus they are classified as orphan nuclear receptors. Human ERRα and ERRβ were first cloned in 1988 in kidney and heart cDNA libraries in a search for new members of ER (4). Similar to ERs, ERRs can bind to the consensus E responsive elements (ERE) and related E/retinoid receptor response element which is recognized by other orphan receptor (steroid factor-1), and thus they could regulate or modulate the transcription of genes, which are also regulated by ERs and other orphan nuclear receptors (5-7). *In vitro* studies show that ERRs are constitutively transactivated without binding to any known physiological ligands in positively expressed cells (8, 9), and they interact with a number of nuclear receptor coactivators (7, 10-12), and also with ER by completing for same coactivators (13). A recent study suggests a mechanism of Ca^{2+}-dependent transactivation of ERR as calmodulin can bind to ERRγ (14). These studies suggest that ERRs could modulate or crosstalk with the classical E-signaling pathway mediated by ERs in regulating physiological activities in normal and cancer cells.

Based on this background, we are interested to investigate the expression pattern of ERRs in various human prostatic cell lines and prostatic tissues in order to reveal whether these receptors have any functional implication and roles in the prostate and prostate cancer.

Materials and Methods

Cell Lines, Xenografts and Human Prostatic Tissues. Four human immortalized prostatic epithelial cell lines (HPr-1, HPr-1AR, PNT1A, PNT2) and seven prostatic cancer lines (CA-HPV-10, CWR22Rv1, LNCaP, C4-2B, DU 145, PC-3, MDA PCa 2b) were used in this study. Total RNA and protein were extracted from cells when they reached 80-90% confluence. Human primary prostate cancer xenograft

CWR22 was grown in male nude mice supplemented with either one slow-release testosterone pellet, while an androgen-independent subline, CWR22R, was grown in castrated nude mice. Benign hyperplastic tissues were obtained from BPH patients by TURP whereas neoplastic tissues were obtained by radical prostatectomy or ultrasound-guided needle biopsy from prostate cancer patients.

RT-PCR. About 5μg DNase I-treated total RNA samples extracted from cultured cells, xenografts and human prostatic tissues were reverse transcribed to cDNAs using Oligo(dT) primer and 1 μl of cDNA samples was used for PCR reaction.

Immunohistochemistry and Western Blot Analysis. A rabbit antiserum against ERRγ, generated by a synthetic peptide of mERRγ conjugated to KLH (15), was used for immunohistochemistry and Western blot analysis.

Results

The results of RT-PCR showed that all immortalized prostatic epithelial cell lines and cancer cell lines, and two CWR22 xenografts expressed one to three isoforms of ERRs (α, β, γ) (Figure 1). Among the three ERR isoforms, ERRα transcripts were expressed in all cell lines and CWR22 xenografts, with weaker expression in DU 145 and PC-3. ERRβ transcripts were weakly detected in the four immortalized cell lines (HPr-1, HPr-1AR, PNT1A, PNT2), three cancer cell lines (LNCaP, PC-3, MDA PCa 2b) and CWR22R xenograft. ERRγ transcripts were expressed variably in different cell lines and CWR22 xenografts, with higher levels in C4-2B, PC-3, CWR22, and CWR22R xenografts.

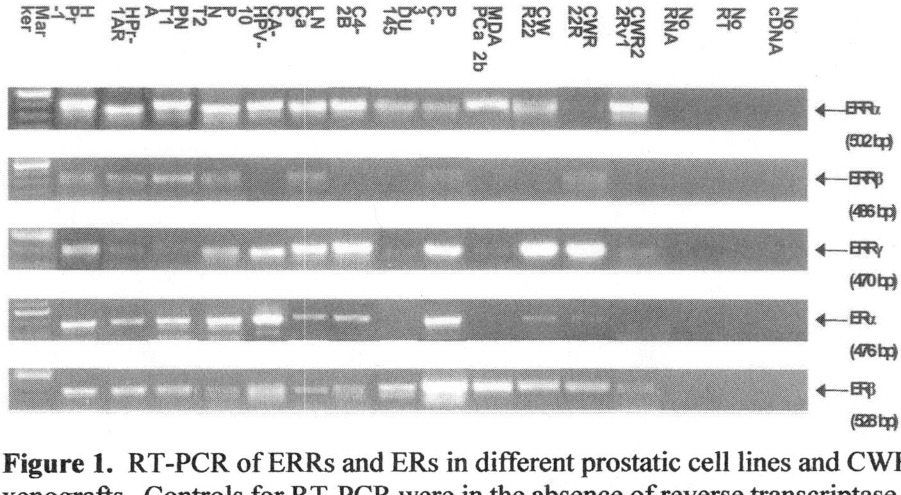

Figure 1. RT-PCR of ERRs and ERs in different prostatic cell lines and CWR22 xenografts. Controls for RT-PCR were in the absence of reverse transcriptase (no RT), total RNA (no RNA) and cDNA templates (no cDNA).

ERRs also exhibited similar patterns in human hyperplastic and neoplastic prostatic tissues as shown in cell lines and xenografts. ERRα transcripts were detected in all hyperplastic and neoplastic samples, whereas ERRβ and ERRγ were variably expressed in hyperplastic tissues but weakly expressed or absent in neoplastic tissues. The gene expression of other steroid hormone receptors in these cell lines and xenografts was also examined and compared to ERRs. In summary, most cell lines and the two CWR22 xenografts positively expressed ERα (except DU 145, MDA PCa 2b, CWR22Rv1), ERβ (Figure 1), androgen receptor (except PNT2, CA-HPV-10, DU 145), progesterone receptor (except HPr-1, HPr-1AR, PNT2, CA-HPV-10, PC-3), and glucocorticoid receptor. The results of gene expression of ERRs and different steroid hormone receptors as detected by RT-PCR are summarized in Table 1.

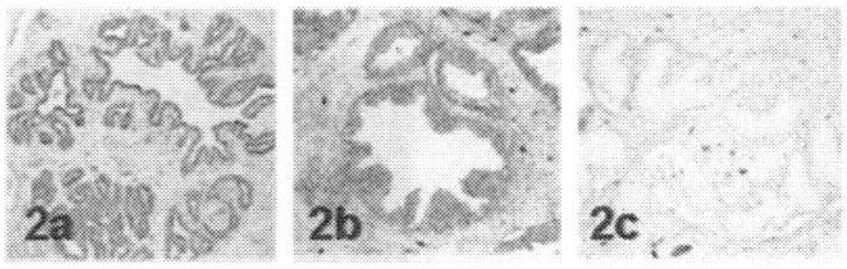

Figure 2. Immunohistochemistry of ERRγ in normal human prostate. **2a and 2b:** The glandular epithelia show positive immunosignals (2a, X100; 2b, X200). **2c:** Control section stained with blocking solution with 3 % BSA (X200).

Table 1: Summary of Gene Expression of ERRs and Steroid Hormone Receptors in Prostatic Cell Lines and Xenografts[1].

Cell lines and tumor xenografts	ERRα	ERRβ	ERRγ	ERα	ERβ	AR	PR	GR
HPr-1	+	+	+	++	++	+	-	+++
HPr-1AR	+	+	+	+	++	+++	-	+++
PNT1A	++	+	+	+	++	+	++	+++
PNT2	++	+	+	++	+	-	-	+++
CA-HPV-10	++	-	++	++	++	-	-	+++
LNCaP	+	+	++	+	+	++	++	++
C4-2B	+	-	+++	+	+	++	+++	+++
DU 145	+	-	-	-	+	-	++	+++
PC-3	+	+	++	++	+	+	-	+++
MDA PCa 2b	++	++	-	-	+++	+++	+	+++
CWR22	+	-	+++	+	++	+++	++	+++
CWR22R	+	+	+++	+	++	+++	+	+++
CWR22Rv1	++	-	-	-	+	++	+	++

[1] Expression levels are graded in arbitrary scales: (-) Negative. (+) Weakly expressed. (++) Moderately expressed. (+++) Highly expressed.

Figure 3. Western blot analysis of ERRγ in prostatic cell lines and tumor xenografts. The antiserum recognizes a band of 51 kDa in LNCaP and two CWR22 xenografts but not in DU 145 and PC-3.

Discussion

In this report, we demonstrated the expression patterns of ERRs in various human prostatic epithelial cell lines, CWR22 xenografts and prostatic tissues. Among the three ERR isoforms, ERRα transcripts were expressed in all tested cell lines, CWR22 xenografts and human prostatic tissues, whereas ERRβ and ERRγ showed variable expression patterns. Immunohistochemistry of ERRγ demonstrated that immunoreacted proteins were mainly localized to the glandular epithelia of normal prostate. The wide expression of ERRα in prostatic cell lines and xenografts suggests that ERRα expression could be important to the prostatic cells. An examination of its relative expression levels as shown by RT-PCR in different cell lines shows that ERRα is expressed at higher levels in the four normal immortalized cell lines and an immortalized primary prostatic carcinoma cell line, CA-HPV-10, and at lower levels in four cancer cell lines, which are established from prostatic carcinoma metastasized to secondary sites. Its expression level was low in two androgen-independent lines, DU 145 and PC-3. Furthermore, RT-PCR of ERRα showed a similar expression pattern in prostatic tissues that ERRα was expressed at higher levels in hyperplastic tissues than in neoplastic tissues, suggesting that ERRα could be down-regulated in the progression of prostate cancer. It is also noted that ERRα transcripts were expressed at high levels in two androgen-independent and bone metastatic prostatic cancer cells, MDA PCa 2b and C4-2B. MDA PCa 2b was established from an androgen-independent bone metastatic prostate cancer patient, whereas C4-2B was derived from LNCaP cells in metastatic lesions in lumbar spine of castrated nude mice, both of which express androgen receptor. However, its expression was low in another bone metastatic and androgen-independent line, PC-3, of which androgen receptor is expressed at low level.

Compared to ERRα and ERRγ, ERRβ showed weak expression in four normal immortalized cell lines and MDA PCa 2b, and barely detected in LNCaP, PC-3 and CWR22R xenografts, but negative in other lines, suggesting that its expression is down-regulated in tumor progression but transcribed in bone metastastic cells. It is uncertain whether the expression of ERRα and ERRβ in two bone metastatic lines, MDA PCa 2 b and PC-3, is related to bone metastasis of prostate cancer.

On the other hand, ERRγ is weakly expressed in the four normal immortalized cell lines and undetected in three androgen-independent lines, CWR22Rv1, DU 145 and MDA PCa 2b. High expression level of ERRγ was detected in two CWR22 xenografts, C4-2B and PC-3. Its variable expression in different prostatic cancer cell lines reflects the clonal and genetic heterogeneity of the tumors.

Since ERRs are closely related to ERs, we also examined and compared their mRNA expression patterns in prostatic cells. We observed that ERβ was positively expressed in all tested cell lines and xenografts, and exhibited a similar expression pattern as that of ERRα. Similarly, ERα showed a similar expression pattern as that of ERRγ in the tested cell lines. The results suggest that members of both nuclear receptors, which are different in their ligand dependence, are co-expressed in the prostatic cells and they could crosstalk between one another or share similar regulatory pathways through competition for DNA binding sites and coregulatory proteins. However, the regulatory roles of ERRs in prostatic cells are still unclear and require further studies.

Acknowledgements

This study was supported by a Direct Grant for Research 2002.1.005 from the Research Grants Council of Hong Kong to FL Chan.

References

1. Griffiths K (2000) Estrogens and prostatic diseases. International Prostate Health Council Study Group. Prostate 45:87-100.
2. Cunha GR, Wang YZ, Hayward SW, et al (2001) Estrogenic effects on prostatic differentiation and carcinogenesis. Reprod Fertil Dev 13:285-296.
3. Jarred RA, McPherson SJ, Bianco JJ, et al (2002) Prostate phenotypes in estrogen modulated transgenic mice. Trends Endocrinol Metab 13:163-168.
4. Giguère V, Yang N, Segui P, et al (1988) Identification of a new class of steroid hormone receptors. Nature 331:91-94.
5. Bonnelye E, Vanacker JM, Dittmar T, et al (1997) The ERR-1 orphan receptor is a transcriptional activator expressed during bone development. Mol Endocrinol 11:905-916.
6. Johnston SD, Liu X, Zuo F, et al (1997) Estrogen-related receptor α-1 functionally binds as a monomer to extended half-site sequences including ones contained within estrogen-response elements. Mol Endocrinol 11:342-352.
7. Zhang Z, Teng CT (2000) Estrogen receptor-related receptor α1 interacts with coactivator and constitutively activates the estrogen response elements of the human lactoferrin gene. J Biol Chem 275:20837-20846.
8. Chen S, Zhou D, Yang C, et al (2001) Molecular basis for the constitutive activity of estrogen-related receptor α-1. J Biol Chem 276:28465-28470.

9. Greschik H, Wurtz JM, Sanglier S, et al (2002) Structural and functional evidence for ligand-independent transcriptional activation by the estrogen-related receptor 3. Mol Cell 9:303-313.
10. Xie W, Hong H, Yang NN, et al (1999) Constitutive activation of transcription and binding of coactivator by estrogen-related receptors 1 and 2. Mol Endocrinol 13:2151-2162.
11. Zhou D, Quach KM, Yang C, et al (2000) PNRC: a proline-rich nuclear receptor coregulatory protein that modulates transcriptional activation of multiple nuclear receptors including orphan receptors SF1 (steriodogenic factor 1) and ERRα1 (estrogen related receptor α-1). Mol Endocrinol 14:986-998.
12. Huss JM, Kopp RP, Kelly DP (2002) Peroxisome proliferator-activator receptor coactivator-1α (PGC-1α) coactivates the cardiac-enriched nuclear receptors estrogen-related receptors-α and -γ. J Biol Chem 277:40265-40274.
13. Zhang Z, Teng CT (2001) Estrogen receptor α and estrogen receptor-related α1 compete for binding and coactivator. Mol Cell Endocrinol 172:223-233.
14. Hentschke M, Schulze C, Süsens U, Borgmeyer U (2003) Characterization of calmodulin binding to the orphan nuclear receptor ERR γ. Biol Chem 384:473-482.
15. Hentschke M, Süsens U, Borgmeyer U (2002) Domains of ERRγ that mediate homodimerization and interaction with factors stimulating DNA binding. Eur J Biochem 269:4086-4097.

Immunohistochemical and *In-Situ* Detection of Sex Hormone-binding Globulin (SHBG) Expression in Breast and Prostate Cancer: Implications for Hormone Regulation

Scott M. Kahn, Daniel J. Hryb, Atif M. Nakhla, Saeed M. Khan, Nicholas A. Romas, and William Rosner

Summary

Sex Hormone Binding Globulin (SHBG), in addition to regulating free concentrations of steroid sex hormones in plasma, also participates in membrane-based steroid signaling in breast and prostate cells. In this study, we address whether the breast and prostate can synthesize their own SHBG. SHBG mRNA and protein were detected in human breast and prostate cancer cell lines, as well as in normal and cancerous breast and prostate tissue where it was prominent in epithelial cells. In cultured human LNCaP prostate cancer cells, SHBG was found both in the cytoplasm, and on the outer cell membrane where it was presumably bound to its high affinity receptor (R_{SHBG}). When LNCaP cells were treated with 2-methoxyestradiol ($2MeOE_2$), an antagonist of SHBG binding to R_{SHBG}, membrane-associated SHBG was eliminated. These results demonstrate that the breast and prostate can synthesize SHBG locally, and that SHBG can bind to, and perhaps participate in autocrine and/or paracrine signaling through R_{SHBG}. Therefore, perturbations in SHBG expression in breast and prostate tumors may affect the regulation of steroid signaling through R_{SHBG} and target SHBG-mediated biologic properties at the cellular level.

Introduction

SHBG is a multifunctional protein that regulates steroid hormone signaling in prostate. Located on chromosome 17p13.1 (Fig. 1), only 30 kb away from the p53 gene (1, 2), the SHBG gene locus undergoes allelic deletions in tumors from many tissue types, including the breast and prostate. SHBG was initially described as a hepatic-derived protein (3), that is secreted into the plasma, where it binds sex steroids such as estradiol (E_2) and testosterone (T) (4, 5). Acting as a chaperone in plasma, SHBG establishes a dynamic equilibrium between free and bound steroids, presumably regulating their bioavailability. Researchers have investigated the relationship between plasma SHBG concentrations and cancer risk for hormone responsive tissues such as the breast and prostate. Certain studies, including data presented in these proceedings, suggest an inverse relationship between plasma SHBG concentration and cancer risk (6-8), although the differences in plasma SHBG are usually slight. This relationship has not been extended to all studies (9-12). If true, plasma SHBG might protect against breast and prostate cancer by sequestering free steroids, or by mechanisms related to additional functions attributed to SHBG.

Figure 1. Close linkage of SHBG and p53 genes on chromosome 17p13.1. (Ref.2)

SHBG is now known to be part of a membrane based steroid signaling system that is found in certain hormone responsive tissues (13, 14), including the breast (15) and prostate (16), where it influences gene transcription and cell growth (17, 18). As shown in Figure 2, unliganded SHBG binds to a specific receptor (R_{SHBG}) on the outer cell membrane (19-22) (steroid-bound SHBG is not an appropriate ligand, presumably due to induced steric changes within the protein). The inactive SHBG-R_{SHBG} complex is activated by subsequent binding of a steroid agonist, such as E_2 or T, which is rapidly followed by the generation of cAMP, and activation of downstream elements of the PKA pathway. R_{SHBG} has yet to be cloned, though data suggest it is a G-protein coupled receptor (23). Downstream effects of activated R_{SHBG} appear to be context dependent. For example, R_{SHBG} signaling in prostate cell lines (LNCaP), induces PSA gene transcription (24) and enhances growth (ALVA-41) (17); in breast, it inhibits growth (MCF-7) (18). The isolation and characterization of R_{SHBG} will allow further elucidation of its biologic properties.

The rate limiting step in R_{SHBG} signaling is not yet known. It could involve R_{SHBG} expression in target cells, local concentrations of free SHBG, or availability of steroid. The known effects of R_{SHBG} signaling (cAMP induction, modulation of cell growth) suggest that activation of this pathway must be rapidly inducible. It is highly unlikely that tissues would rely on changes in plasma SHBG concentrations to induce R_{SHBG}, because plasma SHBG concentrations are relatively stable. Setting aside the immediacy question, we were concerned that global changes in plasma SHBG concentrations could not specifically target only those tissues requiring induction of R_{SHBG}. We therefore addressed the scenario that breast and prostate

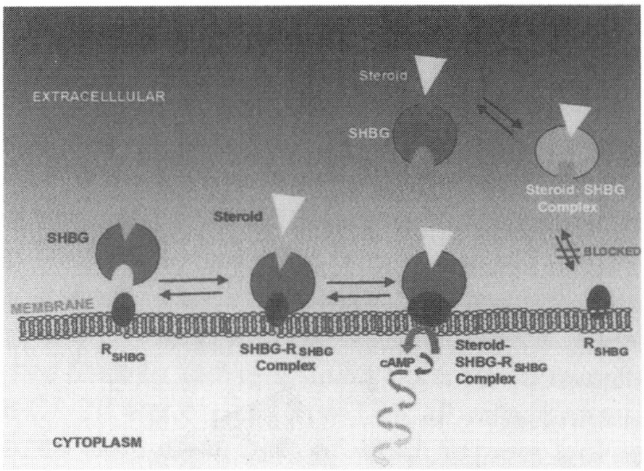

Figure 2. Steroid signaling through the SHBG/R_{SHBG}-Complex – the initial steps.

cells might synthesize their own SHBG for local use. If so, tissues could have direct control over R_{SHBG} activation through regulation of endogenous SHBG gene activity via autocrine or paracrine mechanisms. Here, we report that prostate stromal and epithelial cell explants and breast and prostate cell lines do express SHBG, and that normal human prostate and breast and tumor tissues express both SHBG mRNA and protein (25). Our findings support a model in which the breast and prostate can regulate hormone responses intracellularly and/or through stromal-epithelial cell interactions by the modulation of endogenously produced SHBG. Furthermore, perturbations in SHBG expression in breast and prostate cancer, including frequent intrachromosomal deletions that involve the SHBG gene locus, could contribute to their malignant phenotype.

Results

Breast and Prostate Cells Express SHBG mRNA. SHBG mRNA expression was examined in a panel of human prostate stromal cell explants, epithelial and cancer cell lines (Fig. 3), and in MCF7 breast cancer cell line (data not shown). All expressed SHBG mRNA, was determined by RT-PCR analysis. SHBG mRNA expression was also observed in samples from human liver and prostate, and HepG2 liver cancer cell line (known to express SHBG mRNA). The primers used for RT-PCR spanned exons 5-8 of the SHBG gene, and a smaller SHBG species was detected in all samples (other than TrHPSC 3, a stromal cell explant transiently transfected with a full-length SHBG expression vector). Sequence analysis showed the smaller RT-PCR band lacked exon 7 (data not shown). The exon 5-8 species is consistent with the cDNA for hepatically secreted SHBG. The shorter species is consistent with one previously described for a poorly characterized transcript originally described in testis, and postulated to remain intracellular. These results confirm prior reports on SHBG expression in liver and HepG2 cells, and demonstrate SHBG expression in breast and human prostate cancer cell lines and explants.

Human Breast and Prostate Cancer Cell Lines Express SHBG Protein. To investigate SHBG protein expression, immunohistochemical analysis was performed on LNCaP, PC3, DU145 (Fig. 4), and MCF-7 cells (data not shown). All cell lines showed cytoplasmic SHBG staining. Under nonpermeabilized conditions,

LNCaP and MCF7 cells showed membrane SHBG staining (Fig. 5, top left, and data not shown) (PC3 and DU145 have not yet been tested). Confocal microscopy (Fig. 5) and colocalization studies (data not shown) confirmed the presence of membrane SHBG. Control experiments, not shown, ruled out the medium as a source of SHBG. These results demonstrate that LNCaP, PC3, DU 145 and MCF-7 cells synthesize and secrete their own SHBG.

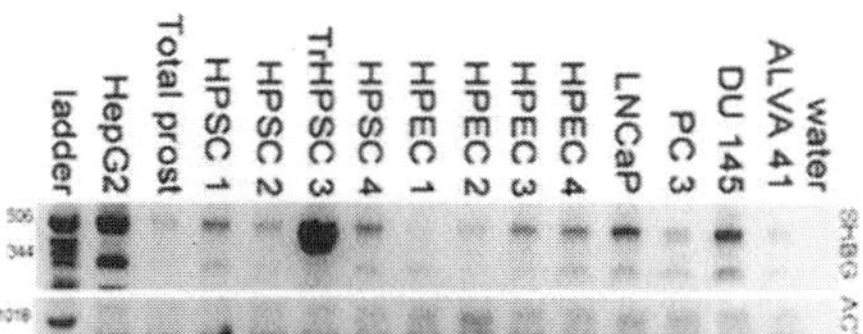

Figure 3. SHBG mRNA expression in prostate explants and cell lines. RT-PCR results are shown for the expression of SHBG and β-actin (25). First strand, cDNAs were prepared from total human liver, HepG2 human liver cancer cells, LNCaP prostate cancer cells, whole human prostate tissue (total prost), human prostate stromal cell explants (HPSC1, 2, 4), TrHPSC 3, a stromal explant transiently transfected with an SHBG expression vector, human prostate epithelial cell lines (HPEC1-4), LNCaP, PC 3, DU 145, and ALVA 41, human prostate cancer cell lines, and the human MCF-7 breast cancer cell line (not shown). The primers, SHBG forward and SHBG reverse, respectively: 5'-ACTCAGGCAGAATTCAATCTC-3' and 5'-CTTTAATGGGAAGCGTCAGT-3', directed amplification of a larger 521 bp fragment containing sequences from exons 5-8 in all samples tested, and a smaller 313 bp $SHBG_T$ fragment lacking exon 7 (30 cycles) in all except for TrHPSC 3 [due to the overexpression of the SHBG cDNA-HPSC 3 itself shows both species (data not shown)].

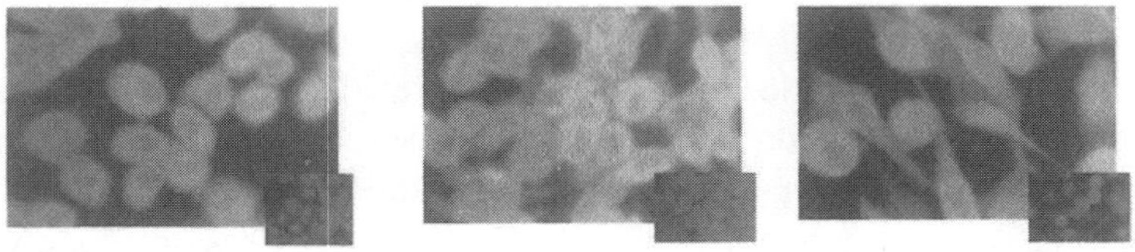

Figure 4. Expression of SHBG protein in DU145, LNCaP, and PC3 cells. Photomicrographs (400 x) of permeabilized DU145 (left), LNCaP (center), and PC3 (right) prostate cancer cells incubated with a monoclonal antibody to SHBG and developed with a rabbit anti-mouse IgG1 linked to the fluor, Alexa-488 (25). Nuclei were stained with DAPI (dark regions inside cells). Similar cytoplasmic staining was seen with MCF-7 cell line (data not shown). Insets- mouse IgG1 control.

2-Methoxyestradiol ($2MeOE_2$), a Steroid Antagonist of SHBG Binding to R_{SHBG}, Prevents SHBG from Associating with LNCaP Cell Membranes. In the current model of R_{SHBG} activation (Fig. 2), SHBG, pre-bound to a steroid, is unable to bind to R_{SHBG}. We tested this model by incubating LNCaP cells with an excess of $2MeOE_2$, an antagonist of SHBG binding to R_{SHBG}. $2MeOE_2$ should prevent both endogenously synthesized SHBG and exogenously added SHBG from localizing to R_{SHBG} on LNCaP membranes. LNCaP cells were grown in serum-free medium for three days on glass chamber slides, and then incubated for 16 h with either carrier alone (EtOH), highly purified SHBG (100 nM), or highly purified SHBG along with $2MeOE_2$ (1 μM). In the absence of added SHBG and steroid, endogenously synthesized SHBG was located both in the cytoplasm and at the membrane (Fig. 5, left). When highly purified SHBG was added, more intense cytoplasmic and membrane staining was seen (Fig. 5, middle). When SHBG and an excess of $2MeOE_2$ were added, membrane binding was eliminated, while little effect was seen on cytoplasmic staining. These results are consistent with the model in Figure 2.

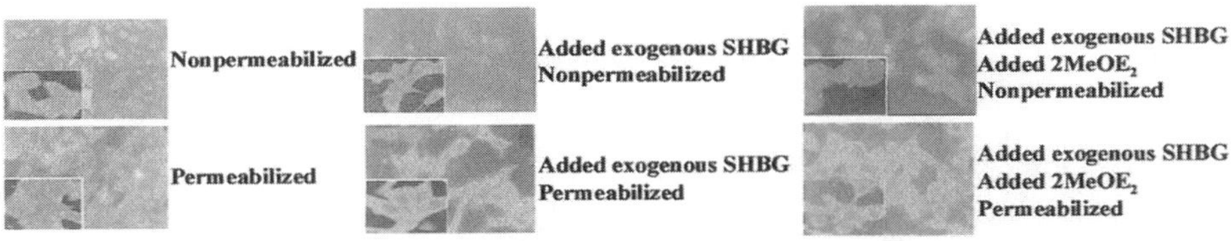

Figure 5. Incubation with steroid displaces SHBG from LNCaP cell membranes. LNCaP cells were plated on glass chamber slides, incubated in serum free medium, then exposed to either EtOH solvent (left panel), highly purified SHBG (100 nM) (middle panel), or SHBG + $2MeOE_2$ (1.0 μM) for 16 h (25). Cells were fixed either in the absence (top row) or presence (bottom row) of 0.1% Triton X-100 to prevent or allow permeabilization of a monoclonal antibody against SHBG (5B2). Insets are confocal micrographs lacking a nuclear counterstain. Staining of permeabilized cells represents both intra- and extracellular SHBG. Staining of non-permeabilized cells represents SHBG outside of the cell, inferentially bound to R_{SHBG}. Free SHBG binds LNCaP cell membranes (top, middle), pre-bound SHBG cannot (top, right). Mouse IgG1 control gave no staining (not shown). Original color figures in reference 25.

SHBG is Expressed Predominantly in Prostate Epithelial Cells in Intact Tissues. *In-situ* hybridization and immunohistochemistry were performed on tissue sections obtained from normal human prostate tissue and prostate tumor tissue (data not shown) (25). In normal prostate, SHBG mRNA expression predominates in luminal epithelial cells, with weak expression in stromal cells. SHBG protein is similarly expressed more intensely, and perhaps with polarity in cells lining the ducts. SHBG protein expression appears to be abundant in cancerous areas of prostate tumor samples that we have examined. Similar SHBG expression patterns were seen in sections from normal human breast tissue and breast tumors.

Conclusions

Our results show breast and prostate cells in culture synthesize and express SHBG. Furthermore, SHBG mRNA and protein are visualized in breast and prostate tissue sections. Taken together, these results suggest that the breast and prostate, two quintessential sex steroid target tissues, can and do synthesize SHBG (the fraction of immunoreactive SHBG in breast and prostate tissue that is internalized from outside sources, if any, is not yet clear). Though its biologic role in breast and prostate is not fully understood, it is reasonable to hypothesize that locally synthesized SHBG binds intracellular steroids, and may (as the major source?) participate in R_{SHBG} signaling. Local SHBG expression must be aptly regulated if it participates in rapid response pathways. These results raise important new questions, does locally expressed SHBG 1) affect intracellular steroid signaling pathways; 2) act in an autocrine/paracrine manner through R_{SHBG}; 3) participate in crosstalk between epithelial and stromal cells, as SHBG is predominantly expressed in the former and R_{SHBG} is predominantly expressed in the latter? Furthermore, do perturbations of SHBG expression in cancer cells, through allelic deletions associated with the SHBG/p53 locus on chromosome 17p13.1 in breast and prostate tumors, contribute to the malignant phenotype? If so, can agonists or antagonists of R_{SHBG} signaling serve as useful therapeutic agents? In summary, regulation of SHBG synthesis in target cells offers new possibilities for the local modification of steroid hormone effects.

Acknowledgments

These studies were supported by awards from the Starr Foundation and the Loewy Foundation, and DOD award #DAMD17-02-1-0572 (to WR).

References

1. Bérubé D, Séralini G, Gagné R, et al (1990) Localization of the human SHBG gene to the short arm of chromosome 17. Cytogenet Cell Genet 54:65-67.
2. April, 2003 UCSC Human Genome Project Working Draft-http://genome.ucsc.edu/cgi-bin/hgTracks?positionX16351&hgsid=22396239
3. Khan MS, Knowles BB, Aden D, Rosner W (1981) Secretion of testosterone-estradiol-binding globulin by a human hepatoma-derived cell line. J Clin Endocrinol Metab 53:448-449.
4. Rosner W, Christy NP, Kelly WG (1966) Evidence for the presence of an estrogen-binding β–globulin in plasma. J Clin End Metab 26:1399-1403.
5. Rosner W, Smith RN (1975) Isolation and characterization of the testosterone-estradiol-binding globulin from human plasma. Use of a novel affinity column. Biochemistry 14:4813-4820.
6. Lipworth L, Adami HO, Trichopoulos D, et al (1996) Serum steroid hormone levels, SHBG, and body mass index in the etiology of postmenopausal breast cancer. Epidemiology 7:96-100.
7. Endogenous Hormones and Breast Cancer Collaborative Group (2002) Endogenous sex hormones and breast cancer in postmenopausal women: reanalysis of nine prospective studies. J Natl Cancer Inst 94:606-616.
8. Gann PH, Hennekens CH, Ma J, et al (1996) Prospective study of sex hormone levels and risk of prostate cancer. J Natl Cancer Inst 88:1118-26.

9. Garland CF, Friedlander NJ, Barrett-Conner E, et al (1992) Sex hormones and postmenopausal breast cancer: a prospective study in an adult community. Am J Epidemiol 135:1220-1230.
10. Helzlsouer KJ, Alberg AJ, Bush TL, et al (1994) A prospective study of endogenous hormones and breast cancer. Cancer Detect Prev 18:79-85.
11. Eaton NE, Reeves GK, Appleby PN, et al (1999) Endogenous sex hormones and prostate cancer. Br J Cancer 80:930-934.
12. Heikkila R, Aho K, Heliovaara M, et al (1999) Serum testosterone and SHBG concentrations and the risk of prostate carcinoma. Cancer 86:312-315.
13. Hryb DJ, Khan MS, Romas NA, et al (1990) The control of the interaction of SHBG with its receptor by steroid hormones. J Biol Chem 265:6048-6054.
14. Kahn S, Hryb DJ, Nakhla AM, et al (2002) SHBG is synthesized in target cells. J Endocrinol 175:113-120.
15. Fissore F, Fortunati N, Comba A, et al (1994) The receptor-mediated action of sex steroid binding protein (SBP, SHBG): accumulation of cAMP in MCF-7 cells under SBP and estradiol treatment. Steroids 59:661-667.
16. Nakhla AM, Khan MS, Rosner W (1990) Biologically active steroids activate receptor-bound human SHBG to cause LNCaP cells to accumulate adenosine 3',5'-monophosphate. J Clin Endo Metab 71:398-404.
17. Nakhla A, Rosner W (1996) Stimulation of prostate cancer growth by androgens & estrogens through intermediacy of SHBG. Endocrinology 137:4126-4129.
18. Fortunati N, Fissore F, Fazzari A, et al (1996) Sex steroid binding protein exerts a negative control on estradiol action in MCF-7 cells through cyclic adenosine 3',5'-monophosphate and PKA. Endocrinology 137:686-692.
19. Hryb DJ, Khan MS, Rosner W (1985) Testosterone-estradiol-binding globulin binds to human prostatic cell membranes. BBRC 128:432-440.
20. Hryb DJ, Khan MS, Romas NA, et al (1990) The control of the interaction of SHBG with its receptor by steroid hormones. J Biol Chem 265:6048-6054.
21. Frairia R, Fortunati N, Berta L, et al (1991) Sex steroid binding protein (SBP) receptors in estrogen sensitive tissues. J Ster Biochem Mol Biol 40:805-812.
22. Porto CS, Musto NA, Bardin CW, et al (1992) Binding of an extracellular steroid-binding globulin to membranes and soluble receptors from human breast cancer cells (MCF-7 cells). Endocrinology 130:2931-2936.
23. Nakhla AM, Leonard J, Hryb DJ, et al (1999) Sex hormone-binding globulin receptor signal transduction proceeds via a G protein. Steroids 64:213-216.
24. Nakhla A, Romas N, Rosner W (1997) Estradiol activates the prostate AR and PSA secretion through intermediacy of SHBG. J Biol Chem 272:6838-6841.
25. Hryb DJ, Nakhla AM, Kahn, S, et al (2002) SHBG in the human prostate is locally synthesized and may act as an autocrine/paracrine effector. J Biol Chem 277:26618-26622.

Activation of Androgen Receptor in Prostate Cancer: Role of Protein Kinase A and Extracellular Signal-regulated Kinases

Yehia Daaka, Elizabeth Kasbohm, and Charles Yowell

Summary

The androgen receptor (AR) is a ligand-controlled transcription factor that is required for normal development of the prostate. It is expressed in both A-dependent and -independent prostate cancer (PCA) cells. Emerging data suggest AR regulates growth of PCA cells even in the absence of As, implying it is activated by factors other than As. Herein, we show that stimulation of A-dependent PCA LNCaP cells with reagents that promote accumulation of cyclic adenosine monophosphate (cAMP) leads to protein kinase A (PKA)-dependent activation of AR. Importantly, dihydrotestosterone (DHT)-regulated activation of AR also required PKA. Stimulation with epidermal growth factor, that robustly activates extracellular signal-regulated kinase pathway, did not promote activation of AR. These data establish a critical role for PKA in AR activation and suggest it may contribute to the growth of A-independent prostate tumors.

Introduction

PCA is the most diagnosed cancer in men, and is the second leading cause of male cancer deaths in the USA (1). PCA is multifocal, and the cancerous gland contains multiple independent tumors, demonstrating disease heterogeneity (2). A unique PCA characteristic is its initial dependence on As for growth. As a result, standard strategies for PCA treatment rely on blockade of A action by: 1. Decreasing circulating A levels via chemical or physical castration; or 2. Treatment with As (*i.e.,* DHT) that compete for binding to the AR (3, 4). The major limitation of this approach, however, is that the PCA frequently develops into A-independent (AI) lesions, characterized by recurrent growth and metastasis even in the continued presence of the hormonal therapies. To date, there is no cure for AI PCA (5).

Mechanisms involved in the transition of PCA from A-dependent (AD) to AI remain poorly defined, although recent findings suggest a role for AR (6). First, AR is overexpressed in up to one third of AI PCAs, suggesting a compensatory mechanism by which the AR adjusts its physiologic rheostat to respond to lower serum levels of As (3, 7). Second, AI lesions exhibit frequent AR mutations, which may allow for its activation by other As or even anti-As (6). Third, AR can be activated by growth factors other than As (8). In all cases, activated AR promotes cell growth and inhibits cell death by regulating expression of specific genes.

In addition to becoming active by binding As, new evidence suggests AR can support transcription in response to growth factor stimulation. For example, in DU145 cells ectopically expressing human AR, stimulation of endogenous insulin-like growth factor 1 (IGF-1) receptors activates transcription of A response element (ARE)-regulated reporter genes (9). Similarly, in LNCaP cells, IGF-1 can stimulate secretion of ARE-regulated prostate specific antigen (PSA) in absence of As (10). Overexpression of the receptor tyrosine kinase (RTK) HER2/neu (11), as well as treatment with forskolin (12, 13) induces PSA expression. Mechanisms involved in the growth factor-regulated activation of AR are not fully understood. Herein, we show that stimulation of LNCaP cells with reagents that increase levels of intracellular 3',5'-cAMP induces protein kinase A (PKA)-dependent activation of AR. Importantly, we show that inhibition of PKA activity abrogates the DHT-mediated activation of AR. We also provide evidence that stimulation of the LNCaP cells with epidermal growth factor (EGF) does not support AR activation.

Materials and Methods

Cell Culture. AD human PCA LNCaP cells (ATCC) were cultured in RPMI 1640 containing 10% fetal bovine serum (FBS) and antibiotics, as described (14). Cells were transiently transfected using DMRIE-C reagent (In Vitrogen) and allowed to recover overnight. Cells were incubated in phenol red-free medium supplemented with 5% charcoal-stripped FBS (starvation medium) and exposed to agonists, as indicated. Where appropriate, cells were pretreated with inhibitors (H89, 30 μM; PD98059, 250 nM; U0126, 1 μM) for 30 min prior to exposure to agonists.

ERK Activation Assay. Cells were incubated overnight in starvation medium, exposed to agonist for 5 min at 37°C, washed once with ice-cold phosphate buffered saline (PBS), and lysed in Laemmli sample buffer. Cell lysates were subjected to protein electrophoresis on 4-20% SDS-polyacrylamide gels and transferred onto nitrocellulose membranes. Phospho-ERK was detected using a 1:2,000 dilution of a rabbit polyclonal phospho-ERK (Cell Signaling) specific antibody, and total ERK2 was detected using a 1:2,000 dilution of ERK2 (Cell Signaling) antibody. Blots were developed with a 1:7,000 dilution of horse-radish peroxidase-conjugated secondary antibody and specific protein bands visualized using enhanced chemiluminescence (Amersham).

Luciferase Assay. LNCaP cells transiently expressing luciferase gene fused to ARE-containing promoter (ARR3-luc/PB3) were grown in phenol red-free RPMI containing 2% charcoal-stripped serum overnight, and lysed in luciferase-lysis buffer (Promega). Protein concentrations were determined by the Bradford method. Relative luciferase units (RLU) were normalized to sample protein concentration. Results are presented as fold induction, the relative luciferase activity of transfected cells over the control cells.

Results and Discussion

Recently, we demonstrated that activated ERK is involved in the agonist-dependent growth of AI PCA cells (15). To begin to assess the potential role of ERK in the mitogenic signaling and growth of AD PCA cells, we used LNCaP cells stimulated with isoproterenol (ISO), EGF, DHT, or forskolin (FSK). The growth factors were chosen because they activate representative members of the three major classes of receptor, namely G protein-coupled receptors (ISO), receptor tyrosine kinases (EGF), and steroid receptors (DHT). Figure 1A shows that while exposure to EGF can robustly induce ERK phosphorylation, treatment with ISO or DHT did not lead to ERK activation. These results are distinct from those reported by Ueda, *et al.*, who showed that stimulation with the synthetic androgen R1881 (16) or FSK (17) induces ERK activation.

The LNCaP cells are dependent for their growth on As. We determined the effect of growth factors on AR activation using a reporter construct made of three congruent rat probasin AREs positioned upstream of the firefly luciferase gene (Prb-luc; 18). Cells were transfected with the cDNA encoding Prb-luc. Luciferase activity was determined after overnight stimulation with ISO, FSK, EGF, or DHT. Figure 1B shows that stimulation with ISO and FSK, but not EGF, induced ~10.0-fold increase in reporter luciferase activity, compared to untreated samples. DHT treatment (used as positive control) promoted a 20.0-fold increase in ARE-reporter activity. Together, these data demonstrate distinct mitogenic effects for ISO/FSK and EGF on the LNCaP cells. Whereas EGF elicits its mitogenic effects by activating ERK, ISO and FSK exert their mitogenic effects by activating the AR.

Recent reports suggest that ERK is required for AR activation (16, 17). However, our data clearly show that ISO and DHT promote AR activation, but do not exert measurable increase in ERK phosphorylation (Figure 1). To directly probe contribution of ERK to AR activation, LNCaP cells were pretreated with ERK pathway inhibitors PD98059 or U0126, followed by overnight stimulation with an agonist. Data shown in Figure 2A demonstrate that inhibition of ERK activation does not interfere with ISO- or DHT-mediated AR activation. Interestingly, we consistently observed that EGT treatment did not activate AR, consistent with published data (17). To ensure that the inhibitor was effective, LNCaP cells were treated with PD98059 for 30 min, exposed to agonists for 5 min, and examined the ERK phosphorylation state. Figure 2B shows that PD98059 was effective because it inhibited the robust EGF-induced phosphorylation of ERK. Together, these data demonstrate ERK does not likely participate in AR activation.

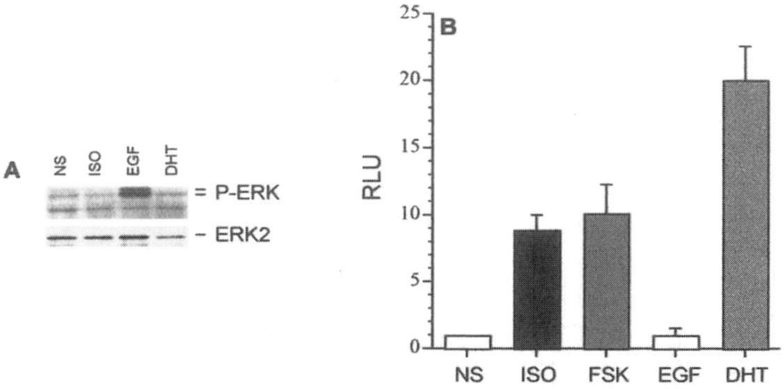

Figure 1. Effect of growth factors on LNCaP cells. **A.** Activation of ERK. LNCaP cells were incubated overnight in starvation medium and stimulated with ISO (10 μM), EGF (10 ng/ml), or DHT (10 nM) for 5 min. Cell lysates were analyzed for phospho-ERK, as described. Filters were stripped of immunoglobulins and reprobed with anti-ERK2 to ensure equal loading of proteins among the samples. **B.** Activation of AR. LNCaP cells were transiently transfected with Prb-luc, incubated overnight in starvation medium with ISO (10 μM), FSK (10 μM), EGF (10 ng/ml), or DHT (10 nM), and luciferase activity determined, as described. Relative luciferase activity was determined as fold increase over unstimulated samples. Data represent the mean ± SD of duplicates from three independent experiments.

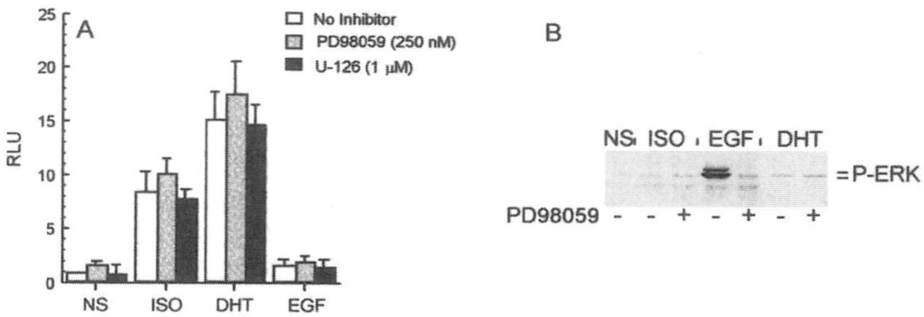

Figure 2. Role of ERK in AR activation. A. Exposure to PD98059 or U0126 does not affect AR activation. LNCaP cells were pretreated with PD98059 (250 nM) or U0126 (1 μM) for 30 min, prior to overnight stimulation with the indicated agonist. Luciferase activity was determined. The data represent mean ± SD of duplicates from three independent experiments. B. PD98059 inhibits EGF-mediated phosphorylation of ERK. LNCaP cells were incubated overnight in starvation medium, exposed to PD98059 for 30 min, treated with the indicated agonist for 5 min. ERK phosphorylation levels were determined, as described.

Recent evidence suggests that mechanisms controlling AD and AI activation of AR converge at the level of AR phosphorylation. Weber, *et al.* (19) identified six A-inducible phospho-serine sites in AR; S16, S81, S256, S308, S424, and S650. They also showed S650 to become phosphorylated in response to stimulation with FSK and EGF. ERK has been suggested to phosphorylate AR *in vitro* and to control its function (20). However, ERK does not appear to directly phosphorylate AR *in vivo* (19). Consistent with this observation is the finding that EGF treatment, which strongly activates ERK, does not cause AR activation (17). Our data support the conclusion that ERK does not play a role in AR activation.

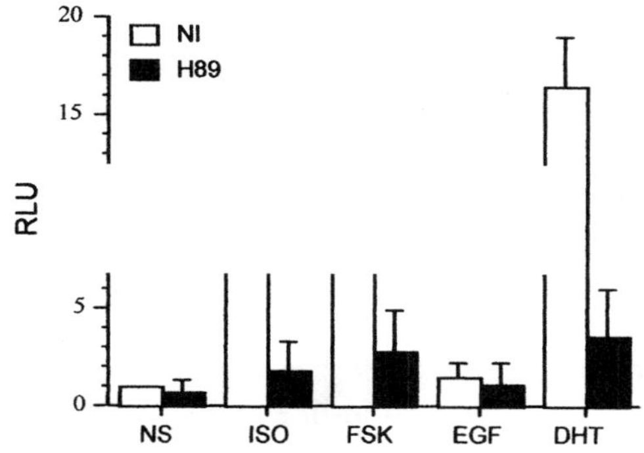

Figure 3. Effect of PKA on AR activation. LNCaP cells were transfected with cDNA encoding the Prb-luc, pretreated with H89 (30 μM) for 30 min, stimulated with ISO (10 μM), FSK (10 μM), EGF (10 ng/ml), or DHT (10 nM) for 16 h. Cells were processed for luciferase activity, as described. Data represent mean ± SD of duplicates from three independent experiments.

ISO and FSK stimulation of LNCaP cells activates AR. ISO induces its effects on target cells by activating its cognate receptor, the β adrenergic receptor (βAR), which belongs to the superfamily of G protein-coupled receptors. Ligand-bound βAR activates heterotrimeric Gαs protein, which activates adenylate cyclase to produce cAMP. FSK is a diterpenoid that directly binds adenylate cyclase and activates it to produce cAMP. Our results show that stimulation of LNCaP cells with ISO and FSK induce accumulation of cAMP.

The best characterized effector of cAMP is PKA. We inhibited PKA with H89 in order to determine its role in the agonist-mediated activation of AR using the LNCaP cells. As shown in Figure 3, treatment of LNCaP cells with 30 μM H89 abrogated the ISO- and FSK-mediated ARE-regulated gene expression. Treatment with H89 also obliterated the DHT-mediated activation of AR. In accordance with these observations is the finding that treatment of monkey kidney CV-1 cells with the PKA inhibitor peptide PKI abrogated FSK-, as well as R1881-induced ARE-regulated gene expression (13).

In conclusion, our data demonstrate a critical role for PKA in the activation of AR. Inhibition of PKA activity abrogates the A- and ISO-mediated AR transactivation. These results suggest that targeted inhibition of PKA, or its activators (*i.e.,* βAR) may prove beneficial in advanced PCA treatment.

References

1. Jemal A, Murray T, Samuels A, et al (2003) Cancer statistics, 2003. CA Cancer J Clin 53:5-26.
2. Bostwick DG (1989) Prostatic intraepithelial neoplasia (pin). Urology 34:16-22.
3. Raj GV, Barki-Harrington L, Kue PF, et al (2002) Guanosine phosphate binding protein coupled receptors in prostate cancer. J Urol 167:1458-1463.
4. Chodak GW, Keane T, Klotz L (2002) Critical evaluation of hormonal therapy for carcinoma of the prostate. Urology 60:201-208.
5. Small EJ (1998) Update on the diagnosis and treatment of prostate cancer. Curr Opin Oncol 10:244-252.
6. Grossmann ME, Huang H, Tindall, DJ (2001) Androgen receptor signaling in androgen-refractory prostate cancer. J Natl Cancer Inst 93:1687-1697.
7. Feldman BJ, Feldman D (2001) The development of androgen-independent prostate cancer. Nat Rev Cancer 1:34-45.
8. Sadar MD, Hussain M, Bruchovsky N (1999) Prostate cancer: Molecular biology of early progression to androgen independence. Endocr Relat Cancer 6:487-502.
9. Culig Z, Hobisch A, Cronauer MV, et al (1994) Androgen receptor activation in prostatic tumor cell lines by insulin-like growth factor-1, keratinocyte growth factor, and epidermal growth factor. Cancer Res 54:5474-5478.
10. Culig Z, Hobisch A, Cronauer MV, et al (1995) Activation of the androgen receptor by polypeptide growth factors and cellular regulators. World J Urol 13:285-289.
11. Craft N, Shostak Y, Carey M, et al (1999) A mechanism for hormone-independent prostate cancer through modulation of androgen receptor signaling by the HER-2/neu tyrosine kinase. Nat Med 5:280-285.
12. Sadar MD (1999) Androgen-independent induction of prostate-specific antigen gene expression via cross-talk between the androgen receptor and protein kinase a signal transduction pathways. J Biol Chem 274:7777-7783.
13. Nazareth LV, Weigel NL (1996) Activation of the human androgen receptor through a protein kinase a signaling pathway. J Biol Chem 271:19900-19907.
14. Taub JS, Guo R, Leeb-Lundberg LM, et al (2003) Bradykinin receptor subtype 1 expression and function in prostate cancer. Cancer Res 63:2037-2041.
15. Barki-Harrington L, Daaka Y (2001) Bradykinin induced mitogenesis of androgen independent prostate cancer cells. J Urol 165:2121-2125.
16. Ueda T, Mawji NR, Bruchovsky N et al (2002) Ligand-independent activation of the androgen receptor by interleukin-6 and the role of steroid receptor coactivator-1 in prostate cancer cells. J Biol Chem 277:38087-38094.

17. Ueda T, Bruchovsky N, Sadar MD (2002) Activation of the androgen receptor n-terminal domain by interleukin-6 via MAPK and STAT3 signal transduction pathways. J Biol Chem 277:7076-7085.
18. Snoek R, Rennie PS, Kasper S et al (1996) Induction of cell-free, in vitro transcription by recombinant androgen receptor peptides. J Steroid Biochem Mol Biol 59:243-250.
19. Gioeli D, Ficarro SB, Kwiek JJ, et al (2002) Androgen receptor phosphorylation. Regulation and identification of the phosphorylation sites. J Biol Chem 277:29304-29314.
20. Yeh S, Lin HK, Kang HY, et al (1999) From HER-2/neu signal cascade to androgen receptor and its coactivators: A novel pathway by induction of androgen target genes through map kinase in prostate cancer cells. Proc Natl Acad Sci USA 96:5458-5463.

Cadmium and Zinc Chloride-induced Preneoplastic Changes in the Rat Ventral Prostate: An Immunohistochemical and Molecular Study

Riánsares Arriazu, José M. Pozuelo, Rosario Rodríguez, Nuno Henriques-Gil, Teresa Perucho, Rocío Martín, and Luis Santamaría

Summary

Cadmium chloride (Cd) is a toxicant that has been implicated in human prostate cancer (PCA). The goal of the present study was to evaluate the immunoexpression of markers for cell proliferation, apoptosis, resistance to apoptosis, and to determine mutations on segments of the *bcl-2* gene, in preneoplastic lesions induced in rat prostate after treatment with Cd alone or in combination with zinc chloride (Zn). We evaluated: 1) The % of cells positively immunostained for the proliferating cell nuclear antigen (PCNA), 2) The % of apoptotic cells (evaluated by TUNEL), 3) The volume fraction of Bcl-2 immunostaining. 4) The mutations on a segment of 253 pb of *bcl-2*, in the ventral prostate lobe of normal and treated rats with Cd alone or in the presence of Zn in the drinking water for 18 mo. Our results indicate that the % of PCNA positive nuclei was significantly increased in preneoplastic prostatic acini of Cd-treated rats alone and in combination with Zn, compared to the normal acini of untreated animals. No significant changes were detected on the apoptotic rate or the volume fraction of Bcl-2. Moreover, no significant changes in the band pattern of the amplified segment of *bcl-2* gene were observed after Cd treatment. In summary, our data indicate that, prostate dysplasia induced in rats by Cd increases proliferative activity, without significant changes in either apoptosis or *bcl-2* immunoreactivity.

Introduction

Prostatic intraepithelial neoplasia (PIN) is widely considered as a premalignant lesion. A number of studies indicate that high-grade PIN (HGPIN) represents a premalignant lesion that could evolve into invasive adenocarcinoma (1,2). Molecular and immunohistochemical markers have been used to characterize the progression of human PIN. Some of these markers include: changes in the level of expression of genes that encode cell proliferation regulators, increase in cell proliferation, genetic instability, and the expression of antiapoptotic oncoproteins (1, 2).

Several rat models for PCA have been developed (2-4). Morphological similarities between human PIN and dysplastic changes experimentally promoted in rodent prostates have been reported (2). These studies have provided substantial evidence implicating Cd as a prostate carcinogen (5), but the epidemiological data linking Cd and PCA are less convincing (5). Cd toxicity may induce changes in cellular homeostasis of essential metal ions, such as Cu, Zn, and Ca. Since there are similarities in the Zn and Cd transport characteristics, it has been suggested that these metals share a common transport mechanism (6). Also, it has been stated that Cd induces premalignant and/or invasive epithelial lesions in rat ventral prostate when administered in drinking water (2, 5).

The present study was undertaken to investigate whether treatment with Cd alone or in combination with Zn would induce changes in cell proliferation (PCNA expression), DNA apoptotic fragmentation, resistance to apoptosis (bcl-2 oncoprotein expression); and mutations on *bcl-2* gene segments (PCR-SSCA).

Materials and Methods

Animals. Male Sprague-Dawley (SD) rats were divided into 3 treatment groups (15 rats/group). In the drinking water, group 1 received 60 ppm Cd; group 2, 50 ppm Zn + Cd; and group 3, water free of these metals. An additional group of 9 rats was used for DNA extraction and PCR-SSCA study. After an 18.0-mo treatment period, all rats were killed by exsanguination after CO_2 narcosis. The ventral lobe of the prostate was dissected and cut into 2 mm-wide slices. The section plane was perpendicular to the sagital axis of the gland. All specimens were fixed in 4% paraformaldehide in phosphate buffered saline (PBS), pH 7.4, for 24 h, and embedded in paraffin.

Sampling Procedure. The ventral lobes from each treatment group were serially sectioned. Sections (5 μm-thick) were used for haematoxylin and eosine and immunohistochemistry staining. Immunostaining evaluation was performed in 20 sections, selected by random systematic sampling (7), from each animal of each group.

Histological Definition of Lesion Changes. The morphological criteria used to classify the lesions was: 1. Dysplastic, enlargement of the epithelial lining of the acinus, usually associated with nuclear pseudostratification. 2. Increase in the basophilic appearance of the epithelium with nuclear crowding. 3. Nuclear pleomorphism. 4. Cribriform pattern.

Immunohistochemistry. For immunostaining, at least 5 selected slides/rat/prostate/antigen were used. The mouse monoclonal antibodies used were PCNA (Biomeda, Foster City, CA, USA) diluted at 1:400, and Bcl-2 protein (Dako, Copenhagen, Denmark) diluted at 1:10. The second antibody used was a biotintilated anti-mouse immunoglobulin (Biomeda, Foster City, CA, USA), diluted at 1:400. After the immunoreactions, the sections were counter-stained with either methyl green. The specificity of the immunohistochemical procedures was determined in the presence of non-immune serum instead of the primary antibodies.

DNA Fragmentation Detection. To detect the apoptotic fragmentation of DNA, a TUNEL (TdT-mediated dUTP-biotin Nick End Labeling) technique (Boehringer Mannheim) was used (8). After TUNEL technique, sections were counter-stained with acetic carmine.

Quantitative Evaluation of Cell Proliferation and Apoptosis. The % PCNA-immunostained nuclei (LI_{PCNA}) (2) and that of labeled apoptotic nuclei (LI_{AP}) (9) were calculated in selected sections from control rats, non-dysplastic acini, and dysplastic acini of treated animals, using the formula: (n) of labeled nuclei x 100/total number of labeled + unlabeled cell nuclei. Measurements were carried out using an Olympus microscope equipped with a 100 x oil immersion lens at a final magnification of 1200 s, and using the stereologic software GRID (Interactivision, Silkeborg, Denmark). PCNA-immunostained nuclei were considered positive regardless of staining intensity. Apoptotic nuclei were considered positive when the stain was uniform and intense.

Quantitative Evaluation of Bcl-2 Immunostaining. To quantify the immunoreactivity of Bcl-2 protein, its volume fraction (V_F) was measured, using the Scion Image Beta 4.02 program, available on the Internet at http://www.scioncorp.com, and expressed as % of immunostained epithelium (2).

DNA Extraction. Fresh tissue samples of ventral prostate from 9 SD rats were separately frozen and stored in liquid N. DNA was extracted from prostate samples following the CTAB (mixed Alkyltrymethyl ammonium bromide) method (10).

Polymerase Chain Reaction-single Strand Conformation Analysis (PCR-SSCA). The primers employed for amplification of *bcl-2* cDNA were: Upper GAT GAC TTC TCT CGT CGC TAC, Lower GCA GAT GCC GGT TCA G; amplified segment of 253 bp and 57°C of annealing temperature. They were selected with the program OLIGO 6.0 (Molecular Biology Insights Inc., CO, USA), based on the sequence of *bcl-2*, GenBank accession L14680; bcl2(a) amplify a 253 base pair (bp) fragment. PCR was performed from 50 ng of template DNA with 0.25 mM of each nucleotide, 0.25 μM of each primer, and 1 unit of *Taq* polymerase (Roche Molecular Biochemicals) in its correspondent buffer. PCR was performed in a

PTC-100 ™ (MJ Research, Waltham, MA, USA) apparatus. For *bcl-2* primers, we used the PCR program consisting of an initial step of 7 min/94 °C, followed by 34 cycles/min/94°C, 1 min/69°C, and 1 min/72°C, with a final step of 7 min/72°C. Electrophoresis was performed with 2 μl of each PCR product plus 5 μl of loading buffer, at 200 V, for 2.5 h, to check that a single product of the expected size was obtained. Band detection was made by silver staining. SSCA was performed by techniques described by Myiers, *et al,* 1997 (11).

Statistical Analysis. For each parameter studied, the mean ± SD was calculated. The differences among different ages, treatments, non-dysplastic and dysplastic groups were evaluated by ANOVA ($p < 0.05$). The variables were normalized by logarithmic transformation.

Results

Histology. Dysplastic changes in the ventral prostate acinar epithelium were detected to the end of the study in 3/9 rats treated either by Cd alone or in the presence of Zn. The control acini showed a columnar monostratified epithelium, whereas the dysplastic acini manifested an irregularly enlarged epithelial lining with occasional polypoid formations. The non-dysplastic acini from treated rats had morphology similar to that of the acini of control rats. When compared to the control acini, the dysplastic glands showed a remarkable enlargement, with a frequent cribriform pattern. The size of nuclei was heterogeneous, and the nucleoli were usually prominent (Figure 1a, b). No tumoral infiltration was observed in the surrounding stroma.

PCNA Immunostaining and Apoptosis. The number of PCNA-immunoreactive nuclei was increased in dysplastic acini, while non-dysplastic acini from Cd ± Zn-treated rats was slightly higher than that of the controls (Figure 1c, d). LI_{PCNA} was significantly higher ($p < 0.05$) in Cd-treated rats treated than in the controls, however the difference was not significant between Cd-treated alone or in combination with Zn (Table 1A). Dysplastic acini LI_{PCNA} was significantly greater than that from non-dysplastic acini (Table 1B). Apoptotic nuclei were evident in all the groups examined (Figure 1e, f). LI_{AP} did not show significant differences among the different treatments (Table 1A) or the presence of dysplastic changes (Table 1B)

Bcl-2 Immunostaining. Bcl-2 immunoreactivity was evident in all groups. It was granular and expressed in the apical border of the epithelium in controls and non-displastic acini (Figure 1g, h). The V_F of Bcl-2 immunostaining did not change in relation to treatment (Table 1A), or the presence of dysplastic acini (Table 1B).

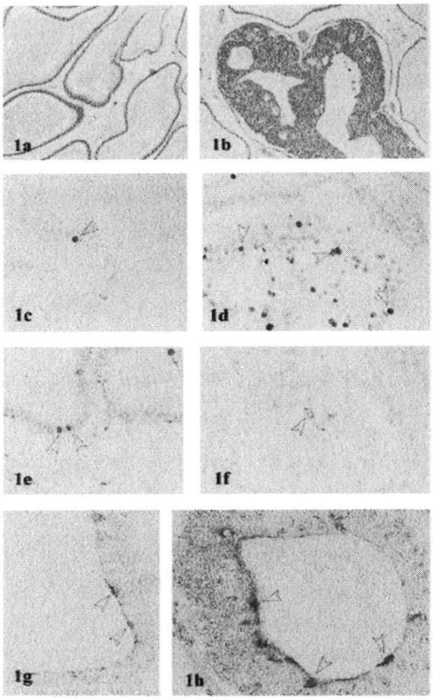

Figure 1. **a:** Normal epithelium from a control rat. **b:** HE of a dysplastic acini from an 18.0-mo Cd + Zn-treated rat (100 x). **c:** PCNA-stained acini from a control rat. Some positive nuclei are observed in the epithelium (arrowhead). **d:** PCNA in a dysplastic acinus of a Cd-exposed rat. The number of positive nuclei is higher than in control acini (arrowheads) (c,d: 400 x). **e:** TUNEL in control acinus, some apoptotic nuclei are seen (arrowheads). **f:** TUNEL in a dysplastic acinus from a Cd-treated rat, an apoptotic nucleus is observed (arrowhead) (e, f x 400). **g:** Bcl-2 staining in a control acinus. Positive cells are observed in apical border of epithelium (arrowheads). **h**: Bcl-2 staining in a Cd + Zn-treated dysplastic acinus in apical border of epithelium (arrowheads) (g,h: 600 x).

Table 1. **A)** Control and Cd ± Zn-treated Values for Ln LI_{PCNA}, LI_{AP}, and V_F bcl-2
B) Non-dysplastic and Dysplastic Acini Values for Ln LI_{PCNA}, LI_{AP}, and V_F bcl-2[1]

A) Variable[2]	**Controls**	**Cd**	**Cd + Zn**
Ln LI_{PCNA}	1.94[3] ± 1.12	2.60[3] ± 0.95	2.18[3] ± 0.94
Ln LI_{AP}	0.62[3] ± 0.79	0.49[3] ± 0.91	0.66[3] ± 0.82
Ln V_F bcl-2	0.80[3] ± 0.51	0.89[3] ± 1.15	1.10[3] ± 0.77
B) Variable[2]	**Non-dysplastic Acini**		**Dysplastic Acini**
Ln LI_{PCNA}	2.26[3] ± 1.08		3.65[3] ± 0.43
Ln LI_{AP}	0.49[3] ± 0.86		0.58[3] ± 0.62
Ln V_F bcl-2	1.35[3] ± 0.70		1.20[3] ± 0.98

1 Values represent the mean ± SD.
2 Logarithmic transformed values. LI_{PCNA}, Labeling index for PCNA (%); LI_{AP}, Labeling index for apoptosis (%); V_F bcl-2, Volume fraction of bcl-2 (%).
3 Values statistically significant ($p < 0.05$).

PCR-SSCA. After Cd ± Zn treatments, the 253 bp fragment of *bcl-2* gene, amplified by PCR-SSCA, did not show any significant changes in band pattern (Figure 2).

Discussion

The present study confirms previous reports regarding the role of Cd in rat prostate tumorigenesis (5). The changes observed resembling dysplasia have been previously described after Cd exposure and other carcinogenic agents (2, 4). The lesions observed were histologically similar to those described as atypical hyperplasia induced in rats by treatment with phenylephrine (12). It should be noted that the changes from untreated acini to those of non-dysplastic and dysplastic acini from Cd-treated rats display a continuum (2). Cd intake may induce initial changes in the prostatic epithelium by increasing cell proliferation. This assumption agrees partially with other descriptions of Cd effects on rat male genital organs because, in contrast to other reports (2), we did not find an increase in apoptosis. Although, it has been reported that Zn may have a potentiating effect on prostatic tumorigenesis (13), in the present study, we did not observed neither an effect on cell proliferation or apoptosis after Zn exposure. Bcl-2 expression was increased in Cd ± Zn-exposed rats, in comparison to untreated animals; however these differences were not significant, contrary to previous studies (2). No differences in *bcl-2* gene band pattern were observed in any of the groups examined indicating that the mutation rate for this gene was lower than expected. Differences in PCNA and Bcl-2 expression and localization between rat and human prostates were remarkable. In normal human prostate, the proliferation compartment is better defined than in rat prostate. It is represented by isolated cells and often is absent in some portions of the acini (the basal layer, PCNA and Bcl-2 positive). Moreover, proliferative activity in the rat is not confined to the basal cells. It has been reported that the main proliferating cells in rat prostate are the columnar cells (14). Cd-induced prostate dysplasia, increased proliferative activity, but did not induced changes on apoptosis, or Bcl-2 immunoreactivity. The addition of Zn did not modify the histological, immunohistochemical, and molecular features observed in the solely Cd-treated rats.

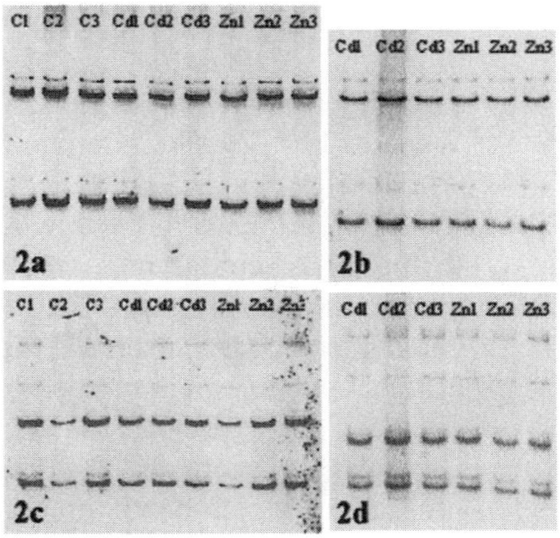

Figure 2. PCR-SSCA gels. There was no variation on the band pattern of the amplified segment of the *bcl-2* gene in relation to the treatment. **a:** Samples after 6.0 mo, **b:** Samples after 18.0 mo. **c:** Samples after 12.0 mo. **d:** Samples after 12.0 mo.

References

1. Bostwick DG (1996) Progression of prostatic intraepithelial neoplasia to early invasive adenocarcinoma. Eur Urol 30:145-152.
2. Martin JJ, Martin R, Codesal J, et al (2001) Cadmium chloride-induced dysplastic changes in the ventral rat prostate: an immunohistochemical and quantitative study. Prostate 46:11-20
3. Morrissey C, Buser A, Scolaro J, et al (2002) Changes in hormone sensitivity in the ventral prostate of aging Sprague-Dawley rats. J Androl 23:341-351.
4. Shirai T, Takahashi S, Cui L, et al (2000) Experimental prostate carcinogenesis-rodent models. Mutat Res 462:219-226.
5. Waalkes MP (2000) Cadmium carcinogenesis in review. J Inorg Biochem 79:241-244.
6. King LM, Banks WA, George WJ (1999) Differences in cadmium transport to the testis, epididymis, and brain in cadmium-sensitive and -resistant murine strains 129/J and A/J. J Pharmacol Exp Ther 289:825-830.
7. Gundersen HJ, Bendtsen TF, Korbo L, et al (1988) Some new, simple and efficient stereological methods and their use in pathological research and diagnosis. APMIS 96:379-394.
8. Gavrieli Y, Sherman Y, Ben Sasson SA (1992) Identification of programmed cell death in situ via specific labeling of nuclear DNA fragmentation. J Cell Biol 119:493-501.
9. Rittmaster RS, Manning AP, Wright AS, et al (1995) Evidence for atrophy and apoptosis in the ventral prostate of rats given the 5 alpha-reductase inhibitor finasteride. Endocrinology 136:741-748.
10. Towner P (1991) Purification of DNA. In: Brown TA (editor), Essential Molecular Biology. Oxford:Oxford University: IRL Press, 47-68.
11. Myers RM, Ellenson LH, Hayashi K (1997) Detection of DNA variation. In: Green Eric D (editor), Genome Analysis II. Detecting Genes. Cold Spring Harbor Laboratory Press, 287-384.
12. Golomb E, Kruglikova A, Duir D, et al (1998) Induction of atypical prostatic hyperplasia in rats by sympathomimetic stimulation. Prostate 34:214-221.
13. Waalkes MP; Rehm S; Riggs LW, et al (1989) Cadmium carcinogenesis in the male Wistar (Crl(WI)BR) rats: Dose-response analysis of effects of zinc on tumor induction in the prostate and in the testes and at the injection site. Cancer Res 49:4282-4288.
14. Sensibar JA (1995) Analysis of cell death and cell proliferation in embryonic stages, normal adult, and aging prostates in human and animals. Microsc Res Tech 30:342-350.

Index

104-R2 cells, 213, 214

A

ACI rat liver, 17β-estradiol metabolism after chronic estradiol treatment in, 367–372, 369f, 370t, 371f
ACI rat mammary gland, 17β-estradiol metabolism after chronic estradiol treatment in, 367–372, 369f, 370t, 371f
ACI rat mammary tumors, 22–24, 23f, 24t, 405–406
  estradiol-induced, dietary genistein on, 405–409, 408t, 409f
  vs. human breast cancers, 28, 28t
ACO:E2, chronic estradiol on, 371, 371f
Adapter proteins, 443–444
  uterine cell Gab-1, hormonal activation of, 443–448, 445f–448f
Adenosis, in prostate cancer, 145
Akt phosphorylation, englitazone on, 480, 484, 484f
Alizarin, 212–213, 213f
α-methylacyl-CoA racemase (AMACR), in prostate cancer, 142, 146
α-zearalanol-induced mutations, and tumorigenesis, 475–478, 476f–478f
Aminoglutethimide, 167–168, 168f. *See also* Aromatase inhibitors
Anastrozole, 167–168, 168f. *See also* Aromatase inhibitors
  in breast cancer treatment, 170–171, 171t
Androgen(s). *See also* specific androgens
  as ancient medicines, 211
  control of action of, 212
  on E2F1 promoter reporter construct transcriptional activity, 241–242, 241f
  and IGF-1, in canine inflammatory mammary carcinoma, 436–441, 439f, 440f
  in mammary growth, normal and neoplastic, 436–437
  race/ethnicity on levels of, 185
  suppression and eradication of androgen-independent prostate tumors in athymic mice by, 215
  suppression of Bcl-2 expression by, 237, 238f
  ZEB-1 regulation by, 455–460, 457f, 460f
Androgen, prostate cancer cells and
  SREBP pathway stimulation via shift in SCAP-retention protein balance by, 357–362, 358f, 360f, 361f
Androgen(s), prostate cancer and, 185–192, 515
  androgen receptor gene in, 186–190, 188t, 189t
  future research on, 191–192
  overview of, 185–186
  steroid 5α-reductase type 2 (SRD5A2) gene in, 190–191, 191t
Androgen(s), prostate cancer cells and
  regulation of Bcl-2 expression in, 237–242, 238f–241f
  suppression of, in culture, 214–215
Androgen receptor (AR)
  CBP and p300 in androgen-independent transactivation of, 494–498, 497f–499f
  in endometrium, normal and neoplastic, 314–320, 317t
  in endometrium in relation to FIGO, expression of, 317, 318f
  in mammary carcinoma, canine inflammatory, 440–441, 440f
  nonsteroidal activator enhancement of function of, 224–225
  p300 in interleukin-6–mediated transaction of, 494–496, 497f, 498, 499f
  structure of, 221
Androgen receptor (AR), in androgen-refractory prostate cancer, 233–242
  LNCaP cell outgrowth with androgen deprivation in, 234–235
  regulation of Bcl-2 expression by androgen in, 237–242, 238f–241f
  regulation of Bcl-2 expression by P13K/PTEN/Akt pathway in, 236–237, 237f
  survival signaling with androgen ablation in LNCaP cells in, 235, 236f
Androgen receptor (AR), in prostate cancer
  androgen-dependent to -independent transformation of, 516
  cytokine regulation of cell growth in, 225–227, 226f, 227f
  growth of, 221
  IL-6 regulation of cell growth in, 225–228, 226f, 227f
  PKA and extracellular signal-regulated kinases in activation of, 515–519, 518f, 519f
  tumor models and clinical specimens of, 222–223
Androgen receptor (AR) activation

androgen-dependent and -independent, phosphorylation in, 519
extracellular signal-regulated kinases in, 517–519, 518f
growth factor–induced, 516
in prostate cancer, protein kinase A and extracellular signal-regulated kinases in, 515–519, 518f, 519f
Androgen receptor (AR) coactivators, in prostate cancer, 227
Androgen receptor (AR) gene, in prostate cancer, 186–190, 188t, 189t
racial/ethnic differences in, 186–188, 188t, 221
Androgen receptor (AR) mutations
in androgen-independent prostate cancer, 336–337
detection of, in prostate cancer, 336–342, 338f–341f
in metastatic prostate cancer, detection of, 338–339, 338f, 339f
in prostate cancer, 223–224
Q640Stop/T877A, 339–341, 340f, 341f
Androgenes, on fatty acid synthase expression, 350
Androstenediol glucorine, race/ethnicity on levels of, 185
Androstenedione, 167–168, 168f. *See also* Aromatase inhibitors
in canine inflammatory mammary carcinoma, 438, 439f, 440
Aneuploidy, 108
in cancer, 108
centrosome amplification and, 108
in estrogen-induced ACI rat mammary neoplasms, 412–417
incomplete DNA replication in, 102, 103f
Animal models, of estrogen-induced oncogenesis, 20–21, 20t, 431. *See also* specific models
mammary tumors in female ACI rats, 22–24, 23f, 24t
mammary tumors in female ACI rats, *vs.* human breast cancers, 28, 28t
Antiandrogens. *See also* specific antiandrogens
on androgen-independent prostate tumors in athymic mice, 215
on prostate tumor growth, 218
transactivation of androgen receptor by, CBP in, 494–499
Antiestrogens. *See* specific antiestrogens
Antioxidant enzyme activity, in ACI rat and mammary gland, after chronic estradiol treatment, 367–369, 369f, 372
Antiprogestins. *See* specific antiprogestins
Apoptosis induction, by E_2 + catechol estrogens with COMT inhibitor, 375, 377–379
ARA70, as coactivator in prostate cancer, 227
Arachidonic acid
on MCF7 breast carcinoma line proliferation, 400, 400f
on reductive and oxidative action of 17β-dehydrogenase, 400, 401f
Aromatase, 166–167
in estrogen hormone synthesis, 130
estrogens and, in breast cancer risk, 172
intratumoral, 130
Aromatase inhibitors, 166–174, 167–168, 168f
for breast cancer, hormone-dependent, 130
in breast cancer prevention, 173–174
in breast cancer treatment, 170–171, 171t
breast endocrinology and, 169–170
as chemopreventive agents, 136–137
classes of, 167, 168f
endocrinology of, 169
potency of, 168, 169t
selective ER modulators *vs.*, 172, 173t
types and structure of, 167, 168f
Aromatase overexpression, 130–137
aromatase inhibitor blocking of, as chemopreventive, 136–137
estrogen on TGFα expression in MMTV-TGFα x MMTV-aromatase transgenic cross and, 135–136
gynecomastia and Leydig tumors in male transgenic mice from, 135
in MG hyperplasia, 134
in transgenic females, biochemical changes in, 132–133, 132f, 133f
in transgenic model, carcinogen susceptibility of, 133–135, 134t
in transgenic model, expression of, 131–132, 131f
Atypical adenomatous hyperplasia, in prostate cancer, 145
Atypical ductal hyperplasia (ADH), 122, 123t
Atypical lobular hyperplasia (ALH), 124–125, 124t
AurA expression in rat mammary carcinogenesis, centrosome amplification and, 9–12. *See also* Centrosome amplification, and AurA...
Aurora A (AurA) kinase, in cancer, 3–8
antibodies against, in centrosomes, 4, 4f
AurA binding at centrosomes and spindles and, 5–6, 6f–8f
as dynamic component of centrosome/mitotic spindle, 6–7, 8f
future directions in, 13–14, 13f
structure of, 5, 5f
targeting domains of, 6, 8f
Aurora A (AurA) kinase family, structure of Aur-A, B, and C kinases in, 5, 5f
Aurora A (AurA) kinase overexpression, 4–5, 4f
cell transformation from, 5

centrosome amplification and chromosome instability from, 110, 111

B

BALB/c p53 null mammary epithelium transplant model, 431–432
hormonal dependence of premalignant progression in, 431–435
hormone-induced tumorigenesis in, 433, 433t
ovarian-dependent growth of normal mammary epithelium in, 432, 433f
tamoxifen inhibition of mammary tumorigenesis in, 433–434, 434f
tumorigenesis in progesterone receptor–deficient mammary epithelium in, 433, 434t
Bcl-2, 233
Bcl-2 expression
androgen receptor on, 237, 239f
P13K/PTEN/Akt pathway regulation of, in prostate cancer cells, 236, 236f
Bcl-2 expression, androgen on, 237, 238f
androgen receptor role in, 237, 238f
gene transcription in, 237–239, 239f
in regulation, 233–234
in regulation, E2F proteins in, 239–240, 239f–241f
in regulation, in prostate cancer cells, 237–242, 238f–241f
Bcl-2 overexpression, in androgen-refractory prostate cancer cell survival, 233
BDII inbred rat, 247–256
chromosome regions with endometrial adenocarcinoma susceptibility in, 255
endometrial adenocarcinoma in, specific genetic changes in, 251–255, 252f, 253t
endometrial adenocarcinoma susceptibility in, 249–250, 250t
RNO4 genetic changes in, 251–254, 252f, 253t
RNO10 genetic changes in, 254–255
tumor heterogeneity in, 256
Benign prostatic hyperplasia (BPH)
estrogens in, 502
Id-1 protein in, 198–199, 199f
prostate cancer and, 145–146
17β-estradiol. *See* E_2 (estradiol)
Bicalutamide, on IL-6-induced androgen receptor activity, 225, 226f
Biomarkers. *See also* specific biomarkers
for metastasis, 463
pituitary tumor-transforming gene-1 (PTTG1) as, 462–467
for prostate cancer (*See* Id-1 protein)
Biopsy, breast cancer risk after, 124t
Bisphenol A–induced mutations, and tumorigenesis, 475–478, 476f–478f
Blc-2 expression, in cadmium and zinc chloride–induced rat ventral prostate, 522, 525–527, 526f, 526t, 527f
Boveri, Theodore, 2, 3f, 106
BRCA1/BRCA2 tumor suppressor genes, in centrosome amplification, 111
Breakage-bridge-fusion cycles, of amplification, 93
Breast
estrogen receptor and normal cell cycle in, 121–122
postmenopausal, endocrinology of, 169–170, 170f
sex hormone–binding globulin synthesis and expression in, 510, 511f, 512–513
Breast cancer. *See also* Inflammatory breast (mammary) carcinoma; specific types
centrosome amplification and, 106–112 (*See also* Centrosome amplification, and breast cancer)
centrosome amplification and AurA expression in, 9–12 (*See also* Centrosome amplification, and AurA expression...)
CHRT in, 50–56 (*See also* Combined estrogen/progestin hormone replacement therapy (CHRT), in breast cancer)
EGCG (green tea) suppression of, 216
epidemiology of, 153
ERα hinge region in, properties of mutants in, 391–396, 393t–396t
estrogen receptor β (ERβ) gene polymorphism in, 287–292 (*See also* Estrogen receptor β (ERβ) gene polymorphism)
in female ACI rat, *vs.* human breast cancers, 28, 28t
histological types of, 47
HRT and, prognosis in, 126
human sporadic, female sex hormones in, 19
intratumoral aromatase in estrogen synthesis by, 130
parity-induced protection from, 126, 420
progesterone receptor (PR) in, 77–82 (*See also* Progesterone receptor (PR), in mammary carcinogenesis)
sex hormone–binding globulin (SHBG) expression in, and hormone regulation, 508–513, 509f, 510f
sex hormones in, history of, 47
soy phytoestrogens on focal adhesion in, 300–306, 303f–305f
tamoxifen prophylaxis and, types of, 126
Breast cancer, in female ACI rat, 22–24, 23f, 24t
cytogenetic analysis of mammary neoplasms in, 412–417 (*See also*

Cytogenetic analysis of ACI rat mammary...)
vs. human breast cancers, 28, 28t
Breast cancer, in p53 null mammary epithelium transplant model, 431–435. *See also* p53 null mammary epithelium transplant model
Breast cancer cells
eicosapentaenoic acid on proliferation and 17β-dehydrogenase activity in, 398–403, 400f, 401f
onapristone regulation of genes of, 308–313, 311t, 312f
soy phytoestrogens on focal adhesion assembly in, 300–306, 303f–305f
Breast cancer prevention, 153
E_2 in, pregnancy levels of, 153–162 (*See also* E_2 (estradiol) at pregnancy levels)
estrogen in, 153–162 (*See also* Estrogen(s), in breast cancer prevention)
hormonal strategies mimicking pregnancy for, 156–157
multiple pregnancies and nursing in, 153
Breast cancer risk
ERβ gene polymorphism in, CA repeats in, 291–292
estrogen in, 288
estrogen in, and aromatase, 172
estrogen replacement therapy in, 122
HRT in, estrogen *vs.* progesterone in, 122
parity on, 126
in premalignant breast disease, after biopsy, 124t
progestin-regulated genes and, 65–72 (*See also* Progestin-regulated genes, and breast cancer risk)
soy phytoestrogens on, 406
Breast carcinoma, inflammatory, 436
genes in, 437
Breast carcinoma, invasive ductal
clinical characteristics of, 48–49
incidence rates of, time trends in, 49–50, 49f
vs. invasive lobular breast carcinoma, 48–49
pathologic, immunohistochemical and genetic characteristics of, 48
Breast carcinoma, invasive lobular, 47–60
clinical characteristics of, 48–49
HRT and risk of, 50–57 (*See also* Hormone replacement therapy (HRT), lobular carcinoma risk and)
incidence rates of, time trends in, 49–50, 49f
vs. invasive ductal breast carcinoma, 48–49
pathologic, immunohistochemical and genetic characteristics of, 48
progesterone in etiology of, 57–59
Breast disease, premalignant. *See* Premalignant breast disease
BTAK/STK15. *See* Aurora A (AurA) kinase

C
c-*Myc* gene, in neoplasia
estrogen-induced, overexpression and amplification of, 25–26, 26f
history of, 88
c-Myc oncoprotein, in neoplasia, 87
deregulation of, 88–89
overexpression of, 88
pathways for, 87
c-Myc oncoprotein deregulation, 88–89
in cancer, 88–92, 89f–91f
in karyotic instability, 89f, 90
in reversible tumorigenesis, 91
c-Myc oncoprotein deregulation, in genomic instability, 87–94
DNA breaks in, 92
illegitimate rounds of replication in, 92–93
locus-specific, 89–90, 89f, 90f
long-range illegitimate recombinations from, 90–92, 91f
mechanisms of, 92–93
overriding functional wild-type p53 in, 93
C75
shortcomings of, 351
tumor cell cytotoxicity of, 343, 345
CAD gene, c-Myc overexpression on stability of, 89–90
Cadmium
in preneoplastic changes in rat ventral prostate, 522–527, 526f, 526t, 527f
in prostate cancer, 523
Cadmium + zinc, in preneoplastic changes in rat ventral prostate, 522–527, 526f, 526t, 527f
CAG repeat, androgen receptor
in prostate cancer risk, 187–190, 188t, 189t
racial/ethnic differences in, 186–189, 188t, 189t
cAMP analogues, androgen receptor stimulation by, 225
Carcinogens. *See also* specific carcinogens
chemical, on aneuploid frequency in mammary tumors, 23–24, 24t
Casodex, on androgen-independent prostate tumors in athymic mice, 215
Catechol estrogens
on apoptosis induction, E_2 and COMT inhibitor on, 375, 377–379
in carcinogenesis, 367–368, 376
carcinogenicity of 2-OH *vs.* 4-OH forms of, 375–376, 378
COMT inactivation of, 376
COMT inhibitor modulation of transforming/clastogenic activities of, 375–379
Catechol-*O*-methyltransferase (COMT) inhibitor

catechol estrogens on apoptosis induction and, 375, 377–379
modulation of transforming/clastogenic activities of catechol estrogens by, 375–379
CBP
in androgen-independent prostate cancer, 494–495, 497–498, 498f, 499f
as coactivator in prostate cancer, 227
CD9, onapristone downregulation of, 311–312, 312f
Cdk6 amplification, in rat endometrial adenocarcinoma, 251–252, 252f, 253t
CDXR cells, 213, 214
Cell cycle
in breast tissue, 121–122
DNA and centrosome cycle coordination with, 109–110
serine/threonine cyclin-dependent protein kinases in, 109
Cell cycle checkpoints, p53, Rb, and Cdk2 on, 110
Centrosome, 1
structure and behavior of, 106–107
Centrosome amplification, 1, 107
aneuploidy and, 108
BRCA1 and BRCA2 tumor suppressor genes in, 111
chromosomal instability and, 109
in genomic instability, 109
as indicator of tumor aggressiveness, 112
pathways of, 4, 13, 13f
pathways of, aberrations in, 13, 13f
pericentrin overexpression in, 111
Centrosome amplification, and AurA expression in rat mammary carcinogenesis, 9–12
after MNU exposure, centrosome counts in, 10, 10f
after MNU exposure, first indications of, 11, 11f
after MNU exposure, mRNA expression in, 11–12, 12f
hormone treatment on, 12, 12f
rat mammary model in, 9, 9f
Centrosome amplification, and breast cancer, 106–112
aneuploidy in, 108–109
chromosomal instability in, 108–109
deregulation of centrosome cycle in, 110–111, 111f
tumor aggressiveness and, 112
Centrosome amplification, in cancer, 107
estrogen-induced oncogenesis of, 26–27, 27f
history of, 2, 3f
recent research on, 2–3
Centrosome anomalies, 1–2
Centrosome anomalies, in cancer
history of, 2, 3f
loss of p53 function and, 110
recent research on, 2–3
Centrosome cycle
deregulation of, in breast cancer, 110–111, 111f
DNA and cell cycle coordination with, 109–110
Centrosome homeostasis, cell cycle checkpoint controls on, 110–111, 111f
Centrosome pathways, 13, 13f
Cerulenin
shortcomings of, 351
tumor cell cytotoxicity of, 343, 345
Checkpoint-signaling proteins, 1
Cholecystokinin (CCK), 468–469
in pituitary and hypothalamus, 469
as proliferative agent in pituitary GH_3 cells, 469
Cholecystokinin A (CCK-A) receptor, 469
Cholecystokinin B (CCK-B) receptor, on GH_3 rat pituitary tumor cells, 471
Cholecystokinin B (CCK-B) receptor mRNA, in human functionless pituitary tumors, 468–473, 471f, 472f
Cholecystokinin (CCK) receptors, 468
forms of, 469
Cholesterol, EGCG (green tea) on, 217
Chromatid exchange model, 93
Chromatid nondisjunction events, incomplete DNA replication in, 102, 103f
Chromosomal aneuploidy, in estrogen-induced oncogenesis, 26f, 27–28
Chromosome instability, 88, 108. *See also* Genomic instability
in breast cancer, centrosome amplification and (*See* Centrosome amplification, and chromosomal...)
in cancer, 108, 417 (*See also* Genomic destabilization)
in carcinogenesis, 106
centrosome amplification and, 109
cyclin E deregulation in, 100, 101f, 102–103, 103f
telomere dysfunction in, 143
Chromosome instability, in estrogen-induced oncogenesis, 19–29
animal models *vs.* human breast cancers and, 28, 28t
c-*Myc* gene overexpression and amplification in, 25–26, 26f
centrosome amplification in, 26–27, 27f
chromosomal instability and aneuploidy in, 26f, 27–28
ectopic uterine tumors in kidney and, 20–22, 21t, 22f
general considerations in, 19–20
mammary tumors in female ACI rats in, 22–24, 23f, 24f
mechanisms of, 26f, 27–28
proposed scheme for, 28–29, 29f
steroid hormone receptors (ERα, PR) in, 24–25, 25f

CHRT. *See* Combined estrogen/progestin hormone replacement therapy (CHRT), in breast cancer
Coactivators. *See also* specific coactivators
  in androgen-independent prostate cancer, 494–498, 497f–499f
  androgen receptor, 494, 498, 499f
  androgen receptor, in prostate cancer, 227–228
Combined estrogen/progestin hormone replacement therapy (CHRT), in breast cancer. *See also* Hormone replacement therapy (HRT)
  current/recent use of, 53–55, 54t
  ever use of, 50–53, 52t–53t
  sequential *vs.* continuous progestin on, 55, 56t
Combined estrogen/progestin hormone replacement therapy (CHRT), in endometrial cancer, progestogen dose on risk of, 273–276, 275f
CREB-binding protein. *See* CBP
Curcumin, 212–213, 213f
Cyclin D1
  in mammary epithelial cell transformation, 79
  on p27 sequestration, 133
  in PR-A transgenic *vs.* wild-type mice mammary glands, 80, 81f
  progesterone receptor regulation of protein levels of, 388, 388f
Cyclin D1 expression
  in aromatase transgenic females, 132
  by progesterone receptor, MAPK in regulation of, 381, 382f, 386–388, 387f
Cyclin D1 promoter, 387
Cyclin D2 gene, c-Myc overexpression on stability of, 89–90
Cyclin-dependent protein kinases (Cdks)
  in cell cycle progression, 109
  in centrosome mitosis and duplication, 109
  DNA replication and centrosome duplication activation of, 109
Cyclin E, in cancer, 98
Cyclin E deregulation, in cancer, 98–104
  in chromosome instability, 100, 101f, 102–103, 103f
  hCDC4 mutation in, relative to cell cycle, 99–100
  mechanisms of, 99
  tumor suppressor mutation synergism with, 101–102
Cyclin E overepxression
  with p53 mutation, in cancer, 110
  regulation of, 98
Cytogenetic analysis of ACI rat mammary neoplasms, 412–417
  frequency of genomic alterations in, 415, 416t
  nonrandom numerical chromosomal alterations in, 414, 415t
  regional genomic alterations in, 415, 415f, 416f
  trisomies and tetrasomies in, 414, 414f

D
Daidzein
  on estrogen receptor, 301
  on FAK activity in ERα cell lines, 303–304, 304f, 305f
  on focal adhesion in breast cancer cells, 300–306, 303f–305f
  on focal adhesions and F-actin, in breast cancer, 302–303, 303f
Deletion plus episome model, 93
DHEA, in canine inflammatory mammary carcinoma, 438, 439f, 440
DHT. *See* Dihydrotestosterone (DHT)
Diet. *See also* Green tea; Soy phytoestrogens
  in prostate cancer, 35–36
Diethylstilbestrol (DES), 450
  endometrial adenocarcinoma in Syrian hamsters treated with, 450–453, 451t, 452f, 453f
  mutations induced by, tumorigenesis and, 475–478, 476f–478f
Dihydrofolate reductase (DHFR) gene, c-Myc overexpression on stability of, 89–90, 90f
5α-Dihydrotestosterone (5α-DHT), 212
  in prostate tumor growth, 215
  suppression of production of, 212
Dihydrotestosterone (DHT)
  androgen receptor activation by, 517, 518f, 519
  protein kinase A in activation of androgen receptors by, 515
DNA breakage, c-*Myc* gene in, 92
DNA cycle, cell and centrosome cycle coordination with, 109–110
DNA hypomethylation, in hepatocarcinogenesis, 294
DNA replication, incomplete, in chromatid nondisjunction events and aneuploidy, 102–103, 103f
Docking proteins, 443–444
  uterine cell Gab-1, hormonal activation of, 443–448, 445f–448f
Ductal carcinomas *in-situ* (DCIS), 122
  estrogen receptor (ERα) in, 124–125

E
E_1 (estron)
  expression of, in aromatase transgenic females, 132
  structure of, 316f
E_2 (estradiol). *See also* Estrogen(s)
  on 17β-estradiol metabolism in ACI rat liver and mammary gland, 367–372, 369f, 370t, 371f

in ACI rat liver and mammary gland, after chronic estradiol treatment, 367–372, 369f, 370t, 371f
in carcinogenesis, 135
on cell survival and invasion signaling proteins, 301
on focal adhesions and F-actin, in breast cancer, 302–303
hinge-region mutant ERα heterodimer hypersensitivity to, 394, 394t
hinge-region mutant ERα–mediated transactivation in AP1 by, inhibition of, 395–396, 396t
metabolism of, after chronic estradiol treatment in ACI rat, 367–372, 369f, 370t, 371f
metabolism of, in hormone-dependent cancers, 367
mutations induced by, and tumorigenesis, 475–478, 476f–478f
short-term high-dose treatment with, in mammary tumor prevention, 159–160 (*See also* E_2 (estradiol) at pregnancy levels)
short-term treatment with, 161–162
structure of, 316f
on tyrosine phosphorylation of FAK in breast cancer cells, 301
uterine cell proliferation and endometrial tumors from, 319
E_2 (estradiol), ACI rat mammary cancer induction by, 407
cytogenetic analysis of genomic destabilization in, 412–417
dietary genistein on, 405–409, 408t, 409f
E_2 (estradiol) + progesterone, high-dose
adverse effects of, 160–161
in mammary tumor prevention, 156–157
short-term in mammary tumor prevention, 159–160
E_2 (estradiol) + progesterone, short-term treatment with, 159–161
in parous *vs.* hormone-protected rats, 160–161
E_2 (estradiol) at pregnancy levels, in breast cancer prevention, 419
with chemical carcinogens, 426–429, 428t, 429t
downregulated genes in, 422–423, 423t
upregulated genes in, 421–422, 422t, 423–424
E2F, on DNA replication and centrosome duplication, 109
E2F proteins, in androgen regulation of Bcl-2 expression, 239–240, 239f–241f
E2F1, in androgen regulation of Bcl-2 expression, 240, 241f
E2F1 promoter reporter, androgen on transcriptional activity of, 241–242, 241f
E_3 (estriol), structure of, 316f
Early growth response (Egr-1) protein
Id-1 anti-sense treatment on expression of, in prostate cancer cell lines, 206, 206f
Id-1 protein ectopic expression on, 203–204, 203f
Id-1 protein interaction with, 202
Ectopic uterine tumors in kidney (EUTK), 20–22, 21t, 22f
in hamsters *vs.* human breast cancers, 28, 28t
tamoxifen prevention of, 21t
EGCG (epigallocatechingallate). *See* Green tea
EGF. *See* Epidermal growth factor (EGF)
EGFR expression, in aromatase transgenic females, 132
Eicosapentaenoic acid
on MCF7 breast carcinoma line proliferation, 400–401, 400f
on proliferation and 17β-dehydrogenase activity in MCF-7 breast carcinoma cells, 398–403, 400f, 401f
on reductive and oxidative action of 17β-dehydrogenase, 400–402, 401f
Endometrial adenocarcinoma
androgen receptor expression in, 314–320, 317t
endometrioid tumors in, 453, 453f
ERβ5 transcript upregulation in, 314–320, 317t
estrogen α and β variant expression in, 314–320, 317t, 318f
MPA (progestogen) dose in postmenopausal combined hormone therapy on, 273–276, 275f
progesterone receptor expression in, 314–320, 317t
progestogen dose in postmenopausal combined hormone therapy on risk of, 273–276, 275f
Endometrial adenocarcinoma, in rat
gene amplification in, 251
inbred, susceptibility for, 249–250, 250t
model for genetic analysis of (*See* BDII inbred rat)
patterns of genetic change in, 250–251
RNO4 genetic changes in, 251–254, 252f, 253t
RNO10 genetic changes in, 254–255
specific genetic changes in development of, 251–255, 252f, 253t
Endometrial adenocarcinoma, in Syrian hamsters, diethylstilbestrol, tamoxifen, and N-ethyl-nitrosurea–treated, 450–453, 451t, 452f, 453f
Endometrial hyperplasia, 451–452, 452f
Englitazone, on fetal growth in late gestation rats, 480–485, 481f, 481t, 484f
Epicatechin, on prostate cancer cell proliferation and survival, 343, 346
Epidermal growth factor (EGF)
activation of p42/p44 MAPKs by, 383, 383f

on Gab-1 docking protein in uterine cells, 447–448, 447f
on ovarian cancer, estrogen and, 262
Epigallocatechingallate. *See* Green tea (EGCG)
ERβ5 transcript upregulation, in endometrial cancer, 314–320, 317t
erbB2 (HER-2/neu), in premalignant breast disease, 125
17β-Estradiol. *See* E_2 (estradiol)
Estradiol 17β-dehydrogenase (OE_2DH)
actions of, 398
in MCF-7 breast carcinoma cells, eicosapentaenoic acid on, 398–403, 400f, 401f
17α-Estradiol-induced mutations, and tumorigenesis, 475–478, 476f–478f
Estriol. *See* E_3 (estriol)
Estrogen(s). *See also* specific estrogens
ACI rat mammary carcinogenesis induction by, cytogenetic analysis of genomic destabilization in, 412–417
aromatase and, in breast cancer risk, 172
aromatase in synthesis of, 130
in benign prostate hyperplasia, 502
biosynthesis of, from cholesterol, 166, 167f
biosynthesis of, pre- *vs.* post-menopause, 166
in breast cancer growth and development, 173, 173f
on breast cancer risk and etiology, 288
in breast growth and development, 77, 121
catechol (*See* Catechol estrogens)
GnRH/GnRH-R crosstalk of, in ovarian cancer, 263–265
in ovarian carcinogenesis, 261–263, 263f, 264f
on phenotypic and biochemical changes in aromatase transgenic mice (*See* Aromatase overexpression)
in progesterone receptor expression, 121–122
in prostate cancer, 322
in prostate growth, normal and neoplastic, 501–502
on reproductive and uterine tissues, 314–315
structures of, 316f
ZEB-1 regulation by, 455–460, 458f, 459f
Estrogen(s), in breast cancer prevention, 153–162
high E_2 + progesterone in, 156–157
microarray analysis of, 419–424, 422t, 423t
mimicking protective effects of pregnancy with, 156–157
multiple pregnancies and nursing in, 153
multiple pregnancies and nursing in, hormonal mimic for, 153–154
parous phenotypes in, human, 154
parous phenotypes in, mouse, 154–155
parous phenotypes in, rat, 155–156
parous rat refractoriness in, cause of, 157–158
parous *vs.* hormone-protected rats and, 159–161
parous *vs.* nulliparous rats and, promotional environment on, 158–159
short-term hormone treatment for, 159–160
Estrogen α and β variant expression, in endometrium, 314–320, 317t, 318f
Estrogen-induced mutations, and tumorigenesis, 475–478, 476f–478f
Estrogen-induced oncogenesis, 20
chromosome instability in, 19–29 (*See also* Chromosome instability, in estrogen-induced oncogenesis)
estrogen concentrations in, 20, 20t
estrogen-induced progesterone expression in, 21, 22f
estrogen receptor expression in, 21, 22f
tamoxifen prevention of, 21t
Estrogen receptor (ER)
in breast cancer, on prognosis, 288
differential expression in normal *vs.* malignant endometrium of, 319–320
isoforms of, 288 (*See also* Estrogen receptor α (ERα); Estrogen receptor β (ERβ))
isoforms of, changes in transcript levels of, in tumorigenesis, 315
structure of, 288
Estrogen receptor α (ERα), 122
actions of, 391
in breast tissue, 121–122
in breast tissue, in premalignant disease, 124–125, 125f
calmodulin interaction with, *vs.* ERβ, 391, 392
crosstalk with GnRH/GnRH-R of, in ovarian cancer, 263–265
in ductal carcinomas *in-situ* (DCIS), 124–125
in estrogen-induced oncogenesis, 20, 21, 22f
hinge region mutants of, and breast cancer, 391–396, 393t–396t
hyperplastic and atypical *in-situ* breast lesions and, 121
in mammary development, 77
in ovarian carcinogenesis, 261–263, 263f, 264f
in PR-A transgenic *vs.* wild-type mice mammary glands, 80, 81f
progestins on, 68
in prostate cancer, 502
selective inhibitors of, *vs.* of β form of, 392
Estrogen receptor α (ERα) blocking agents. *See also* specific agents
on high-grade prostatic intraepithelial neoplasia, 321–327, 324t, 325t
Estrogen receptor α (ERα) expression
in aromatase transgenic mice, 134

estrogen-induced oncogenesis on, 24–25, 25f
Estrogen receptor α (ERα) expression, in endometrium
normal and neoplastic, 314–320, 317t
pre- and postmenopausal, 317, 318f
in relation to FIGO, 317, 318f
Estrogen receptor α (ERα) hinge-region mutants
calmodulin binding of, 391–393
E_2 inhibition of transactivation in AP1 by, 395–396, 396t
hypersensitivity to E_2 of, 394, 394t
hypersensitivity to E_2 of heterodimers of, 395, 395t
W7 sensitivity of, 391, 393, 393t
Estrogen receptor β (ERβ)
actions of, 391
calmodulin interaction with, *vs.* ERα, 391, 392
in ovarian cancer, 261–262
potential roles of, 288
in prostate cancer, 502
selective inhibitors of, *vs.* of α form of, 392
Estrogen receptor β (ERβ) expression, in endometrium
pre- and postmenopausal, 317, 318f
in relation to FIGO, 317, 318f
Estrogen receptor β (ERβ) gene polymorphism, in breast cancer, 287–292
allele lengths and, *vs.* controls, 291, 291t
CA repeats and breast cancer risk in, 291–292
CA repeats in alleles of, 289, 290f, 290t
ERβ genotype variant length in, 287, 289
potential effects of, 288
SS, SL and LL genotype distribution in, 290, 291f, 291t
Estrogen receptor–related receptors (ERRs), 502
expression of, in human prostate cancer cells, 501–506, 503f–505f, 504t
Estrogen replacement therapy (ERT), breast cancer risk with, 122
Estrogen replacement therapy (ERT), lobular carcinoma risk and. *See also* Hormone replacement therapy (HRT), lobular...
current/recent use of, 53–55, 54t
ever use of, 50–53, 52t–53t
Estron. *See* E_1 (estron)
Exemestane, 167–168, 168f. *See also* Aromatase inhibitors
in breast cancer treatment, 170–171, 171t
Experimental results interpretation, dominant hypothesis on, 71
Extracellular signal-regulated kinases, in androgen receptor activation, 517–519, 518f
in prostate cancer, 515–519, 518f, 519f

F

Fatty acid synthase (FAS), 68
as antineoplastic therapy target, 344–345
clinical significance of, 69–70
expression of, in normal *vs.* neoplastic cells, 351
functions of, 351
progestin regulation of, 68–69
upregulation of, in tumorigenesis, 344–345
Fatty acid synthase (FAS) gene silencing by RNA interference, on LNCaP prostate cancer cell growth and apoptosis, 350–354, 352f, 353f, 354t
Fatty acid synthase (FAS) inhibitors. *See also* specific agents
green tea (EGCG) as, in cancer prevention, 343–347, 346t
tumor cell cytotoxicity of, 343, 345
Fatty acid synthase (FAS) overexpression
in cancer cells, 70–71, 350, 351
in prostate cancer, 358
Fatty acids
on enzyme activity, 401–402
on proliferation and 17β-dehydrogenase activity in MCF-7 breast carcinoma line, 398–403, 400f, 401f
Finasteride, 212–213
on androgen-independent prostate tumors in athymic mice, 215
on prostate tumor growth, 218
Focal adhesion assembly in breast cancer cells, soy phytoestrogens on, 300–306, 303f–305f
Focal adhesion kinase (FAK), 301
Focal adhesion kinase (FAK) activity
in breast cancer cells, soy phytoestrogens on, 300–306, 303f–305f (*See also* Soy phytoestrogens)
genistein on, 301 (*See also* Genistein)
Focal adhesions, 301
Formestane, 167–168, 168f. *See also* Aromatase inhibitors
Forskolin, androgen receptor activation by, 225, 517, 518f, 519

G

Gab-1 docking protein in uterine cells
hormonal activation of, 443–448, 445f–448f
hormonal activation of, growth factor, 447–448, 447f
hormonal activation of, ovarian steroid hormone, 445–447, 446f
hormonal activation of, uterine in estrous cycle, 445, 445f
immunolocalization of, 447–448, 448f
GADD45, on cell cycle checkpoint and centrosome cycle, 111
γ-linolenic acid, 212–213, 213f

Gene amplification. *See also* specific genes
mechanisms of, 92–93
in rat endometrial cancer, 251
Genistein
on estrogen receptor, 301
on FAK activity in ERα cell lines, 303–304, 304f, 305f
on focal adhesion in breast cancer cells, 300–306, 303f–305f
on focal adhesion kinase activity, 301
on focal adhesions and F-actin, in breast cancer, 302–303, 303f
Genistein, dietary
on estradiol-induced mammary carcinogenesis in ACI rat, 405–409, 408t, 409f
on mammary carcinogenesis in rodent models, 407
Genome alterations, somatic, in prostate cancer cells, 36–38, 36f, 37f
Genomic destabilization
in estrogen-induced ACI rat mammary neoplasms, cytogenetic analysis of, 412–417
in estrogen-induced oncogenesis, 412–413
Genomic instability, 87–88. *See also* Chromosome instability
in aneuploid tumors, 108
c-Myc deregulation in complex network of, 87–92 (*See also* c-*Myc* oncoprotein deregulation, in genomic instability)
centrosome amplification in, 109
locus-specific, in c-*Myc* gene, 89–90, 89f, 90f
Genomics-related internet sites, rat, 248t
GFP-AurA
cell cycle localization of, 6, 7f
in centrosome/mitotic spindle, 6–7, 8f
chromosomes marked by, in G_2 phase of HeLa cell cycle, 6, 7f
regulated expression of, 6, 6f
Glucose, EGCG (green tea) on, 217
Glutathione S-transferase (GSTP1), in prostate cancer, 142, 144–145
Glycine N-methyltransferase (GNMT)
in carcinogenesis, 293–294
expression and function of, 293
Glycine N-methyltransferase *(GNMT)* gene, 294
polymorphisms in, 294, 294f
Glycine N-methyltransferase *(GNMT)* gene polymorphism, in hepatocellular carcinoma, 293–298
cancer cell lines and polymorphisms in, 295, 296t
HBV infection & GNMT polymorphism in, 297–298, 298t
INS/DEL & SNP1 genotype frequencies in, 295–297, 297t
Gonadotropin-releasing hormone (GnRH), in cancer, 259
Gonadotropin-releasing hormone (GnRH), on ovarian cancer
estrogen/estrogen receptor crosstalk of, 263–265
inhibitory actions of, 258–260
inhibitory actions of, mechanism of, 260–261
as therapy, 265
Gonadotropin-releasing hormone II (GnRH-II), 259–260
Gonadotropin-releasing hormone receptor (GnRH-R)
estrogen/ER crosstalk of, in ovarian cancer, 263–265
inhibition of ovarian cancer by, 258–261
Grb-2-associated binder-1 (Gab-1), 444. *See also* Gab-1 docking protein in uterine cells
Green tea (EGCG)
as ancient medicine, 211
body weight reduction by, via reduced food intake, 216
cancer prevention by, 343–344
cancer prevention by, via EGCG-mediated fatty acid synthase inhibition, 343–347, 346t
cell death induction in prostate cancer but not normal fibroblasts by, 346
endocrine system modulation by, 216–217
fatty acid synthase inhibition by, 345, 346t
fatty acid synthase inhibition by, in cancer prevention, 343–347, 346t
fatty acid synthase inhibition by, mechanism of, 346–347
on proliferation and apoptosis in prostate cancer cells, 345–346, 346t
on serum glucose, lipids, triglycerides, and cholesterol, 217
structure of EGCG in, 344, 344f
suppression of prostate and breast tumors by, 216
on testosterone level, 217
Growth factors. *See also* specific growth factors
androgen receptor activation by, 224–225, 516
in androgen-refractory progression of prostate cancer, 234
on Gab-1 docking protein in uterine cells, 447–448, 447f
on MAPKs, 383, 383f
GSTA1, 34
GSTP1, 34
in proliferative inflammatory atrophy, 38–39
GSTP1, in prostate cancer prevention and pathogenesis, 34–35, 38
GSTP1 CpG island hypermethylation, in prostate cancer, 36, 37t, 39
Gynecomastia, in male transgenic mice from aromatase overexpression, 135

H

hCDC4 mutation, in cyclin E deregulation in cancer, relative to cell cycle, 99–100
Hepatitis B virus infection, and glycine N-methyltransferase gene polymorphism in hepatocellular cancer, 297–298, 298t
Hepatoblastoma, glycine N-methyltransferase gene polymorphism in, 293–298
Hepatocellular carcinoma, glycine N-methyltransferase gene polymorphism in, 293–298. *See also* Glycine N-methyltransferase (GNMT) gene polymorphism
Hepatocyte proliferation, 480
High-grade prostatic intraepithelial neoplasia (HBPIN). *See* Prostatic intraepithelial neoplasia, high grade
Hormone replacement therapy (HRT)
  combined (*See* Combined estrogen/progestin hormone replacement therapy (CHRT))
  estrogen (*See* Estrogen replacement therapy (ERT))
  estrogen *vs.* progesterone on breast cancer risk in, 122
  prognosis of carcinomas with, 126
Hormone replacement therapy (HRT), lobular carcinoma risk and, 50–57
  conclusions on, 59–60
  invasive lobular *vs.* invasive ductal cancer risk in, 50, 51t
  sequential *vs.* continuous progestin in CHRT and, 55, 56t
  summary of, 55–57
Hormones. *See* specific hormones
HRT. *See* Hormone replacement therapy
Human chorionic gonadotrophin (hCG), in mammary tumor prevention, 157
Hydroxyfultamide, on IL-6-induced androgen receptor activity, 225, 226f

I

Id-1 protein, 197–207
  in benign prostatic hypertrophy, 198–199, 199f
  early growth response-1 (Egr-1) interaction with, 202
  ectopic expression of, on Raf-1, MEK1/2, and Egr-1 expression, 203–204, 203f
  functional studies of, 199–202
  functional studies of: DNA synthesis and cell cycle distribution, 200, 201f
  functional studies of: ectopic expression on prostate cancer growth, 199, 200f
  functional studies of: Rb/p16^{INK4a} pathway, 201–202, 202f
  in human prostate cancer, 198–199, 199f
  Id-1 anti-sense treatment on Id-1/Egr-1 protein expression in prostate cancer cell lines expressing, 206, 206f
  MAPK signaling pathway and, 202–206
  MEK 1/2 inhibitor PD098059 on serum-independent prostate cancer growth induced by, 205–206, 205f, 206f
  MEK1/2 inhibitor PD098059 on MEK1/2 and Egr-1 expression in transfectant clones of, 204, 204f
  nature of, 198
  as oncogene, 202
  overexpression of, in rat prostate cancer model, 197–198, 198f
Inbred animal models, 247. *See also* BDII inbred rat
Inflammation, prostate, in prostate cancer, 35–36
Inflammatory breast (mammary) carcinoma, 436
  canine, IGF-1 and androgens in, 436–441, 439f, 440f
  genes in, 437
INS/DEL genotypes, of glycine N-methyltransferase in hepatocellular carcinoma, 293–298
Insulin-like growth factor (IGF)
  in cancer, 437, 441
  in canine inflammatory mammary carcinoma, 436–441, 439f, 440f
  englitazone on, 480–485, 481f, 481t, 484f
  on Gab-1 docking protein in uterine cells, 447–448, 447f
  on ovarian cancer, estrogen and, 262
Interferon-inducible protein 9-27 gene, onapristone upregulation of, 311, 311t
Interleukin-6, transactivation of androgen receptor by, 496
  p300 in, 494–499
Interleukin-6 signaling, in prostate cancer, androgen receptor and, 225–228, 226f, 227f
Internet sites, rat genomics-related, 248t
Intestinal trefoil factor gene, onapristone upregulation of, 311, 311t
Isoproterenol, androgen receptor activation by, 517, 519

K

Karyotypic instability, 88
  in c-*Myc* gene in cancer, 89f, 90
Kennedy's disease, 186–187
Kidney, ectopic uterine tumors in, 20–22, 21t, 22f
Kinases. *See* specific kinases

L

Letrozole, 167–168, 168f. *See also* Aromatase inhibitors

on aromatase-induced preneoplastic changes and breast hyperplasia, 136
in breast cancer treatment, 170–171, 171t
Leydig tumors, in male transgenic mice from aromatase overexpression, 135
Lipids, serum. *See also* specific lipids
EGCG (green tea) on, 217
Lipogenesis. *See also* Fatty acid synthase (FAS)
increased, in prostate cancer, 358
LNCaP cells
104-R1, 213, 214
104-S, 214
activation of survival signaling under androgen ablation conditions in, 235
androgen-refractory outgrowth of, with androgen deprivation, 234–235, 234f
androgens on SCAP in, 359, 360f
androgens on SREBPs in, 358–359
epidermal growth factor on androgen receptors in, 515–519, 518f, 519f
lipogenic effects of androgens in, 358
protein kinase A–dependent activation of androgen receptors in, 515–519, 518f, 519f
RNA interference silencing of fatty acid synthase gene on, 350–354, 352f, 353f, 354t
LNCaP-IL-6+ subline, 226, 227f
Lobular neoplasia, 124–125, 125t
LOH acceleration, cyclin E deregulation in, 100, 101f
Lost in inflammatory breast carcinoma (LIBC) gene, 437

M

Mammary carcinoma, inflammatory, 436
canine, IGF-1 and androgens in, 436–441, 439f, 440f
genes in, 437
Mammary model, Medina laboratory rat, 9, 9f
Mammary premalignant progression, hormonal dependence of, 431–435, 433f, 433t, 434f
Mammary tumors, in female ACI rats, 22–24, 23f, 24t
vs. human breast cancers, 28, 28t
Mammographic densities (MD)
estrogen metabolic pathways on, 278
hormone replacement therapy on, 278
tamoxifen on, 278
Mammographic densities (MD), and urinary hormones in healthy women, 277–284
age on, 281, 282t
BMI on, 281, 282t
by categories of percent MD, 282–283, 283f
ethnicity on, 280–281, 281t
hormones *vs.* percent densities, age, and BMI in, 282t
HRT on, 281–283
pre- *vs.* postmenopausal, 281–282, 281t
study population in, 281t
MAPK(s). *See also* specific MAPKs
in GnRH antiproliferative effect in ovarian cancer, 261
phosphorylation of progesterone receptor by, 383–384, 383f
in progesterone receptor regulation of cyclin D1 expression, 381, 382f, 386–388, 387f
MAPK activation
by liganded progesterone receptor, 382f, 385–386
in progesterone receptor actions, 382, 382f
in progesterone receptor localization, 382f, 383, 383f
in progesterone receptor transcriptional activity, 384, 385f
MAPK pathway
in breast cancer cell models, bidirectional regulation of progesterone receptors and, 381–388
Id-1 protein as prostate cancer biomarker and, 202–206
Medroxyprogesterone acetate (MPA)
dose in postmenopausal CHT of, on endometrial cancer risk, 273–276, 275f
human breast cancer cell line assays of, 67
on mammary carcinogenesis, 65–66, 82
MEK 1/2 inhibitor PD098059
on MEK1/2 and Egr-1 expression in transfectant clones of Id-1 protein, 204, 204f
on serum-independent prostate cancer growth induced by Id-1 protein, 205–206, 205f, 206f
MEK1/2, Id-1 protein ectopic expression on, 203–204, 203f
Met amplification, in rat endometrial adenocarcinoma, 251–252, 252f, 253t
Metaphase-anaphase transition, cyclin E deregulation on, 103
Microarray analysis, of estrogen-induced protection against breast cancer, 419–424, 422t, 423t
Micrometastasis, PTTG1 detection of, 462–467. *See also* Pituitary tumor-transforming gene-1 (PTTG1)
Microtubule organizing center (MTOC), 1
Mifepristone, on mammary dysplasias in PR-A transgenic mice, 81
Mitogen activated protein kinases (MAPKs). *See* MAPK; specific protein kinases
Monosomy, 87
MSR1, in prostate cancer, 35
Myc proteins, mutant, in genomic instability, 91–92, 91f
Myristoleic acid, 212–213, 213f

N
N-ethyl-nitrosurea, 450, 451
endometrial adenocarcinoma in Syrian hamsters treated with, 450–453, 451t, 452f, 453f
NADPH-dependent hepatic oxidation of estradiol, chronic estradiol on, 367, 370, 370t
Nuclear receptors, 455–456
Nullisomy, 87

O
ODC gene, c-Myc overexpression on, 89–90
Onapristone
antiprogesterone and antitumor activity of, 308–309
breast cancer cell genes regulated by, 308–313, 311t, 312f
Oncogene. *See also* specific oncogenes
Id-1 protein as, 202
Oncogenesis. *See also* Tumorigenesis; specific cancers
estrogen-induced, chromosome instability in, 19–29 (*See also* Chromosome instability, in estrogen-induced oncogenesis)
Onion skin model, 93
ORC, 102, 103f
Orphan nuclear receptors, 501
Ovarian cancer, 258–265
estrogen/ER and gonadotropin-releasing hormone/GnRH receptor crosstalk in, 263–265
estrogen/ER in carcinogenesis of, 261–263, 263f, 264f
gonadotropin-releasing hormone and GnRH receptor as therapy for, 265
gonadotropin-releasing hormone and GnRH receptor inhibition of, 258–260
repeated ovulation in, 258

P
P13K/AKT, in androgen-refractory prostate cancer
LNCaP cell outgrowth with androgen deprivation in, 234–235
survival signaling with androgen ablation in, 235, 236f
P13K/Akt, in androgen-refractory prostate cancer cell survival, 233–242
P13K/PTEN/Akt, in regulation of Bcl-2 expression by prostate cancer cells, 236–237, 237f
p21
estrogenic stimulation of BC cell lines on, 133
in mammary epithelial cell transformation, 79
in PR-A transgenic *vs.* wild-type mice mammary glands, 80, 81f
p21/waf1
on centrosome duplication, 109, 110–111
p53 regulation of, 110–111
p27
on centrosome duplication, 109
on cyclin/CDK in cell cycle progression, 132
cyclin D1 on sequestration of, 133
estrogenic stimulation of BC cell lines on, 133
$p27^{KIP1}$, in prostate cancer, 144
p42 MAPK. *See also* MAPK(s)
growth factors as activators of, 383, 383f
p44 MAPK. *See also* MAPK(s)
growth factors as activators of, 383, 383f
p53
abnormalities of, on centrosomes in cancer, 110
on centrosome cycle, 110–111
in premalignant breast disease, 125
p53 null mammary epithelium transplant model
hormonal dependence of premalignant progression in, 431–435
hormone-induced tumorigenesis in, 433, 433t
ovarian-dependent growth of normal mammary epithelium in, 432, 433f
tamoxifen inhibition of mammary tumorigenesis in, 433–434, 434f
tumorigenesis in progesterone receptor–deficient mammary epithelium in, 433, 434t
p160 coactivators, on androgen receptor, 494
p300, in androgen-independent prostate cancer, 494–496, 497f, 498, 499f
PCNA expression
in aromatase transgenic females, 132
in cadmium and zinc chloride–induced rat ventral prostate, 522, 525, 526f, 526t
PD098059
on Id-1 protein–induced serum-independent prostate cancer growth, 205–206, 205f, 206f
on MEK1/2 and Egr-1 expression in Id-1 protein transfectant clones, 204, 204f
Pericentrin overexpression, centrosome amplification by, 111
Perictriolar material, 106–107
Peroxisome proliferator-activated receptor gamma ligands
on fetal growth in late gestation rats, 480–485, 481f, 481t, 484f
roles of, 480–481
Phenylbutyrate, androgen receptor activation by, 224–225
Phosphatase and tension homology deleted on chromosome ten. *See* PTEN
Phosphatidylinositol (PI)
in CCK-B receptor transduction, 471, 472–473, 472f

as intracellular second messenger, 472
Phosphorylation
Akt, englitazone on, 480, 484, 484f
of androgen receptors, 519
on centrosomes in cell cycle, 110
of progesterone receptor by MAPK, 383–384, 383f
STAT3, in prostate tumor differentiation, 226
tyrosine, of FAK in breast cancer, E_2 on, 301
Phytoestrogens. *See also* Soy phytoestrogens
on estradiol-induced mammary carcinogenesis in ACI rat, 405–409, 408t, 409f
estrogenic activity of, 322
Pituitary tumor, human functionless, CCK-B receptor mRNA in, 468–473, 471f, 472f
Pituitary tumor-transforming gene-1 (PTTG1), 462–463, 463t
in chromosomal instability, 462–463
Pituitary tumor-transforming gene-1 (PTTG1) mRNA, as micrometastasis marker, 462–467
cell lines developed for assay with, 463, 464t
human tumor cell line expression of, 465, 465f
primer characteristics in, 464, 465t
specificity of analysis with, 465–466, 466f, 466t, 467f
Polyunsaturated fatty acids (PUFAs)
on enzyme activity, 401–402
on proliferation & 17β-dehydrogenase activity in MCF-7 breast carcinoma line, 398–403, 400f, 401f
Pre-replication complexes, assembly of, 102, 103f
Precancerous lesions. *See also* specific lesions
rationale for treatment of, 325
Pregnancy, in breast cancer prevention, 126, 420
Premalignant breast disease, 121–127
atypical ductal hyperplasia (ADH) in, 122, 123t
atypical lobular hyperplasia (ALH) in, 124–125, 125t
breast cancer risk after biopsy and, 124t
ductal carcinomas *in-situ* (DCIS) in, 122
estrogen receptor evidence in breast tissue in, 124–125, 125f
hormonal dependence of, 431–435, 433f, 433t, 434f
hyperplastic and premalignant lesions in, 122, 123t
lobular carcinoma *in-situ* (LCIS) in, 124–125, 125t
subsequent invasive carcinoma in, 125–126
Prevention. *See* specific cancers
Progesterone. *See also* Progestin
in breast cancer etiology, 57–59
on breast lobular alveolar structures in pregnancy, 58
on breast lobular epithelial cell proliferations, 58
estrogen-induced expression of, in oncogenesis, 21, 22f
high, in mammary tumor prevention, 156
in mammary development, 77
as mitogen in breast cancer, 381–382
in normal breast proliferation and development, 58
Progesterone receptor (PR)
A and B isoforms of, 59, 78
bidirectional regulation of, and MAPK pathway in breast cancer cell models, 381–388
direct phosphorylation of, by MAPKs, 383–384, 383f
MAPK activation by liganded modules of, 382f, 385–386
MAPK activation in receptor localization of, 382f, 383, 383f
MAPK activation in transcriptional activity of, 384, 385f
MAPK in actions of, 382, 382f
MAPK in regulation of cyclin D1 expression by, 381, 382f, 386–388, 387f
in pregnancy and lactation, 78
regulation of cyclin D1 protein levels by, 388, 388f
Progesterone receptor (PR), in mammary carcinogenesis, 77–82
isoform imbalance in, 79–80, 79f–81f, 82
mifepristone on mammary dysplasias in PR-A transgenic mice and, 81
overexpression of PR-A in mammary cell transformation and, 78–81, 79f–81f
signaling of, 81–82
Progesterone receptor (PR), in mammary development, 77–78
in PR-A transgenic *vs.* wild-type mice, 80, 80f
in PR-B transgenic mice, 80
in transgenic mice with altered A:B ratio, 79–80, 79f–81f, 82
Progesterone receptor (PR) antagonists, 308–309. *See also* specific agents
Progesterone receptor (PR) expression, 121–122
in endometrium in relation to FIGO, 317, 318f
estrogen-induced oncogenesis on, 24–25, 25f
excess, of A form, 78–81, 79f–81f
in normal and neoplastic endometrium, 314–320, 317t
regulation of, 77–78
Progesterone receptor (PR) subtype, 59, 78
differential expression of, in normal *vs.* malignant endometrium, 319
Progesterone signaling, in mammary carcinogenesis, 81–82
Progestin. *See also* Progesterone
antiestrogenic activity of, 67–68
on breast cancer, recent study on, 71
on breast cancer cells, 66, 66f
on estrogen, 65
fatty acid synthase regulation by, 68–69

as mitogens in breast cancer, 381–382
on postmenopausal mammary carcinogenesis, 65
on postmenopausal mammary glands, 66–67, 66f
Progestin-regulated genes, and breast cancer risk, 65–72
antiestrogenic activity of progestins in, 67–68
background on, 65
experimental results interpretation in, 71
fatty acid synthase in, 68
fatty acid synthase in, clinical significance of, 69–70
fatty acid synthase in, overexpression of, 70–71
fatty acid synthase in, progestin regulation of, 68–69
human breast cancer cell line detection of receptor-induced proteins in, 67
progestins on mammary glands in, 66–67, 66f
recent study on, 71
therapeutic implications of, 71–72
Progestogen dose, in postmenopausal CHT, on endometrial cancer risk, 273–276, 275f
Prolactin-inducible protein, onapristone downregulation of, 311–312, 312f
Proliferative inflammatory atrophy, in prostate cancer, 34, 38–39, 143–145
Prophylaxis. *See* specific cancers
Prostate
mean nuclear volume weighted by volume in, 329–334, 332t, 333f
morphology of prostatic acini in, 331, 331f, 333f
sex hormone–binding globulin synthesis and expression in, 510, 511f, 512–513
Prostate cancer
adenosis in, 145
α-methylacyl-CoA racemase (AMACR) in, 142, 146
androgen growth suppression of cultured cells of, 214–215
androgen receptor in (*See* Androgen receptor (AR))
androgens and (*See* Androgen(s), prostate cancer and)
ARA70 as coactivator in, 227
atypical adenomatous hyperplasia (adenosis) in, 145
benign prostatic hyperplasia and, 145–146
biomarker for (*See* Id-1 protein)
cadmium in, 523
cell progression of, models for, 213–214, 214f
diet in, 35–36
epidemiology and etiology of, 141, 183–184
estrogen receptor α (ERα) in, 502
estrogen receptor β (ERβ) in, 502
estrogens in, 322, 501–502
fatty acid synthase overexpression in, 358
glutathione S-transferase (GSTP1) in, 142, 144–145
green tea on (*See* Green tea)
GSTP1 CpG island hypermethylation in, 36, 37t, 39
GSTP1 in pathogenesis of, 34–35, 38
Id-1 protein in (*See* Id-1 protein)
lipogenesis in, increased, 358
mean nuclear volume weighted by volume in, 329–334, 332t
MSR1 in, 35
oncostatin M growth stimulation in, 225–226, 226f
p27^{KIP1} in, 144
progression and treatment of, four stages of, 214, 214f
proliferative inflammatory atrophy in, 34, 38–39, 143–145
prostatic acini morphology in, 331, 331f, 333f
prostatic intraepithelial neoplasia in, 142–143
prostatic intraepithelial neoplasia in, in pathogenesis model, 141, 142f
prostatitis in, 35–36
race/ethnicity in, 183–185
risk factors for, 184–185, 185f
RNASEL in, 35
sex hormone–binding globulin (SHBG) expression in, and hormone regulation, 508–513, 509f–512f
soy phytoestrogens and, 322 (*See also* Soy phytoestrogens)
steroid 5α-reductase type 2 (SRD5A2) gene in, 190–191, 191t
yeast-based functional assay to detect mutant androgen receptors in, 336, 337–338, 338f
Prostate cancer, androgen-independent, 515
CBP in, 494–495, 497–498, 498f, 499f
coactivators of, 494–498, 497f–499f
coactivators of, androgen receptor as, 227–228
coactivators of, CBP as, 227, 494–498, 497f–499f
coactivators of, p300 as, 494–498, 497f–499f
Prostate cancer, androgen-refractory
growth factors in progression of, 234
p13K/AKT in growth of, 234–235, 236f
Prostate cancer, molecular pathogenesis of, 34–39
background on, 34–35
GSTP1 CpG island hypermethylation in, 36, 37t, 39
inflammation and diet in epidemiology of, 35–36
model of, 141, 142f
prevention of, new opportunities for, 39
proliferative inflammatory atrophy in, 38–39

somatic genome alterations in cells in, 36–38, 36f, 37f
Prostate cancer cells. *See also* specific cell lines, e.g., LNCaP cells
androgen on Bcl-2 expression in, 237–242, 238f–241f
androgen-refractory, Bcl-2 overexpression in survival of, 233
androgen-refractory, P13K/Akt in survival of, 233–242
early growth response (Egr-1) protein expression in, Id-1 anti-sense treatment on, 206, 206f
epicatechin on proliferation and survival of, 343, 346
estrogen receptor–related receptors (ERRs) in, 501–506, 503f–505f, 504t
fatty acid synthase gene silencing by RNA interference in, on growth and apoptosis, 350–354, 352f, 353f, 354t
genome alterations in, somatic, 36–38, 36f, 37f
green tea (EGCG) cell death induction in, but not in normal fibroblasts, 346
green tea (EGCG) on proliferation and apoptosis in, 345–346, 346t
P13K/PTEN/Akt in regulation of Bcl-2 expression by, 236–237, 237f
PTEN in, cell cycle regulation and apoptosis of, 487–492, 489t, 490f, 491f
PTEN in, regulation of Bcl-2 expression by, 236–237, 237f
RNA interference fatty acid synthase gene silencing on, 350–354, 352f, 353f, 354t
steroid hormone receptor expression in, 504, 504t
Prostate cancer prevention, 39
green tea in (*See* Green tea (EGCG))
GSTP1 in, 34–35, 38
new opportunities for, 39
Prostate cancer risk, with high-grade prostatic intraepithelial neoplasia, 322
Prostate epithelial carcinogenesis, 141–146
Prostate specific antigen (PSA)
high-grade prostatic intraepithelial neoplasia on, 325–326
toremifene on, 321, 324, 324t
Prostatic intraepithelial neoplasia, 142–143, 321
cadmium and zinc chloride–induced, 522–527, 526f, 526t, 527f
mean nuclear volume weighted by volume in, 329–334, 332t
in molecular pathogenesis model, 141, 142f
prostatic acini morphology in, 331, 331f, 333f
Prostatic intraepithelial neoplasia, high grade, 142–143, 321
epidemiology of, 321–322
prostate cancer risk with, 322
on PSA, 325–326
toremifene on, 321–327, 324t, 325t
treatment of, rationale for, 325
Prostatitis, prostate cancer and, 35–36
Protein kinase A, androgen receptor activation by, 225
in prostate cancer, 515–519, 519f
Protein phosphorylation. *See* Phosphorylation
PTEN, 487
in cell cycle regulation and apoptosis in prostate cancer, 487–492, 489t, 490f, 491f
expression and action of, 488
mutations of, 487–488
overexpression of, on cell cycle and apoptosis, 490, 491f
in prostate cancer cells, regulation of Bcl-2 expression by, 236–237, 237f
PTEN/MMAC, 487
PTTG1. *See* Pituitary tumor-transforming gene-1 (PTTG1)

Q

Q640Stop/T877A mutant androgen receptor, 339–341, 340f, 341f

R

R5020, human breast cancer cell line assays of, 67
Raf-1, Id-1 protein ectopic expression on, 203–204, 203f
RAPD-PCR fingerprinting, detecting of estrogen-induced mutations by, 475–478, 476f–478f
Rat mammary model. *See also* specific models
Medina laboratory, 9, 9f
Rb/p16^{INK4a} pathway, Id-1 protein on, 201–202, 202f
Receptor tyrosine kinases (RTKs)
in androgen receptor activation, 517–519, 518f
docking proteins on signals from, 443–444
Receptors, steroid hormone. *See also* specific receptors
estrogen-induced oncogenesis on expression of, 24–25, 25f
5α-Reductase inhibitors, 212–213
on androgen action, 212
on prostate tumor growth, 218
Replication-driven model, 93
Retinoblastoma tumor suppressor (Rb), DNA replication and centrosome duplication control by, 109
Reversible tumorigenesis, c-*Myc* gene deregulation in, 91
Ribonucleotide reductase R2 gene, 93
c-Myc overexpression on stability of, 89–90

RNA interference, 351, 352f
fatty acid synthase gene silencing by, on LNCaP prostate cancer cell, 350–354, 352f, 353f, 354t
RNASEL, in prostate cancer, 35
Ro41-0960, modulation of transforming/clastogenic activities of catechol estrogens by, 375–379
R_{SHBG} signaling, 509–510, 510f

S

SCAP-retention protein balance, in SREBP pathway in prostate cancer, androgens on, 357–362, 358f, 360f, 361f
Securin and tumor transforming protein gene 1. *See* Pituitary tumor-transforming gene-1 (PTTG1)
Segregation-driven model, 93
Selective ER modulators (SERMs). *See also* specific agents
vs. aromatase inhibitors, 172, 173t
Serine/theronine kinases, aurora family of, 3–8. *See also* Aurora A (AurA) kinase, in cancer
Serine/threonine cyclin-dependent protein kinases, 109
Sex hormone–binding globulin (SHBG)
cancer risk and, 509
expression in breast/prostate cancer of, and hormone regulation, 508–513, 509f–512f
steroid signaling by, 509, 510f
structure and function of, 508–509, 509f
Short-term E_2 + progesterone treatment (STEPT), 159–161. *See also* Estrogen(s), in breast cancer prevention
in parous *vs.* hormone-protected rats, 161–162
Short-term E_2 treatment (STET), 161–162. *See also* Estrogen(s), in breast cancer prevention
in parous *vs.* hormone-protected rats, 161–162
Somatic mutation idea, 2
Soy phytoestrogens
on breast cancer risk, 406
on cell survival and invasion signaling proteins, 301
on estrogen receptor, 301
estrogenic activity of, 322
on FAK activity in ERα cell lines, 303–304, 304f, 305f
on focal adhesion in breast cancer cells, 300–306, 303f–305f
on focal adhesions and F-actin, in breast cancer, 302–303, 303f
on mammary carcinogenesis in ACI rat, estradiol-induced, 405–409, 408t, 409f
on mammary carcinogenesis in rodent models, 407
prostate cancer and, 322
SREBP-cleavage activating protein (SCAP), 358
in LNCaP cells, androgens on, 359, 360f
SREBP pathway activation via overexpression of, 360–361, 361f
in SREBP pathway in prostate cancer, androgens on, 357–362, 358f, 360f, 361f
STAT3 phosphorylation, in prostate tumor differentiation, 226
Steroid 5α-reductase type 2 (SRD5A2) gene
in prostate cancer, 190–191, 191t
racial/ethnic differences in, 190–191, 191t
Steroid hormone receptor expression. *See also* specific receptors
in human prostatic cells, 504, 504t
Steroid hormone receptor–growth factor crosstalk, in progesterone receptors in breast cancer cell models, 381–388
Sterol regulatory element binding protein (SREBP) pathway, 345
androgen stimulation of, in prostate cancer, 357–362, 358f, 360f, 361f
androgen stimulation of, sterol-regulating structural elements in, 359, 360f
SCAP overexpression in activation of, 359, 360f
Sterol regulatory element binding proteins (SREBPs)
on fatty acid synthase overexpression, 351
structural and sterol-mediated activation mechanism of, 358, 358f
synthesis and functions of, 358
STK15 allele, mutational status of, 4f, 5
Sulfotransferase activity, in ACI rat and mammary gland after chronic estradiol treatment, 367, 369–370, 369f, 370t
SULT1A1 activity, in ACI rat and mammary gland after chronic estradiol treatment, 367, 369–370, 369f, 370t, 371
SULT2A1 activity, in ACI rat and mammary gland after chronic estradiol treatment, 367, 369–370, 369f, 370t
Syrian hamster embryo fibroblasts
COMT inhibitor modulation of transformation/clastogenic activities of catechol estrogens in, 375–379
as model for estrogen-induced transformation, 376

T

Tamoxifen, 450, 451
vs. aromatase inhibitors, 172, 173t
on BALB/c p53 null mammary epithelium transplant model, 433–434, 434f
endometrial adenocarcinoma in Syrian hamsters treated with, 450–453, 451t, 452f, 453f
ERα *vs.* ERβ response to, 319–320
on ovarian cancer, estrogen and, 262–263

uterine cell proliferation and endometrial tumors from, 319
Tamoxifen, prophylactic
in breast cancer prevention, 153
breast cancers in women on, 126
on estrogen-induced oncogenesis, 21t
Tea, green. *See* Green tea (EGCG)
Telomere dysfunction, in chromosomal instability, 143
Terminal end buds (TEBs), in pregnancy protection against breast cancer, 155, 157
Testosterone. *See also* Androgen(s)
on androgen-independent prostate tumors in athymic mice, 215
in canine inflammatory mammary carcinoma, 438, 439f, 440
race/ethnicity on levels of, 185
Testosterone-repressed prostate message-2 (TRPM-2) gene, onapristone upregulation of, 311, 311t
Tetrasomy, 87
TGBα expression, in aromatase mice, 134
TGBβ1 expression, in aromatase transgenic females, 132
Thiazolidinediones, 481
Toremifene, on high grade prostatic intraepithelial neoplasia, 321–327, 324t, 325t
Transforming growth factor α (TGFα)
estrogen on expression of, in MMTV-TGFα x MMTV-aromatase transgenic cross, 135–136
on ovarian cancer, estrogen and, 262
Triglycerides, EGCG (green tea) on, 217
Tumor suppressor mutation, cyclin E deregulation synergisms with, 101–102
Tumorigenesis. *See also* Oncogenesis; specific cancers
reversible, c-*Myc* gene deregulation in, 91

U
Uterine cells, hormonal activation of Gab-1 docking protein in, 443–448, 445f–448f

W
W7
hinge-region mutant ERα sensitivity to, 391, 393, 393t
inhibition of ERα-mediated transcriptional activation by, 391, 393, 393t

Y
Yeast-based functional assay, to detect mutant androgen receptors in prostate cancer, 336, 337–338, 338f

Z
α-Zearalanol-induced mutations, and tumorigenesis, 475–478, 476f–478f
ZEB-1, 455–456, 456f
hormonal regulation of, and reproductive cancer progression, 455–460, 456f–460f
Zinc-α2-glycoprotein gene, onapristone downregulation of, 311–312, 312f
Zinc chloride, in preneoplastic changes in rat ventral prostate, 522–527, 526f, 526t, 527f
Zinc finger E-box binding (ZEB)-1 protein, 455–456, 456f. *See also* ZEB-1